PERSPECTIVES IN PEDIATRIC CARDIOLOGY

Series Editor
Robert H. Anderson, M.D.

Perspectives in
Pediatric Cardiology   Volume 5

Genetic and Environmental Risk Factors of Major Cardiovascular Malformations:

The Baltimore-Washington Infant Study: 1981–1989

by

Charlotte Ferencz, M.D., M.P.H.
Research Professor of Epidemiology
  and Preventive Medicine and of Pediatrics
University of Maryland
School of Medicine
Baltimore, Maryland

Adolfo Correa-Villaseñor, M.D., Ph.D.
Associate Professor of Epidemiology
Johns Hopkins University
School of Hygiene and Public Health
Baltimore, Maryland

Christopher A. Loffredo, Ph.D.
Research Assistant Professor
  of Epidemiology and Preventive Medicine
University of Maryland
School of Medicine
Baltimore, Maryland

P. David Wilson, Ph.D.
Associate Professor of Epidemiology
University of Maryland
School of Medicine
Baltimore, Maryland

With contributions in

Pediatric Epidemiology:

Judith D. Rubin, M.D., M.P.H.
*Associate Professor of Epidemiology
  and Preventive Medicine and of Pediatrics
University of Maryland
School of Medicine
Baltimore, Maryland*

Dysmorphology/Genetics:

Iosif W. Lurie, M.D., Ph.D.
*Adjunct Professor of Epidemiology
  and Preventive Medicine
University of Maryland
School of Medicine
Baltimore, Maryland*

Environmental Epidemiology:

Carol A. Magee, PhD., M.P.H.
*Research Assistant Professor of
  Epidemiology and Preventive Medicine
University of Maryland
School of Medicine
Baltimore, Maryland*

Pediatric Cardiology:

Karen S. Kuehl, M.D.
*Associate Professor of Pediatrics
Georgetown University
Washington, D.C.*

**Futura Publishing
Company, Inc.**
Armonk, NY

Futura Publishing Company, Inc.
135 Bedford Road
Armonk, New York 10504
ISBN#: 0-87993-675-4

Perspectives in Pediatric Cardiology ISSN#: 1044-4157

Every effort has been made to ensure that the information in this book is as up to date and accurate as possible at the time of publication. However, due to the constant developments in medicine, neither the authors, nor the editor, nor the publisher can accept any legal or any other responsibility for any errors or omissions that may occur.

Printed in the United States of America.

**DEDICATED
TO THE MEMORY OF
TOMAŠ PEXIEDER, M.D.**

THE RESEARCH GRANT

Genetic and Environmental Risk Factors of Major Cardiovascular Malformations:
The Baltimore-Washington Infant Study: 1981–1989

Principal Investigator: Charlotte Ferencz, M.D., M.P.H.

National Heart, Lung & Blood Institute: R37 HL25629
December 1, 1980–November 30, 1992

Meritorious Extension in Time (MERIT) Award
December 1, 1992–November 30, 1997

Overview

Cardiovascular malformations constitute a major segment of birth defects with profound medical, psychosocial, and economic consequences. Previous research was principally directed to clinical methods of diagnosis and treatment, but the need for prediction, prenatal counseling, and preventive interventions requires further knowledge of familial and environmental risk factors.

The Baltimore-Washington Infant Study is a large population-based etiologic study of congenital heart disease. With the collaboration of six pediatric cardiology centers, 53 area hospitals, and more than 800 physicians, 4390 cases and 3572 controls were enrolled between 1981 and 1989.

Cases are liveborn infants with cardiovascular malformations ascertained through multiple sources in the defined geographic area of about 100,000 annual births. Control infants are representative of the regional livebirth cohort (N = 906,626). Detailed diagnostic information on cardiac and noncardiac malformations was obtained on case probands; a questionnaire administered by trained interviewers in home visits to case and control mothers recorded genetic, sociodemographic, and medical/obstetric data, and parental exposures to potentially harmful substances in lifestyle, medical therapies, and in-home and occupational activities. Interview data were obtained on 3377 cases (90% of eligible) and on 3572 controls (95% first or second choices from the random listing).

A comprehensive evaluation of familial and infant characteristics was completed for all cardiovascular malformations[152] and for 17 anatomic types of major malformations. Multivariate analyses of genetic and environmental factors have generated etiologic hypotheses for future studies by multidisciplinary teams. The identification of vulnerable populations should eventually lead to improved preventive strategies.

Research Personnel

PRINCIPAL INVESTIGATOR

Charlotte Ferencz, M.D., M.P.H.
Pediatric Cardiologist/Epidemiologist

RESEARCH CENTER INVESTIGATORS

Judith Rubin, M.D., M.P.H.
Pediatric Epidemiologist

Adolfo Correa-Villaseñor, M.D., Ph.D.
Epidemiologist, Pediatrician

P. David Wilson, Ph.D.
Biostatistician

Carol A. Magee, Ph.D., M.P.H.
Environmental Epidemiologist

Christopher A. Loffredo, Ph.D.
Data Analyst/Manager

Robert J. McCarter, Sc.D.
Epidemiologist

Joann A. Boughman, Ph.D.
Genetic Epidemiologist

Iosif W. Lurie, M.D., Ph.D.
Geneticist, Dysmorphologist

PEDIATRIC CARDIOLOGISTS

Joel I. Brenner, M.D.
University of Maryland
Medical Systems

Gerard M. Martin, M.D.
Children's National Medical Center
George Washington University

Catherine A. Neill, M.D.
The Johns Hopkins
Medical Institutions

John W. Downing, M.D.
Howard University
Hospital

Lowell W. Perry, M.D.
Georgetown University
Children's Medical Center

Mohamed K. Mardini, M.D.
Seymour I. Hepner, M.D.
Fairfax Hospital

Karen S. Kuehl, M.D.
Georgetown University

Research Staff

Research Associates
Linda Reich, M.S.—Biostatistics
Anne Damiano, M.S.—Biostatistics
Cynthia Maier, M.Ed.—Scientific Editing

Study Coordinators
Glenn Heartwell (1981–1985)
Kimberly Hofmeister-Ripley (1985–1991)

Field Supervisors
Alice Good (1980–1987)
Joan Cantori (1987–1990)

Registration
Linda Finazzo
Beverly Law
S. Patricia Chiraboga-Roby
Rosemarie Cherry
Leslie Schinella
Crystal Stroud
Michelle Wade
Lori Gorschboth

Interviewers
Jeanne Schlossberg
Gertrude Rosen
Kathleen Sadler
Betty Concha
Shirley Kramer
Constance Nunnamaker
Norma Lutes

Data Editing
Denise Edmonds
Allen Arnold
Betty Williams

Data Entry
Karen Volpini
Colleen Shaffer

Secretaries
Beulah Maultsby
Jennifer Aronson
Wilhelmena Smith
Chandra Brown-Staten

Programming
Jacqueline Astemborksi, M.S.
Nittaya Suppapanya, M.S.
Mary Tang, M.S.

Participating Hospitals

Baltimore, Maryland
Francis Scott Key Medical Center, Franklin Square Hospital, Greater Baltimore Medical Center, Johns Hopkins Hospital, Liberty Medical Center, Maryland General Hospital, Mercy Medical Center, St. Agnes Hospital, Union Memorial Hospital, University of Maryland Medical System

Maryland Counties
Anne Arundel General Hospital, Calvert Memorial Hospital, Carroll County General Hospital, Dorchester General Hospital, Edward W. McCready Memorial Hospital, Frederick Memorial Hospital, Garrett County Memorial Hospital, Harford Memorial Hospital, Holy Cross Hospital of Silver Spring, Howard County General Hospital, Kent and Queen Anne's Hospital, Memorial Hospital and Medical Center of Cumberland, Memorial Hospital at Easton, Montgomery General Hospital, Peninsula General Hospital and Medical Center, Physicians Memorial Hospital, Prince George's General Hospital and Medical Center, Sacred Heart Hospital, St. Mary's Hospital, Shady Grove Adventist Hospital, Southern Maryland Hospital Center, Union Hospital of Cecil County, Washington Adventist Hospital, Washington County Hospital Association

Washington, D.C.
Children's Hospital National Medical Center, District of Columbia General Hospital, Georgetown University Hospital, George Washington University Medical Center, Greater Southeast Community Hospital, Howard University Hospital, Providence Hospital, Sibley Memorial Hospital, Washington Hospital Center

Virginia
Alexandria Hospital, Arlington Hospital, Fairfax Hospital, Fauquier Hospital, Loudon Memorial Hospital, Potomac Hospital, Prince William Hospital

Maryland Medical Examiner's Office
John E. Smialek, MD., Chief
Baltimore, Maryland

Directors of Vital Statistics
Frances J. Warthen, Chief
Division of Vital Statistics
Maryland Center for Health Statistics

Grover Chamberlain, Chief
Olive Shisana, Sc.D., Acting Chief
Research and Statistics
Government of the District of Columbia

Dorothy Harshberger, Assistant Director
Russell Booker, State Registrar
Virginia Center for Health Statistics

ACKNOWLEDGMENTS

The Baltimore-Washington Infant Study was funded by the National Heart Lung and Blood Institute (R37-HL25629); we express our gratitude to the Institute for granting us this unique research opportunity; we thank Dr. Barbara Packard for initial support, Dr. Millicent Higgins, and Dr. Thomas Blaszkowski for their continued advice and counsel over many years, and Dr. James Norman, who has helped us to bring the study to a successful end.

A great debt is due to Dr. Genevieve M. Matanoski, Professor of Epidemiology, The Johns Hopkins University School of Hygiene and Public Health, for her major role in the design of this study, as well as for her willing help in solving many problems.

A crucial contribution to this study was the committed collaboration of the Pediatric Cardiology community of the region—especially the "study cardiologists" Drs. Joel I. Brenner, Catherine A. Neill, Lowell W. Perry, Gerard Martin, John W. Downing, Mohamed Mardini, and Seymour Hepner, who enrolled the cases, established diagnostic accuracy and uniformity of coding practices, and fostered the family contacts, which resulted in the remarkable response rate. The cardiologists attended many planning and monitoring meetings, participated in the preparation of publications, and made many presentations at pediatric and pediatric cardiology meetings. This intensive collaboration of clinicians in an epidemiologic study was exemplary, and constituted a great strength of the Baltimore-Washington Infant Study.

The value of an epidemiologic study depends on the quality of the information. We owe a great debt to Dr. Judith D. Rubin, whose conscientious oversight of the field and data work is now rewarding us with confidence in our data; her detailed attention to the coding and accuracy of reported information on pharmaceuticals and other drug exposures must be unique in an epidemiologic study.

Dr. Joann A. Boughman opened new horizons of population genetics in the evaluation of case and family information, and was the principal author of the Genetics chapter in our first book. Dr. Iosif W. Lurie contributed his encyclopedic knowledge of genetics and dysmorphology to greatly enhance the quality of this work. By his reassessment and updating of all case information on noncardiac anomalies he created the Glossary on Dysmorphology (Appendix), and his reviews and discussions enriched this volume immeasurably.

Dr. Carol A. Magee and Dr. Karen S. Kuehl generously provided their expertise in the environmental and cardiological assessments respectively.

Dr. Wolfgang Mergner, gave generously of his time to achieve the illustrations of the normal cardiac architecture.

Dr. Robert McCarter implemented data management, quality control procedures, and the computerization of management procedures. Computer time for this project was supported through the facilities of Academic Computing/Health Informatics of the University of Maryland at Baltimore.

Administrative officers of the Department of Epidemiology and Preventive Medicine, Ms. Wendy Cohan, Roxanne Zaghab, Mary Kristin Schlack, Doris Farrell, and Donna Ferger provided much needed help throughout the years.

We thank the Administrators and Record Room staff of the 53 regional hospitals for their collaboration, and the many obstetricians, pediatricians, and pathologists who aided in case and control enrollment; more than 700 practicing physicians in the community supported enrollment of study subjects and helped us to achieve diagnostic accuracy.

Rapid advances in embryology and in cardiac morphogenesis over the years of the study provided extensive new information; Dr. Edward B. Clark and Dr. Tomaš Pexieder contributed greatly to our considerations on developmental, and toxicological aspects of cardiac development. We used Dr. Clark's mechanistic categorization of congenital heart disease, and the definition of "pure" diagnoses was initiated by him. Dr. Clark promoted the presentation of our study results in national and international meetings, helping to link epidemiologic data with biologic advances. The ongoing interest and helpful considerations of Dr. Tomaš Pexieder were an important source of encouragement; all Baltimore-Washington Infant Study investigators keenly feel the tragedy of his untimely death (see Tribute).

We thank Dr. Walter C. Willett, Professor of Nutrition, Harvard School of Public Health, for allowing us to use his Food Frequency Questionnaire during the final 3 years of the study. The analyses of these nutritional data are now in progress in collaboration with the Birth Defects and the Nutrition Branches of the Centers for Disease Control in Atlanta, Georgia; we are grateful for the interest of Drs. Kelley Scanlon, Lorenzo Botto, and Muin Khoury.

We owe a great debt to Dr. Robert H. Anderson of the National Heart Institute in London, England, for his ongoing encouragement and many suggestions concerning anatomic and developmental considerations, and appropriate terminology. We are deeply grateful for his interest and help in including this research monograph in the Perspectives in Pediatric Cardiology Series.

Our very devoted research staff maintained their unfailing commitment to excellence, each in a specific area of the study; special thanks go to Beulah Maultsby, our secretary for many years; Glenn Heartwell and Kimberly Hofmeister-Ripley, study coordinators; and Cynthia Maier, academic assistant who created a slide file and reprint library that have served us so well! Updated literature entries were added to these documents in recent years by Jean Schlossberg and Betty Jean Williams. Jenifer Aronson and Chandra Brown-Staten converted the old bibliographic card files into a modern, computerized Reference Manager System.

A great debt is due to the Health Sciences Library, University of Maryland: Mrs. Frieda Weise, Director; Mr. Richard Behles, Historical Librarian; and the Reference Librarians who never tired of our many requests. Special thanks go to the Osler Library at McGill University, Montreal, Canada—for permitting access to the files of Dr. Maude E. Abbott and to the famous monograph of Dr. Carl von Rokitansky. We thank Mr. Wayne LeBel, Assistant Osler Librarian, and Ms. E. Phebe Chartrand, McGill University Archivist, for their patient search for helpful documents and for photographs.

We acknowledge the outstanding support of the Division of Medical Illustrations: the creative artwork was prepared by Mary Grindell-Donhouser, BFA, the graphs by Stephanie Czech, and the photographs by Mark Teske, Thomas Wynn, Laura Steber, Tom Jenski, and Ursula Goldman. In the preparation of this manuscript we relied on the untiring commitment of Betty Jean Williams, research assistant, and Chandra Brown-Staten, secretary, who gave their enthusiastic support and talents to every page of every chapter. We are greatly indebted to Futura Publishing, Inc. and Mr. Steven Korn, in particular, for his enthusiastic support.

Finally, but most importantly, our thanks and those of our readers, go to the 7000 families of case and control infants who willingly gave their time and effort to participate in a lengthy interview that gathered the substance of a new body of knowledge. Together we hope that this work will, in time, create stepping stones toward the prevention of congenital heart disease.

TRIBUTE

Tomaš Pexieder, M.D. 1941–1995

The Baltimore-Washington Infant Study mourns an advisor and friend, whose thoughts, reflections, and advice are integral to the understanding of congenital heart disease as revealed in these chapters.

In the planning phase of the study, embryologist Mary Lou Oster-Granite familiarized us with Pexieder's important studies on the categorization of cardiovascular malformations and we noted his emerging authority in cardiac embryology. Then, at the Second World Congress of Pediatric Cardiology in New York in 1985, our poster on early epidemiologic results awoke the intense interest of a Congress participant with whom we engaged in an animated discussion. Suddenly I noticed his name tag. "Are you *the* Pexieder, the great embryologist??" I asked, and received an amused affirmative reply while I transferred my imagined "great" embryologist to this smiling, slender young man. Soon we received some reprints from him, and in a rapid sequence of letters and encounters we crystallized areas of common interest and a sense of mutual understanding.

Within a year of meeting we began an exchange of prevalence data, comparing Baltimore-Washington Infant Study to the EUROCAT and Swiss studies. We developed common diagnostic definitions for comparative evaluations, and the findings on these "sentinel" cases were demonstrated in joint poster presentations, EUROCAT publications, and the final discussion presented by Tomaš in Tokyo in December 1993.[346]

Over the year Tomaš' visits to Baltimore were inserted into his various USA tours related to his collaborative studies with Ed Clark, Don Patterson, Roger Markwald, and many others, as well as his participation in meetings of the Teratology Society, the American Heart Association, and special sessions on morphogenesis. Three Baltimore-Washington Infant Study investigators (Boughman, Correa, and Ferencz) were part of the 1992 Wiley-Liss Symposium at the Teratology Society meeting, which he organized with Ed Clark. In 1988, I had the great pleasure of visiting Lausanne and spending some days seeing the outstanding work carried on by Pexieder's research group on various experimental studies on mice and rats with genetic and teratogenic interventions, as well as the epidemiologic studies and his constantly updated Teratologic Information System, which provided up-to-date data to clinicians. The warm hospitality of Tomaš and his family will remain as treasured memories.

At the Baltimore-Washington Infant Study 10th Anniversary Celebration on December 13, 1990, Tomaš Pexieder presented the keynote address. It was remarkable to note his global perceptions and his overview of so many frontiers of knowledge.[434] His vision, like that of an archeologist, was always on the *whole target* but never neglecting any small detail, any possible clue to a new direction. He never stopped acquiring new ideas and new study methods. He was a true scholar who could lead, join, interrelate, and collaborate. He stood at the center point of a remarkable network of like-minded scientists: sincere, sensible, generous in giving, sharing, and giving back much more than he received.

This book, the culmination of the Baltimore-Washington Infant Study's evaluation of genetic and environmental risk factors, owes much to him and he awaited it keenly, sending congratulations to us on the book's progress in his last letter, written on October 16, 1995, just 12 days before his death. Then his bright and warm light was extinguished. He left us in our eleventh hour. He was to have been the first and most important reader of this book. He would have known how to blow away the chaff and transplant the seeds into the domain of his generation and then perhaps the next. Twenty years his senior, I watched his luminous career, saw him in the center of the coming new world of knowledge which, as he hoped, *"will realize a new opportunity for integration of cellular and molecular biology as well as molecular genetics with cardiac embryology and teratology, each partner taking advantage of the other partner's expertise so that one can consider with sufficient confidence the development of first experimental and later clinical primary prevention."*[344]

Tomaš Pexieder left us a purpose, a way, and an example. How thankful we must be that for a whole decade we shared the road with him and were encouraged by his interests and his visions! *"Do not worry"*, Tomaš wrote on February 27, 1995, *"about burdening me with what you call 'Baltimore-Washington Infant Study junk,'* (small items) *"at all times I am pleased to review it because it challenges my own thinking, beliefs, and understanding of the mystery of normal and abnormal cardiac development."*

We miss him greatly.

Charlotte Ferencz

PREFACE

Congenital heart disease encompasses a wide range of abnormalities of the heart and great vessels. Some of these are mild and can be well tolerated throughout a normal lifetime, but others present serious problems in early infancy and constitute a major cause of infant death, either as single anomalies or as part of multiple malformations. Over the past 50 years the introduction and then the rapid expansion of surgical methods has been the centerpiece of professional and public attention: defects in the heart have been corrected and the circulation has been rearranged to provide relief of symptoms and an improved outlook for life and health. With the spectacular rise of medical technology, the diagnosis of cardiovascular anomalies was accomplished earlier and earlier, and finally focused on the infancy period; recently it has extended also to the fetus by obstetric ultrasound studies.

It has been said that in the field of Pediatric Cardiology "the surgeons have shown the way," and untold millions of dollars are readily available for even a partial or temporary amelioration of the clinical course of sick babies. The possibility of *preventing* congenital heart disease has been viewed as a remote hope, even though the question continues to be raised by the parents of infants with cardiovascular malformations and by adults whose own cardiac malformation has been successfully treated and who realize that their risk of having a child with a cardiac malformation is considerably increased.

Now society has, at last, begun to consider strategies to prevent birth defects by addressing nutritional inadequacies and by providing public education regarding the risks of using alcohol and recreational drugs.[88] Moreover, new legal issues have arisen concerning the possible teratogenic effects of herbicide exposures in veterans of the Vietnam War, chemical exposures in the Persian Gulf War, and domestic exposures at toxic waste sites such as Love Canal, while the earlier epidemics of birth defects due to rubella infection and thalidomide have now faded in memory.

The Baltimore-Washington Infant Study, an etiologic study of congenital heart disease, was designed to address the serious lack of knowledge of the epidemiology of congenital heart disease that became evident in the 1970s, a time of extensive legal activity concerning the possible teratogenic effect of maternal hormone therapy, which was thought to result in various malformations, including those of the heart. Smaller studies, including our own,[147] that were funded to test this hypothesis, could not answer the research question because of the absence of any information on possible correlates of congenital heart disease. It was clear that an exploratory hypothesis-generating study was needed to document the epidemiology of cardiovascular malformation. The Baltimore-Washington Infant Study was designed to respond to this need.

At the beginning, the knowledge about congenital heart disease appeared to be a gigantic jigsaw puzzle of disconnected pieces. There were historical pieces of information and pieces representing a wide variety of experimental morphogenetic studies, which offered some insights into the nature and developmental mechanisms of the major cardiovascular malformation groups. However, this information failed to reveal strategies by which the malformations could be prevented. It was expected that the identification of genetic and environmental risk factors would fill in the missing pieces of the puzzle, and define the risk profiles of specific cardiovascular phenotypes.

A serious commitment to define the epidemiology of cardiovascular malformations became a shared objective of a multidisciplinary team of investigators who, in 1978, created a research alliance of clinicians, epidemiologists, and basic scientists, known as the Baltimore-Washington Infant Study Group. This commitment challenged the view that congenital heart disease occurs equally everywhere and among all kinds of people. The Mid-Atlantic region was considered to be a microcosm of the multiethnic American population, in which rich and poor live in highly urbanized as well as in remote rural areas, with a wide variety of lifestyles and occupational activities, providing an appropriate "community laboratory" in which the distribution of congenital heart disease could be effectively studied.[220]

Such an exploratory case-control study requires a large study population for the evaluation of multiple factors. Through multicenter participation over nearly a decade, the Baltimore-Washington Infant Study enrolled a regional case pool of infants with congenital heart disease from a livebirth cohort of almost 1,000,000 livebirths, and selected a representative sample of births as controls from the defined study area. The study methods, instruments, and the findings on the aggregate case and control groups that demonstrated potential risks in genetic and environmental factors, were published in 1993 in *The Epidemiology of Congenital Heart Disease: The Baltimore-Washington Infant Study, 1981–1989.*[152] That book presented for the first time a systematic description of the epidemiology of congenital heart disease by time, place, and person. It established the framework for the present book on the selected diagnosis-specific evaluations of more homogenous case subsets.

Part I of this new book lays the groundwork for the subsequent sections by detailing the historical background, categorization of defects, and epidemiologic and statistical methods. Part II is a compilation of information on major anatomic phenotypes, presented chapter by chapter in uniformly designed tables, figures, and text. The set of summary chapters (Part III) provides a synthesis and discussion of the diagnosis-specific material. Thus, this book provides a documentary of the systematic evaluation of possible genetic and environmental risk factors of cardiovascular malformations. This book should serve as a scientific basis for future studies. It will permit the consideration of preventive strategies to protect vulnerable individuals and groups and to lower the frequency of occurrence of congenital heart disease.

There is reason for optimism because the comparative evaluation of various anatomic types of cardiovascular malformations has revealed a striking coherence with the concepts of recent morphogenetic studies. Thus we can open a dialogue with colleagues in the basic sciences who can elucidate the mechanisms by which the abnormalities arise, through an understanding of the actions of teratogens identified in this epidemiologic study.

A major finding of this epidemiologic study is the etiologic categorization of cardiovascular malformations into abnormalities of primary and of secondary cardiogenesis, respectively. For *primary* defects, two etiologic pathways were identified. In one of these pathways, maternal diabetes was the principal risk factor, and other risk factors included past reproductive failures and exposures to certain therapeutic drugs. The other pathway leading to the same major anomalies was through trisomies of chromosomes 13,18, and 21; for these cases, only the well-known effect of increasing maternal age was found to be a strong risk factor, a notable finding, since we analyzed over 200 potential risk variables. Malformations of *secondary cardiogenesis* include those in which the fundamental design of the heart is normal. Those with obstructive lesions of the right or left side of the heart showed a strong genetic etiology and evidence of a role of environmental toxicants, especially solvent exposures.

The findings of the Baltimore-Washington Infant Study reveal answers to many questions that have long been asked about cardiovascular malformations, but the etiology of congenital heart disease remains elusive. Nora[324] aptly expresses the paradox of his research: *"Although we have opened many doors to learn about the causes of congenital heart disease, we have found that the doors lead to new corridors flanked by many more doors."* The use of epidemiologic methods has allowed us to open some of those doors. We are confident that a new generation of researchers will make use of these findings and open the remaining doors that lead to the prevention of congenital heart disease and other major malformations.

CONTENTS

THE RESEARCH GRANT vii

ACKNOWLEDGMENTS xi

TRIBUTE TO TOMAŚ PEXIEDER xv

PREFACE xvii

PART I BACKGROUND AND METHODS

CHAPTER 1 HISTORICAL PERSPECTIVES *Charlotte Ferencz* 3

    Community Concern About Patients with Heart Disease 4

    Witness to the New Era 4

    Maude E. Seymour Abbott (1869–1940) 5

    Helen Brooke Taussig (1898–1986) 8

    John D. Keith (1908–1988) 9

    Richard D. Rowe (1923–1988) 10

    Alexander S. Nadas (1913–) 10

    Maurice Campbell (1891–1973) 11

**CHAPTER 2 CATEGORIZATION OF CARDIOVASCULAR
MALFORMATIONS FOR RISK FACTOR ANALYSIS** 13

    THE NORMAL HEART 14

    THE MALFORMED HEART 15

    PREVIOUS CLASSIFICATIONS OF CARDIOVASCULAR

        MALFORMATIONS 24

    CATEGORIZATION OF CARDIOVASCULAR MALFORMATIONS

        FOR RISK FACTOR ANALYSIS 26

        A Hierarchical Order 26

        Mechanistic Grouping 26

        "Pure" Cardiac Diagnoses 26

    ASSOCIATED NONCARDIAC ANOMALIES 27

**CHAPTER 3 EPIDEMIOLOGIC CASE-CONTROL DESIGN AND
STATISTICAL METHODS** 29

    Cases 30

        Length of Follow-up 31

        Sampling of Mild Defects for Interview 31

    Controls 31

    Data Collection and Quality Control 33

    Variables Analyzed 35

Statistical Analysis 35
    Evaluation of Genetic Component 36

PART II RISK FACTOR ANALYSIS OF DIAGNOSTIC GROUPS

CHAPTER 4 DEFECTS OF LATERALITY AND LOOPING.............................. 41
    STUDY POPULATION.. 42
        Cardiac Abnormalities... 42
        Noncardiac Anomalies 44
    DESCRIPTIVE ANALYSES.. 46
        Prevalence by Time, Season, and Area of Residence 46
        Diagnosis and Course 46
        Gender, Race, and Twinning...................................... 46
        Birthweight and Gestational Age 48
        Sociodemographic Characteristics............................ 48
    POTENTIAL RISK FACTORS 50
        Familial Cardiac and Noncardiac Anomalies........................... 50
        Genetic and Environmental Factors 50
            Univariate Analyses 50
            Multivariate Analyses 53
    DISCUSSION 53
        Prevalence.. 54
        Cardiovascular Malformations and Noncardiac Anomalies.... 54
        Genetic Heterogeneity... 54
        Maternal Diabetes 55
        Environmental Factors 56
        Socioeconomic Status 56
    CONCLUSIONS... 57

CHAPTER 5 MALFORMATIONS OF THE CARDIAC OUTFLOW TRACT 59
        Clinical Perspectives 60
        Developmental Perspectives 60
        Epidemiologic Perspectives....................................... 62
    STUDY POPULATION.. 63
        Cardiac Abnormalities... 64
        Noncardiac Anomalies 64
    DESCRIPTIVE ANALYSES.. 68
        Prevalence by Time, Season, and Area of Residence 68
        Diagnosis and Course 69
        Gender, Race, and Twinning...................................... 72
        Birthweight and Gestational Age 72
        Sociodemographic Characteristics............................ 72
    POTENTIAL RISK FACTORS 79
        Familial Cardiac and Noncardiac Anomalies........................... 79
        Genetic and Environmental Factors 79
            Univariate Analysis 79
            Maternal Reproductive History 79
            Maternal Illnesses 84
            Maternal Medications... 84
            Paternal Medical Exposures 88

Lifestyle Exposures.. 88
Maternal Home and Occupational Exposures................. 88
Multivariate Analysis.. 88
The Major Groups of Transposition and Normal
    Great Artery Outflow Tract Defects............................. 90
Diagnosis-Specific Analyses: Transposition with Intact
    Ventricular Septum and Tetralogy of Fallot 94
Summary of Findings 96
DISCUSSION 96
Genetic Risk Factors 96
Proband... 99
Family History 99
Heritable Blood Disorders 99
Maternal Diabetes 101
Environmental Factors 101
CONCLUSIONS... 102

**CHAPTER 6 ATRIOVENTRICULAR SEPTAL DEFECTS WITH
AND WITHOUT DOWN SYNDROME 103**
STUDY POPULATION... 105
Cardiac Abnormalities.. 105
Noncardiac Anomalies 106
Chromosomal and Mendelian Disorders 107
Non-Mendelian Associations and Other Noncardiac Defects . 107
DESCRIPTIVE ANALYSES.. 108
Prevalence by Time, Season, and Area of Residence 108
Diagnosis and Course 109
Gender, Race, and Twinning.. 109
Birthweight and Gestational Age 111
Sociodemographic Characteristics................................. 111
POTENTIAL RISK FACTORS 114
Familial Cardiac and Noncardiac Anomalies 114
Genetic and Environmental Factors 114
Univariate Analyses 114
Maternal Age and Reproductive History 114
Maternal Illnesses and Medications 118
Lifestyle Exposures.. 119
Parental Home and Occupational Exposures................. 119
Multivariate Analysis 119
DISCUSSION 120

CHAPTER 7 VENTRICULAR SEPTAL DEFECTS
SECTION A: OVERVIEW... 124

SECTION B: MEMBRANOUS TYPE 128
STUDY POPULATION... 128
Cardiac Abnormalities.. 128
Noncardiac Anomalies 128
DESCRIPTIVE ANALYSES.. 132
Prevalence by Time, Season, and Area of Residence 132

Diagnosis and Course 132
Gender, Race, and Twinning.. 133
Birthweight and Gestational Age 135
Similarity of Small and Moderate/Large Defects...................... 135
Sociodemographic Characteristics.. 136
POTENTIAL RISK FACTORS 137
Familial Cardiac and Noncardiac Anomalies............................ 137
Genetic and Environmental Factors 137
Univariate Analysis 137
Multivariate Analysis... 140
DISCUSSION 145
SUMMARY 147
SECTION C: MUSCULAR TYPE.. 149
STUDY POPULATION... 149
Cardiac Abnormalities... 149
Associated Cardiac Abnormalities... 150
Noncardiac Anomalies 151
DESCRIPTIVE ANALYSES... 151
Prevalence by Time, Season, and Area of Residence 151
Diagnosis and Course 154
Gender, Race, and Twinning.. 155
Birthweight and Gestational Age 155
Sociodemographic Characteristics.. 157
POTENTIAL RISK FACTORS 158
Familial Cardiac and Noncardiac Anomalies............................ 158
Genetic and Environmental Factors 158
Univariate Analysis 159
Multivariate Analysis... 159
DISCUSSION 159
Possible Genetic Risk Factors 162
Changes in Prevalence and Risk Factors 162
Reported Epidemiologic Studies... 163

CHAPTER 8 LEFT-SIDED OBSTRUCTIVE LESIONS
SECTION A: OVERVIEW.. 166
Historical and Diagnostic Considerations 166
Analysis Plan... 167
RISK FACTOR ANALYSIS OF THE TOTAL GROUP 168
Diagnostic Definition 168
Genetic Relationships Among Left-Sided Heart Defects 168
Infant Characteristics 171
POTENTIAL RISK FACTORS 172
Univariate Analysis 172
Multivariate Analysis... 174
Isolated/Simplex Subset... 174
Multiple Anomalies/Multiplex Subset 176
SUMMARY 176
SECTION B: HYPOPLASTIC LEFT HEART SYNDROME..................... 178
STUDY POPULATION... 179

Cardiac Abnormalities.. 179
Noncardiac Anomalies 180
DESCRIPTIVE ANALYSES.. 180
Prevalence by Time, Season, and Area of Residence 180
Diagnosis and Course 181
Gender, Race, and Twinning.. 181
Birthweight and Gestational Age 183
Sociodemographic Factors 183
POTENTIAL RISK FACTORS 183
Familial Cardiac and Noncardiac Anomalies............................ 183
Genetic and Environmental Factors 185
Univariate Analysis.. 185
Multivariate Analysis.. 187
DISCUSSION 187
Family History.. 188
Maternal Illnesses 188
Environmental Risk Factors... 188
Variability of Anatomic Features 189
SECTION C: COARCTATION OF AORTA................................ 190
STUDY POPULATION.. 191
Cardiac Abnormalities.. 192
Noncardiac Anomalies 192
DESCRIPTIVE ANALYSIS.. 193
Prevalence by Time, Season, and Area of Residence 193
Diagnosis and Course 193
Gender, Race, and Twinning.. 193
Birthweight and Gestational Age 195
Sociodemographic Characteristics... 195
POTENTIAL RISK FACTORS 196
Familial Cardiac and Noncardiac Anomalies............................ 196
Genetic and Environmental Factors 197
Univariate Analysis.. 197
Multivariate Analysis.. 199
DISCUSSION 200
SECTION D: AORTIC STENOSIS.. 202
STUDY POPULATION.. 203
Noncardiac Anomalies 203
DESCRIPTIVE ANALYSES.. 204
Prevalence by Time, Season, and Area of Residence 204
Diagnosis and Course 204
Gender, Race, and Twinning.. 204
Birthweight and Gestational Age 206
Sociodemographic Characteristics... 206
POTENTIAL RISK FACTORS 208
Familial Cardiac and Noncardiac Anomalies............................ 208
Genetic and Environmental Factors 208
Univariate Analysis.. 208
Multivariate Analysis.. 210
DISCUSSION 210

SECTION E: BICUSPID AORTIC VALVE 212
STUDY POPULATION... 212
    Cardiac Abnormalities... 212
    Noncardiac Anomalies 212
DESCRIPTIVE ANALYSES.. 212
    Prevalence by Time, Season, and Area of Residence 212
    Diagnosis and Course 213
    Gender, Race, and Twinning... 214
    Birthweight and Gestational Age 214
    Sociodemographic Characteristics.................................... 214
POTENTIAL RISK FACTORS 216
    Familial Cardiac and Noncardiac Anomalies........................ 216
    Genetic and Environmental Factors 217
    Univariate Analysis 217
    Multivariate Analysis 218
DISCUSSION 219
SECTION F: SUMMARY AND CONCLUSIONS 220
DESCRIPTIVE ANALYSES.. 220
    Infant Characteristics 220
    Race and Gender: Epidemiologic Considerations.................... 220
POTENTIAL RISK FACTORS 224
    Evaluation of the Effect of Doubly Ascertained Families 224
CONCLUSIONS 225

CHAPTER 9 RIGHT-SIDED OBSTRUCTIVE DEFECTS
    SECTION A: OVERVIEW 228

SECTION B: PULMONARY VALVE STENOSIS 230
    Recognition: Changes By Time... 232
    Categorization of Severity... 233
STUDY POPULATION... 233
    Cardiac Abnormalities... 233
    Noncardiac Anomalies 234
DESCRIPTIVE ANALYSES.. 235
    Prevalence by Time, Season, and Area of Residence 235
    Diagnosis and Course 236
    Gender, Race, and Twinning... 237
    Birthweight and Gestational Age 240
    Sociodemographic Characteristics.................................... 240
POTENTIAL RISK FACTORS 242
    Familial Cardiac and Noncardiac Anomalies........................ 242
    Genetic and Environmental Factors 244
        Univariate Analysis 244
        Multivariate Analysis 245
DISCUSSION 245

SECTION C: PULMONARY VALVE ATRESIA WITH INTACT
VENTRICULAR SEPTUM... 249
STUDY POPULATION... 249

Cardiac Abnormalities.. 249
Noncardiac Anomalies 250
DESCRIPTIVE ANALYSES.. 250
Prevalence by Time, Season, and Area of Residence 250
Diagnosis and Course 251
Gender, Race, and Twinning.. 251
Birthweight and Gestational Age 252
Sociodemographic Characteristics... 252
POTENTIAL RISK FACTORS 254
Familial Cardiac and Noncardiac Anomalies........................... 254
Genetic and Environmental Factors 254
Exposures In the Second and Third Trimesters 257
DISCUSSION 257
SECTION D: TRICUSPID ATRESIA WITH NORMALLY
RELATED GREAT ARTERIES... **258**
STUDY POPULATION.. 259
Cardiac Abnormalities.. 259
Noncardiac Anomalies 259
DESCRIPTIVE ANALYSES.. 259
Prevalence by Time, Season, and Area of Residence 259
Diagnosis and Course 260
Gender, Race, and Twinning.. 260
Birthweight and Gestational Age 260
Sociodemographic Characteristics... 260
POTENTIAL RISK FACTORS 262
Familial Cardiac and Noncardiac Anomalies........................... 262
Genetic and Environmental Factors 262
DISCUSSION 265

CHAPTER 10 ATRIAL SEPTAL DEFECT **267**
STUDY POPULATION.. 269
Cardiac Abnormalities.. 269
Noncardiac Anomalies 270
DESCRIPTIVE ANALYSES.. 271
Prevalence by Time, Season, and Area of Residence 271
Diagnosis and Course 271
Gender, Race, and Twinning.. 274
Birthweight and Gestational Age 274
Sociodemographic Characteristics... 274
POTENTIAL RISK FACTORS 275
Familial Cardiac and Noncardiac Anomalies........................... 275
Genetic and Environmental Factors 278
Univariate Analysis 278
Multivariate Analysis 280
DISCUSSION 281
Diagnosis and Prevalence 281
Infant Characteristics 282
Genetic Factors.. 282
Environmental Factors 283

CHAPTER 11 PATENT ARTERIAL DUCT **285**
   INTRODUCTION.. 286
   STUDY POPULATION.. 287
      Noncardiac Anomalies 287
   DESCRIPTIVE ANALYSES.. 288
      Prevalence by Time, Season, and Area of Residence 288
      Diagnosis and Course 288
      Gender, Race, and Twinning...................................... 290
      Birthweight and Gestational Age 290
   POTENTIAL RISK FACTORS 293
      Familial Cardiac and Noncardiac Anomalies........................... 293
      Sociodemographic Characteristics........................... 293
      Genetic and Environmental Factors 294
         Univariate Analysis.. 294
         Multivariate Analysis... 296
      Maternal Exposures During the Second and
         Third Trimesters 296
   DISCUSSION 297
      Prevalence and Relative Frequency........................... 297
      Potential Risk Factors 297

CHAPTER 12 UPDATES AND COMMENTS ON PREVIOUS
PUBLICATIONS **301**
   SECTION A: TOTAL ANOMALOUS PULMONARY
   VENOUS RETURN.. 303
   STUDY POPULATION.. 303
      Cardiac Abnormalities.. 304
      Noncardiac Anomalies 304
   DESCRIPTIVE ANALYSES.. 306
      Prevalence by Time, Season, and Area of Residence 306
      Diagnosis and Course 306
      Gender, Race, and Twinning...................................... 307
      Birthweight and Gestational Age 307
   POTENTIAL RISK FACTORS 309
      Familial Cardiac and Noncardiac Anomalies 309
      Genetic and Environmental Factors 309
   DISCUSSION 312

   SECTION B: CARDIOMYOPATHY... 313
   STUDY POPULATION.. 313
   Noncardiac Anomalies 314
   DESCRIPTIVE ANALYSES.. 316
      Prevalence by Time, Season, and Area of Residence 316
      Diagnosis and Course 316
      Gender, Race, and Twinning...................................... 317
      Birthweight and Gestational Age 317
      Sociodemographic Characteristics........................... 317
   POTENTIAL RISK FACTORS 320
      Familial Cardiac and Noncardiac Anomalies........................... 320

Genetic and Environmental Factors 320
    Univariate Analysis 320
    Multivariate Analysis 321
DISCUSSION 323

*SECTION C: EBSTEIN'S MALFORMATION OF
THE TRICUSPID VALVE* 325
STUDY POPULATION 325
    Cardiac Abnormalities 325
    Noncardiac Anomalies 325
DESCRIPTIVE ANALYSES 326
    Prevalence by Time, Season and Area of Residence 326
    Diagnosis and Course 326
    Gender, Race, and Twinning 326
    Birthweight and Gestational Age 328
    Sociodemographic Characteristics 328
POTENTIAL RISK FACTORS 329
    Familial Cardiac and Noncardiac Anomalies 329
        Univariate and Multivariate Analysis 329
DISCUSSION 332

PART III SYNTHESIS

CHAPTER 13 THE INFANT WITH CONGENITAL HEART DISEASE 337
    INTRODUCTION.. 338
    PREVALENCE 339
    THE MALFORMATION TRIAD 344
        Cardiovascular Malformations 344
        Noncardiac Anomalies 345
            Chromosomal Syndromes 348
        Small Size at Birth: Premature and Growth Retarded 351
    IMPLICATIONS OF THE MALFORMATION TRIAD 354

CHAPTER 14 RISK FACTOR ANALYSIS: A SYNTHESIS 359
    INTRODUCTION.. 360
    FAMILY HISTORY OF CARDIOVASCULAR MALFORMATIONS............. 361
        Familial Aggregation of Cardiac and
            Noncardiac Abnormalities 364
        Genetic Heterogeneity and Genetic Liability............. 366
    ENVIRONMENTAL RISK FACTORS 367
        Maternal Diabetes 368
        Maternal Influenza and Fever 370
        Maternal Reproductive History............................... 371
        Maternal Use of Prescription and Nonprescription
        Pharmaceuticals... 372
            Antitussives 372
            Clomiphene... 372
            Corticosteroids 374
            Ibuprofen.. 374

Diazepam, Benzodiazepines 374
Metronidazole 374
Salicylates 375
Lifestyle Factors 375
Alcohol 376
Cigarette Smoking.. 376
Recreational Drugs 376
Caffeine... 376
Hair Dyes... 377
Home and Occupational Exposures 377
Solvents.. 377
Pesticides... 379
Lead and Other Metals.. 379
Paternal Factors 379
Comments on Environmental Factors..................................... 380

CHAPTER 15 RESEARCH IMPLICATIONS **383**
INTRODUCTION... 384
CARDIOLOGY AND CARDIAC MORPHOGENESIS 386
DYSMORPHOLOGY AND GENETICS *Iosif W. Lurie, M.D., Ph.D.* 387
MUTATIONS AND MICRODELETIONS 388
EPIDEMIOLOGY 391
NEED FOR MULTIDISCIPLINARY PARTNERSHIPS................................ 393

APPENDICES **395**
A. GLOSSARY OF MALFORMATION SYNDROMES: *I. W. Lurie, MD, PhD*
   Multiple Malformations Syndromes Found in the
   Baltimore-Washington Infant Study: 1981–1989 396
B. VARIABLES
   Categories of Variables Examined in the
   Case-Control Analyses... 404
C. CASE LISTINGS
   Identification Numbers of Infants with Outflow
   Tract Anomalies and Associated Noncardiac Anomalies 406
   Identification Numbers of Down Syndrome
   Cases with Atrial or Ventricular Septal Defects 411
D. CONTROL LISTINGS
   Distribution of Potential Risk Factor Variables
   Among 3572 Controls... 413
   Cardiovascular Malformations Among First-Degree
   Relatives of Control Infants.. 420
   Noncardiac Malformations Among First-Degree
   Relatives of Control Infants.. 422

REFERENCES **429**

INDEX **449**

PART I

BACKGROUND AND METHODS

Chapter 1

Historical Perspectives

Charlotte Ferencz

From: Ferencz C, Loffredo CA, Correa-Villaseñor A, Wilson PD, eds: *Genetic & Environmental Risk Factors of Major Cardiovascular Malformations: The Baltimore-Washington Infant Study 1981–1989.* Armonk, NY: Futura Publishing Co., Inc; ©1997.

Historical perspectives serve as a useful background to an etiologic study as we recognize the foundations laid by our predecessors in the description of the disease under investigation. A prime example is the work of Cheadle, who in 1900 separated childhood rheumatism from the rheumatic diseases of adults; he thereby characterized a specific disease and opened the way for specific remedies and preventive therapies.[79] For congenital malformations of the heart, such a categorization has not yet been achieved and there is no clear view today of either etiology or prevention.

Community Concern About Patients with Heart Disease

The foundations of the discipline of cardiology for adults and children were the heart clinics. Segall[390] reminds us that these clinics originated in response to the urging of Social Service workers who were concerned about the patients' lives in their communities.

At the Massachusetts General Hospital, a Children's Heart Clinic was founded in 1910; this was the first pediatric heart clinic, and was established to study acute endocarditis in children with rheumatic heart disease. In New York City, Dr. Haven Emerson, the Commissioner of Health from 1915 to 1917, noted that in the school system there were 20,000 children with heart disease who needed organized medical care and public health interventions. The American Heart Association was formed in 1924 as a voluntary agency, with Dr. Lewis A. Connor as its first president.[390] National legislation [Shepherd Towner Act (1921) and Social Security Act, Title V (1935)] provided support for the care and protection of children with heart disease due to rheumatic fever and for those with orthopedic disabilities due to poliomyelitis. The comprehensive Crippled Children's Services were available also to children with congenital heart disease when, in 1930, Dr. Edwards A. Park, the third Chief of Pediatrics at the Johns Hopkins Hospital, formed the Cardiac Clinic at the Harriet Lane Home for Invalid Children[408] and chose Helen Brooke Taussig as its Director. The Cardiac Clinic at the Harriet Lane Home became the site of dramatic innovations in the operative care of children with cyanotic congenital heart disease, which opened the "new surgical" era.

Witness to the New Era

The year was 1945, the historic year that marked the end of World War II. At McGill University in Montreal, Canada, the wartime acceleration of medical training was concluded and the University, under its visionary principal, F. Cyril James,[167] was planning the reeducation of veterans of the Canadian Armed Forces returning from overseas. Medical students who had received concurrent military training were "demobilized," and competed for hospital positions with colleagues whose clinical experiences had been in bombed cities and battlefields.

In the Spring of 1945 the Women's Medical Society honored the graduating women at an afternoon tea, made memorable by the announcement that *a woman in Baltimore helped to devise an operation which turns blue children pink!* This was almost incredible: tying off a patent arterial duct was impressive enough, but rearranging the circulation was truly revolutionary; where could this lead in the future? Now, half a century later, we can report on "miracles" of technology and human skills, which signaled the rise of pediatric cardiology and pediatric cardiac surgery and the later emergence of two further areas of specialization: the adult with congenital heart disease and fetal cardiology, an obstetric-pediatric interface.

Pediatric cardiology as a discipline has been characterized by the medical-surgical teamwork that achieved the treatment of cyanotic congenital heart disease by the Blalock-Taussig shunting operation for tetralogy of Fallot in 1945.[45,134] The previous and very first surgical intervention, ligation of a patent arterial duct, was an anatomic correction carried out in 1938 by Robert Gross of Boston; evidence exists in an exchange of letters on the thoughtful medical-surgical considerations[139] that preceded that event. As a medical student I watched the surgeons at the Children's Hospital of Montreal perform this operation, perhaps my first introduction to my future career. Thus the new era was in my time: I write this history in personal terms, recognizing the clinicians and scientists whose innovative work clarified the nature of cardiovascular malformations and laid the foundations of our etiologic considerations. My contemporaries have already begun to document their own work and their reflections.[43,317,356]

Maude E. Seymour Abbott (1869–1940)

Maude Abbott has been the most important of the pioneers in establishing congenital heart disease as a living part of clinical medicine, wrote Paul Dudley White in his introduction to her *Atlas of Congenital Heart Disease.*[6]

Abbott, curator of the Medical Museum of Pathologic Specimens at McGill University, had died before I entered the medical school, but I was one of the fortunate students to gain support from the Maude E. Abbott Scholarship founded in her honor by an anonymous donor. Abbott's scholarly influence was transmitted to us by her close associate, our beloved teacher, Harold N. Segall (1898–1992), who had worked with her and still followed some of the patients she had seen. I well remember a woman with tetralogy of Fallot who carried on a demanding job in a war industry in spite of cyanosis, fatigue, and shortness of breath, while attempting to disguise her blue lips and nailbeds with thick lipstick and nail polish. Her hands were photographed in color in her youth to show cyanosis and clubbing in Maude Abbott's monograph *Congenital Heart Disease.*[7] When Dr. Segall presented this patient to the Pediatric Grand Rounds of the Montreal Children's Hospital, she was in her forties, unable to work, deeply cyanotic and dyspneic at rest, but she still wanted to live; a Blalock-Taussig shunt was attempted, and she died at operation—she had come just too late!

Maude Abbott was a scholar who pursued in detail every piece of information she could gather on the origin, genesis, pathology, and clinical consequences of cardiovascular malformations. I came to know her work well because it was I who, as a fellow in pediatric cardiology in 1948, unsealed Dr. Abbott's scientific papers, which came to the Children's Hospital in Montreal 8 years after her death. While our pathologist, Dr. F. W. Wiglesworth, attended to the historical specimens, I sorted and catalogued the reprints and filed her letters in alphabetical order as they can still be found today in the historic Osler Library of McGill University. Through these letters I gained an insight into the professional network that expanded and crystallized her thoughts on cardiac malformations and their clinical consequences.

Maude Abbott descended from a prominent Canadian family and was one of the first women admitted to McGill University. She obtained her bachelor of arts degree in 1890, earning a Gold Medal, but as women were still excluded from medical studies, she went on to Bishops College,[287,440] which had many of the same teachers and used the same English teaching hospitals of Montreal. She received her doctor of medicine and surgery degree in 1894, earning the Anatomy and Chancellor's Prizes. Two travel years to Europe followed, where she attended clinics and courses from the famous teachers in Vienna, Heidelberg, Zürich, London, and Edinburgh. Returning to Montreal, she opened a medical practice and in 1898 she became Curator of the McGill Medical Museum (the pathology collection).

> *Museum teaching was quite a spontaneous development,* she is quoted by McDermot, *as the students began dropping in asking questions, and very soon the entire final year had enrolled itself in groups, which came weekly in rotation.*[287]

In December, 1898 the McGill Faculty of Medicine sent Abbott to Washington DC to see the Army Medical Museum "and other institutions on route." This is how Abbott came to meet William Osler in Baltimore. William Osler (1849–1919) was the second of the "Great Four Physicians" appointed to the new Johns Hopkins School of Medicine a decade earlier in 1889. His reputation as the "consummate clinician and teacher" was already well established in his native Canada and especially at McGill University in Montreal (1875–1884), where he established a lasting heritage of bedside instruction and the clinical-pathologic evaluation of disease. His notes on more than 1000 autopsies performed at the Montreal General Hospital formed the nucleus of the McGill Medical Museum and were used in the class teachings of Abbott. When she finally met her famous and charismatic predecessor, she was greatly inspired by him as he praised the McGill Medical Museum in his now famous words: *Pictures of life and death together. Wonderful.*

And so he gently dropped a seed that dominated all my future work, Abbott wrote. However, McDermot notes, *it is hardly possible to think of Maude working with much more enthusiasm and vigor, than before her contact with Osler.*[287]

Abbott's interest in cardiovascular malformations is dated to 1900, when she came across the very unusual trilocular heart, which had been reported in 1824 by Dr. Andrew Holmes, founder of the McGill School of Medicine. Abbott republished the case in the Montreal Medical Journal in 1901. Within a few years her expertise was recognized, and in 1905 Osler invited her to write a section in his textbook of medicine, suggesting that she treat the subject statistically.

The resultant detailed treatise on 412 cases grouped into 23 categories[1] astonished even Osler, whose letter of appreciation is worthy of quote:

> *Dear Dr. Abbott:*

> *I knew you would write a good article but, I did not expect one of such extraordinary merit. It is by far and away the best thing written on the subject*

in English—possibly in any language. I cannot begin to tell you how much I appreciate the care and trouble you have taken, but I know you will find it to have been worth while. For years it will be the standard work on the subject, and it is articles of this sort—and there are not many of them—that make a system of medicine. Then too, the credit which such a contribution brings to the school is very great. Many, many, thanks.

Sincerely yours,

Wm. Osler

P.S.—I have but one regret, that Rokitansky and Peacock are not alive to see it. Your tribute to R. is splendid.

In 1906 the International Association of Medical Museum Curators was organized in Washington, DC, and in 1907 Maude Abbott became the permanent international secretary, and later the editor of its journal. She realized every advantage of this role, as it gave her worldwide contacts. Letters went to and from Montreal discussing clinical cases and pathologic specimens of congenital heart disease.[139] It could have only been through such a very extensive ascertainment system that Maude Abbott could recognize the wide variety of malformations and clinical courses that enabled her to systematize and categorize the findings. In 1924, working with Dawes, a physiologist, she created a clinicopathologic classification of congenital heart disease[8] that took into account the anatomy and the physiologic consequences of each malformation as well as the potential influence of a changing pulmonary-vascular resistance, a concept well ahead of its time. Maude Abbott's pathophysiologic classification of congenital heart disease became acutely relevant as it paved the way for the rise of cardiac surgery. Her classification, still valid today, represents a timeless example of her true and honest scholarship:

It is a well recognized fact that in clinical medicine the intimate personal knowledge of a relatively small number of individual cases is likely to yield a richer harvest in the understanding of disease conditions, than wider generalizations covering a more vast material. In the intriguing subject before us, the study of each individual case is indeed the sole key to the comprehension of these obscure conditions.[6]

Maude Abbott was honored worldwide and her picture appears on the famous mural of the History of Cardiology by Diego Rivera in the Institute of Cardiology in Mexico City.[78]

In the last decade of her life Abbott presented a framework for etiologic inquiry by recognizing potential variations in fetal pathology and hence the heterogeneity of cardiovascular malformations:

Is it a fault in the germ plasm? Is it true inheritance? Is it altered environment of the embryo? Is it a disease state of maternal tissues or secretions? or fetal trauma? or exhaustion of a parturient uterus? or disease of the early embryo itself?[7]

These questions are so well focused that they can guide us today and be extended to the current research horizons on the developmental determinants of the biologic processes that control fetal life and the development of anomalies.[140] However important and challenging these questions were in 1932, they were quickly eclipsed by attention to the surgical alterations of the circulation,[44] which until the recent advances in developmental molecular studies, formed the "raison d'être" of pediatric cardiology.

Helen Brooke Taussig (1898–1986)

The Harriet Lane Home Cardiac Clinic initially served patients with rheumatic fever and rheumatic heart disease, but Helen Taussig was also caring for infants and children with congenital heart disease. Once, after returning from a visit to Montreal, she saw an infant in her own clinic with clinical and radiological findings similar to those of an anomaly shown to her by Dr. Abbott. To the astonishment of many, she made the same diagnosis in life and realized, herself, that the systematic description of findings, and especially the observation of the heart and great vessels under the fluoroscope, could identify different diagnostic entities among similarly cyanotic patients. The use of the fluoroscope and Taussig's exquisitely sensitive clinical evaluation did for clinicians what Abbott had done for pathologists: she brought order and a definition of diagnostic entities. Taussig's fabulous memory of every case she had seen made it possible for her to create a book on *Congenital Malformations of the Heart* (1947), describing her own experience in the clinical, physiologic, and pathologic picture of each anatomic diagnosis.[406] In this volume, and in the later greatly expanded second edition (1960) of her book,[407] the descriptions of the diagnoses of her patients were enhanced by beautifully executed drawings. These contributions of the Department of Medical Illustrations, founded at The Johns Hopkins Hospital by Max Brödel,[63] represented the essential characteristics of each malformation as interpreted by the author to the artist.

The Blalock-Taussig operation[45] came at a fortuitous time and caught the imagination of the medical and lay world. The dramatic relief brought to suffering children, adolescents, and adults in that critical year of 1945 was a cause for joy and international collaboration, which blurred the memory of the divisiveness and destruction of World War II. The operation did not stand alone as a medical achievement. Technological advances accelerated during World War II found application in peace-time medical practice. Instruments improved in kind and in quality, and invasive tests such as cardiac catheterization and angiocardiography became possible and acceptable.[75,97] The Johns Hopkins laboratory for the physiologic and diagnostic study of congenital heart disease, under the direction of Richard J. Bing, soon produced a series of papers that described new physiologic assessments and clinical data, malformation by malformation.[43,44] A fellowship in "the cath-lab" complemented a fellowship in "the Clinic" and established competence to carry the new practice of pediatric cardiology to other centers.

From the time I entered the Clinic as a fellow in 1949 to the very end of her life in 1986, Helen Taussig was a great teacher and friend to me as well as to her many other young colleagues. Following her sudden accidental death, her former fellows came together under the leadership of Dr. Dan McNamara to share our memories and to recount a brief history of her accomplishments[289] and of her epoch-making role in the surgical treatment of cyanotic children in partnership with Dr. Alfred Blalock.

John D. Keith
(1908–1988)

John Keith became interested in children's heart diseases during his pediatric training at the Strong Memorial Hospital in Rochester, NY and at the Birmingham Children's Hospital in England. In 1938 a cardiac department had already been established at the Hospital for Sick Children in Toronto, with John D. Keith as physician-in-charge. He wrote his own story in the *History of Cardiology in Canada:*[390]

In 1938, congenital heart was as often a diagnosis in itself and little attempt was made to differentiate the various subdivisions. The only people interested in the different types of congenital heart disease were a few pathologists such as Dr. Maude Abbott. Interest in specific diagnoses was awakened by Dr. Gross' operation for patent ductus in 1938. From then on it became extremely important to decide what specific lesion was present.

In World War II Keith served in the Canadian Navy from 1942–1946. Before his return to civilian life he visited Helen Taussig, *to acquaint myself with their techniques, surgical, medical, and diagnostic. This was a most interesting visit and I always tell Helen that I regard myself as her first fellow.* In visits to André Cournand in New York, he became acquainted with the intracardiac catheter technique to study hemodynamics,[97] and he studied the methods of angiocardiography developed by Castellanos[75] in Cuba, which seemed *a very obvious technique to apply to young infants and children . . . we purchased a wartime American aerial photography machine that was able to take several pictures a second. Soon we were in business taking excellent pictures of contrast media going through infant hearts.* At about the same time Arnold L. Johnson, also a medical officer in the Royal Canadian Navy, studied with Paul D. White in Boston, then with Taussig in Baltimore, and subsequently organized a cardiology service at the Children's Hospital in Montreal. As his first associate, I well remember the crude angio-machine and a box spring film holder from which subsequent films jumped out rapidly delivered by the push of a broom stick! Of special interest to our Baltimore-Washington Infant Study is Keith's *little pioneering work in the data collecting field,* the "zebra sheet" dreaded by all his fellows, on which systematic information was collected on clinical, hemodynamic, surgical, and pathologic data on over 10,000 cases. This material formed the subject of Keith's textbook on *Heart Disease in Infancy and Childhood,*[218] written in collaboration with his two close associates, Richard D. Rowe and Peter Vlad. John Keith also helped to gain support for research in this new field. He was a co-founder of the Ontario Heart Foundation and later the Canadian Heart Association. The Hospital for Sick Children in Toronto became a leading international training center.

Richard D. Rowe
(1923–1988)

A native of New Zealand, Rowe joined John Keith at the Toronto Hospital for Sick Children at about the same time as Peter Vlad, a native of Romania, and they became the closest of friends. Their cooperative work explored many new frontiers, with a special focus on diagnostic studies and treatment of infants. Early pediatric cardiology had dealt with survivors: the many children, adolescents, and adults who came for help in various phases of their disease. Babies were considered "too small" or "too sick" to do any procedures. Rowe and Vlad made the too small/too sick concept a thing of the past. It was Richard Rowe who systematically described the distinguishing clinical manifestations of major cardiovascular malformations in this vulnerable patient group; from him we learned to recognize the urgency for action in the "sick baby" with vague symptoms of cardiorespiratory distress. This understanding was fundamental to the establishment of the Regional Infant Transport Systems, which bring infants and high-risk mothers to tertiary care centers to benefit from the newly developed technologies and miniaturized instruments that altered traditional practices in obstetrics and neonatology.

The Neonate with Congenital Heart Disease[376] (1968) was the textbook that launched the era of infant cardiology. It was written during Rowe's years at the Johns Hopkins Hospital (1965–1973), where he followed Helen Taussig as Director of the Cardiac Clinic. Subsequently, he succeeded John Keith at the Toronto Hospital for Sick Children as Chairman of Cardiology; *he filled the shoes of two giants, a great and gentle man, the Father of Neonatal Cardiology.*[424]

Just as he was to enter a long-anticipated period of readings and writing in retirement, his life was ended abruptly by a swift and malignant illness. Stella Van-Praagh spoke eloquently at the memorial service and placed a plaque on his portrait: "Graduated with Honors."[376]

Alexander S. Nadas
(1913–   )

A great training Center of Pediatric Cardiology arose at the Boston Children's Hospital—it was third only in timing (1949), but unequaled in its eventual scope and strength, encompassing besides clinical cardiology, a constellation of subspecialties of many related fields of knowledge. Only a strong, forceful, and visionary leader could make this happen, and that was Alexander Nadas!

Born in Hungary, and a graduate of the famous Pázmány Péter University Medical School in Budapest, Nadas complied with the USA requirements for foreign medical graduates through an internship in pediatrics at the Boston Children's Hospital,[172] and then, as he assumed the leadership of the cardiac division, he naturally stepped forward to become also a national and international leader.

I first met Nadas in Old Point Comfort, VA at the very tip of land where the James River passes through Hampton Roads into the Atlantic, at a meeting of the American Society for Pediatric Research. The Great Hotel is no more, but in 1951 (or 1952) a few young pediatricians, recent fellows in cardiology, were exchanging ideas on a sun porch when *the new man from Boston* arrived (clearly my country-man with the unmistakable Hungarian accent), and said *we must do this systematically and organize to meet as cardiologists.* Nadas became a leader in the organization of the Section of Cardiology of the American Academy of Pediatrics, and even more importantly, of the Sub-board of Pediatric Cardiology for specialty certification in 1961. Himself a great clinician, he promoted and preached science. This was, I think, the salient difference between Baltimore and Boston: as "Taussig fellows" dispersed they became leaders of clinics in many lands.[316] As "Nadas fellows" dispersed, research programs were founded. Nadas, himself, was at the forefront of systematic organized research. He fully supported the idea of Don Fyler to create the New England Regional Infant Cardiac Program (NERICP).[17] He played a leading role in the Natural History Study,[312] and the randomized clinical trial of indomethazine in premature infants with patent ductus,[179] as major examples. To all of us in the Taussig Group he was a faithful friend. Each summer when Dr. Taussig vacationed on Cape Cod, she was invited to spend a day at the Boston Children's Hospital with the Nadas fellows and students. When Dr. Taussig had her famous reunions in Baltimore, Alex Nadas was an active participant (the picture above was taken at the 1976 Reunion in Dr. Taussig's garden). Now living in retirement, he was honored on his 80th birthday in the Alexander S. Nadas Lecture of the American Heart Association; the lecturer was James Moller of Minneapolis, MN, who, in an incisive review, recounted the many important events in pediatric cardiology that marked Dr. Nadas' active years.[303]

Maurice Campbell
(1891–1973)

Among many national and international leaders of cardiovascular medicine who visited the Johns Hopkins Hospital in the early years, it was a privilege to meet Maurice Campbell of London, England. Greatly admired by Dr. Taussig, this kind and wise physician visited the Harriet Lane Home Cardiac Clinic on several occasions. Always interested in seeing the recent advances in diagnoses and therapies, he posed questions to us on causation and on the possible role of heredity in the genesis of cardiac anomalies. Our generation-though quite aware of the familial occurrences of anomalies in patients under our own observations and those reported by Polani and Campbell,[350] Lamy,[239] and others-was not moved to pursue this line of investigation.

As soon as he turned his attention to congenital heart disease in 1946, Campbell's first target of interest was the origin of malformations of the heart. A truly classic treatise (1965) on the first concise epidemiologic discussion of incidence, parental factors, seasonal distributions, and genetic and environmental factors was entitled *Causes of Malformations of the Heart.*[71] This paper presented research guidelines for genetic and epidemiologic studies, and we may well be embarrassed 30 years later in attempting to fill the gaps of knowledge that he had already delineated!

The life history of this great physician presents no insights into his precise and purposeful etiologic interest in birth defects; far from it! Hardly out of medical school, he served in the field with an ambulance unit in Mesopotamia and Persia in World War I, receiving recognition by the prestigious Order of the British Empire (OBE) while still in his twenties. He became a member of the Royal College of Physicians in 1921. After some years in physiologic research and interest in the circulatory system, he was appointed to the National Heart Hospital under Dr. John Parkinson. He was the 14th member of the Cardiac Club (founded in 1922), which became the Cardiac Society of Great Britain and Ireland, and of which he was a founding member, its first secretary, and later its president.

In World War II Dr. Campbell directed an emergency medicine hospital that cared for casualties and evacuees from the London air raids. It was only subsequent to this experience that he became interested in the study of congenital heart disease, after Dr. Alfred Blalock visited Guy's Hospital, and Mr. Russell Brock (later Lord Brock) initiated a direct surgical approach to relieve stenoses of the mitral and pulmonic valves. His partnership with Brock made Guy's Hospital world renowned for cardiology and cardiac surgery, but one could not imagine the contrast between the leaders: when "Mr. Brock" visited Hopkins we avoided him best we could, while Dr. Campbell was sought out by young and old alike! Regrettably we did not respond to his message—if we had, a comprehensive epidemiologic study would have taken place decades earlier!

Acknowledgments   Credits for Photographs:
Maude E. Abbott: Courtesy of the Osler Library, McGill University, Montreal, Canada.
Richard D. Rowe: Courtesy of Peter Rowe, M.D., Associate Professor of Pediatrics, Johns Hopkins University School of Medicine.
Maurice Campbell: Courtesy of Horst Kolo Photographers, London, England.
Helen B. Taussig, John D. Keith, and Alexander S. Nadas: photographs were taken by the author (CF).

Chapter 2

Categorization of Cardiovascular Malformations for Risk Factor Analysis

From: Ferencz C, Loffredo CA, Correa-Villaseñor A, Wilson PD, eds: *Genetic & Environmental Risk Factors of Major Cardiovascular Malformations: The Baltimore-Washington Infant Study 1981–1989.* Armonk, NY: Futura Publishing Co., Inc; ©1997.

Cardiovascular defects encompass a wide range of abnormalities of the heart and of the great arteries and veins. Such abnormalities, clinically recognizable in liveborn infants, must have been compatible with an adequate performance of the fetal circulation. The recognition of congenital heart disease by clinicians over the years depended on indirect signs and symptoms: *heart murmurs,* arising from increased turbulence in the bloodstream or from bloodflow across a narrowed area; *cyanosis* (blueness of the mucous membranes, nailbeds, and skin), indicating inadequate oxygenation of the arterial blood; and *respiratory distress,* due to either inadequate oxygenation (hypoxia) or congestive heart failure. Ultrasound technology developed over the past 15 to 20 years made it possible to visualize the internal structure of the heart very accurately in fetal and in postnatal life.

The Normal Heart

Some details of normal cardiovascular anatomy and of the specific cardiovascular malformations evaluated in this epidemiologic study must be understood by all readers who are concerned with potential risk factors: this includes the families of patients who seek counsel, the genetic counselors, and the scientists in multiple disciplines that relate to birth defects. *Geneticists* will find this information useful in evaluating cardiac and noncardiac anomalies; *toxicologists* who evaluate the teratogenic effects of environmental agents can better refine comparisons of experimental and human anomalies; and *epidemiologists* who conduct birth defects surveillance programs and etiologic studies can appropriately prioritize attention to single or multiple malformations. The undifferentiated or erroneous characterization of heart defects in epidemiologic studies will introduce serious errors in the evaluation of presumed associations.

The heart is a pump that delivers blood to the tissues to supply them with oxygen and nutrients, and then redirects the oxygen and nutrient-depleted blood from the tissues to be replenished either in the placenta during fetal life or in the lungs in postnatal life. Life cannot exist without this circulation, so that a remarkable interdependence of form and function is a unique and very early feature of cardiac development.[84,219] The task of the heart begins when it is only a single tube, and continues as the tube loops, rotates, and differentiates into a four-chambered heart with uniquely formed valves that control bloodflow from the atria to the ventricles and from the ventricles into the great arteries. Some abnormalities seriously disrupt the essential functions of the heart, but others have no significant hemodynamic consequences in either fetal or postnatal life.

Normal cardiovascular anatomy is the basis of reference for all comparisons to abnormal hearts. Certain salient anatomic features of the heart are relevant to etiologic considerations, and we have therefore prepared a series of illustrations to emphasize these features: the position of the heart in the thorax; the sequence of the cardiac chambers and great vessels; the interior structure of the right and left ventricle; and the central position of the membranous portion of the ventricular septum, which is a critical site in the developmental completion of the normal cardiovascular architecture.

Single organs such as the heart, the liver, the stomach, and the spleen are asymmetrically placed (Figure 2.1A): the heart, stomach, and spleen on the left side, the liver on the right. The systemic great veins (superior and inferior vena cava) carry venous blood from the body into the right atrium, and the pulmonary veins (lying behind the heart) carry oxygenated blood from the lungs into the left atrium. From each atrium, blood passes to the appropriate ventricle, the muscular pumping portion of the heart, and then into the great arteries, the pulmonary artery and aorta respectively. An anterior view of the heart indicates the anatomic terminology of the cardiac structures (Figure 2.1B).

Stylized drawings illustrate the interior structure of the heart (Figures 2.2A and 2.2B). The *right ventricular view* shows the inflow portion through the tricuspid valve and the muscular outflow portion, which leads to the pulmonary artery through the infundibulum or "conus." The ventricular septum is in a curved plane and the membranous portion of the ventricular septum is in a posterior position behind the septal leaflet of the tricuspid valve (illustrated by dotted lines).

The *left ventricular view* illustrates a different relationship of the inflow and outflow valves: the mitral valve is immediately related to the aorta, with fibrous continuity between the mitral and the aortic valves. This four-chamber view of the heart shows the membranous portion of the ventricular septum lying just below the aortic valve and in apposition to the right atrium and tricuspid valve as well as both ventricles.

Photographs of the opened heart were prepared with transillumination from one to the other ventricle to further highlight the delicate nature and critical position of the membranous portion of the ventricular septum, which is the final closure point in cardiac septation (Figures 2.3A and 2.3B).

Remarkable studies are now in progress utilizing molecular techniques that will elucidate the initiation, alteration, and "turning off" of various embryonic events. Thus we must avoid traditional concepts of cardiac morphogenesis, which are constantly superseded by new information.[181,278,344,345]

The Malformed Heart

The major forms of cardiovascular malformations are represented in illustrations taken from the publications of the great clinicians and anatomists who formed the foundation of our knowledge of congenital heart disease (Figures 2.4 to 2.8). This format is chosen for four reasons:

- to introduce at least some of the classic sources of information;
- to emphasize that there has been a long standing and precise understanding of certain specific cardiac defects;
- to avoid discussion of the many recently described anatomic variations;
- to avoid the problem of terminology regarding the fine points of specific malformations, a constant subject of disagreement among experts.

The cardiovascular malformations illustrated present at least one example of each of the major categories of defects that will be discussed in subsequent chapters. Salient features are noted in the figure legends and details of the malformations are described in the appropriate diagnostic chapters.

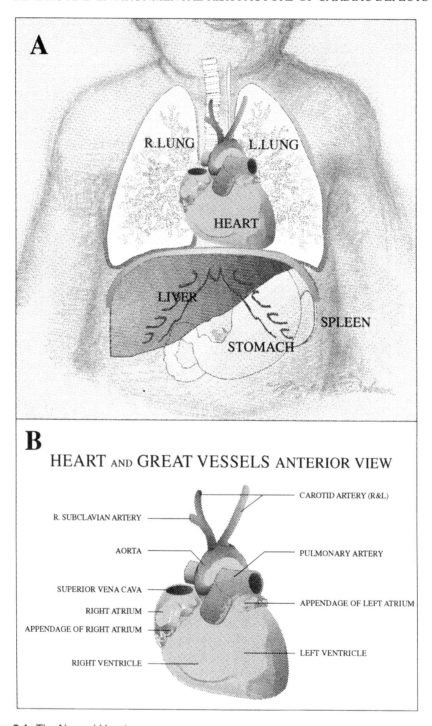

Figure 2.1: The Normal Heart
A. Normal positions of the heart and abdominal organs, anterior view; the apex of the heart points to the left side of the chest; the liver is in the right upper quadrant of the abdominal cavity; the stomach and spleen are on the left side.
B. Normal heart with identification of the chambers of the heart, the great arteries, and the superior vena cava, seen in cross section; the inferior vena cava is posterior and inferior and not visible in this view.

NORMAL HEART

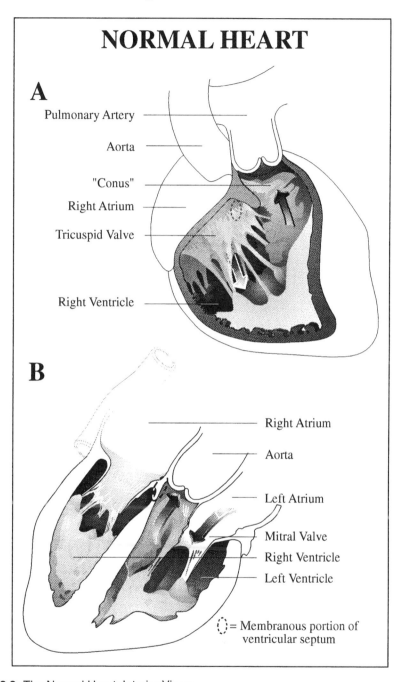

Figure 2.2: The Normal Heart: Interior Views
A. Right ventricle: This anterior view shows that blood that enters through the tricuspid valve (light arrow) is ejected into the pulmonary artery (dark arrow) through the muscular outflow portion of the right ventricle called the infundibulum or "conus". The membranous portion of the ventricular septum (dotted outline) lies behind the tricuspid valve.
B. Four chamber view of the heart: This longitudinal posterior section shows the septum between the right and left ventricles and the relationship of the membranous part of the septum (dotted outline) to the right atrium, tricuspid valve, and aortic valve. Blood entering through the mitral valve (arrow) is directed into the aorta (arrow) with membranous continuity of the mitral and aortic valves.

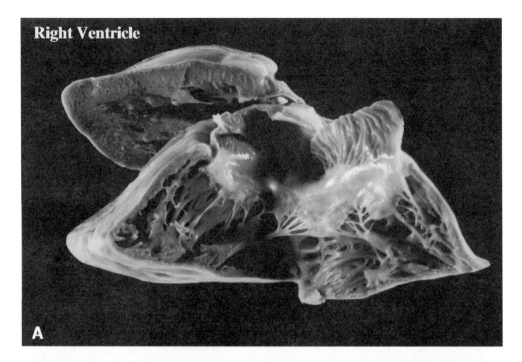

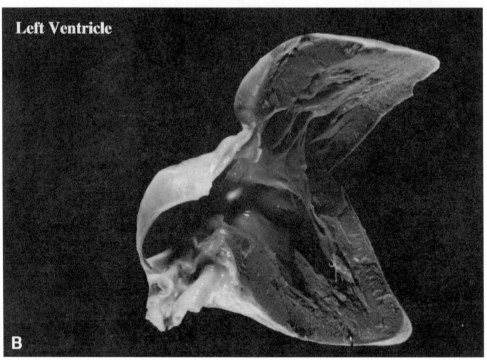

Figure 2.3: The Normal Heart: Photographs of the opened right and left ventricles with transillumination; the delicate structure and location of the membranous ventricular septum is evident.
A. Right ventricular view
B. Left ventricular view

LATERALITY LOOPING DEFECT

MEMBRANOUS VENTRICULAR SEPTAL DEFECT

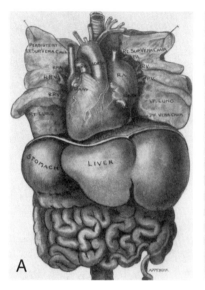

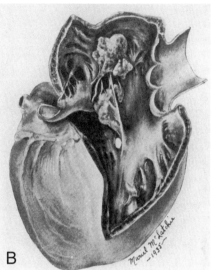

COARCTATION OF AORTA

HYPOPLASTIC LEFT HEART SYNDROME

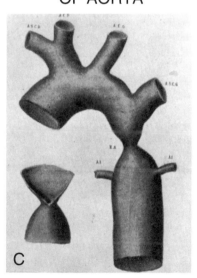

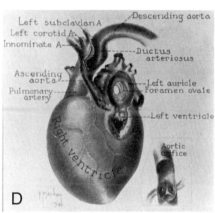

Figure 2.4: From ABBOTT ME. (1936) *Atlas of Congenital Heart Disease*. Historical Illustrations.
A. Laterality/Looping Defect: Complete inversion of thoracic and abdominal organs; heart and stomach on the right, liver and appendix on the left.
B. Defect of the Membranous Ventricular Septum, lying behind the tricuspid valve.
C. Coarctation of the Aorta: Marked constriction of the aorta below the left subclavian artery at the level of the arterial duct (not shown); dilated branches of the aortic arch and first pair of intercostal arteries.
D. Hypoplastic Left Heart Syndrome: Described by Abbott as a case of "Congenital Atresia of the Aortic Orifice from Foetal Myocarditis".
Copyright of Abbott's Atlas belonged to its author (deceased 1940); she had no descendants.

TETRALOGY OF FALLOT

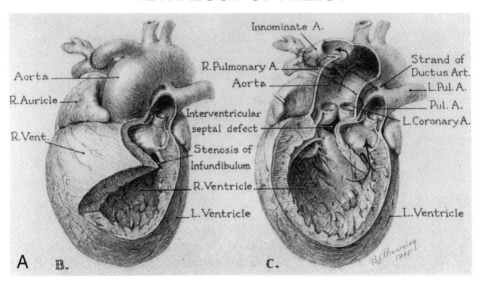

TRANSPOSITION OF GREAT ARTERIES

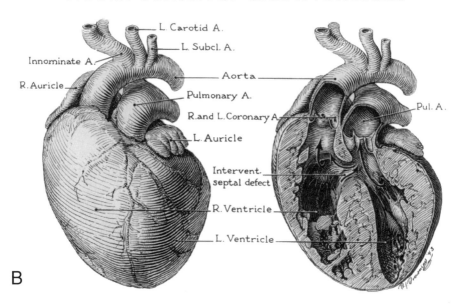

Figure 2.5: From TAUSSIG HB (1960) *Congenital Malformations of the Heart.*
A. Tetralogy of Fallot: Drawings show a large right ventricle, severe stenosis of the in-fundibulum, small pulmonary valve and pulmonary artery, and a large aorta which "overrides" a ventricular septal defect. The innominate artery has been anastomosed to the right pul-monary artery (Blalock-Taussig shunt).
B. Transposition of the Great Arteries: The aorta arises from the right ventricle and the pul-monary artery from the left. A defect in the ventricular septum is shown to have a bidirectional shunt (arrows).
From Taussig HB. *Congenital Malformations of the Heart.* (Copyright © 1947, 1960) The Com-monwealth Fund. Reprinted by permission of Harvard University Press.

ATRIOVENTRICULAR SEPTAL DEFECT

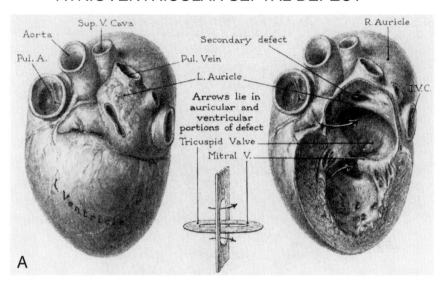

A

TRICUSPID VALVE ATRESIA

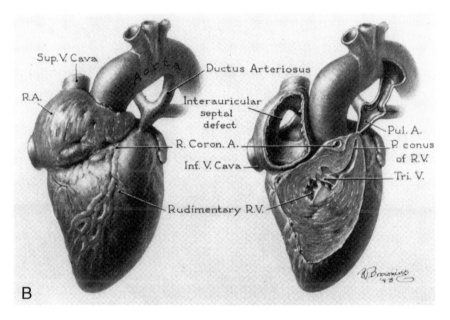

B

Figure 2.6: From TAUSSIG HB. (1960) *Congenital Malformations of the Heart.*
A. Atrioventricular Septal Defect: The exterior of the heart is viewed from the back. The interior view shows a large defect of the lower atrial septum (septum primum), a large ventricular septal defect and a common atrioventricular valve. The schematic planes illustrate the communications between all four chambers.
B. Tricuspid Atresia: Anterior view of the heart shows a large right atrium, small pulmonary artery and a large patent arterial duct. There is no communication through a tricuspid valve between the right atrium and right ventricle.
From Taussig HB. *Congenital Malformations of the Heart,* (Copyright © 1947, 1960) The Commonwealth Fund. Reprinted by permission of Harvard University Press.

PULMONIC VALVE STENOSIS

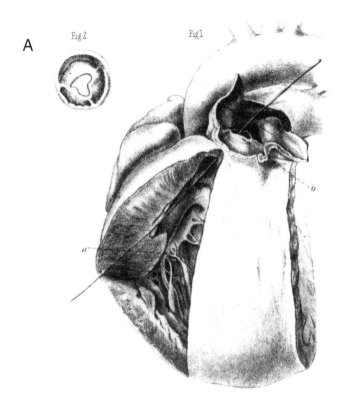

BICUSPID AORTIC VALVE

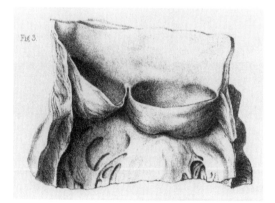

Figure 2.7: From PEACOCK T. (1858) *On Malformations of the Human Heart*.
A. Pulmonary Valve Stenosis: The dome-shaped, obstructed pulmonary valve (probe from right ventricle to pulmonary artery) is associated with severe hypertrophy of the right ventricular muscle; the fused valve cusps are seen in the inset.
B. Bicuspid Aortic Valve: The two cusps (instead of the normal three cusps) may function normally, but impart a risk for bacterial infection and calcification.

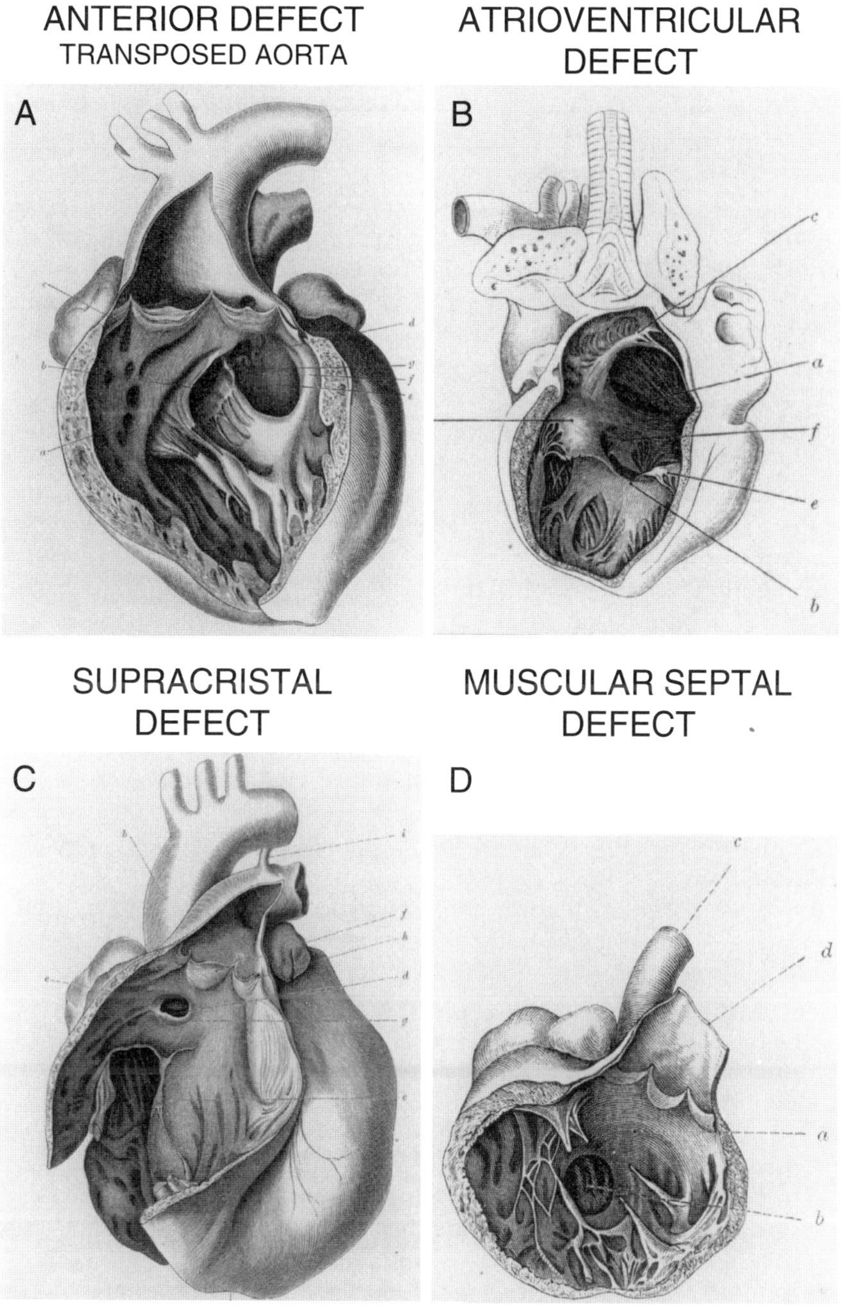

ANTERIOR DEFECT
TRANSPOSED AORTA

ATRIOVENTRICULAR DEFECT

SUPRACRISTAL DEFECT

MUSCULAR SEPTAL DEFECT

Figure 2.8: From ROKITANSKY C. (1875) *Defects of Cardiac Septation*.
Ventricular Septal Defects:
A. Large anterior defect just below the valve of a transposed aorta.
B. Atrioventricular septal defect, posterior view (similar to Figure 2.6A): The large defect of the lower atrial septum permits a view of a common atrioventricular valve; the ventricular part of the defect is small.
C. Supracristal defect in the pulmonary outflow tract, with a fibrous edge.
D. Large defect of the muscular septum, seen from the right ventricle.
Photographs of the original woodcut prints by courtesy of the Osler Library of McGill University, Montreal, Canada.

Previous Classifications of Cardiovascular Malformations

Attempts to classify cardiovascular malformations constitute hallmarks of the expanding knowledge of congenital heart disease. Initial attempts at groupings searched for a framework by which the great variety of abnormalities could be viewed.[27,366]

Rokitansky, in his famous *Manual of Pathologic Anatomy,* described virtually the entire range of anomalies we recognize today under four categories: 1) deficiency of formation (ie, arrested development), 2) anomalies of form, 3) anomalies of position, including the heart as well as its components, and 4) anomalies of size in comparison to a list of standard measurements of mean values for men and for women, of the external as well as internal dimensions of all component parts. These measurements were derived from the extensive autopsy material assembled in the Imperial Pathologic Museum of the University of Vienna, which was the magnet for medical studies, *a conventional Mecca for American practitioners where the amount of pathologic material was so great that a man might see in a few weeks as many cases of a disease as he would see in America in years.*[155]

Bamberger was Professor of the Medical Clinic in Wurzburg, Germany. In his *Textbook of Heart Disease,* he devoted a chapter to congenital malformations of the heart characterized by *genetic and anatomic characteristics* with consideration of the possible origin as: 1) arrest at a certain fetal stage, 2) abnormal formation, and 3) products of inflammatory changes. He states that direct inheritance seems to be rare, and *other origins are yet obscure leaving us in doubt whether alterations of the maternal organism during gestation or independent developmental events in the embryo are principally responsible.* These same questions about the possible origin of congenital heart disease were also clearly formulated by Abbott.[7,140]

Maude Abbott searched for a taxonomy that would reflect the principles by which the malformations should be grouped; hence her classification of congenital cardiac malformations was based on pathologic and physiologic alterations. This was a *clinical* classification.[8] Abbott was proud of this work; she wrote to Paul Dudley White, *it is my magnum opus.*[4] She created an impressive exhibit for the 100th Anniversary of the British Medical Society,[7] which illustrated the three major categories of cardiovascular malformations:

- group 1: *the acyanotic group,* in which no abnormal intracardiac communication exists;
- group 2: the cases with *arterial venous shunting* and possible terminal or transient reversal of flow *(cyanose tardive);* and
- group 3: cases of *venous-arterial shunt* or "*true morbus coeruleus*".

The circulatory diagrams, drawings, and photographs became famous as they were reproduced in the Atlas of Congenital Cardiac Disease,[6] which was intensively studied by the members of the first generation of pediatric cardiologists.

An incomparable classic work on cardiovascular malformations was recently published by Bharati and Lev.[40] Their two-volume compendium, *The Pathology of Congenital Heart Disease,* describes the terminology and classification of more than 6300 malformed hearts, with excellent illustrations. The variations and exquisite details of each anomaly represent a treasure trove to pathologists and surgeons; readers of this epidemiologic study should be aware of this unusual resource.

Morphogenesis studies sought to order malformations according to developmental events considering the changes of cells and tissues and the movements of embryonic structures, which determine the segmental arrangement of venous inflow into the right and left atria and ventricles and the appropriate origins of the great arteries.[423]

Clark[82] brought together information from experimental studies in a mechanistic classification of congenital heart disease. This classification was useful in our epidemiologic study because it separated the various anatomic types of septal defects. However, some mechanisms are as yet unknown (eg, the extent and effect of mesenchymal cell migration) and some categories (eg, hemodynamic factors) are too broad and include many anatomic types of abnormal hearts.

Thus far, every attempt to categorize cardiovascular defects has failed to satisfy etiologic considerations because each of them has only rearranged the phenotypic entities in various ways (Figure 2.9). Our epidemiologic study hopes to find clues that will regroup the malformations according to their origin

Cardiovascular malformations occur in endless variations. After decades of caring for patients with congenital heart disease, Dr. Taussig once remarked that she was seeing new malformations all the time. *It would seem to me that after all these years I should have seen at least one of every kind,* she exclaimed, expressing very

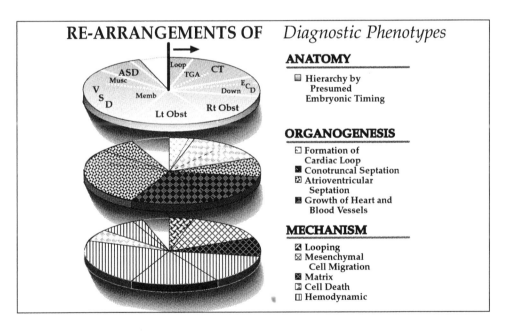

Figure 2.9: Rearrangements of Diagnostic Phenotypes:
Clinical diagnoses were ordered by embryonic timing beginning with abnormalities of the cardiac loop and ending with simple defects of the atrial septum. Consideration of organogenetic events, and of presumed embryonic mechanisms resulted in rearrangements of the original diagnostic phenotypes. Our etiologic search aimed to discover regrouping of the phenotypes by risk factors.

graphically the individuality of cardiovascular anomalies. This wide variation occurring within each diagnostic entity was also highlighted as the *most intriguing aspect of congenital heart disease,* by Bharati and Lev.[40] As epidemiologists, how then should we consider these variations, which must indicate a multiplicity of factors responsible for their occurrence? How should we proceed to define case groups that are anatomically homogeneous but of adequate size for risk factor analyses?

Categorization of Cardiovascular Malformations for Risk Factor Analysis

In the Baltimore-Washington Infant Study data we attempted the best of both possible worlds: we established a morphogenetic sequence (hierarchy) by embryonic timing, and envisioned for analysis "pure" defects within each of Clark's mechanistic subgroups.[82]

A Hierarchical Order

Defects were considered by their presumed embryonic timing to establish a *single principal* diagnosis for each case (Table 2.1). This order was based on anatomic criteria without regard to clinical or hemodynamic changes. This practice followed the example of the New England Regional Infant Cardiac Program.[174]

Mechanistic Grouping

We then placed each diagnosis into the various mechanistic categories described by Clark[82] to allow for the analysis of groups of related malformations; in addition we established the "pure" diagnoses which would be targeted for separate case-control analyses (Table 2.2).

"Pure" Cardiac Diagnoses

Because of the tremendous variation in anatomic phenotypes there could be no simple criterion by which a "pure" diagnosis could be defined, nor was the expected number of cases known. As the case collection grew, we could perceive the directions we could take. For instance, the outflow tract defects group was large enough to separately examine the cases of transposition of the great arteries with intact ventricular septum in which the fundamental architecture of the heart is normal, and also those of tetralogy of Fallot, the classic malformation of the *maladie bleu* recognized in early times in children and adults.

In cardiac malformations of the lower hierarchies, ie, stenoses and septal defects, we limited the "pure" group to single defects. Such a limitation had been made for standardized diagnostic subsets for prevalence comparisons with the EUROCAT data,[346] but clearly, the sample size demands of multivariate analyses did not permit very narrow definitions. Diagnostic flow sheets in each chapter indicate all component defects and the anatomic subgroups that were examined. Subgroups that remained too small, and/or had multiple or nonspecific diagnoses or could not be allocated to larger groups, were omitted from the analyses.

Table 2.1
Cardiovascular Malformations: Hierarchical Allocation
by Presumed Embryological Timing
Baltimore-Washington Infant Study (1981–1989)

I. *Defects of Laterality and Cardiac Looping*
    Malposition of the heart/cardio-visceral discordance
    "Corrected" (Levo) transposition of the great arteries
II. *Defects of Outflow and Inflow and of the Atrioventricular Junction*
    Complete (Dextro) transposition of the great arteries
    Common arterial trunk
    Double-outlet right ventricle
    Tetralogy of Fallot
    Double-inlet left ventricle ("single ventricle")
    Total anomalous pulmonary venous return
    Atrioventricular septal defects
III. *Severe Obstructions* (*Hypoplasias and Atresias*)
    Pulmonary atresia with intact ventricular septum
    Tricuspid atresia
    Divided left atrium
    Hypoplastic left heart syndrome
    Interrupted aortic arch
    Ebstein's anomaly
IV. *Obstructive Lesions of Valves and Vessels*
    Pulmonic stenosis
    Aortic stenosis
    Coarctation of the aorta
    Bicuspid aortic valve
    Pulmonary artery branch stenosis
V. *Simple Defects of Septation and Patent Arterial Duct*
    Ventricular septal defects, membranous and muscular
    Atrial septal defects
    Patent arterial duct
VI. *Other Great Vessel and Coronary Artery Lesions*
    Aortic arch anomalies
    Coronary artery anomalies
    Partial anomalous pulmonary venous return
    Pulmonary artery sling
VII. *Cardiomyopathies*

Associated Noncardiac Anomalies

An important group of cases are infants with multiple malformations; that is, anomalies of the heart and other organs. Anomalies associated with known etiologic agents (eg, rubella, fetal alcohol, diphenylhydantoin, and retinoic acid embryopathies) involve multiple organs and often show a typical abnormal facial appearance; however only a few teratogen-specific syndromes are known.[213] Conversely, the allocation of cases to a specific multiple malformation syndrome in the absence of a known teratogen has created a wide range of eponymic entities, some of which occur in families and are presumed to be of genetic origin. A great number of such syndromes occurred among our cases, indicating the common origin of cardiac and noncardiac anomalies. We are indebted to Dr. Iosif W. Lurie for the glossary of the syndromes that were encountered in this study (See Appendix).

Table 2.2

Classification of Cardiovascular Malformation by
Presumed Developmental Mechanism*
Baltimore-Washington Infant Study (1981–1989)

Presumed Developmental Mechanism	Diagnostic Types
Lateralization and Looping	Laterality disturbances, complex defects, and corrected (L) transposition
Malformations of the Ventricular Outlets and Arterial Trunks	
I. Mesenchymal Cell Migration	Normally related (spiral) great arteries: Tetralogy of Fallot Double-outlet right ventricle Common arterial trunk Ventricular septal defect, supracristal type Aortic-pulmonary window
II. Complete Transposition	Transposed (parallel) great arteries: with intact ventricular septum with ventricular septal defect with double-outlet right ventricle with pulmonary/tricuspid atresia
Extracellar Matrix Defects	Atrioventricular septal defects: Complete and partial types Ventricular septal defect type
Hemodynamic (Flow) Defects	Left-Sided Obstruction Defects Hypoplastic left heart syndrome Coarctation of the aorta Aortic valve stenosis Bicuspid aortic valve Right-Sided Obstruction Defects Pulmonic valve stenosis Pulomnary atresia with intact ventricular septum Tricuspid atresia Septation defects and Patent Duct Ventricular septal defect, membranous type Atrial septal defect Patent arterial duct
Cell Death	Ventricular septal defect, muscular type Ebstein's anomaly
Targeted Growth	Anomalous pulmonary venous return: total and partial types
Cardiomyopathy	Cardiomyopathies

*From Clark (1987).

In the categorization of cases we considered only the principal cardiovascular diagnosis, however, in the analyses we considered also the noncardiac anomalies of the proband and the congenital abnormalities in the families. Thus, the evaluation of the presumably genetic segments of the case population was accomplished by subdivisions of the diagnostic group and different analytic models for isolated/simplex and multiple/multiplex families. These analytic methods will be described in Chapter 3.

Chapter 3

Epidemiologic Case-Control Design and Statistical Methods

From: Ferencz C, Loffredo CA, Correa-Villaseñor A, Wilson PD, eds: *Genetic & Environmental Risk Factors of Major Cardiovascular Malformations: The Baltimore-Washington Infant Study 1981–1989.* Armonk, NY: Futura Publishing Co., Inc; ©1997.

The Baltimore-Washington Infant Study is a collaborative population-based case-control study of cardiovascular malformations among the resident livebirths of the State of Maryland, the District of Columbia, and six counties of Northern Virginia, during the years 1981 to 1989 (Figure 3.1). The details of the design and methods have been reported in detail.[151] Salient features of the case ascertainment, control selection, and data quality control methods are described below, followed by a detailed description of our general method of statistical analysis. The data collection represented a comprehensive inquiry into potential risk factors: (a) hereditary disorders and major malformations in the nuclear families of the proband and both parents, (b) medical and lifestyle exposures of both parents, and home and workplace exposures. Statistical analysis estimated case-control odds ratios for multiple variables simultaneously and corrected for effect modification and confounding.

Cases

The study cases were found by a systematic search to achieve complete ascertainment of all liveborn infants with cardiovascular malformation in whom the diagnosis was confirmed before 1 year of age by echocardiography, cardiac catheterization, surgical inspection, or autopsy. The cases were ascertained from multiple sources, including the six pediatric cardiology centers of the region, infant death certificates, and pathology records in 52 obstetrical hospitals. Detailed

Baltimore-Washington Infant Study Area

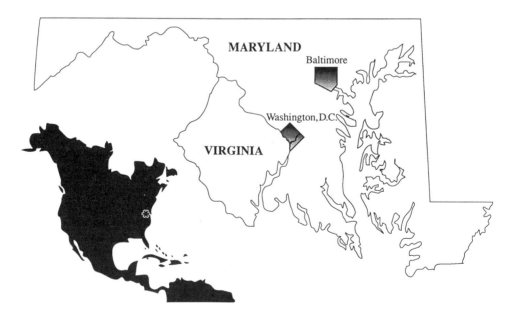

Figure 3.1: Map of the study region, encompassing the State of Maryland, the District of Columbia, and adjacent counties of Virginia, in the United States of America.
Reprinted from Ferencz C, et al. (1993) *Epidemiology of Congenital Heart Disease,* with permission from Futura Publishing Company, Inc.

diagnostic codes were assigned by pediatric cardiologists using the International Society of Cardiology (ISC) coding system[206] (expanded by Dr. Lowell Perry). A diagnostic update was obtained from the cardiologists' follow-up report or from autopsy findings. Precise descriptions of cardiac and noncardiac anomalies were recorded. A hierarchial order assigned a primary diagnosis to each case, giving the highest priority to structural malformations of the earliest embryonic origin. The hierarchial allocation was made by the regional pediatric cardiologists and confirmed by a cardiologist at the research center.

For the purpose of etiologic analyses, we reviewed the diagnostic codes for each type of defect to remove any misclassified cases and to select cases that represent a "pure" anatomic diagnosis of the specific cardiovascular malformation. The "pure" diagnostic group was obtained by removing cases with multiple defects and those with rare variations of the specified cardiac defect. This selection process is illustrated in diagnostic flow charts in the reports on each specific defect. Each flow chart shows the total number of cases enrolled, the exclusions, and the final number of cases analyzed in the chapter. Bold type indicates the subgroups, which were analyzed by multivariate logistic regression.

Length of Follow-Up

The protocol for case registration in this study required an update form on cardiac and noncardiac diagnoses from the cardiology centers on the first birthday of each surviving infant. Thus our intention was to have 12 months of follow-up time in order to provide a final diagnosis for the risk factor analysis. The actual lengths of follow-up for all cases as a group and for individual diagnostic groups is shown in Table 3.1. Among the 4390 total cases, 7.4% were lost to follow-up or had mild lesions that were not intended to be followed by the Cardiology Centers. Of those who had update reports, 19.6% died in the first year of life. Among the cases who were followed, the median length of follow-up was 10 months (mean = 10.1, standard error = 0.1). The great majority of the diagnostic groups each achieved 10 to 12 months of follow-up; the exceptions were mild lesions such as muscular ventricular septal defect and the less severe forms of pulmonic stenosis, with 8 and 9 months of follow-up, respectively. Thus, in each of the diagnosis-specific chapters when we refer to events during the first year of life we recognize that there is some variation in the follow-up period, which is also diagnosis-specific.

Sampling of Mild Defects for Interview

The final number of cases enrolled and interviewed is shown in Figure 3.2. Due to funding limitations, not all families could be interviewed, and 553 infants with mild lesions (all with isolated heart defects) were removed by sampling. Of the families eligible for interview, 90% participated.

Controls

Controls were 3572 infants without cardiovascular malformations who were selected from the 52 area hospitals by computer-generated random sampling to achieve a representative sample of the livebirth cohort of the region from which

Table 3.1
Length of Follow-Up
Baltimore-Washington Infant Study (1981–1989)

Diagnostic Group	Total N	Not Followed	(%)	Number Followed	Died	(%)	Cases Who Were Followed Alive		Median Age (months) at Last Visit
					n	(%)	n	(%)	
All Cases	4390	323	(7.4)	4067	798	(19.6)	3269	(80.4)	10
Analytic Group:									
Laterality/Looping	131	3	(2.3)	128	59	(46.1)	69	(53.9)	11
Transposition Group	239	1	(0.4)	238	70	(29.4)	168	(70.6)	11
Normal Great Arteries Group	400	4	(1.0)	396	120	(30.3)	276	(69.7)	11
AVSD-Down Syndrome	210	3	(1.4)	207	50	(24.2)	157	(75.8)	11
AVSD, Nonsyndromic	88	4	(4.5)	84	28	(33.3)	56	(66.7)	12
VSD Membranous	895	75	(8.4)	820	56	(6.8)	764	(93.2)	10
VSD Muscular	429	81	(18.9)	348	4	(1.1)	344	(98.9)	8
Atrial Septal Defect	291	36	(12.4)	255	18	(7.1)	237	(92.9)	10
Patent Arterial Duct	80	7	(8.8)	73	4	(5.5)	69	(94.5)	10
Hypoplastic Left Heart	162	1	(0.6)	161	140	(87.0)	21	(13.0)	11
Coarctation of Aorta	126	4	(3.2)	122	9	(7.4)	113	(92.6)	11
Aortic Stenosis	74	2	(2.7)	72	10	(13.9)	62	(86.1)	10
Bicuspid Aortic Valve	67	12	(17.9)	55	2	(3.6)	53	(96.4)	9
Pulmonic Valve Stenosis:									
all	341	38	(11.1)	303	1	(0.3)	302	(99.7)	10
severe	64	0		64	1	(1.5)	63	(98.5)	11
moderate	125	10	(8.0)	115	0	(0)	115	(100)	9
mild	152	28	(18.4)	124	0	(0)	124	(100)	9
Pulmonary Atresia	53	1	(1.9)	52	24	(46.1)	28	(53.9)	12
Tricuspid Atresia	33	1	(3.0)	32	11	(34.4)	21	(65.6)	11
TAPVR	60	0		60	24	(40.0)	36	(60.0)	9
Ebstein's	47	3	(6.4)	44	14	(31.8)	30	(68.2)	11
Cardiomyopathy	86	5	(5.8)	81	21	(25.9)	60	(74.1)	11

AVSD = atrioventricular septal defect; TAPVR = total anomalous pulmonary venous return; VSD = ventricular septal defect.

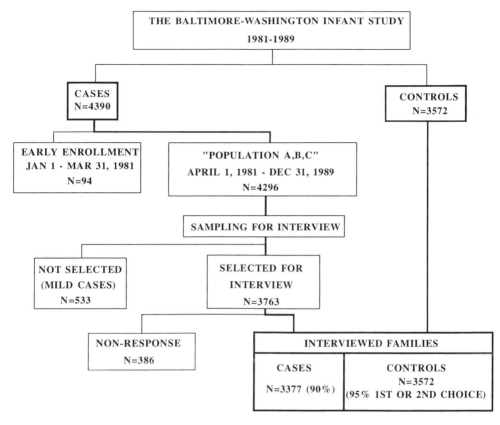

Figure 3.2: Flow sheet demonstrating the enrolled study population and the selection process for family interviews with selective sampling of mild defects. Response rates in cases and controls are shown in parentheses.

the cases were ascertained during the years of the case births (1981–1989). Random control selection was performed annually in each of these years, and was independent of the enrollment of cases. The targeted number of controls was achieved by replacing nonrespondents with alternates from the same birth hospital. We examined the representativeness of the controls by comparing maternal and infant characteristics of the 9-year control sample to those of the regional cohort of 906,626 livebirths of the same period. Characteristics studied were infant gender, race, birthweight, plurality, season of birth, and maternal age. We found the control group to be nearly identical to the regional birth cohort in these characteristics.[152]

Data Collection and Quality Control

Home interviews were conducted with parents following written consent. A structured questionnaire administered by trained interviewers obtained information on sociodemographic, medical, reproductive, genetic, lifestyle, and environmental factors. The family history of the first- and second-degree relatives included cardiac and noncardiac anomalies with details of diagnostic tests and treatments; validation by medical record review depended on the consent of the appropriate relative or proxy. When two siblings that are both affected were in the same diagnostic case group, we did not include that family in the analysis. However when two affected siblings were in different diagnostic case groups, they were both kept in the analysis. The environmental inquiry included reports on parental smoking, alcohol and caffeine intake, recreational and therapeutic drug use, diagnostic radiography, occupational history, and questions on exposures to pesticides, dyes, metals, ionizing radiation, and solvents.

More than 90% of the interviews for both cases and controls were conducted within 12 months of the birth of the study subject, and almost 100% were conducted within 18 months (Figure 3.3). Pregnancies subsequent to those with the study infants occurred before the interview in only 4.2% of case mothers and 2.2% of control mothers. The mean duration of the interview was 76 minutes for cases and 67 minutes for controls; the interviews ranged from 25 to 240 minutes.

During the years 1987 to 1989, with the collaboration of Dr. Walter Willet and his research team, we also collected information on maternal diets using the Harvard Food Frequency Questionnaire.[449] From these data, macro- and micronutrient scores were calculated as measures of daily nutritional intake. The statistical analysis of this information on nutritional factors is in progress in cooperation with researchers at the Centers for Disease Control and Prevention.[51]

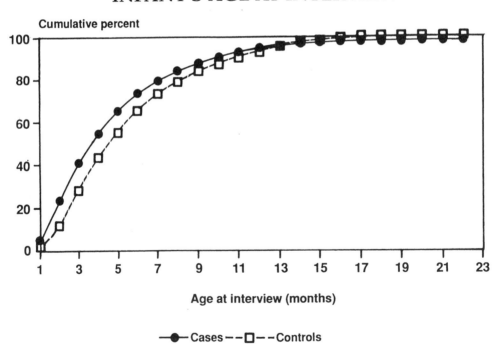

Baltimore-Washington Infant Study (1981-1989)

INFANT'S AGE AT INTERVIEW

Figure 3.3: The cumulative percent distributions of the infant's age at which the interview took place reveals a close similarity between cases and controls.
Reprinted from Ferencz C, et al. (1993) *Epidemiology of Congenital Heart Disease,* with permission from Futura Publishing Company, Inc.

Variables Analyzed

An extensive set of variables was analyzed. Categories of variables, the specific variables themselves, and scores derived from the variables are shown in Appendix 1. Several scores were constructed, including: a family history score (defined in Appendix 1); a score of family socioeconomic status (SES) based on maternal education, head-of-household occupation, and annual household income (defined in Appendix 1); and summary scores for lead and solvent exposures from all sources.[257] Race of the infant was defined as a dichotomous variable, white and nonwhite.[95] Total dose of xanthines and salicylates (mg/day) were each calculated from all reported sources (defined in Rubin and Loffredo 1993).[380]

For exposure, the time window of interest (the "critical period") was defined as the first trimester of pregnancy and the preceding 3 months. For some malformations, we also evaluated exposures in the second and third trimesters.

The distribution of the variables among the controls is shown in Appendix 2, which shows the number and percent of subjects exposed.

Statistical Analysis

For each diagnostic group, possible associations with known and potential environmental risk factors were evaluated by univariate and multivariate analyses. As the first step, all of the variables were screened by univariate analysis separately for each diagnostic group. Using controls as the reference group, the univariate analyses estimated the unadjusted (crude) odds ratio (OR) and 95% confidence intervals (CI) for each variable. In the univariate analyses, each level of any variable shown with more than two levels (eg, number of cigarettes) was analyzed separately. Variables were retained for evaluation as potential risk factors if the lower limit of the 95% confidence interval exceeded 1.0.

For each diagnostic group separately, *multiple logistic regression methods*[384] were used to select the final set of risk factors and detect possible effect modifiers and confounders. As potential risk factors we considered all variables detected as significant in the univariate analysis, and we also considered their first-order interactions. Potential effect modifiers and confounders included all other risk factors selected by the univariate analyses, as well as the following *a priori* risk factors: family history of congenital heart disease, maternal diabetes, maternal age, maternal smoking, maternal alcohol consumption, maternal ionizing radiation exposure, race of infant, and socioeconomic score. Any ambiguity in model selection (between 2 models with different sets of effect modifiers and/or confounding variables) was resolved by the Bayes information criterion of Schwarz,[389] which leads to the selection of the most parsimonious and best-fitting model. Adjusted odds ratios and confidence intervals were obtained from maximum likelihood estimates for the regression coefficients and their covariance matrix.

To construct the multiple logistic regression model for each subgroup, we proceeded in four steps, using the candidate risk factor variables found in univariate analysis:

1. We created the first-order interactions of all of the candidates and then used step-wise selection to select all of the candidates and interac-

tions that would remain in the model with significance level at 0.05. If an interaction was selected, we arbitrarily included its main effects (hierarchical rule).

2. Any *a priori* risk factors and the other candidates not included in step 1 were considered for effect modification of any variable of the step 1 model as follows: the interaction of each potential effect modifier with each variable in the step 1 model was tested for significance (0.05). All significant interactions and their corresponding main effects were included in the model if they remained significant when included together.

3. The other *a priori* factors and candidates not included in steps 1 and 2 were considered for confounding of any of the variables of the step 1 model and any of their interactions found in step 2. These other variables were all included in the model together and removed one at a time to select the minimal set responsible for any confounding. The criterion for confounding was a change of 30% or more in the OR for any risk factor in step 2 model. Confounding variables were kept in the model.

4. Variables not found to be confounders in step 3 were included in the model if their presence reduced the standard error of the OR of any variable in the step 1 model or any interaction found in step 2, and did not increase the standard error of any of these.

Any exceptions to this four-step procedure are described in the chapters for the individual diagnostic groups. In addition, the following additional analyses were performed in several diagnostic groups.

To evaluate the role of maternal reproductive history in those diagnostic groups in which there was a univariate association of one or more of reproductive history variables with case/control status, the four-step procedure described above was carried out first in the absence of any reproductive history variables, and then repeated using those reproductive history variables that showed univariate associations, while stratifying on the number of previous pregnancies. The reproductive history variables were: 1) number of previous preterm births, 2) number of previous miscarriages, 3) number of previous stillbirths, 4) number of induced abortions, 5) bleeding during study-baby pregnancy (yes/no), 6) history of subfertility (yes/no), 7) use of fertility drugs for study-baby pregnancy (yes/no).

Evaluation of Genetic Component

To provide separate analyses (by the above 4-step procedure) for cases with a possible genetic basis for malformations and those without such basis, we compared the pattern of risk factors in these two case strata. Although we do not have the data to determine a genetic basis for malformations, we attempted an approximation by classifying each case as *isolated/simplex* or not *isolated/simplex*. A case was classified as isolated/simplex if it had the specific cardiac defect without any noncardiac defects, and if no nuclear family member (parents, siblings, and half-siblings) had any congenital defect, cardiac or noncardiac (Figure 3.4). Those infants not classified as isolated/simplex are called *multiple anomalies/multiplex,*

Subdivision of cases and controls for multivariate analysis
Baltimore-Washington Infant Study (1981-1989)

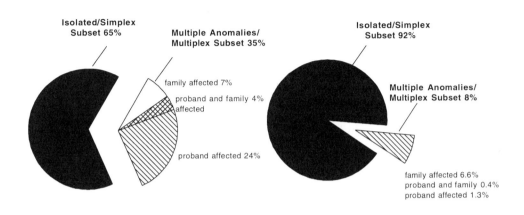

Figure 3.4: Subdivision of the case and control groups for the analysis of risk factors: The isolated/simplex subset of cases comprised infants free of noncardiac anomalies whose families were reportedly free of any birth defect. For control infants, the corresponding subset had no birth defects in proband and family members; multiple anomalies/multiplex subsets were those in which either the proband or family members had additional birth defects.

and this classification is our best approximation of a possible genetic basis for malformations. The isolated/simplex cases would then be nongenetic cases. For analysis in each of these strata, the control group consisted of all controls except those 245 controls with either a noncardiac defect or a nuclear family member with a cardiac or noncardiac defect.

Details of the affected family members of case infants are shown in each diagnosis-specific chapter. The affected first-degree relatives of controls are described in Appendix 3 for cardiovascular malformations and in Appendix 4 for noncardiac anomalies.

PART II

RISK FACTOR ANALYSIS OF DIAGNOSTIC GROUPS

Chapter 4

Defects of Laterality and Looping

Malformations of the human heart can be divided by embryonic timing into those that are initiated in the earliest phase of cardiogenesis, those that are initiated in later organogenesis, and those that occur during later fetal cardiac growth.

In the human embryo a single cardiac tube appears on the 22nd day. It bends to form a loop, which then achieves its destined rotation by the 28th day of embryonic life.[382] This looping and rotation of the primary heart tube establishes the final cardio-visceral relationships, ie, the lateralization of the atria in relation to the great veins and the establishment of the segmental order of the cardiac chambers.[423,427] This anatomic process is known to be gene driven, and highly conserved across species.[278] Recent developmental studies indicate that cells that will eventually constitute the ventricles and the conotruncus have been spatially specified in the initial heart fields.[64] Thus abnormal lateralization increases the potential for atrioventricular and conotruncal malformations.[204,345]

Abnormalities of cardio-visceral positions are associated with defects of embryonic laterality in other organs such as the bronchial tree. When abnormal cardiac defects are associated with asplenia or polysplenia, the eponym "Ivemark syndrome" has been used.[207] However, since the various forms of lateralization defects with and without splenic abnormality may have a common etiology, we have not subdivided cases of abnormal situs into "Ivemark" and "non-Ivemark" subgroups in this etiologic study. Detailed pathologic studies of lateralization defects have explored the relationships of cardiac and noncardiac defects, and specifically the state of the spleen as a correlate of certain morphological defects of the heart,[32,426,428,429] but the complexity of lateralization abnormalities has not yet been resolved in agreement among the experts.

In the Baltimore-Washington Infant Study a hierarchical allocation of cases assigns the highest priority to laterality and looping abnormalities, defined by the presence of cardiac and/or visceral malposition, complex multiple major cardiac defects with or without asplenia/polysplenia, and corrected (levo) transposition of the great arteries, a specific defect in looping. Positional abnormalities of the heart due to skeletal, diaphragmatic, or pulmonary anomalies were not included as cases. The research records of all cases were individually reviewed for this report in order to remove any misclassified cases, to reevaluate syndromic associations, and to reassess the appropriateness of coding.

Study Population

Between January 1, 1981 and December 31, 1989, the Baltimore-Washington Infant Study enrolled 4390 cases with cardiovascular malformations, and 3572 controls. Among the cases, 141 infants were originally diagnosed with abnormalities of cardiac situs and looping. Ten cases were excluded, including three conjoint twins and one case of ectopia cordis with a structurally normal heart. Thus enrollment data on 131 cases permitted the estimation of regional prevalence and the diagnostic descriptions of cardiac and noncardiac morphology. Family interviews were completed for 112 of these infants for evaluation of potential genetic and environmental risk factors.

Cardiac Abnormalities

The complexity of laterality and cardiac looping abnormalities is shown in Figure 4.1. Initially, we considered two diagnostic categories: heterotaxy (malpo-

Spectrum of Laterality and Cardiac Looping Defects, N=141

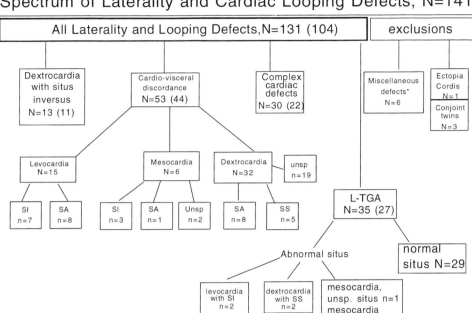

*D-transposition of the great arteries with discordant atrioventricular connections (n=3)
 L-looping with tricuspid atresia (n=3, one with dextrocardia in SS)
 numbers in parentheses indicate interviewed families

Figure 4.1: A wide spectrum of laterality and cardiac looping defects is encompassed in the diagnoses of 131 cases. There was no homogeneous subgroup suitable for separate analysis. Abbreviations: SA = situs ambiguous; SI = situs inversus; SS = situs solitus; Unsp = situs unspecified; L-TGA = levo-transposition of great arteries.

sitions) and L-transposition, but the malposition group included 30 patients coded as too complex without specification of visceral situs, and also 13 patients with mirror-image dextrocardia with situs inversus, in whom there is no cardio-visceral discordance. Among 35 cases with left (L) transposition, six also had situs anomalies and the remaining group was too small for separate analysis. Finally, we considered it best to use the total group as a diagnostic entity for risk factor analysis.

A reassessment of the diagnostic codes of all cardiac defects by pediatric cardiologist (KSK) did support the potential of a delineation of two separate groups: cardiac defects associated with laterality abnormalities and those associated with levo-transposition of the great arteries (L-TGA) alone. Principles that appeared to govern this dichotomy are as follows:

- complex intracardiac defects are features of the laterality group;
- major anomalies of pulmonary and systemic venous return occur only with disturbed laterality;
- atrioventricular septal defects and positional abnormalities of the great vessels are common in the laterality group in the presence of normal cardio-visceral situs; L-TGA is not associated with these same complex defects, but it is overwhelmingly associated with double-inlet left ventricle (22 of 29 cases);

- when the relationship of the heart and viscera is normal (as in complete situs inversus with dextrocardia), cardiac defects are absent or minor.

Associations of cardiac malformations were also evaluated in relation to spleen status. The associations with spleen status could be evaluated in 63 cases in whom this information was coded. In cases with coded spleen status, concordance with the clinically defined categories was found to be fairly good, but not complete. (Table 4.1)

Noncardiac Anomalies

Noncardiac defects in cases with malformations of cardiac looping are unusual from the embryological point of view, as the abnormality of the heart itself is a part of a complex visceral lateralization defect. Thus the separation of cardiac and noncardiac anomalies is not distinct as in other diagnostic groups. Bilateral left or bilateral right lungs, asplenia or polysplenia, left-sided liver, and intestinal heterotaxy are not independent morphological defects, but only different manifestations of a single developmental field defect.[331] Abnormalities of any field may be isolated (if defects of other fields are not involved) or multiple (when defects of other fields are also present). Twenty-six cases had multiple defects, which here are considered noncardiac anomalies (Table 4.2).

Chromosome pathology was found in three cases: 1 with trisomy 13 and 2 with deletions of 10q or 13q. Five patients had recognizable monogenic syndromes: 3 had Kartagener syndrome, 1 had Williams syndrome, and 1 had Larsen syndrome.

Of the total group of 131 patients, 105 had only isolated laterality field defects. The other 18 patients had defects in other fields, including 2 with VACTERL association, 2 with Goldenhar syndrome, and 1 with CHARGE association. The oto-palato-digital syndrome is suspected in a male (#1860) with cleft palate, syn-

Table 4.1
Laterality and Looping Defects:
Clinically Defined Categories of Cardiovascular Malformations
in relation to Coded Spleen Status
Baltimore-Washington Infant Study (1981–1989)

Clinical Categories	No.	Coded Spleen Status		
		Asplenic	Polysplenic	Normal
Asplenia heart	20	16	2	2
Polysplenia heart	22	4	15	3
Complex heart	16	10	3	3
Minor or no cardiac defects	5	2	1	2
Total	63	32	21	10

Asplenia heart = total anomalous pulmonary venous return, atrial and/or ventricular or atrioventricular septal defects, pulmonary stenosis or atresia; Polysplenia heart = abnormal pulmonary venous return, usually partial, absence of the inferior vena cava or inferior vena cava-azygous vein continuation, atrial and/or ventricular septal defects; Complex heart = cases with multiple abnormal features but not obviously in the asplenia or polysplenia group; Minor or no cardiac defects = cases with simple abnormalities such as isolated ventricular septal defect or isolated pulmonic stenosis, and cases with no structural cardiac abnormality.

Table 4.2
Laterality and Looping Defects (N=131)
Noncardiac Anomalies*
Baltimore-Washington Infant Study (1981–1989)

Chromosomal Abnormalities:	N=3 (2.3%)	
Trisomy 13	1	[# 1585]
Deletion 10q	1	[# 2907]
Deletion 13q	1	[# 10909]
Syndromes	N=12 (9.2%)	
a. Mendelian Syndromes:	5	
Kartagener	3	[# 1253, 1943, 2365]
Williams	1	[# 3018]
Larsen	1	[# 3760]
b. Non-Mendelian Syndromes:	7	
Goldenhar	2	[# 3157, 11448]
VACTERL	2	[# 2372, 2529]
CHARGE	1	[# 10702]
Microgastria-limb deficiency (see text)	1	[# 11011]
Ambiguous genitalia, renal dysfunction (see text)	1	[# 2780]
Multiple, Nonclassified Anomalies:	N=2 (1.5%)	
Cleft palate, ear defect, syndactyly	1	[# 1860]
Tracheal stenosis, lung hypoplasia	1	[# 3677]
Organ Defects:	N=9 (6.9%)	
Microcephaly	1	[# 1979]
Hypoplastic cerebellar vermis	1	[# 2841]
Esophageal stenosis	1	[# 10068]
Absence of diaphragm	1	[# 2411]
Intestinal atresia	2	[# 2964, 11059]
Omphalocele	1	[# 2129]
Hydronephrosis	1	[# 2568]
Polydactyly	1	[# 10086]
No Associated Anomalies	N=105 (80.1%)	

* Excludes defects of the laterality field, eg, biliary atresia (5 cases).

dactyly, and ear defects, who died and in whom the clinical description was not adequate to confirm the diagnosis.

Two patients had unusual findings. In case #2780, a girl with dextrocardia, tetralogy of Fallot, and aneurysm of the pulmonary artery with hypoplasia of its branch, also had ambiguous genitalia and renal dysfunction. In her sister, a "new" autosomal recessive syndrome of heart defects, oligomeganephronic renal hypoplasia and clitoromegaly was segregating, and has been reported under the proposed acronym of HOMAGE (Heart defects, OligoMeganephronia, Abnormal Genitalia).[265]

The other unusual patient, case #11011, was the second of male twins (co-twin had a large hemangioma on the chest wall but no other abnormalities). This case with multiple noncardiac defects had been considered to be a case of VACTERL association, but the findings of a missing 4th finger and microgastria allows reassignment of this case to the microgastria-limb reduction complex.[100] Details of this case have been reported.[266] Among the remaining cases with associated single malformations, several could be considered as forme fruste of the VACTERL association.

Descriptive Analyses

Prevalence by Time, Season, and Area of Residence

During the study period, the overall prevalence of laterality and looping anomalies was 1.44 per 10,000 livebirths (Table 4.3). Prevalence was slightly higher between 1984 and 1986 (1.96/10,000) than in other years of the study (1.27 between 1981 and 1983 and 1.10 from 1987 to 1989). There was no significant variation by season of birth. Prevalence was slightly higher among residents of urban areas (2.21), as compared to residents of suburban or rural areas (1.20 and 1.33, respectively).

Diagnosis and Course

The clinical presentation and course of the patients with cardiac laterality and looping defects is in accord with the severity of cardiovascular malformations and the associated defects (Table 4.4). More than half were identified in the first week of life (in contrast to 28% of all other cases in the study). Most infants with laterality and looping defects had complicated diagnoses, and the diagnosis was rarely accomplished by echocardiography alone, ie, without cardiac catheterization, surgery, or autopsy (18.3% in contrast to 49% in all other cases). Half of the patients underwent a cardiac surgical procedure in the first year of life. In this group of patients with complex, severe disease poorly palliated by surgery, mortality at 1 year was high (46%—all other cases 16%).

Gender, Race, and Twinning

Among cases with laterality and looping defects there was a slight excess of females (Table 4.5); the male/female ratio was 0.93 compared to 1.04 in controls.

Table 4.3
Prevalence of Laterality and Looping Defects
Baltimore-Washington Infant Study (1981–1989)

	Cases	Area Births	Prevalence per 10,000
Total Subjects	131	906,626	1.44
Year of Birth:			
1981–1983	35	274,558	1.27
1984–1986	58	295,845	1.96
1987–1989	37	336,223	1.10
Birth Quarter:			
1st (Jan–Mar)	29	219,145	1.32
2nd (Apr–Jun)	33	231,777	1.42
3rd (Jul–Sep)	41	233,626	1.75
4th (Oct–Dec)	28	222,078	1.26
Area of Residence:			
urban	46	208,568	2.21
suburban	70	584,022	1.20
rural	15	112,318	1.33

Table 4.4
Laterality and Looping Defects: Clinical Data
Baltimore-Washington Infant Study (1981–1989)

	Cases N=131 no.	(%)
Age at Diagnosis:		
< 1 week	70	(53.4)
1–4 weeks	39	(29.8)
5–24 weeks	19	(14.5)
25–52 weeks	3	(2.3)
Method of Diagnosis:		
echocardiography only	24	(18.3)
echocardiography and cardiac catheterization	29	(22.1)
echocardiography and surgery	3	(2.3)
echocardiography, cardiac catheterization and surgery	40*	(30.5)
autopsy following other methods	32	(24.4)
autopsy only	3	(2.3)
Surgery During First Year of Life	65	(49.6)
Follow-up:		
not followed	3	(2.3)
followed:	128	
died in the first year of life	59	(46.1)
alive	69	(53.9)
median age at last visit	11 months	

* includes 1 cardiac catheterization only, 1 cardiac catheterization and surgery.

Table 4.5
Laterality and Looping Defects: Infant Characteristics
Baltimore-Washington Infant Study (1981–1989)

Characteristic	Cases N=131 no.	(%)	Controls N=3572 no.	(%)
Gender:				
male	63	(48.1)	1,817	(50.9)
female	68	(51.9)	1,755	(49.1)
Race:*				
white	75	(59.1)	2,362	(66.1)
black	42	(33.1)	1,109	(31.0)
other	10	(7.9)	98	(2.7)
Twin births**	4	(3.1)	53	(1.5)

*missing race in 4 cases and 3 controls; "other" race includes infants with parents from each of the following countries: Laos, Cambodia, Korea, China, Japan, Iran, Pakistan, Guatemala, El Salvador.
**conjoint twinning (n=3) is not included in case group.

The racial distribution of cases was slightly different from controls. There were four twin infants among cases (3.1%), a proportion twice as great as that among controls (1.5%). All the co-twins were normal. Of four case infants from like-sex pairs, three were reported by their parents to be identical. The fourth case, the male infant with microgastria-limb reduction complex differed markedly from his co-twin in birthweight (950 g versus 2580 g). Multiple births among cases also included three sets of thoracopagus conjoint twins, but these unique single-ovum twins were excluded from the clinical and risk factor analyses.

Birthweight and Gestational Age

Infants with laterality and looping defects were characterized by prematurity, low birthweight, and intrauterine growth retardation (Table 4.6). The proportion of case infants born under 38 weeks of gestational age (25%) was more than twice as great as that in controls (9.5%); accordingly the mean birthweight of cases was low relative to controls. Intrauterine growth retardation was defined by the 10th percentile of a standard birthweight distribution for each week of gestation.[61] Cases were smaller than controls (odds ratio = 3.5). A separate evaluation of the 13 cases of mirror-image dextrocardia without structural cardiac defects showed a normal mean birthweight for the eight full-term infants, but a reduced birthweight for gestational age for the five preterm infants.

Sociodemographic Characteristics

Case-control differences in the distributions of sociodemographic characteristics are notable, as shown in Table 4.7. Compared to controls, mothers were

Table 4.6

**Laterality and Looping Defects: Fetal Growth Characteristics
Baltimore-Washington Infant Study (1981–1989)**

Characteristic	Cases no.	Cases (%)	Controls no.	Controls (%)
Number of Families Interviewed	112		3572	
Birthweight (grams)				
<2500	23	(20.5)	252	(7.1)
2500–3500	66	(58.9)	1853	(51.9)
>3500	23	(20.5)	1467	(41.0)
mean±standard error	2999±68		3351±10	
range	848–4649		340–5273	
Gestational Age (weeks)				
<38	28	(25.0)	339	(9.5)
38+	84	(75.0)	3233	(90.5)
mean±standard error	38.8±0.3		39.6±0.1	
range	28–44		20–47	
Size for Gestational Age				
small (SGA)	20	(17.9)	211	(5.9)
normal	80	(71.4)	2712	(75.9)
large	12	(10.7)	649	(18.2)
Odds Ratio for SGA (95% CI)	3.5	(2.1–5.7)	1.0	(reference)

Table 4.7

Laterality and Looping Defects:

Analysis of Sociodemographic Characteristics

Baltimore-Washington Infant Study (1981–1989)

Interviewed Families	Cases N=104 no.	(%)	Controls N=3572 no.	(%)	Odds Ratio (95% CI)	
Maternal Marital Status:						
not married	43	(41.4)	990	(27.7)	**1.8**	**(1.2–2.7)**
married	61	(58.6)	2582	(72.3)	1.0	(reference)
Maternal Education:						
<high school	32	(30.8)	659	(18.4)	**2.0**	**(1.2–3.1)**
high school	31	(29.8)	1265	(35.4)	1.0	(0.6–1.6)
college	41	(39.4)	1648	(46.1)	1.0	(reference)
Paternal Education:						
<high school	28	(26.9)	650	(18.2)	**1.9**	**(1.2–3.2)**
high school	40	(38.5)	1298	(36.3)	1.4	(0.9–2.5)
college	36	(34.6)	1624	(45.5)	1.0	(reference)
Annual Household Income:						
<$10,000	27	(27.0)	686	(19.2)	**1.7**	**(1.0–2.8)**
$10,000–$19,999	24	(24.0)	699	(19.6)	1.5	(0.9–2.5)
$20,000–$29,999	17	(17.0)	737	(20.6)	1.0	(0.5–1.8)
$30,000+	32	(32.0)	1373	(38.4)	1.0	(reference)
Maternal Occupation:						
not working	33	(31.7)	1142	(32.0)	1.4	(0.8–2.6)
clerical/sales	25	(24.0)	1119	(31.3)	1.1	(0.6–2.1)
service	22	(21.2)	444	(12.4)	**2.4**	**(1.2–4.7)**
factory	9	(8.7)	137	(3.8)	**3.2**	**(1.4–7.5)**
professional	15	(14.4)	730	(20.4)	1.0	(reference)
Paternal Occupation:						
not working	12	(11.5)	246	(6.9)	**2.4**	**(1.2–5.1)**
clerical/sales	13	(12.5)	618	(17.3)	1.0	(0.5–2.1)
service	12	(11.5)	480	(13.4)	1.2	(0.6–2.6)
factory	48	(46.2)	1279	(35.8)	**1.9**	**(1.1–3.2)**
professional	19	(18.3)	947	(26.5)	1.0	(reference)
Month of Pregnancy Confirmation:						
1st month	18	(17.3)	511	(14.3)	1.0	(reference)
2nd month	46	(44.2)	1955	(54.7)	0.7	(0.4–1.2)
3rd month	20	(19.2)	663	(18.6)	0.9	(0.4–1.6)
4th month or later	20	(19.2)	416	(11.6)	1.4	(0.7–2.6)

Data are missing for income (4 cases, 77 controls), paternal occupation (2 controls), and pregnancy confirmation month (27 controls).

more likely to be unmarried, poorly educated, and employed in service and factory occupations. Fathers of cases were also more likely to have low educational attainment and factory-related occupations than were fathers of controls, and there was a case excess of nonworking fathers. Household income was also lower among the case families than among controls, with 27% of cases and 19% of controls reporting household incomes below $10,000 per year. There was no significant case-control difference in the month of pregnancy confirmation.

Potential Risk Factors

Familial Cardiac and Noncardiac Anomalies

Information about birth defects and hereditary disorders was available in 112 interviewed families of cases. Infants with cardiac laterality and looping defects were more likely than controls to have first-degree relatives with congenital heart disease (4.8% versus 1.2%) and familial noncardiac anomalies (10.6% versus 4.6%).

Cardiac and noncardiac malformations in the families are shown in Table 4.8. Of four siblings among the three cases of Kartagener syndrome, one brother was similarly affected and the cardiac malposition was an intrinsic part of the genetic entity. Cardiovascular malformations were found in siblings of four additional probands. The sisters with the new HOMAGE syndrome, mentioned above, were concordant for tetralogy of Fallot, but one also had dextrocardia. Two patients with heterotaxy had sisters with ventricular septal defects and patent ductus arteriosus, respectively. One female with dextrocardia and absent diaphragm had a brother with a single ventricle; the parents of these siblings were first cousins. Although there are reports of presumably autosomal recessive inheritance of the laterality field defect,[103,115,462] neither absence of the diaphragm nor absence of the ventricular septum are typical components of this complex.

In six families, different forms of noncardiac defects were found in relatives. In case #10702 the baby had L-transposition, coarctation of the aorta, malformed ear, choanal atresia, and cleft lip, characteristic for the CHARGE association; his mother reportedly had Treacher-Collins syndrome. It may be presumed that the maternal diagnosis was not correct, and that this might be an autosomal dominant form of the CHARGE association,[254] but it is possible that a child with CHARGE association was born to a woman with Treacher-Collins syndrome. In five more families different disorders in relatives were reported. In one family (case #10184) an unusual association of birth defects, presumably multifactorial in origin, included the proband with heterotaxy complex, his mother with cleft lip and palate, and a maternal uncle with pyloric stenosis. Of five Mendelian disorders three were hematologic abnormalities.

Genetic and Environmental Factors

Univariate Analyses

Family history of cardiovascular defects and of noncardiac anomalies were significantly associated with case status (Table 4.9). Among maternal illnesses, overt diabetes was significantly associated, but there were only five affected case mothers. Maternal urinary tract infections were more common in the case groups and a case-control excess was found for those mothers who received combination sulfonamide therapy (sulfamethoxazole/trimethoprim). The occurrence of this treatment showed more than a seven-fold greater rate than in controls, but was represented by only three mothers. Reports of taking antitussive preparations containing dextromethorphan for influenza were more common among case than among control mothers, but there was only a small number of exposed cases. Among lifestyle exposures, both maternal and paternal cocaine use showed sig-

Table 4.8
Laterality and Looping Defects:
Congenital Defects and Hereditary Disorders in First-Degree Relatives
The Baltimore-Washington Infant Study (1981–1989)

Proband	Sex	Diagnosis	Relative	Diagnosis
		A. Congenital Heart Defects		
2780	F	Dextrocardia and atrial/ abdominal SS Tetralogy of Fallot Kidney disease Hypertrophic clitoris (HOMAGE syndrome)	Sister Father	Tetralogy of Fallot Oligomeganephronia Hypertrophic clitoris (HOMAGE syndrome) Branchial cyst
1253	M	Dextrocardia and atrial/ abdominal SI* Kartagener syndrome	Brother	Detrocardia and atrial/abdominal SI* Kartagener syndrome
2411	F	Dextrocardia and atrial/ abdominal SI* Absence of diaphragm	Brother	Single ventricle Parents are first cousins
1684	F	Levocardia and abdominal SI	Sister	Ventricular septal defect
3004	F	Dextrocardia, unspecified Atrial septal defect	Sister	Parent ductus arteriosus
		B. Congenital Defects of Other Organs		
10702	M	L-Transposition CHARGE-association	Mother	Treacher-Collins syndrome
10184	M	Complex heart with asplenia	Mother Mat. Uncle	Cleft lip and palate Pyloric stenosis
3677	F	Dextrocardia, unspecified Tracheal stenosis Hypoplastic lung	Brother	Limb reduction defect
1585	F	Dextrocardia and atrial/ abdominal SA Trisomy 13	Mat. Half- Brother	Pyloric stenosis
2692	M	Heterotaxia–too complex to allocate	Mat. Half- Sister	Hydrocephaly
1447	M	L-Transposition	Sister Mat. Uncle	Coloboma of the iris Cleft lip
		C. Mendelian Disorders (except malformations)		
2430	M	Mesocardia and atrial/ abdominal SA	Mother	Congenital deafness
3487	M	Dextrocardia, unspecified	Father and Grandfather	Friedreich ataxia
2964	M	Levocardia and abdominal SA	Father	Rhesus incompatibility
10553	F	L-Transposition	Father	Rhesus incompatibility
1951	M	Dextrocardia and atrial/ abdominal SA	Sister	Sickle cell anemia

SA = situs ambiguous, SI = Situs inversus, SS = situs solitus; *Normal structure of mirror-image heart.

nificantly elevated case/control odds ratios. Maternal cigarette use showed a significantly elevated odds ratio for smoking more than 20 cigarettes per day. Findings were similar for paternal cigarette exposure.

In an evaluation of the subset of 13 cases of mirror-image dextrocardia with-

Table 4.9
Laterality and Looping Defects:
Univariate Analysis of Risk Factors
Baltimore-Wasington Infant Study (1981–1989)

Variable	Cases N=104 no. (%)		Controls N=3572 no. (%)		Odds Ratio (95% CI)	
Family History (first-degree relatives)						
congenital heart disease	5	(4.8)	43	(1.2)	**4.1**	**(1.6–10.7)**
noncardiac anomalies	11	(10.6)	165	(4.6)	**2.4**	**(1.3–4.7)**
Parental Age						
maternal age:						
<20	17	(16.3)	507	(14.2)	1.3	(0.7–2.2)
20–29	53	(51.0)	2009	(56.2)	1.0	(reference)
30+	34	(32.7)	1056	(29.6)	1.2	(0.8–1.9)
paternal age:						
<20	8	(7.7)	226	(6.3)	1.4	(0.6–3.0)
20–29	44	(42.3)	1724	(48.3)	1.0	(reference)
30+	52	(50.0)	1622	(45.4)	1.3	(0.8–1.9)
Maternal Reproductive History						
number of previous pregnancies:						
none	33	(31.7)	1159	(32.4)	1.0	(reference)
one	29	(27.8)	1097	(30.7)	0.9	(0.6–1.5)
two	24	(23.1)	709	(19.9)	1.2	(0.7–2.0)
three or more	18	(17.3)	607	(17.0)	1.0	(0.6–1.9)
previous miscarriage(s)	20	(19.2)	681	(19.1)	1.0	(0.6–1.7)
Illnesses and Medications						
diabetes:						
overt	5	(4.8)	23	(0.6)	**7.8**	**(2.9–20.9)**
gestational	4	(3.9)	115	(3.2)	1.2	(0.4–3.3)
influenza:						
total	8	(7.7)	278	(7.8)	1.0	(0.5–2.1)
treated with antitussives	3	(2.9)	23	(0.6)	**4.6**	**(1.4–15.5)**
urinary tract infection:						
total	21	(20.2)	467	(13.1)	**1.7**	**(1.0–2.7)**
treated with sulfonamide*	3	(2.9)	14	(0.4)	**7.5**	**(2.1–26.6)**
Lifestyle Exposures						
cocaine:						
mother	5	(4.8)	42	(1.2)	**4.2**	**(1.6–11.0)**
father	7	(6.7)	111	(3.1)	**2.3**	**(1.0–5.0)**
alcohol (any amount)	56	(53.8)	2101	(58.8)	0.8	(0.6–1.2)
smoking (cigarettes/day):						
none	62	(59.6)	2302	(64.5)	1.0	(reference)
1–10	20	(19.2)	565	(15.8)	1.3	(0.8–2.2)
11–20	12	(11.5)	529	(14.8)	0.8	(0.5–1.6)
>20	10	(9.7)	176	(4.9)	**2.1**	**(1.1–4.2)**
paternal smoking (cigarettes/day):						
none	49	(48.0)	2053	(58.0)	1.0	(reference)
1–10	16	(15.7)	510	(14.4)	1.3	(0.7–2.3)
11–20	17	(16.7)	638	(18.0)	1.1	(0.6–2.0)
>20	20	(19.6)	342	(9.6)	**2.5**	**(1.4–4.2)**
Home and Occupational Exposures						
ionizing radiation (occupational)	1	(1.0)	48	(1.3)	0.7	(0.1–5.2)
Race of Infant						
white	60	(58.3)	2362	(66.2)	0.7	(0.5–1.1)
nonwhite	43	(41.7)	1207	(33.8)	1.0	(reference)

*trimethoprim-sulfamethoxazole combination; risk factors are *maternal* unless labelled paternal.

out structural malformations of the heart, the following exposures were reported: overt diabetes, untreated urinary tract infection, paternal and maternal cocaine use, and paternal and maternal smoking. Exclusion of these cases revealed no change in the odds ratios on the remaining cases. Accordingly, these subjects without structural malformations of the heart were retained in further analyses.

Multivariate Analyses

Multiple logistic regression analysis of the total group of cardiac laterality and looping defects identified six possible risk factors: family history of heart defects, family history of noncardiac anomalies, maternal diabetes, antitussive use, paternal smoking, and low socioeconomic status (Table 4.10). There was no evidence of interaction between socioeconomic status and any of the above variables. Overall, paternal cigarette smoking (greater than 20 cigarettes per day) was associated with a two-fold increase in risk, however, the magnitude of this effect depended on whether the father was present when the mother provided data. In the absence of the father at the interview the odds ratio for paternal smoking was 1.6, while there was a marked increase in risk (odds ratio 5.6) if the father was present at the interview and reported his own smoking habits. The risk of laterality and looping defects was increased among infants of families classified as having middle and lower socioeconomic status. These results did not change materially with adjustment for race or after exclusion of mirror-image dextrocardia cases or twins from the dataset.

Discussion

Abnormalities of cardiac looping are rare, with only a single case expected among 7000 livebirths. As the clinical manifestations of abnormal embryonic lat-

Table 4.10
Laterality and Looping Defects (N=104)
Multivariate Analysis of Potential Risk Factors
Baltimore-Washington Infant Study (1981–1989)

Variable	Odds Ratio	95% CI	99% CI
Family history of CVM	4.5	1.7–11.8	1.5–13.5
Family history of NCA	2.5	1.3–4.8	1.0–5.9
Diabetes	8.3	3.0–23.0	2.2–31.5
Antitussives	6.3	1.8–21.6	1.3–31.5
Paternal smoking*, father not present at interview	1.6	0.9–2.9	0.7–3.6
Paternal smoking*, father present at interview	5.6	2.5–12.9	1.9–16.6
SES score:			
high	1.0	(reference)	
medium	1.4	0.9–2.2	0.8–2.4
low	3.7	2.0–6.8	1.6–8.3

CVM = cardiovascular malformations; NCA = noncardiac anomalies.
* Paternal smoking = 20 or more cigarettes (1 pack) per day.

erality vary greatly, clinical reports have given attention to many specific diagnostic, pathologic, and surgical aspects of the component defects.

Prevalence

No comprehensive epidemiologic studies of infants with cardiac laterality and looping abnormalities have been reported, and few population-based prevalence studies of cardiovascular malformations have considered the various forms of this abnormality. The prevalence of L-transposition in our study (0.45 per 10,000 livebirths) was three times greater than that reported by Fyler et al[174] for the New England Regional Infant Care Program, which included only hospitalized infants. In contrast, prevalence of heterotaxy in these two studies (1.06 and 0.88 per 10,000 respectively) was similar. The prevalence of dextrocardia in our study (48 cases) was 0.53, similar to that reported by Kidd et al[223] in Australia (0.41), but half the rate reported by Stoll et al[403] in Strasbourg (1.04) in a study that included stillbirths.

Cardiovascular Malformations and Noncardiac Anomalies

At the planning stage of the Baltimore-Washington Infant Study there was no available etiologic information to guide decisions regarding the creation of etiologic subgroups of cardiovascular malformations for risk factor analysis. The decision to enroll all apparent laterality and looping cases provides information on the population frequency of these diagnoses in the first year of life, and encompasses the full range of severity of defects including infants with structurally normal hearts and those with multiple cardiac and noncardiac anomalies too complex to be allocated into specified cardiac subentities. This diagnostic composition, along with ascertainment of clinically referred infants as well as those found in a community search for infant deaths, renders this study population unlike clinical series, which include patients referred to pediatric cardiologists or surgically treated or identified in autopsy series.*

Clinical reports have documented normal hearts in patients with polysplenia, and morphogenetic studies have emphasized the role of atrial isomerism in identifying patients with intracardiac pathology.[15,159] However, in our study of liveborn patients, atrial isomerism was not consistently identified and only rarely reported. Among patients with clinical and pathologic designations of asplenic heart or polysplenic heart, these typical heart complexes were found in patients with normal spleens or with spleen status unlike the expected; these findings are consistent with previous case observations.[112,174]

Genetic Heterogeneity

Abnormalities of laterality of embryonic structures, including cardiac and noncardiac malformations, represent a distinct field defect. But in addition these cardiac defects also occurred as components of other multiple congenital anom-

---

*References 15, 39, 112, 159, 240, 341, 401, 422, 430.

alies in 26 (20%) of our 131 cases: 5 had monogenic syndromes, chromosome abnormalities were present in 3, and 4 had known associations. Among cases with nonsyndromal anomalies, we identified one (#2780) in which the association of defects may constitute a Mendelian genetic syndrome, the HOMAGE syndrome. A second case (#11011) provided new information not only on the origin of the microgastria-limb reduction complex, but also on the significance of twinning as a mechanism of origin of some components of the heterotaxy complex.[66]

Among the remaining associated noncardiac abnormalities, two main groups were of special interest: abnormalities of eyes, ears, lip, and palate, which may be related to abnormal neural crest cell migration, and abnormalities involving the gastrointestinal and urinary systems, which, in conjunction with cardiovascular malformations, may be forme fruste of the VACTERL association.

Heterotaxy has previously been described in chromosomal abnormalities such as trisomy 13, trisomy 22, trisomy 1q, trisomy 2p, trisomy 6q, deletion 18p, and apparently balanced rearrangements with breakpoints in 7q22, 8q12, 8q24, 12q13.[196,229,318,413,433,450] It has also been described in monogenic syndromes such as Meckel syndrome, Smith-Lemli-Opitz syndrome, Fryns syndrome, Kartagener syndrome, Saldino-Noonan syndrome, Beemer-Langer syndrome, Carpenter syndrome, reno-hepato-pancreatic dysplasia, and in associations such as VACTERL, agnathia-holoprosencephaly complex, Goldenhar syndrome, and others.* There are reports of autosomal recessive,[68] X-linked recessive,[293] and even autosomal-dominant inheritance[74] of isolated heterotaxy field defect, but only a small number of the cases can be explained in this way. The rare occurrence of affected siblings is evidence against recessive inheritance, both autosomal and X-linked, and the many associations of defects confirm the heterogeneity of the abnormal laterality complex with multiple congenital abnormalities.

The association of twinning with laterality defects has been described; looping and situs anomalies may be a co-outcome of the twinning process and due to the same etiologic factors.[66]

Maternal Diabetes

The strong association of overt maternal diabetes with heterotaxy indicates an early embryonic teratogenic impact of diabetes.[295] Mothers with diabetes have a well-established increased risk of major malformations in their offspring, with increased prevalence of multisystem anomalies and prominence of cardiovascular malformations.[81,285,296,302,340,377] The association with laterality and looping defects draws further attention to such an early effect of maternal diabetes on precardiac structures. An association of maternal diabetes, polysplenia, and extrahepatic biliary atresia is reported by Davenport et al.[108] This is of interest in that extrahepatic biliary atresia also occurred in five of our looping cases, two of whom had maternal diabetes; these were the only cases of biliary atresia in the Baltimore-Washington Infant Study.[73] The specific teratogenic mechanism of diabetes mellitus remains unknown. With postulated mechanisms of hyper- or hypoglycemia, exogenous insulin, and other associated metabolic derangements,[9,137,297,333] the teratogenic role of insulin is thought to be indirect, as insulin

---

References 107, 117, 291, 349, 355, 395, 421, 438.

does not cross the human placenta during the embryonic period.[9,295] In our small number of mothers with overt diabetes, all were taking insulin prior to and throughout the pregnancy, and no assessment could be made of possible independent effects.

Environmental Factors

Our findings suggest a possible association of maternal and paternal exposures to environmental factors during pregnancies in which laterality and looping defects develop. These exposures are varied, and range from medications taken for maternal respiratory infections to cigarette smoking by fathers. However, these associations are based on small numbers of exposed cases, and await confirmation.

Maternal use of multi-ingredient antitussive medications during the first trimester represents one of these associations, but no similar associations have been reported in the literature. The antitussive preparations used by Baltimore-Washington Infant Study mothers contained dextromethorphan in combination with acetaminophen, pseudoephedrine, and various antihistamines. Dextromethorphan, the d-isomer of the codeine analog of levorphanol, may have teratogenic potential as it acts on the central nervous system,[209] but unlike other opioids the drug has low systemic toxicity.

Another possible association found only in univariate analysis, is with the sulfonamide combination *sulfamethoxazole-trimethoprim*. Sulfonamide drugs have been reported as potential risk factors for cardiovascular malformations in at least two previous reports: a case-control study of all cardiovascular malformations[394] with a univariate odds ratio of 1.6, and a clinical report of three mothers exposed to the drug during pregnancy who gave birth to infants with multiple congenital anomalies, including two infants with congenital heart disease.[320] A clinical trial of sulfamethoxazole-trimethoprim in 186 pregnancies revealed no significant differences in congenital abnormalities between drug and placebo groups;[65] however, statistical power of such a small prospective study is too low to detect significant case-control differences. Since trimethoprim, a folic acid antagonist, is teratogenic in rats,[234] this association should be further evaluated.

Socioeconomic Status

The association of laterality and looping defects with lower socioeconomic status is of interest. The increasing trend in risk of these defects with lower socioeconomic status suggests an association with conditions of poverty, such as nutritional deficiencies and environmental exposures. The association remained when we took into account race and possible case-control differences in various medical, lifestyle, and environmental factors. It is possible that socioeconomic status represents a constellation of factors whose association with disease can only be detected in the aggregate, due to their interactive, rather than individual, effects. Further research on this association is warranted, as this may lead to the identification of high-risk maternal groups in need of better access to detection and treatment facilities.

Conclusions

The complexity and heterogeneity of laterality and cardiac looping defects is strikingly manifest in the findings of this epidemiologic study. The cardiovascular malformations are components of a composite field defect of abnormal lateralization, but are also associated with anomalies of other fields. Chromosomal and monogenic disorders in the infants, and cardiac and noncardiac abnormalities in the families provide evidence of genetic disturbances in the origin of these multisystem disorders. Maternal diabetes, coupled with environmental exposures and a possible role of socioeconomic disadvantage, emerges as the most important risk factor for these severe early embryonic defects.

Chapter 5

Malformations of the Cardiac Outflow Tract

From: Ferencz C, Loffredo CA, Correa-Villaseñor A, Wilson PD, eds: *Genetic & Environmental Risk Factors of Major Cardiovascular Malformations: The Baltimore-Washington Infant Study 1981–1989.* Armonk, NY: Futura Publishing Co., Inc; ©1997.

Clinical Perspectives

Malformations of the cardiac outflow tract and of the origins of the great arteries represent severe forms of congenital heart disease associated with early cyanosis and many distressing clinical manifestations of hypoxia. Case histories of children and young adults with "darkness of the face," shortness of breath, paroxysms of dyspnea, faintness, and convulsions were assembled in the first text book on *Malformations of the Human Heart,* by Thomas Peacock[338] (1858), and explained by the presence of intracardiac defects with pulmonary outflow obstruction. Abbott dramatically described the *typical cases of morbus coeruleus* as *unfortunate infants in whom each separate cell is strangled.*[8] Nevertheless, survival was possible sometimes even to a full life span.[447] The surgical relief of hypoxia by the creation of a systemic-pulmonary communication that bypassed pulmonary obstructive lesions[45] was a landmark event, which brought into the limelight children and adolescents with tetralogy of Fallot and other similar defects. Within 4 years of the first operation in 1945 more than 1000 patients had traveled to Baltimore for creation of a "Blalock-Taussig shunt" at The Johns Hopkins Hospital.[409,412] Their dramatic and often long persistent clinical improvement[410,411] was not equaled by any new therapy for almost 20 years. By then the focus of attention had shifted to infants as the target population of the rising discipline of pediatric cardiology, and transposition of the great arteries became a common diagnosis in cyanotic neonates. Interventions by balloon septostomy[358] and venous and arterial methods of surgical repair[210,309,391] once again transformed a grave prognosis into a favorable outlook for long-term survival.[162,222,263,308]

In addition to tetralogy of Fallot and transposition of the great arteries, a third severe anomaly, persistence of a single arterial trunk (truncus arteriosus), has recently attracted special attention, as it occurs as part of the DiGeorge syndrome and in association with a submicroscopic chromosomal deletion.[118] Thus clinical interest in outflow tract anomalies must now encompass medical, surgical, and genetic considerations.

Developmental Perspectives

The three malformations mentioned above represent the major anatomic phenotypes of outflow tract anomalies which are often collectively referred to as conotruncal defects. These are disturbances in the ventriculoarterial portion of the ascending limb of the primitive S-shaped cardiac loop—the conus or bulbus cordis—[382] which will become septated by ridges derived from the endocardial cushions and by the aorticopulmonary septum respectively, to form the divided arterial outflow from the right and left ventricles and of the pulmonary artery and aorta.

A wide range of anomalies occur in this portion of the developing heart. Speculations on morphogenetic and pathogenetic processes have been ongoing for more than a century. In his classic monograph, *Die Defekte der Scheidewände des Herzens* (Defects of Cardiac Septation), Rokitansky[367] recognized the unitary nature of anomalies that involve intracardiac defects and alterations in the division of the primitive arterial trunk.

In transposition of the great arteries, the aorta arises from the right ventricle (instead of the left) and the pulmonary artery arises from the left ventricle (instead of the right). The curiosity of this anomaly in humans lies in the fact that such a reversed ventricular-great vessel relationship has never been described in phylogenetic or ontogenetic sequences. Maude Abbott described transposition as a *bizarre condition which seems at first sight to contradict every known principle of development.*[7] Warkany contrasted transpositions with other malformations, which *can be explained by an arrest of development, thus representing a profound disturbance of the basic architecture of the heart.*[435]

He reviewed the many embryological and etiologic theories previously presented, including the imaginative theory of Spitzer (1914), who believed that the abnormally placed aorta may be an embryonic remnant of the primitive right aorta of the reptilian heart.[435] Abbott[7] considered this *the most logical and satisfactory explanation as yet available.* However other investigators considered transposition as the extreme form of aortic displacement in a spectrum of related conotruncal anomalies in which lesser degrees of aortic malpositions are represented by tetralogy of Fallot and double-outlet right ventricle, and at the mild extreme, a simple ventricular septal defect without great vessel disturbances. Some support exists for this unifying view in the familial occurrence of the various outflow tract defects.[91]

A totally different teratologic horizon was opened up by the discovery that migrating neural crest cells form part of the aorticopulmonary septum and the cardiac outflow tract.[225,286] Thus certain anomalies of the heart may represent anomalous ectomesenchymal development. Excising portions of the neural crest in experimental animals enabled Kirby[225,226] to identify the responsible portion of the neural crest (cardiac neural crest), the ablation of which produced a spectrum of anomalies including double-outlet right ventricle, tetralogy of Fallot, and common arterial trunk. In a mechanistic classification of congenital heart disease, Clark grouped outflow tract anomalies as defects of mesenchymal cell migration.[82] This perception is supported by the association of outflow tract anomalies with maternal diabetes,[153,296,299,340,377] thalidomide,[244] and retinoid acid exposure[123,267] -all factors associated with multiple abnormal fetal outcomes that are consistent with abnormal neural crest cell migration.

In very rapid succession, a range of new study methods has expanded our knowledge on embryonic development and maldevelopment beyond the classic perceptions of cardiac pathogenesis.[28,29,109,110,187] These changing horizons have been succinctly described in Pexieder's brilliant overview, *The Conotruncus and Its Septation at the Advent of the Molecular Biology Era.*[345]

Multidisciplinary new information is now becoming available every month, which needs international coordination to advance the multilevel analysis of cardiac pathogenesis.[344] Using as an example the experimental induction of transposition of great arteries by retinoic acid administration in mice, Pexieder describes the results of timed sampling of embryos of treated mice at multiple levels:

- at the *organ* level the embryos showed early alterations in looping and in aorticopulmonary septation
- at the *tissue* level there was hypoplasia of the myocardial wall and hypoplasia of the early septal tissues
- at the *cellular* level there was altered cell migration and proliferation and extracellular matrix formation demonstrated by immunohistoclinical studies

- at the *molecular* level there were abnormal protein profiles and specific patterns of retinoic acid binding proteins, which could lead to a rational interpretation of the timing and action of specific teratogens[347]
- at the *functional* level there was an increased contractile response to calcium ions.

Illustrated in this example is the biologic complexity of developmental systems, the existence of subtle pathogenetic chains in the activation and regulation of cellular processes, and the functional adaptions to hemodynamic requirements of the functioning developing heart. *This complexity,* wrote Tomaš Pexieder, *is somewhere in conflict with the natural trend of the human mind which, since the Renaissance, searches for the single key which would open all the doors. However, this is not likely in the new and challenging frontier of teratogenesis.*[345]

Epidemiologic Perspectives

The rarity of congenital heart disease creates difficulties in the evaluation of anatomic subtypes. Therefore, some studies of cardiovascular malformations have considered outflow tract defects as a single case group.[10,417] To date, no epidemiologic research has addressed potential etiologic heterogeneity within the group of outflow tract anomalies, although we recognized that a contrast between transposition of the great arteries and tetralogy of Fallot is evident in experimental and clinical data.

The Baltimore-Washington Infant Study (1981–1989), included 639 infants with outflow tract defects. It is hypothesized that etiologic differences exist between the following groups and subgroups:

- outflow tract defects with and without transposition ie, parallel versus spiral (normally related) great arteries;
- transposition of the great arteries with intact ventricular septum where the heart is essentially normal, and transposition of the great arteries with ventricular septal defect where the heart itself is abnormal;
- tetralogy of Fallot and other outflow defects, common arterial trunk, and double-outlet right ventricle with spiral (normally related) great arteries.

The separate descriptions of potential risk factors by anatomic diagnoses would evaluate the concept that outflow tract defects consist of a spectrum of abnormal great artery positions.[29,171,247,323,336,441]

This report describes the occurrence of outflow tract defects and the distributions of diagnostic subgroups with evaluations of infant and parental characteristics, and of potential familial and environmental risk factors in the Baltimore-Washington Infant Study region. Multivariate analyses of potential risk factors, including interaction and confounding analysis, could be performed only in the larger case groups. The findings are evaluated regarding the relatedness of the anatomic subgroups according to shared risk factors and the possible etiologic specificity of diagnostic subtypes that have unique associations.

Study Population

Between January 1, 1981 and December 31, 1989 the Baltimore-Washington Infant Study enrolled 4390 cases with cardiovascular malformations, among whom there were 639 cases with outflow tract anomalies (14.6%); family interviews were conducted in 555 (87%).

The diagnostic subdivisions and the frequencies of diagnostic subsets among enrolled subjects and among those with interviewed families are shown in Figure 5.1.

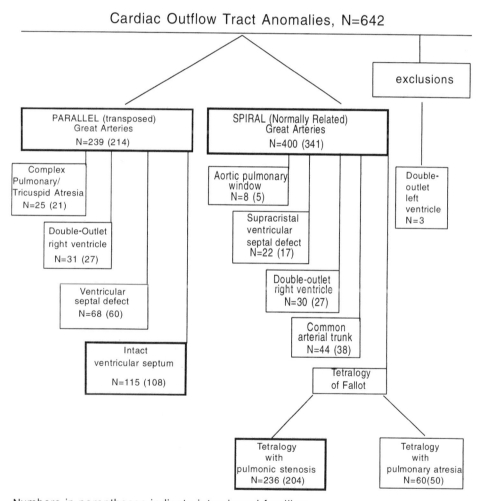

Numbers in parentheses indicate interviewed families.

Figure 5.1: Cases with transposed great arteries were divided into four component phenotypes; complex transposition includes cases with tricuspid or pulmonary atresia. Outflow tract defects with normally related great arteries encompassed five phenotypes; tetralogy of Fallot was further subdivided by the type of pulmonary obstruction.

Cardiac Abnormalities

The outflow tract anomalies considered here had normal cardio-visceral relationships (situs solitus). They were divided into two major categories defined solely by normal or abnormal great artery relationships. Anomalies with aortic origin from the right ventricle are characterized as parallel great arteries, and include transposition of the great arteries with any of the following: intact ventricular septum, ventricular septal defect, pulmonary or tricuspid atresia (complex cases), and double-outlet right ventricle with an anterior aorta or side-by-side great arteries. This group will also be referred to as the transposition defects.

Anomalies with normal ventriculoarterial relationships are characterized as spiral great arteries, and include the diagnoses of tetralogy of Fallot, tetralogy with pulmonary atresia, double-outlet right ventricle without transposition, common arterial trunk (truncus), aortic-pulmonary window, and supracristal ventricular septal defect. This group of cases will also be referred to as the nontransposition or normally related great artery defect group.

Noncardiac Anomalies

Genetic disorders and noncardiac anomalies rarely occurred in the transposition group, but frequently occurred in those groups without transposition; the cases of complex transposition occupied an intermediate position in this comparison. In a detailed review of noncardiac abnormalities in each case, Lurie[265] provided an updated dysmorphology categorization. Associated defects were found in one third of the patients in whom the great arteries were in a normal (spiral) relationship, but in only 10% of patients with transposed great arteries, and were especially rare among those with intact ventricular septum (8.7%) and with ventricular septal defect (4.4%). A listing of these defects is shown in Table 5.1. Among 46 chromosomal defects only two occurred in the transposition group (trisomy 8 and trisomy 18); among the 44 with normally related great arteries there was a wide range of cytogenetic abnormalities. Of the 22 infants with Down syndrome, 20 had tetralogy of Fallot.

Townes-Brooks and Schinzel-Gideon syndromes occurred in both groups, pseudo trisomy 13 occurred only with transposition, and the Holt-Oram, BBB, Noonan, Smith-Lemli-Opitz, Roberts, and Holtzgreve syndromes occurred only in infants with normally related great arteries. Additionally a new autosomal recessive syndrome of oligomeganephronic renal hypoplasia and clitoromegaly was found in two siblings with tetralogy of Fallot, one of whom also had dextrocardia, and was described with the acronym of HOMAGE (Heart defect OligoMeganephronic, Abnormal Genitalia).

DiGeorge syndrome occurred in 17 cases; only one in the transposition group (0.4%), and 16 in the normally related great arteries group (4.0%). VACTERL association was diagnosed in both groups with approximately the same frequency (1.3% in the transposition group and 1.8% in those with normally related great arteries).

In the total group multiple noncardiac malformations outnumbered single associated defects by 2:1. This finding is demonstrated in the diagnosis-specific frequencies of single, multiple, syndromic, and chromosomal malformations in Figure 5.2. Appendix 5 lists the affected cases by identification number and cardiac diagnosis.

Table 5.1
Noncardiac Anomalies In Infants with Outflow Tract Anomalies • Baltimore-Washington Infant Study (1981–1989)

Noncardiac Anomaly	Transposed Great Arteries (TGA)					Normally Related Great Arteries (NGA)						
	TGA Total N=239	TGA-IVS N=115	TGA-VSD N=68	TGA-DORV N=31	TGA-Complex N=25	NGA Total N=400	TOF-PS N=236	TOF-PAtr N=60	NGA-DORV N=30	Truncus N=44	ScVSD N=22	AP Window N=8
Chromosome abnormalities	**2 (0.8%)**	**0**	**1 (1.5%)**	**1 (3.2%)**	**0**	**43 (10.8%)**	**28 (11.9%)**	**5 (8.3%)**	**6 (20%)**	**1 (2.3%)**	**2 (9.1%)**	**1 (12.5%)**
Trisomy 8	1	0	1	0	0	0	0	0	0	0	0	0
Trisomy 13	0	0	0	0	0	10	5	2	1	1	0	1
Trisomy 18	1	0	0	1	0	7	4	0	2	0	1	0
Trisomy 21	0	0	0	0	0	22	18	2	1	0	1	0
Trisomy 1q	0	0	0	0	0	1	0	0	1	0	0	0
Tetrasomy 8p	0	0	0	0	0	1	0	0	1	0	0	0
Deletion 5p	0	0	0	0	0	1	0	1	0	0	0	0
Partial trisomy, uncharacterized	0	0	0	0	0	1	1	0	0	0	0	0
Syndromes	**8 (3.3%)**	**2 (1.7%)**	**1 (1.5%)**	**3 (9.7%)**	**2 (8%)**	**44 (11%)**	**17 (7.2%)**	**7 (11.7%)**	**3 (10%)**	**13 (29.5%)**	**4 (18.2%)**	**0**
a. Mendelian	**3**	**1**	**0**	**1**	**1**	**13**	**6**	**1**	**2**	**3**	**1**	**0**
Townes-Brocks	1	0	0	1	0	2	2	0	0	0	0	0
Holt-Oram	0	0	0	0	0	2	2	0	0	2	0	0
BBB	0	0	0	0	0	1	0	0	0	0	0	0
Noonan	0	0	0	0	0	1	1	0	0	0	0	0
Smith-Lemli-Opitz	0	0	0	0	0	3	1	0	1	0	1	0
Schinzel-Giedion	1	1	0	0	0	1	0	1	1	0	0	0
Roberts	0	0	0	0	0	0	0	0	0	0	0	0
Pseudotrisomy 13	1	0	0	0	1	1	0	0	0	1	0	0
Holzgreve	0	0	0	0	0	0	0	0	0	0	0	0
HOMAGE	0	0	0	0	0	1	1	0	0	0	0	0
b. Non-Mendelian Associations	**5**	**1**	**1**	**2**	**1**	**31**	**11**	**6**	**1**	**10**	**3**	**0**
DiGeorge	1	0	0	1	0	16	2	2	0	9	3	0
VACTERL	3	1	0	1	1	7	5	2	0	0	0	0
Cantrell	0	0	0	0	0	3	2	1	0	0	0	0

Table 5.1—Continued
Noncardiac Anomalies In Infants with Outflow Tract Anomalies
Baltimore-Washington Infant Study (1981–1989)

Noncardiac Anomaly	Transposed Great Arteries (TGA)					Normally Related Great Arteries (NGA)						
	TGA Total N=239	TGA-IVS N=115	TGA-VSD N=68	TGA-DORV N=31	TGA-Complex N=25	NGA Total N=400	TOF-PS N=236	TOF-PAtr N=60	NGA-DORV N=30	Truncus N=44	ScVSD N=22	AP Window N=8
Goldenhar	0	0	0	0	0	2	1	0	0	1	0	0
CHARGE	1	0	1	0	0	1	0	1	0	0	0	0
Klippel-Feil	0	0	0	0	0	2	1	0	1	0	0	0
Multiple Anomalies, Nonclassified	3 (1.3%)	1 (0.9%)	1 (1.5%)	1 (3.2%)	0	9 (2.3%)	5 (2.1%)	0	2 (6.7%)	0	2 (9.1%)	0
Organ Defects	11 (4.6%)	7 (6.1%)	2 (6.5%)	2 (6.5%)	1 (4%)	39 (9.8%)	26 (11.4%)	4 (6.7%)	4 (13.3%)	4 (9.1%)	1 (4.5%)	0
Brain, Skull, and Eye												
microcephaly	0	0	0	0	0	2	1	0	0	1	0	0
crainosynostosis	0	0	0	0	0	2	2	0	0	0	0	0
hydrocephaly	0	0	0	0	0	1	1	0	0	0	0	0
iris defects	0	0	0	0	0	1	0	0	1	0	0	0
Gastrointestinal												
duodenal atresia	0	0	0	0	0	3	3	0	0	0	0	0
Pyloric stenosis	1	1	0	0	0	1	1	0	0	0	0	0
omphalocele	0	0	0	0	0	2	2	0	0	0	0	0
malrotation	1	0	0	0	1	0	0	0	0	0	0	0
Meckel's diverticulum	0	0	0	0	0	2	0	1	1	1	0	0
anal atresia	0	0	0	0	0	1	0	1	0	0	0	0

Genitourinary												
kidney agenesis	1	0	0	0	0	3	2	1	0	0	0	0
polycystic kidney	0	0	0	0	0	1	1	0	0	0	0	0
hydronephrosis/ureter	2	2	0	0	0	1	0	1	0	0	0	0
hypospadias	2	2	0	0	0	2	1	1	0	0	0	0
ambiguous genitalia	0	0	0	0	0	1	1	0	0	0	0	0
Skeletal Defects												
polydactyly	3	1	1	1	0	2	0	0	0	2	0	0
vertebral defects	0	0	0	0	0	2	0	0	1	0	1	0
Other Systems												
cleft lip/palate	0	0	0	0	0	3	3	0	0	0	0	0
choanal atresia	0	0	0	0	0	1	1	0	0	0	0	0
absence of ear	0	0	0	0	0	1	1	0	0	0	0	0
bronchial defect	0	0	0	0	0	1	1	0	0	0	0	0
thrombocytopenia	0	0	0	0	0	1	1	0	0	0	0	0
anemia	0	0	0	0	0	1	1	0	0	0	0	0
tuberous sclerosis	0	0	0	0	0	1	1	0	0	0	0	0
accessory spleen	1	0	0	1	0	2	1	0	1	0	0	0
diaphragmatic hernia	0	0	0	0	0	1	1	0	0	0	0	0
No Associated Anomalies	215 (90%)	105 (91%)	64 (94.1%)	24 (77.4%)	22 (88%)	265 (66.3%)	160 (67.8%)	44 (73.3%)	15 (50%)	26 (59.1%)	13 (59.1%)	7 (87.5%)

Note: identification numbers for these infants are listed in Appendix C.

Abbreviations: A-P-aortic-pulmonary; DORV-double-outlet right ventricle; IVS-intact ventricular septum; PAtr-pulmonary atresia; PS-pulmonic valve stenosis; Sc-supracristal; TOF-tetralogy of Fallot; VSD-ventricular septal defect.

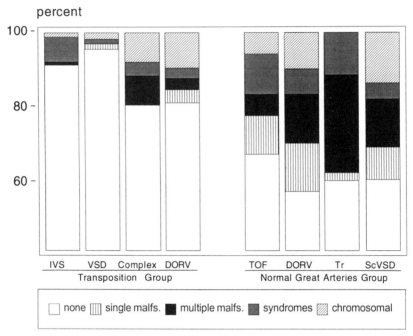

Outflow Tract Anomalies: Associated Noncardiac Malformations
Baltimore-Washington Infant Study (1981 - 1989)

Figure 5.2: Differential distribution of associated noncardiac malformations among outflow tract defects with and without transposition of the great arteries. Multiple malformations, syndromes, and chromosomal anomalies combined were more common than single associated malformations in both the major groups.
Abbreviations: IVS = intact ventricular septum; VSD = ventricular septum defect; DORV = double-outlet right ventricle; TOF = tetralogy of Fallot; Tr = common arterial trunk; ScVSD = supracristal ventricular septal defect.

Descriptive Analyses

Prevalence by Time, Season, and Area of Residence

The regional prevalence of all cardiovascular malformations over the 9 study years was 48.4 per 10,000 livebirths; that of outflow tract anomalies was 7.05 per 10,000. The prevalence of the group with parallel (transposed) great arteries and of the group with normally related (spiral) great arteries is shown in Figure 5.3 for the entire 9-year study period, for each 3-year time segment, for birth quarters, and by the area of residence (urban, suburban, rural).

There were no changes in the prevalence of transposition defects by year of birth but there was a slight increase over time in the defects with normally related great arteries. Both groups of defects showed a slight seasonal change: prevalence was increased in the 3rd and 4th birth quarters relative to the earlier quarters. Prevalence by area of residence varied little for the transposition group but was increased in urban areas among infants with normally related great arteries.

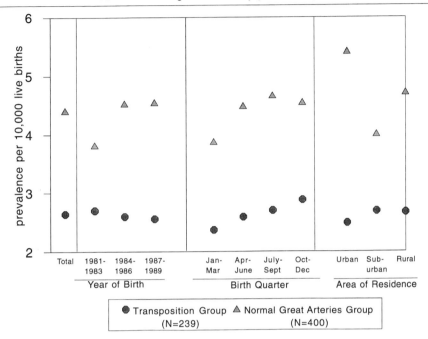

Figure 5.3: Arising from the regional birth cohort of 906,626 liveborn infants, cases with transposed great arteries occurred more rarely and with less variability by time and place than cases of outflow tract defects with normally related great arteries.

Diagnosis and Course

Enrollment and outcome characteristics in infants with transposed and with normal great artery relationships are shown in Table 5.2 for all subtypes in each major group. All of these severe cardiovascular malformations were recognized early in the infants' lives. The role of cyanosis leading them to medical attention is evident among infants with transposition and intact ventricular septum, who are the most deeply cyanotic, and were thus diagnosed in the first week of life. The proportion of neonatal diagnoses diminished among those with septal defects, which allow intracardiac communications. Among the infants with normally related great arteries, cardiac diagnoses occurred slightly later; the greatest proportion of first week diagnosis was in tetralogy with pulmonary atresia. Very few defects were diagnosed in the second half of the first year of life, suggesting that these infants are representative of the total liveborn group with these malformations.

In the transposition group, almost every infant underwent cardiac catheterization and most underwent surgery in the first year of life with a remarkable survival rate of 80% or more in the two major forms of transposition. Infants with outflow tract defects with normally related great arteries were also subjected to cardiac catheterization in a very high proportion of cases, and only six infants with tetralogy of Fallot and only one infant with supracristal ventricular septal defect were diagnosed by echocardiography only. The rate of surgery was also

Table 5.2
Outflow Tract Anomalies: Clinical Data
Baltimore-Washington Infant Study (1981–1989)

A. Infants with Transposed Great Arteries (TGA)

	All TGA N=239 no. (%)	TGA-IVS N=115 no. (%)	TGA-VSD N=68 no. (%)	TGA-DORV N=31 no. (%)	TGA-Complex N=25 no. (%)
Age at Diagnosis					
<1 week	188 (78.7)	102 (88.7)	53 (77.9)	16 (51.6)	17 (68.0)
1–4 weeks	34 (14.2)	8 (7.0)	11 (16.2)	9 (29.0)	6 (24.0)
5–24 weeks	14 (5.9)	4 (3.5)	3 (4.4)	5 (16.1)	2 (8.0)
25–52 weeks	3 (1.3)	1 (0.8)	1 (1.5)	1 (3.2)	0
Method of Diagnosis					
echocardiography only	2 (0.8)	2 (1.7)	0	0	0
echocardiography and cardiac catheterization	34 (14.2)	9 (7.8)	13 (19.1)	8 (25.8)	4 (16.0)
echocardiography and surgery	18 (7.5)	16 (13.9)	0	2 (6.5)	0
echocardiography, cardiac catheterization and surgery	148* (61.9)	70 (60.9)	51 (75.0)	14 (45.2)	13 (52.0)
autopsy following other methods	31 (13.0)	14 (12.2)	4 (5.9)	6 (19.4)	7 (28.0)
autopsy only	6 (2.5)	4 (3.5)	0	1 (3.2)	1 (4.0)
Cardiac Surgery in 1st year	202 (84.5)	97 (84.3)	65 (94.2)	23 (74.2)	17 (68.0)

Alive at diagnostic

B. Infants with Spiral (Normally Related) Great Arteries (NGA)

update	All NGA N=400 no. (%)		TOF/PS N=236 no. (%)		TOF/PAtr N=60 no. (%)		NGA-DORV N=30 no. (%)		Truncus N=44 no. (%)		Sc VSD N=22 no. (%)		A-P Window N=8 no. (%)	
	169	(70.7)	84	(73.0)	52	(75.4)	18	(58.1)	15	(60.0)				
Age at Diagnosis														
<1 week	174	(43.5)	86	(36.4)	41	(68.3)	15	(50.0)	26	(59.1)	4	(18.2)	2	(25.0)
1–4 weeks	115	(28.8)	76	(32.2)	8	(13.3)	9	(30.0)	10	(22.7)	7	(31.8)	5	(62.5)
5–24 weeks	90	(22.5)	60	(25.4)	9	(15.0)	4	(13.3)	8	(18.2)	8	(36.4)	1	(12.5)
25–52 weeks	21	(5.2)	14	(5.9)	2	(3.3)	2	(6.7)	0		3	(13.6)	0	
Method of Diagnosis														
echocardiography only	65	(16.3)	50	(21.2)	4	(6.7)	4	(13.3)	2	(4.6)	4	(18.2)	1	(12.5)
echocardiography and cardiac catheterization	70	(17.5)	42	(17.8)	13	(21.7)	6	(20.0)	7	(15.9)	2	(9.1)	0	
echocardiography and surgery	9	(2.3)	7	(3.0)	1	(1.7)	1	(3.3)	0		0		0	
echocardiography, cardiac catheterization and surgery	198**	(49.5)	118	(50.0)	28	(46.7)	10	(33.3)	22	(50.0)	15	(68.2)	5	(62.5)
autopsy following other methods	42	(10.5)	14	(5.9)	10	(16.7)	9	(30.0)	8	(18.2)	0		1	(12.5)
autopsy only	16	(4.0)	5	(2.1)	4	(6.7)	0		5	(11.4)	1	(4.6)	1	(12.5)
Cardiac Surgery in 1st year	237	(59.3)	132	(55.9)	42	(70.0)	16	(53.3)	27	(61.4)	15	(68.2)	5	(62.5)
Alive at diagnostic update	281	(70.3)	198	(83.9)	35	(58.3)	11	(36.7)	13	(29.6)	19	(86.4)	5	(62.5)

*includes 1 infant with TGA-IVS who had cardiac catheterization and surgery; **includes 1 infant with cardiac catheterization only and 3 with cardiac catheterization and surgery (all with TOF/PS); IVS = intact ventricular septum; VSD = ventricular septal defect; DORV = double-outlet right ventricle; TOF = tetralogy of Fallot; PS = pulmonic valve stenosis; PAtr = pulmonary atresia; Sc = supracristal, A-P = aortic-pulmonary.
For the TGA group, only one infant was not followed after registration. Mortality as 29% overall in the first year of life; survivors were followed for a median of 11 months. For the NGA group, 5 infants were not followed; overall mortality was 30%, and survivors were followed for a median of 11 months.

lower, as was survival, except for tetralogy with pulmonary stenosis and supracristal ventricular septal defects. The highest mortality was among infants with truncus arteriosus. All-cause mortality in the first year of life was 29% for infants with transposed great arteries, and 30% for infants with normally related great arteries.

Gender, Race, and Twinning

Male predominance was found in almost all diagnoses (Table 5.3, Figure 5.4), and was greatest in transposition with intact ventricular septum (73% versus 51% of controls); this was the only difference of statistical significance (case-control odds ratio of 2.6) (Figure 5.4). The race distributions of cases in both major diagnostic groups do not differ statistically from that of controls. Among 639 cases with outflow tract defects, 12 occurred in multiple births (1.9%), a frequency similar to controls (1.5%): 2 were in the transposition group and 10 had normally related great arteries; all co-twins were normal.

Birthweight and Gestational Age

The distribution of low birthweight, gestational age, and small size-for-gestational-age are shown for all subjects in each diagnostic subgroup and for subjects with isolated congenital heart disease (Table 5.4). Infants in the transposition group had normal birthweight except for those with complex cardiac anomalies, in whom the odds ratios for low birthweight and for small size-for-gestational-age were elevated in comparison to controls. These findings, however, were not significant when subjects with noncardiac anomalies were excluded. In contrast, in the group with normally related great arteries, all subgroups showed high odds ratios for low birthweight as well as for premature birth and for small size-for-gestational-age, findings which persisted also for most diagnoses occurring as isolated anomalies. These differences are graphically presented in Figure 5.5. Of note is the difference between the two groups of double-outlet right ventricle, those with and without transposition, respectively. A morphogenetic evaluation by Männer et al[274] suggests that double-outlet right ventricle (DORV) with normal great arteries may be an inflow, rather than an outflow defect.

Sociodemographic Characteristics

Sociodemographic characteristics of cases and controls are shown in Table 5.5 for cases with parallel great arteries (the transposition group) and in Table 5.6 for cases with spiral (normally related) great arteries. There were no significant case-control differences in maternal marital status, parental educational levels, household income, parental occupational categories, or the month of pregnancy confirmation in the analysis of infants in the transposition group (Table 5.5). Among infants with spiral great arteries (Table 5.6), case mothers were somewhat more likely than controls to have jobs classified as clerical/sales-related, or factory-related, but there were no other significant sociodemographic differences.

Table 5.3
Characteristics of Infants with Outflow Tract Anomalies
Baltimore-Washington Infant Study (1981–1989)

A. Infants with Transposed Great Arteries (TGA)

	All TGA N=239 no. (%)		TGA IVS N=115 no. (%)		TGA VSD N=68 no. (%)		TGA DORV N=31 no. (%)		TGA Complex N=25 no. (%)		Controls N=3572 no. (%)	
Gender:												
male	158	(66.1)	84	(73.0)	39	(57.4)	19	(61.3)	16	(64.0)	1,817	(50.9)
female	81	(33.9)	31	(27.0)	29	(42.6)	12	(38.7)	9	(36.0)	1,755	(49.1)
Race*												
white	164	(68.6)	83	(72.1)	47	(69.1)	20	(66.7)	14	(56.0)	2,362	(66.1)
black	53	(22.2)	24	(20.9)	14	(20.6)	8	(26.7)	7	(28.0)	1,109	(31.0)
other	15	(6.3)	7	(6.1)	4	(5.9)	2	(6.6)	2	(8.0)	98	(2.7)
Twin births	2	(0.8)	0		1	(1.5)	0		1	(4.0)	53	(1.5)

B. Infants with Spiral (Normally Related) Great Arteries (NGA)

	All NGA N=400 no. (%)		TOF/PS N=236 no. (%)		TOF/PAtr N=60 no. (%)		NGA-DORV N=30 no. (%)		Truncus N=44 no. (%)		ScVSD N=22 no. (%)		A-P Window N=8 no. (%)		Controls N=3572 no. (%)	
Gender:																
male	217	(54.3)	133	(56.4)	30	(50.0)	15	(50.0)	22	(50.0)	13	(59.1)	4	(50.0)	1,817	(50.9)
female	183	(45.7)	103	(43.6)	30	(50.0)	15	(50.0)	22	(50.0)	9	(40.9)	4	(50.0)	1,755	(49.1)
Race:*																
white	245	(62.2)	144	(62.3)	37	(62.7)	16	(53.3)	34	(77.3)	9	(40.9)	5	(62.5)	2,362	(66.1)
black	124	(31.5)	73	(31.6)	17	(28.8)	12	(40.0)	8	(18.2)	11	(50.0)	3	(37.5)	1,109	(31.0)
other	25	(6.3)	14	(6.1)	5	(8.5)	2	(6.7)	2	(4.5)	2	(9.1)	0		98	(2.7)
Twin births	10	(2.5)	7	(3.0)	2	(3.3)	0		1	(2.3)	0		0		53	(1.5)

*Data are missing for race: 1 TGA-IVS, 3 TGA-VSD, 1 TGA-DORV, 2 TGA-Complex, 5 TOF/PS, 1 TOF/PAtr, 3 controls.
A-P = aortic-pulmonary; DORV = double-outlet right ventricle; IVS = intact ventricular septum; PS = pulmonic valve stenosis; Sc = supracristal; TOF = tetralogy of Fallot; VSD = ventricular septal defect.

Outflow Tract Anomalies: Infant Gender
Baltimore-Washington Infant Study (1981 - 1989)

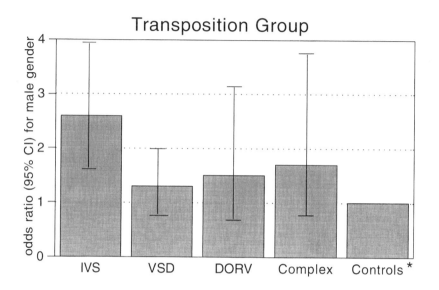

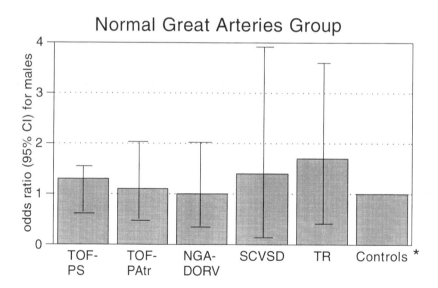

*Reference group

Figure 5.4: A significant excess of males is noted only among cases with transposed great arteries with intact ventricular septum (odds ratio 2.6; 95% confidence interval 1.7 to 4.0). The evaluated odds ratios for male gender in all other subgroups are nonsignificant.

Abbreviations: IVS = intact ventricular septum; VSD = ventricular septal defect; DORV = double-outlet right ventricle; TOF = tetralogy of Fallot; NGA = normally related great arteries; Tr = common arterial trunk; ScVSD = supracristal ventricular septal defect; PS = pulmonic stenosis; PAtr = pulmonary atresia.

Table 5.4
Outflow Tract Anomalies: Fetal Growth Characteristics in all Diagnostic Subsets in the Presence and Absence of Noncardiac Anomalies
Baltimore-Washington Infant Study (1981–1989)

Diagnostic Group	N	Low Birthweight (<2500 g)		Prematurity (Gest. Age <38 wks.)		Small for Gestational Age*	
		no. (%)	Odds Ratio 95% CI	no. (%)	Odds Ratio 95% CI	no. (%)	Odds Ratio 95% CI
A) All Subjects							
TGA Group	**214**	14 (6.5)	0.9 (0.5–1.6)	20 (9.3)	1.0 (0.6–1.6)	17 (7.9)	1.4 (0.8–2.3)
TGA-IVS	106	4 (3.8)	0.5 (0.2–1.4)	9 (8.5)	0.9 (0.4–1.8)	5 (4.7)	0.8 (0.3–2.0)
TGA-VSD	60	4 (6.7)	0.9 (0.3–2.6)	5 (8.3)	0.9 (0.3–2.2)	6 (10.0)	1.8 (0.8–4.2)
TGA-Complex	21	4 (19.0)	**3.1 (1.0–9.3)**	3 (14.3)	1.6 (0.5–5.4)	4 (19.0)	**3.7 (1.3–11.2)**
TGA DORV	27	2 (7.4)	1.1 (0.2–4.5)	3 (11.1)	1.2 (0.4–4.0)	2 (7.4)	1.3 (0.3–5.4)
NGA Group	**341**	96 (28.2)	**5.2 (3.9–6.8)**	84 (24.6)	**3.1 (2.4–4.1)**	86 (25.2)	**5.4 (4.1–7.1)**
TOF/PS	204	52 (25.5)	**4.5 (3.2–6.3)**	43 (21.1)	**2.5 (1.8–3.6)**	49 (24.0)	**5.0 (3.5–7.1)**
TOF/PAtr	50	20 (40.0)	**8.8 (4.9–15.7)**	22 (44.0)	**7.5 (4.2–13.2)**	11 (22.0)	**4.5 (2.3–8.9)**
NGA-DORV	27	12 (44.4)	**10.5 (4.9–22.8)**	5 (18.5)	**2.2 (0.8–5.8)**	15 (55.6)	**19.9 (9.2–43.1)**
Truncus	38	8 (21.1)	**3.5 (1.6–7.7)**	10 (26.3)	**3.4 (1.6–7.1)**	7 (18.4)	**3.6 (1.6–8.3)**
ScVSD	17	4 (23.5)	**4.1 (1.3–12.5)**	4 (23.5)	**2.9 (1.0–9.0)**	4 (23.5)	**4.9 (1.6–15.1)**
Controls	3572	252 (7.1)	1.0 (reference)	339 (9.5)	1.0 (reference)	211 (5.9)	1.0 (reference)
B) Subjects without Noncardiac Anomalies (isolated)							
TGA Group	**194**	9 (4.6)	0.6 (0.3–1.2)	16 (8.2)	0.9 (0.5–1.4)	12 (6.2)	1.0 (0.6–1.9)
TGA-IVS	96	3 (3.1)	0.4 (0.1–1.3)	8 (8.3)	0.9 (0.4–1.8)	4 (4.2)	0.7 (0.3–1.9)
TGA-VSD	57	3 (5.3)	0.7 (0.2–2.3)	4 (7.0)	0.7 (0.3–2.0)	5 (8.8)	1.5 (0.6–3.9)
TGA-Complex	19	3 (15.8)	2.5 (0.7–8.5)	2 (10.5)	1.1 (0.3–4.9)	3 (15.8)	3.0 (0.9–10.3)
TGA-DORV	22	0	—	2 (9.1)	1.0 (0.2–4.1)	0	—
NGA Group	**225**	44 (19.6)	**3.2 (2.3–4.5)**	41 (18.2)	**2.1 (1.5–3.0)**	42 (18.7)	**3.7 (2.5–5.3)**
TOF/PS	139	23 (16.6)	**2.6 (1.6–4.1)**	20 (14.4)	**1.6 (1.0–2.6)**	22 (15.8)	**3.0 (1.9–4.8)**
TOF/PAtr	36	13 (36.1)	**7.4 (3.7–14.8)**	13 (36.1)	**5.4 (2.7–10.7)**	7 (19.4)	**3.9 (1.7–8.9)**
NGA-DORV	14	5 (35.7)	**7.3 (2.4–21.9)**	3 (21.4)	**2.6 (0.7–9.4)**	6 (42.9)	**12.0 (4.1–34.8)**
Truncus	22	2 (9.1)	1.3 (0.3–5.6)	4 (18.2)	2.1 (0.7–6.3)	6 (27.3)	**6.0 (2.3–15.4)**
ScVSD	10	1 (10.0)	1.5 (0.2–11.5)	1 (10.0)	1.1 (0.1–8.4)	1 (10.0)	1.8 (0.2–14.1)
Controls	3509	249 (7.1)	1.0 (reference)	333 (9.5)	1.0 (reference)	207 (5.9)	1.0 (reference)

*Birthweight below 10th percentile at given week of gestation (Brenner et al. 1976).

TGA = transposed great arteries; IVS = intact ventricular septum; VSD = ventricular septal defect; DORV = double-outlet right ventricle; TOF = tetralogy of Fallot; PS = pulmonic valve stenosis, PAtr = pulmonary atresia; NGA = normally related great arteries; Sc = supracristal.

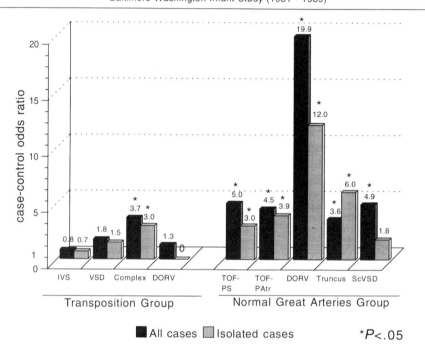

Outflow Tract Anomalies: Small for Gestational Age
Baltimore-Washington Infant Study (1981 - 1989)

Figure 5.5: An important difference between outflow tract defects with and without transposition of the great arteries is evident in the case-control odds ratios for small for gestational age (odds ratios are shown above the bars) for all cases and for isolated cases.
Abbreviations: IVS = intact ventricular septum; VSD = ventricular septal defect; DORV = double-outlet right ventricle; TOF = tetralogy of Fallot; Tr = common arterial trunk; ScVSD = supracristal ventricular septal defect; PS = pulmonic stenosis; Patr = pulmonary atresia.

Table 5.5

Infants with Parallel Great Arteries *(Transposition Group)*:
Analysis of Sociodemographic Characteristics
Baltimore-Washington Infant Study (1981–1989)

Interviewed Families	Cases N=214 no.	(%)	Controls N=3572 no.	(%)	Odds Ratio (95% CI)	
Maternal Marital Status:						
not married	49	(22.9)	990	(27.7)	1.0	(reference)
married	165	(77.1)	2582	(72.3)	0.8	(0.6–1.5)
Maternal Education:						
<high school	29	(13.6)	659	(18.4)	0.7	(0.5–1.1)
high school	88	(41.1)	1265	(35.4)	1.2	(0.9–1.6)
college	97	(45.3)	1648	(46.1)	1.0	(reference)
Paternal Education:						
<high school	32	(15.0)	650	(18.2)	0.8	(0.5–1.2)
high school	82	(38.3)	1298	(36.3)	1.0	(0.8–1.4)
college	100	(46.7)	1624	(45.5)	1.0	(reference)
Annual Household Income:						
<$10,000	36	(17.0)	686	(19.2)	0.8	(0.5–1.1)
$10,000–$19,999	40	(18.9)	699	(19.6)	0.8	(0.6–1.2)
$20,000–$29,999	42	(19.8)	737	(20.6)	0.8	(0.6–1.2)
$30,000+	94	(44.3)	1373	(38.4)	1.0	(reference)
Maternal Occupation:						
not working	60	(28.0)	1142	(32.0)	0.9	(0.6–1.3)
clerical/sales	80	(37.4)	1119	(31.3)	1.2	(0.8–1.7)
service	23	(10.7)	444	(12.4)	0.8	(0.5–1.4)
factory	6	(2.8)	137	(3.8)	0.7	(0.3–1.7)
professional	45	(21.0)	730	(20.4)	1.0	(reference)
Paternal Occupation:						
not working	12	(5.7)	246	(6.9)	0.8	(0.4–1.5)
clerical/sales	42	(19.8)	618	(17.3)	1.1	(0.7–1.6)
service	28	(13.2)	480	(13.4)	0.9	(0.6–1.5)
factory	70	(33.0)	1279	(35.8)	0.9	(0.6–1.2)
professional	60	(28.3)	947	(26.5)	1.0	(reference)
Month of Pregnancy Confirmation:						
1st month	32	(15.0)	511	(14.3)	1.0	(reference)
2nd month	123	(57.5)	1955	(54.7)	1.0	(0.4–1.5)
3rd month	43	(20.1)	663	(18.6)	1.0	(0.6–1.7)
4th month or later	16	(7.4)	416	(11.6)	0.6	(0.3–1.1)

Data are missing for income (2 cases, 77 controls), paternal occupation (2 cases, 2 controls) and pregnancy confirmation month (27 controls).

Table 5.6

Infants with Spiral Great Arteries (*Normally Related Great Arteries Group*): Analysis of Sociodemographic Characteristics Baltimore-Washington Infant Study (1981–1989)

Interviewed Families	Cases N= 341 no.	(%)	Controls N=3572 no.	(%)	Odds Ratio (95% CI)	
Maternal Marital Status:						
not married	103	(30.2)	990	(27.7)	1.1	(0.9–1.4)
married	238	(69.8)	2582	(72.3)	1.0	(reference)
Maternal Education:						
<high school	57	(16.7)	659	(18.4)	0.9	(0.7–1.2)
high school	126	(37.0)	1265	(35.4)	1.0	(0.8–1.3)
college	158	(46.3)	1648	(46.1)	1.0	(reference)
Paternal Education:						
<high school	53	(15.5)	650	(18.2)	0.9	(0.6–1.2)
high school	133	(39.0)	1298	(36.3)	1.1	(0.8–1.4)
college	155	(45.5)	1624	(45.5)	1.0	(reference)
Annual Household Income:						
<$10,000	63	(18.7)	686	(19.2)	0.9	(0.7–1.3)
$10,000–$19,999	76	(22.6)	699	(19.6)	1.1	(0.8–1.5)
$20,000–$29,999	63	(18.7)	737	(20.6)	0.9	(0.6–1.2)
$30,000+	135	(40.0)	1373	(38.4)	1.0	(reference)
Maternal Occupation:						
not working	112	(32.8)	1142	(32.0)	1.3	(0.9–1.8)
clerical/sales	116	(34.0)	1119	(31.3)	**1.4**	**(1.0–1.9)**
service	38	(11.1)	444	(12.4)	1.1	(0.7–1.7)
factor	19	(5.6)	137	(3.8)	**1.8**	**(1.0–3.1)**
professional	56	(16.4)	730	(20.4)	1.0	(reference)
Paternal Occupation:						
not working	22	(6.5)	246	(6.9)	1.0	(0.6–1.6)
clerical/sales	61	(18.0)	618	(17.3)	1.1	(0.8–1.5)
service	39	(11.5)	480	(13.4)	0.9	(0.6–1.3)
factory	130	(38.3)	1279	(35.8)	1.1	(0.8–1.5)
professional	87	(25.7)	947	(26.5)	1.0	(reference)
Month of Pregnancy Confirmation:						
1st month	68	(20.0)	511	(14.3)	1.0	(reference)
2nd month	168	(49.4)	1955	(54.7)	**0.6**	**(0.5–0.9)**
3rd month	66	(19.4)	663	(18.6)	0.7	(0.7–1.1)
4th month or later	38	(11.2)	416	(11.6)	0.7	(0.5–1.0)

Data are missing for income (4 cases, 77 controls), paternal occupation (2 cases, 2 controls) and pregnancy confirmation month (1 case, 27 controls).

Potential Risk Factors

Familial Cardiac and Noncardiac Anomalies

The frequency of congenital heart disease in first-degree relatives of infants with outflow tract anomalies was similar to that in the control families (1.2% versus 1.0%), as was the frequency of familial noncardiac malformations (5.1% versus 4.7%). Details of the familial abnormalities in first-degree relatives (noting also affected second-degree relatives within the same families) are shown in Table 5.7. Although familial aggregation of congenital heart disease was rare, the concordance of outflow tract defects is notable in both major groups, but the cardiac defects were not identical in the two brother-sister pairs with transposition or in the sisters of two female infants with tetralogy of Fallot.

Hematologic abnormalities were the most notable noncardiac familial disorders. Hemophilia was present in parents of two transposition cases; a third had a father with congenital anemia treated since birth at one of the University Hospitals. Of infants with normally related great arteries, a mother, 2 fathers, and 2 half-siblings had blood disorders including sickle cell disease, hemolytic anemia, thalassemia, and thrombocytopenia, respectively.

For all outflow tract defects as a whole, case-control comparisons of all familial hematologic disorders grouped together revealed a case excess for defects in first-degree relatives (odds ratio 2.4; 95% confidence 1.1 to 5.3) and a case excess for defects in fathers (odds ratio 4.3; 95% confidence interval 1.2 to 15.3). Case-control odds ratios were elevated, but not significantly, for defects in siblings (odds ratio 2.6; 95% confidence interval 0.5 to 13.3) and defects in mothers (odds ratio 1.2; 95% confidence interval 0.3 to 5.3). An etiologic role of heritable blood disorders was suspected early in the study,[154] but so far the relationship has not been clarified.

Genetic and Environmental Factors

Univariate Analysis

These analyses describe the case and control populations and define possible risk factors for further evaluation by multivariate analyses. Unless otherwise indicated, all variables are binary, and the period in which the exposure occurred is the critical (periconception) period (defined in Chapter 3). The two major categories of defects and their subsets are described in the tables, and additional salient findings are noted for each potential risk factor:

Maternal Reproductive History

Case-control differences in maternal reproductive characteristics in the normal great arteries group were evident in reported subfertility (Table 5.8), and history of adverse pregnancy outcomes in (Table 5.9). No other set of maternal variables so clearly distinguished the normal great artery group, which was strongly associated with adverse reproductive history, from the transposition group, which lacked statistical associations with these variables.

Table 5.7

Outflow Tract Anomalies: Familial Cardiac and Noncardiac Anomalies and Hematologic Disorders in First-Degree Relatives of Infants in Each Major Diagnostic Group Baltimore-Washington Infant Study (1981–1989)

		Proband		Affected Family Members	
ID	Gender	Cardiac Anomaly	Noncardiac Anomaly	Relative	Congenital Anomalies
Transposition Group					
(a) Congenital Heart Disease					
2503	F	Intact ventricular septum	—	Brother	Transposition with ventricular septal defect, dextrocardia
11160	M	Double-outlet right ventricle	—	Sister (#3430)	Transposition with tricuspid atresia
(b) Noncardiac Anomalies					
2978	F	Intact ventricular septum	Meningomyelocele, Cleft palate, limb Anomalies	Father Mat. Aunt	Congenital hypoplastic anemia Transposition with ventricular septal defect, pulmonary atresia
1343	F	Ventricular septal defect, AV canal type	—	Father Pat. Uncle	Hemophilia, Factor VIII Hemophilia, Factor VIII
11291	F	Ventricular septal defect	—	Mother	Hemophilia, Factors IX, X
3691	F	Double-outlet right ventricle	Di George	Mother 3 Mat. Aunts	Darier-White disease Darier-White disease
3113	M	Intact ventricular septum	Polydactyly	Mother Brother Father	Polydactyly Polydactyly Polydactyly
2563	F	Ventricular septal defect	Polydactyly		
Normal Great Arteries Group					
(a) Congenital Heart Disease					
10207	M	Tetralogy, Pulmonary atresia	Di George	Sister	Tetralogy
1833	F	Tetralogy	—	Mother	Ventricular septal defect

ID	Sex	Diagnosis	Syndrome	Relative	Condition
10816	M	Tetralogy	—	Sister (#3198)	Common arterial trunk
3566	M	Tetralogy	—	Mother	Bicuspid aortic valve

(b) Noncardiac Anomalies

ID	Sex	Diagnosis	Syndrome	Relative	Condition
10115	M	Supracristal ventricular septal defect	hemivertebrae	Father	Hemolytic anemia
11548	M	Double-outlet right ventricle	VACTERL	Father	Thalassemia
11462	F	Common arterial trunk	—	Half-Brother	Thrombocytopenia
2621	F	Tetralogy	—	Mother	Sickle cell disease
11367	F	Double-outlet right ventricle	—	Half-Sister	Sickle cell disease
1728	F	Tetralogy	—	Mother	Spina bifida occulta
1539	M	Tetralogy	—	Father	Spina Bifida occulta
3767	M	Tetralogy	—	Mother	Cleft palate
				Mat. Aunt	Polydactyly
10832	F	Tetralogy	—	Father	Cleft palate
2732	M	Common arterial trunk	Di George	Half-Brother	Cleft palate
2779	F	Supracristal ventricular septal defect	—	Half-Brother	Abnormal genitalia
3709	M	Tetralogy	—	Mother	Pilonidal cyst
11416	M	Double-outlet right ventricle	—	Father	Fanconi (kidney) syndrome
				Mat. Cousin	Pectus excavatum
1279	F	Double-outlet right ventricle	Di George	Father	Balanced translocation

Table 5.8

Outflow Tract Anomalies: Maternal Subfertility

Baltimore-Washington Infant Study (1981–1989)

Group	N	Subfertility*		Any Fertility Medication		Clomiphene Citrate	
		no. (%)	Odds Ratio 95% CI	no. (%)	Odds Ratio 95% CI	no. (%)	Odds Ratio 95% CI
TRANSPOSITION GROUP	214	20 (9.3)	0.9 (0.6–1.4)	5 (2.3)	2.0 (0.8–5.1)	2 (0.9)	0.8 (0.2–3.2)
Intact ventricular septum	106	13 (12.3)	1.2 (0.7–2.2)	4 (3.8)	2.5 (0.9–6.9)	2 (1.9)	1.6 (0.6–4.5)
Ventricular septal defect	60	4 (6.7)	0.6 (0.2–1.7)	1 (1.7)	1.1 (0.1–7.8)	0	—
Complex	21	1 (4.8)	0.4 (0.1–3.2)	0	—	0	—
Double-outlet right ventricle	27	2 (7.4)	0.7 (0.2–2.9)	0	—	0	—
NORMAL GREAT ARTERIES GROUP	341	48 (14.1)	**1.4 (1.0–1.9)**	16 (4.7)	**3.1 (1.8–5.5)**	11 (3.2)	**2.7 (1.4–5.4)**
Tetralogy	204	32 (15.7)	**1.6 (1.1–2.4)**	11 (5.4)	**3.6 (1.9–6.9)**	8 (3.9)	**3.4 (1.6–7.2)**
Tetralogy with pulmonary atresia	50	6 (12.0)	1.2 (0.5–2.8)	1 (2.0)	1.3 (0.2–9.4)	0	—
Double outlet right ventricle	27	4 (14.8)	1.5 (0.5–4.3)	1 (3.7)	2.4 (0.3–18.1)	0	—
Truncus	38	5 (13.2)	1.3 (0.5–3.3)	3 (7.9)	**5.4 (1.6–18.0)**	3 (7.9)	**7.0 (2.0–23.7)**
Supracristal VSD	17	1 (5.9)	0.5 (0.1–4.0)	0	—	0	—
CONTROLS	3572	374 (10.5)	1.0 (reference)	56 (1.6)	1.0 (reference)	43 (1.2)	1.0 (reference)

*Trying to conceive 12 months, fertility tests, or treatments; **significant odds ratios are shown in bold type.**

Table 5.9

Outflow Tract Anomalies: Maternal Reproductive History

Baltimore-Washington Infant Study (1981–1989)

Group	N	Previous Miscarriage		Previous Preterm Birth		Bleeding During Pregnancy	
		no. (%)	Odds Ratio 95% CI	no. (%)	Odds Ratio 95% CI	no. (%)	Odds Ratio 95% CI
TRANSPOSITION GROUP	214	39 (18.2)	0.9 (0.6–1.4)	10 (4.7)	0.9 (0.5–1.8)	34 (15.9)	1.1 (0.7–1.6)
Intact ventricular septum	106	21 (19.8)	1.0 (0.6–1.7)	5 (4.7)	0.9 (0.4–2.3)	11 (10.4)	0.7 (0.4–1.2)
Ventricular septal defect	60	10 (16.7)	0.8 (0.4–1.7)	3 (5.0)	1.0 (0.3–3.2)	13 (21.7)	1.6 (0.8–2.9)
Complex	21	2 (9.5)	0.4 (0.1–1.9)	1 (4.8)	0.9 (0.1–7.0)	5 (23.8)	1.8 (0.6–4.9)
Double-outlet right ventricule	27	6 (22.2)	1.2 (0.5–3.1)	1 (3.7)	0.7 (0.1–5.4)	5 (18.5)	1.3 (0.5–3.4)
NORMAL GREAT ARTERIES GROUP	341	101 (29.6)	**1.8 (1.4–2.3)**	32 (9.4)	**2.0 (1.3–2.9)**	76 (22.3)	**1.6 (1.2–2.1)**
Tetralogy	204	59 (28.9)	**1.7 (1.3–2.4)**	17 (8.3)	1.7 (1.0–2.9)	43 (21.1)	**1.5 (1.1–2.2)**
Tetralogy with pulmonary atresia	50	12 (24.0)	1.3 (0.7–2.6)	8 (16.0)	**3.6 (1.7–7.8)**	16 (32.0)	**2.7 (1.5–4.9)**
Double-outlet right ventricle	27	7 (25.9)	1.5 (0.6–3.5)	2 (7.4)	1.5 (0.4–6.4)	4 (14.8)	1.0 (0.3–2.9)
Truncus	38	13 (34.2)	**2.2 (1.1–4.3)**	4 (10.5)	2.2 (0.8–6.3)	9 (23.7)	1.8 (0.8–3.8)
Supracristal VSD	17	7 (41.2)	**3.0 (1.1–7.8)**	0	—	4 (23.5)	1.8 (0.6–5.4)
CONTROLS	3572	681 (19.1)	1.0 (reference)	180 (5.0)	1.0 (reference)	534 (15.0)	1.0 (reference)

Significant odds ratios are shown in bold type.

Subfertility, defined as trying to conceive for more than 12 months or undergoing fertility tests or treatments, was associated with tetralogy of Fallot but not with other diagnostic subgroups (Table 5.8). Use of fertility medications, chiefly clomiphene citrate, was strongly associated with tetralogy of Fallot and with common arterial trunk. No significant associations were detected within the transposition group.

The contrast between the two main outflow tract groupings was further evident in the greater case-control odds ratios for mothers reporting previous miscarriages, preterm births, and bleeding during pregnancy among those with normally related great arteries than among the transposition group (Table 5.9). In the total normal great arteries group and in tetralogy, the odds ratios for 3 or more previous miscarriages were: 5.4 (95% confidence interval 2.7 to 10.7) and 5.2 (95% confidence interval 2.2 to 12.1) respectively (not shown).

Multiparity, three or more previous pregnancies (not shown), was also more common in the group with normally related great arteries (25%) than among the transposition group (14%) and controls (17%). Bleeding during the index pregnancy also occurred in excess in the total group and in tetralogy, especially in the small subgroup with pulmonary atresia.

Maternal Illnesses

Diabetes, whether considered overt, gestational, or diabetes with insulin therapy, was associated with both groups of outflow tract anomalies. In particular, overt diabetes and diabetes with insulin use showed very strong odds ratios in almost all of the subgroups of both major groups. The association with gestational diabetes was significant only in the normal great artery group (Table 5.10). No other maternal illness exhibited such an association.

Thyroid disease (hypothyroidism) was absent among mothers in the transposition group, but occurred in all but one of the diagnostic groups with normally related great arteries (Table 5.11). Two mothers reported both diabetes and hypothyroidism. Significant associations were detected for influenza and transposition with intact ventricular septum, and for tetralogy of Fallot with pulmonary atresia. Urinary tract infections were associated with the small case groups of double-outlet right ventricle with transposition and supracristal ventricular septal defects.

Maternal Medications

Benzodiazepines, ibuprofen, and progesterone use was more frequently reported by mothers of the transposition group than by mothers of controls (Table 5.12). Among mothers in the normal artery group, benzodiazepines showed a significant case excess only in the large group of tetralogy of Fallot; progesterone had a case excess in truncus. Among reported benzodiazepines, the predominant medication was diazepam in both cases (11 of 16 users) and controls (18 of 35 users). Additionally, in the normally related great artery group as a whole, four mothers reported taking metronidazole for urinary tract infection (odds ratio 5.3; 95% confidence interval 1.6 to 17.6) (not shown).

Table 5.10
Outflow Tract Anomalies: Maternal Diabetes
Baltimore-Washington Infant Study (1981–1989)

Group	N	Overt Diabetes no. (%)	Overt Diabetes Odds Ratio 95% CI	Diabetes/Insulin Use no. (%)	Diabetes/Insulin Use Odds Ratio 95% CI	Gestational Diabetes no. (%)	Gestational Diabetes Odds Ratio 95% CI
TRANSPOSITION GROUP	214	5 (2.3)	**3.7** **(1.4–9.8)**	3 (1.4)	3.2 (0.9–10.9)	8 (3.7)	1.2 (0.6–2.4)
Intact ventricular septum	106	1 (0.9)	1.5 (0.2–11.0)	0	—	2 (1.9)	0.6 (0.1–2.4)
Ventricular septal defect	60	0	—	0	—	4 (6.7)	2.2 (0.8–6.0)
Complex	21	2 (9.5)	**16.2** **(3.6–73.8)**	1 (4.8)	**11.4** **(1.4–88.1)**	0	—
Double-outlet right ventricle	27	2 (7.4)	**12.3** **(2.8–55.2)**	2 (7.4)	**17.8** **(3.9–81.4)**	2 (7.4)	2.4 (0.6–10.3)
NORMAL GREAT ARTERIES GROUP	341	12 (3.5)	**5.6** **(2.8–11.4)**	9 (2.6)	**6.0** **(2.6–13.7)**	18 (5.3)	**1.7** **(1.0–2.8)**
Tetralogy	204	4 (2.0)	**3.1** **(1.1–9.0)**	2 (1.0)	2.2 (0.5–9.6)	10 (4.9)	1.6 (0.8–3.0)
Tetralogy with pulmonary atresia	50	3 (6.0)	**9.9** **(2.9–33.9)**	4 (8.0)	**19.3** **(6.2–60.0)**	5 (10.0)	**3.3** **(1.3–8.6)**
Double-outlet right ventricle	27	1 (3.7)	5.9 (0.8–45.6)	1 (3.7)	**8.6** **(1.1–66.9)**	2 (7.4)	2.4 (0.6–10.3)
Truncus	38	3 (7.9)	**13.2** **(3.8–46.1)**	2 (5.4)	**12.4** **(2.7–55.7)**	1 (2.6)	0.8 (0.1–6.0)
Supracristal VSD	17	0	—	0	—	0	—
CONTROLS	3572	23 (0.6)	1.0 (reference)	16 (0.5)	1.0 (reference)	115 (3.2)	1.0 (reference)

Table 5.11
Outflow Tract Anomalies: Maternal Illnesses
Baltimore-Washington Infant Study (1981–1989)

Group	N	Thyroid Disease no. (%)	Thyroid Disease Odds Ratio 95% CI	Influenza no. (%)	Influenza Odds Ratio 95% CI	Urinary Tract Infection no. (%)	Urinary Tract Infection Odds Ratio 95% CI
TRANSPOSITION GROUP	214	0	—	21 (9.8)	1.3 (0.8–2.1)	34 (15.9)	1.3 (0.9–1.8)
Intact ventricular septum	106	0	—	16 (15.0)	**2.1** (**1.2–3.6**)	14 (13.2)	1.0 (0.6–1.8)
Ventricular septal defect	60	0	—	3 (5.0)	0.6 (0.2–2.0)	9 (15.0)	1.2 (0.6–2.4)
Complex	21	0	—	0	—	2 (9.5)	0.7 (0.2–3.0)
Double-outlet right ventricle	27	0	—	2 (7.4)	0.9 (0.2–4.0)	8 (29.6)	**2.8** (**1.2–6.4**)
NORMAL GREAT ARTERIES GROUP	341	8 (2.3)	1.6 (0.8–3.5)	29 (8.5)	1.1 (0.7–1.6)	50 (14.7)	1.1 (0.8–1.6)
Tetralogy	204	3 (1.5)	1.0 (0.3–3.1)	13 (6.4)	0.8 (0.5–1.4)	32 (15.7)	1.2 (0.8–1.8)
Tetralogy with pulmonary atresia	50	0	—	8 (16.0)	**2.3** (**1.0–4.9**)	3 (6.0)	0.4 (0.1–1.4)
Double-outlet right ventricle	27	3 (11.1)	**8.5** (**2.5–29.0**)	4 (14.8)	2.1 (0.8–6.0)	4 (14.8)	1.0 (0.3–2.8)
Truncus	38	1 (2.6)	1.7 (0.2–12.8)	2 (5.3)	0.7 (0.2–2.7)	5 (13.2)	1.0 (0.4–2.6)
Supracristal VSD	17	1 (5.9)	4.0 (0.5–30.7)	2 (11.7)	1.6 (0.4–6.9)	6 (35.3)	**3.6** (**1.3–9.9**)
CONTROLS	3572	52 (1.5)	1.0 (reference)	278 (7.8)	1.0 (reference)	467 (13.1)	1.0 (reference)

Significant odds ratios are shown in bold type.

Table 5.12
Outflow Tract Anomalies: Maternal Medications
Baltimore-Washington Infant Study (1981–1989)

Group	N	Benzodiazepines no. (%)	Odds Ratio 95% CI	Ibuprofen no. (%)	Odds Ratio 95% CI	Progesterone no. (%)	Odds Ratio 95% CI
TRANSPOSITION GROUP	214	8 (3.7)	**3.9 (1.8–8.6)**	13 (6.1)	**1.9 (1.0–3.4)**	7 (3.3)	**2.8 (1.3–6.4)**
Intact ventricular septum	106	3 (2.8)	2.9 (0.9–9.7)	10 (9.4)	**3.0 (1.5–6.0)**	4 (3.8)	**3.3 (1.2–9.4)**
Ventricular septal defect	60	0	—	0	—	2 (3.3)	2.9 (0.7–12.3)
Complex	21	2 (9.5)	**10.6 (2.4–47.4)**	0	—	0	—
Double-outlet right ventricle	27	3 (11.1)	**12.6 (3.6–43.9)**	3 (11.1)	**3.6 (1.1–12.2)**	1 (3.7)	3.2 (0.4–24.4)
NORMAL GREAT ARTERIES GROUP	341	8 (2.3)	**2.4 (1.1–5.3)**	14 (4.1)	1.2 (0.7–2.2)	6 (1.9)	1.5 (0.7–3.6)
Tetralogy	204	5 (2.5)	**2.5 (1.0–6.6)**	10 (4.8)	1.5 (0.8–2.9)	3 (1.5)	1.2 (0.4–3.9)
Tetralogy with pulmonary atresia	50	1 (2.0)	2.1 (0.3–15.4)	3 (6.0)	1.9 (0.6–6.0)	1 (2.0)	1.6 (0.2–12.1)
Double-outlet right ventricle	27	1 (3.7)	3.9 (0.5–29.4)	0	—	0	—
Truncus	38	0	—	1 (2.6)	0.8 (0.1–5.8)	2 (5.3)	**4.5 (1.0–19.1)**
Supracristal VSD	17	1 (5.9)	6.3 (0.8–48.9)	0	—	0	—
CONTROLS	3572	35 (1.0)	(reference)	119 (3.3)	(reference)	44 (1.2)	1.0 (reference)

Significant odds ratios are shown in bold type.

Paternal Medical Exposures

Paternal exposures to medical x-rays and general anesthesia during the period of 1 month to 6 months prior to conception of the proband were associated only with tetralogy of Fallot (not shown). In this subgroup, 7 fathers reported having abdominal x-rays (odds ratio 2.6; 95% confidence interval 1.1 to 5.7) and 14 reported general anesthesia for a variety of surgical procedures (odds ratio 2.6; 95% confidence interval 1.5 to 4.7); in 3 cases both exposures occurred simultaneously (odds ratio 7.6; 95% confidence interval 2.0 to 29.6).

Lifestyle Exposures

Case-control differences in parental use of recreational drugs were found only within certain transposition subgroups (not shown). Paternal use of marijuana (odds ratio 1.8; 95% confidence interval 1.1 to 2.8) and cocaine (odds ratio 2.6; 95% confidence interval 1.2 to 5.4) were each associated with transposition with intact ventricular septum. Excess cigarette use was associated only with transposition with ventricular septal defect. The odds ratios for mothers who smoked between 21 and 39, and 40+ cigarettes daily were 2.3 (95% confidence interval 1.4 to 3.8) and 5.2 (95% confidence interval 1.9 to 14.5), respectively, compared to the reference group of nonsmokers and mothers who smoked fewer than 21 cigarettes. This was the only environmental risk factor detected for transposition with ventricular septal defect. Because heavy smokers tend to have other exposures, we examined the possible coteratogenic influences of alcohol and caffeine intake, but found no interaction or independent effects of those two variables.

Maternal Home and Occupational Exposures

Two home and occupational exposures exhibited a consistent case-control difference for the transposition group: solvents and ionizing radiation (Table 5.13). Solvent exposures at home and at the workplace (paint thinners, printing inks, and photographic solvents) as well as occupational exposures to ionizing radiation (reported by x-ray technicians and laboratory workers) were each statistically associated with the transposition group and with transposition with intact septum. Two of the smaller subgroups with normally related great arteries, double-outlet right ventricle and supracristal ventricular septal defect, also showed associations with solvent exposures. Maternal use of hair dyes at home was weakly associated with the normal great artery group (odds ratio 1.6; 95% confidence interval 1.1 to 2.3) (not shown). In addition, four mothers of infants in the normal great arteries group (not shown) reported auto body repair work (odds ratio 4.2; 95% confidence interval 1.3 to 13.6).

Multivariate Analysis

Associations identified by univariate analysis were examined further with several logistic regression models. Multivariate analyses focused initially on the major subgroups. Logistic regression analyses were conducted on all transposition cases

Table 5.13
Outflow Tract Anomalies: Maternal Home and Occupational Exposures
Baltimore-Washington Infant Study (1981–1989)

Group	N	Solvents no. (%)	Solvents Odds Ratio 95% CI		Ionizing Radiation no. (%)	Ionizing Radiation Odds Ratio 95% CI	
TRANSPOSITION GROUP	214	11 (5.1)	**3.0**	**(1.5–5.7)**	8 (3.7)	**2.9**	**(1.3–6.1)**
Intact ventricular septum	106	8 (7.6)	**4.5**	**(2.1–9.6)**	5 (4.7)	**3.6**	**(1.4–9.3)**
Ventricular septal defect	60	3 (5.0)	2.9	(0.9–9.5)	0	—	—
Complex	21	0	—	—	2 (9.5)	**7.7**	**(1.8–34.1)**
Double-outlet right ventricle	27	0	—	—	1 (3.7)	2.8	(0.4–21.2)
NORMAL GREAT ARTERIES GROUP	341	6 (1.8)	1.0	(0.4–2.3)	7 (2.1)	1.5	(0.7–3.4)
Tetralogy	204	1 (0.5)	0.3	(0.1–1.9)	4 (2.0)	1.5	(0.5–4.1)
Tetralogy with pulmonary atresia	50	1 (2.0)	1.1	(0.2–8.2)	1 (2.0)	1.5	(0.2–11.1)
Double-outlet right ventricle	27	2 (7.4)	**4.4**	**(1.0–18.9)**	0	—	—
Truncus	38	0	—	—	1 (2.6)	2.0	(0.3–14.8)
Supracristal VSD	17	2 (11.8)	**7.3**	**(1.6–32.6)**	1 (5.9)	4.6	(0.6–35.3)
CONTROLS	3572	64 (1.8)	1.0	(reference)	48 (1.3)	1.0	(reference)

Significant odds ratios are shown in bold type.

and separately on all cases with normally related great arteries (Table 5.14 and Table 5.15). For each of these groups, associations with environmental factors, alone and in the presence of maternal reproductive factors, were examined first.

The Major Groups of Transposition and Normal Great Artery Outflow Tract Defects

For the total group of transposition cases, the input variables for logistic regression were: overt diabetes, influenza, ibuprofen, benzodiazepines, progesterone, solvents, ionizing radiation, smoking, marijuana (paternal), cocaine (paternal), maternal age, alcohol, xanthine score, salicylate score, race, and socioeconomic status score. The final regression model for environmental factors included influenza, ibuprofen, diabetes, benzodiazepines, progesterone, solvents, ionizing radiation, paternal use of marijuana, and maternal age (Table 5.14A). The additive effect of influenza and ibuprofen was also examined in this model.

For the total group of cases with normally related great arteries, the input variables for logistic regression analysis were: diabetes, benzodiazepines, metronidazole, hair dyes, auto body repair work, ionizing radiation, general anesthesia (paternal), abdominal x-rays (paternal), family history of cardiovascular malformations, previous preterm birth, previous miscarriage, fertility medications, bleeding during pregnancy, alcohol, smoking, maternal age, race, xanthine score, salicylate score, and socioeconomic status score. The final regression model included diabetes, benzodiazepines, metronidazole, hair dyes, auto body repair, and paternal exposure to general anesthesia (Table 5.14A).

Inclusion of maternal reproductive factors for all transposition (Table 5.14B) did not alter the final regression model shown in Table 5.14A. On the other hand, when maternal reproductive factors were included for the total group of normal great arteries, the final model of Table 5.14A was enhanced, as shown in Table 5.14B, by inclusion of previous miscarriages, fertility medications, and bleeding during pregnancy and interaction between maternal smoking and previous preterm births. The interactions showed a five-fold odds ratio if the mother had a history of preterm births and smoked 20 cigarettes or more per day during the pregnancy. These enhancements to the Table 5.14B did not importantly confound the associations of Table 5.14A.

To examine the role of potential genetic factors, all transposition cases and normally related great artery cases were each stratified into two new data subsets. A nongenetic group, was defined by the absence of associated noncardiac anomalies and heritable disorders in probands and family members, ie, the proband has isolated congenital heart disease and is the only affected person in the nuclear family (simplex family). In the remaining subjects with a noncardiac abnormality and/or affected family members (multiplex family), a genetic susceptibility may exist.[56] Separate multivariate analyses were conducted on these two subsets (Table 5.15). A further analysis was conducted on the multiplex cases after excluding chromosomal defects from both cases and controls.

For the transposition group, the regression of environmental factors remained somewhat the same for the isolated/simplex subset (Table 5.15A) compared to the results obtained with the entire dataset (Table 5.14A). However many of the associations are weaker and less precise. The associations with diabetes,

Table 5.14
Outflow Tract Anomalies: Multivariate Models for Transposition and Normal Great Artery Groups
Environmental and Reproductive Factors • Baltimore-Washington Infant Study (1981–1989)

A. Environmental Factors

Variable	O.R.	95% CI	99% CI
Transposition Group (N=214)			
influenza	1.2	0.8–2.0	0.7–2.3
ibuprofen	1.6	0.9–2.9	0.7–3.5
influenza and ibuprofen[1]	2.0	0.9–4.1	0.7–5.2
diabetes	3.7	1.4–10.0	1.0–13.6
benzodiazepines	3.3	1.6–6.6	1.3–8.2
progesterone	2.5	1.1–5.8	0.9–7.4
solvents	2.4	1.2–4.7	1.0–5.7
ionizing radiation	2.3	1.1–5.1	0.8–6.4
marijuana (paternal)	1.6	1.1–2.3	1.0–2.6
maternal age:			
<20	1.0	(reference)	
20–29	1.3	1.0–1.6	0.9–1.8
30+	1.7	1.1–2.7	0.9–3.1
Normal Great Artery Group (N=341)			
diabetes	5.8	2.8–11.8	2.3–14.7
benzodiazepines	2.4	1.1–5.3	0.9–6.7
metronidazole	5.5	1.6–18.4	1.1–26.8
hair dyes	1.5	1.0–2.2	0.9–2.5
auto body repair	3.8	1.1–12.7	0.8–18.4
general anesthesia (pat.)	2.0	1.2–3.3	0.9–3.9

B. Environmental Factors and Reproductive History

Variable	O.R.	95% CI	99% CI
Transposition Group (N=214)			
(same as part A)			
Normal Great Artery Group (N=341)			
diabetes	5.4	2.5–10.8	2.0–13.6
benzodiazepines	2.3	1.0–5.0	0.8–6.5
metronidazole	6.0	1.8–20.7	1.2–30.4
hair dyes	1.4	0.8–1.9	0.7–2.3
auto body repair	3.5	1.1–12.9	0.7–19.0
general anesthesia (pat.)	1.8	1.1–3.1	0.9–3.7
prev. miscarriage	1.5	1.1–2.1	1.0–2.3
fertility medications	3.0	1.7–5.3	1.4–6.4
bleeding during pregnancy	1.5	1.1–2.0	1.0–2.1
smoker* without prev. preterm births	0.8	0.5–1.4	0.4–1.7
nonsmoker with prev. preterm births	1.4	0.8–2.2	0.7–2.5
smoker* with prev. preterm births	5.1	1.8–14.6	1.3–20.3

[1] odds ratio for exposure to *both* influenza and ibuprofen relative to exposure to neither (additive effect)
* smoker = 20 or more cigarettes per day.

wTable 5.15

Outflow Tract Anomalies: Multivariate Models for Subdivisions of Transposition and Nontransposition Groups

Baltimore-Washington Infant Study (1981–1989)

	A: Isolated/ Simplex Subset			B1: Multiple Anomalies/ Multiplex Subset				B2: Multiple/Multiplex Subset Excluding Chromosome Defects	
Variable	O.R.	95% C.I.	99% C.I.	Variable	O.R.	95% C.I.	99% C.I.	O.R. 95% C.I.	99% C.I.
Transposition Group (N=189)				**Transposition Group (N=25)** (too few subjects for analysis)				**Transposition Group (N=23)** (too few subjects for analysis)	
influenza[1]	1.2	0.6–2.3	0.5–2.8						
ibuprofen[2]	1.4	0.7–2.9	0.5–3.7						
influenza and ibuprofen[3]	4.8	0.9–24.3	0.6–40.3						
diabetes	**5.1**	**1.8–13.9**	**1.4–19.1**						
benzodiazepines	**4.1**	**1.9–8.6**	**1.5–10.8**						
progesterone	1.8	0.6–5.3	0.5–7.3						
solvents	1.6	0.7–3.9	0.5–5.2						
ionizing radiation	**2.6**	**1.1–6.0**	0.9–7.8						
marijuana (paternal)	**1.6**	**1.1–2.3**	0.9–2.6						
maternal age:									
<20	1.0	(reference)							
20–29	**1.3**	**1.0–1.6**	0.9–1.7						
30+	1.6	0.9–2.6	0.8–3.0						

Normal Great Arteries Group (N=209)

diabetes	5.0	2.0–12.0	1.5–15.8
benzodiazepines	2.5	0.9–5.9	0.6–7.9
metronidazole*	7.8	1.9–30.4	1.2–46.8
hair dyes	1.7	1.1–2.8	0.9–3.2
auto body repair	1.9	0.2–13.6	0.1–24.8
general anesthesia (pat.)	1.8	0.9–3.6	0.8–4.4
prev. miscarriages	1.3	0.9–2.0	0.8–2.3
fertility medications	1.1	0.4–3.1	0.3–4.2
bleeding during pregnancy	1.4	1.0–2.0	0.9–2.3
ionizing radiation	2.4	1.1–5.5	0.8–7.1
smoker without prev. preterm births	0.6	0.4–1.2	0.3–1.7
nonsmoker with prev. preterm births	1.3	0.8–1.5	0.7–2.1
smoker with prev. preterm births	7.5	2.1–26.6	1.4–39.6

Normal Great Arteries Group (N=132)

diabetes	6.3	2.3–17.3	1.7–23.8
benzodiazepines	2.2	0.6–20.1	0.3–35.9
metronidazole*	—	—	—
hair dyes	1.0	0.5–1.9	0.2–2.8
auto body repair*	—	—	—
general anesthesia (pat.)	1.6	0.7–3.7	0.5–4.8
previous miscarriages	1.7	1.1–2.7	0.9–3.1
fertility medications	6.0	3.1–12.2	2.4–15.2
bleeding during pregnancy	1.5	1.0–2.6	0.9–3.1
ionizing radiation*	—	—	—
smoker without prev. preterm births	1.3	0.8–3.1	0.7–3.8
nonsmoker with prev. preterm births	1.6	0.6–2.5	0.3–3.2
smoker with prev. preterm births	6.1	1.2–30.6	0.7–49.4

Normal Great Arteries Group (N=94)

diabetes	8.9	3.2–24.5	2.3–33.8
benzodiazepines	2.1	0.5–8.9	0.3–14.0
metronidazole*	—	—	—
hair dyes	0.7	0.3–1.7	0.2–2.2
auto body repair*	—	—	—
general anesthesia (pat.)	0.6	0.1–2.6	0.1–4.1
previous miscarriages	1.4	0.8–2.4	0.7–2.8
fertility medications	6.7	2.9–15.2	2.2–19.7
bleeding during pregnancy	1.4	0.9–2.4	0.7–2.8
ionizing radiation*	—	—	—
smoker without prev. preterm births	1.4	0.6–3.3	0.4–4.3
nonsmoker with prev. preterm births	1.5	0.7–3.3	0.5–4.2
smoker with prev. preterm births	5.0	0.6–42.5	0.1–81.5

*no reported exposure;
[1] odds ratio for exposure to influenza in the absence of exposures to ibuprofen;
[2] odds ratio for exposure to ibuprofen in the absence of influenza exposure;
[3] odds ratio for exposure to both influenza and ibuprofen relative to exposure to neither (interaction).

benzodiazepines, ionizing radiation, and paternal marijuana remained significant, with those of diabetes and benzodiazepines strengthened. The multiplex subset was too small for this kind of analysis. For the isolated/simplex group with normal great arteries (Table 5.15A), a new variable, ionizing radiation exposure, was found to be associated with the disease. Perhaps because of the decreased sample size, some associations of Table 5.15A became less precise (ie, wider confidence intervals); others became weaker. Associations remained significant for diabetes, metronidazole, hair dyes, bleeding during pregnancy, and association of smoker with previous preterm births.

Analysis of the multiplex subset of normal great arteries cases shows significant associations with maternal diabetes, previous miscarriages, fertility medications, bleeding during pregnancy, and interaction of smoking with previous preterm births. There was a decrease in the precision of the other associations (Table 5.15B1). In addition, the exclusion of 38 case and control subjects with chromosome abnormalities resulted in several important changes in the magnitude of the associations (Table 5.15 column B-2 versus column B-1). Only the associations with diabetes and fertility medications remained significant, and the odds ratio of diabetes was increased.

Diagnosis-Specific Analyses: Transposition with Intact Ventricular Septum and Tetralogy of Fallot

Next we used multivariate analyses to examine the associations of potential risk factors with each of the diagnostic subgroups that had adequate sample size to support multivariate analysis: transposition with intact ventricular septum and tetralogy of Fallot. For each of these subgroups separately, associations with environmental factors alone and in the presence of maternal reproductive factors were examined (Table 5.16).

For transposition with intact ventricular septum, the following input variables were used in the logistic regression analysis of environmental factors: influenza, ibuprofen, benzodiazepines, progesterone, painting, solvents, ionizing radiation, marijuana (paternal), cocaine (paternal), maternal age, xanthine score, salicylate score, smoking, alcohol, race, and socioeconomic status score. The final regression model of environmental factors included influenza, ibuprofen, benzodiazepines, progesterone, solvents, ionizing radiation, and paternal use of marijuana (Table 5.16A).

For tetralogy with pulmonic stenosis, the input variables for analysis of environmental factors were: diabetes, benzodiazepines, hair dyes, painting (both parents), general anesthesia (paternal), abdominal x-rays (paternal), previous miscarriage, clomiphene citrate, family history of cardiovascular malfunction (CVM), bleeding during pregnancy, maternal age, xanthine score, salicylate score, smoking, alcohol, race, and socioeconomic status score. The final model included diabetes, benzodiazepines, hair dyes, painting (both parents), and paternal exposure to general anesthesia.

For transposition with intact ventricular septum, inclusion of the maternal reproductive factors did not lead to any new significant associations or any evidence of confounding of the environmental factors in the analysis. On the other hand, after such inclusion for tetralogy with pulmonary stenosis, two reproductive factors

Table 5.16

Outflow Tract Anomalies: Diagnosis-Specific Multivariate Models for Transposition with Intact Ventricular Septum and for Tetralogy with Pulmonic Stenosis Baltimore-Washington Infant Study (1981–1989)

A. Environmental Factors

Variable	O.R.	95% CI	99% CI
Transposition/Intact Septum (N=106)			
influenza	2.2	1.2–4.1	1.0–4.9
ibuprofen	2.5	1.2–4.5	1.0–6.1
influenza and ibuprofen[1]	5.5	2.2–13.6	1.7–18.2
benzodiazepines	3.0	1.1–7.9	0.8–10.8
progesterone	3.0	1.0–8.6	0.7–12.0
solvents	3.2	1.4–7.1	1.1–9.2
ionizing radiation	2.8	1.1–7.6	0.8–10.4
marijuana (paternal)	1.7	1.1–2.7	0.9–3.2
Tetralogy/Pulmonic Stenosis (N=204)			
diabetes	1.9	1.1–3.4	0.9–4.1
benzodiazepines	2.7	1.2–6.2	0.9–7.9
hair dyes	1.6	1.0–2.6	0.9–3.0
painting (both parents)	1.8	1.1–3.1	0.9–3.7
general anesthesia (pat.)	2.7	1.5–4.8	1.3–5.7

B: Environmental Factors and Reproductive History

Variable	O.R.	95% CI	99% CI
Transposition/Intact Septum (N=106)			
(same as part A)			
Tetralogy/Pulmonic Stenosis (N=204)			
diabetes	1.8	1.0–3.2	0.8–3.9
benzodiazepines	2.8	1.3–6.4	0.9–8.3
hair dyes	1.5	0.9–2.7	0.7–3.0
painting (both)	1.7	1.0–3.0	0.8–3.5
general anesthesia (pat.)	2.5	1.4–4.5	1.2–5.4
previous miscarriage	1.5	1.0–2.2	0.9–2.4
clomiphene citrate	3.2	1.6–6.3	1.3–7.8

[1] Odds ratio for exposure to *both* influenza and ibuprofen relative to absence of both (additive effect).

had significant associations: the number of previous miscarriages and the use of clomiphene citrate. However, there was no appreciable confounding of the other variables in the analysis (Table 5.16B).

Next, these two diagnostic subgroups were each stratified into subjects without and with potential genetic factors (simplex and multiplex as above) (Table 5.17). The final regression model of environmental factors shown in Table 5.16A remained somewhat the same here for the isolated/simplex subset of transposition with intact ventricular septum (Table 5.17A). However many of the associations were weaker and less precise. The associations for benzodiazepines was strengthened. The multiplex subset of transposition with intact ventricular septum was too small for multivariate analysis. In the case of the isolated/simplex cases of tetralogy with pulmonic stenosis, only three variables were found to be associated with disease: hair dyes, painting (both parents), and paternal exposure to general anesthesia (Table 5.17A). Analysis of the multiplex subset showed no pattern of associations consistent with that shown in the simplex subset. Significant associations were detected for general anesthesia, previous miscarriages, and clomiphene citrate (Table 5.17B).

Summary of Findings

The significant associations detected in the analyses are shown in Table 5.18, which summarizes, in four categories, the risk factors by diagnoses: those associated only with transposition, those associated with only normal great arteries, those associated with only both transposition and normal great arteries, and those possibly associated both. It is apparent that unique risk factors are associated with the transposition group, and that unique risk factors are found also for the group with normally related great arteries. However, two important risk factors occur in common among both groups: maternal overt diabetes and exposure to benzodiazepines. Additionally, a fourth category shows three risk factors (progesterone, solvents, and marijuana) that were significantly associated in multivariate analyses of transposition cases, but were only associated in univariate results among cases with normal great arteries.

Discussion

This report on infants with outflow tract anomalies of the heart, in comparison to nonaffected infants in the birth cohort, categorizes the component cardiovascular malformations into subgroups according to common or unique genetic and environmental risk factors. The findings support the proposed hypothesis of a dichotomy by great artery relationships—transposed (parallel) and normal (spiral)—and of etiologically distinct anatomic phenotypes.

Genetic Risk Factors

The study revealed three indications of the importance of genetic factors in the origin of outflow tract anomalies: the associations of noncardiac anomalies in probands, family history of concordant malformations, and the occurrence of heritable blood disorders among parents.

Table 5.17

Outflow Tract Anomalies: Multivariant Models for Subdivisions of Transposition with Intact Ventricular Septum and for Tetralogy with Pulmonic Stenosis

Baltimore-Washington Infant Study (1981–1989)

A. Isolated/Simplex Subset				B: Multiple Anomalies/Multiplex Subset			
Variable	O.R.	95% CI	99% CI	Variable	O.R.	95% CI	99% CI
Transposition/Intact Septum (N=85)				**Transposition/Intact Septum (N=21)**			
influenza	2.0	1.1–3.6	0.9–4.3	(too few subjects for analysis)			
ibuprofen	2.6	1.3–5.4	1.0–6.8				
influenza and ibuprofen[1]	5.1	2.1–12.9	1.5–17.2				
benzodiazepines	3.6	1.4–9.6	1.0–12.9				
progesterone	2.6	0.8–8.8	0.5–12.8				
solvents	2.3	0.9–6.1	0.6–8.3				
ionizing radiation	2.6	0.9–7.6	0.6–10.7				
marijuana (paternal)	1.7	1.0–2.6	0.8–3.1				
Tetralogy/Pulmonic Stenosis (N=128)				**Tetralogy/Pulmonic Stenosis (N=76)**			
diabetes	1.2	0.5–2.9	0.4–3.7	diabetes	1.9	0.7–5.2	0.5–7.1
benzodiazepines	2.3	0.8–6.5	0.6–9.0	benzodiazepines	4.8	0.9–20.3	0.8–34.1
hair dyes	1.9	1.1–3.3	0.9–3.9	hair dyes	0.9	0.4–2.3	0.3–3.8
painting (both)	2.7	1.5–4.8	1.2–5.7	painting (both)	0.5	0.1–2.4	0.1–4.0
general anesthesia (pat.)	2.3	1.1–5.0	0.9–6.3	general anesthesia (pat.)	3.6	1.2–14.2	0.9–20.9
previous miscarriage	1.2	0.7–1.9	0.6–2.2	previous miscarriages	2.6	1.2–5.3	1.0–6.7
clomiphene citrate	0.6	0.2–4.5	0.1–8.2	clomiphene citrate	6.0	1.5–27.7	0.9–43.8

[1]Odds ratio for exposure to *both* influenza and ibuprofen relative to exposure to neither (additive model).

Table 5.18

Summary of Risk Factor Analysis for Outflow Tract Anomalies • Baltimore-Washington Infant Study (1981–1989)

Risk Factor	Transposition Group					Normal Great Arteries Group					
	All Cases N=214	IVS N=108	VSD N=60	Complex N=21	DORV N=27	All Cases N=341	TOF-PS N=204	TOF-PAtr N=50	DORV N=27	Tr N=38	SeVSD N=17
Associated Only with Transposition											
influenza/ibuprofen	**2.0**[1]	**5.5**[1]	—	—	3.6[2]	—	—	—	—	—	—
ionizing radiation	2.3	2.8	—	7.7	—	—	—	—	—	—	—
Associated Only with Normal Great Arteries											
previous miscarriage	—	—	—	—	—	**1.5**	1.5	3.6(p)	—	2.2	3.0
fertility medications	—	—	—	—	—	**3.0**	3.2	—	—	7.0	—
bleeding in pregnancy	—	—	—	—	—	**1.5**	—	2.7	—	—	—
metronidazole	—	—	—	—	—	**6.0**	—	—	—	—	—
auto body repair	—	—	—	—	—	**3.5**	—	—	—	—	—
painting	—	—	—	—	—	—	1.7	—	—	—	—
gen. anesthesia (pat.)	—	—	—	—	—	**1.8**	2.5	—	—	—	—
Associated With Both Groups											
overt diabetes	**3.7**	—	—	16.2	12.3	**5.4**	1.8	9.9	—	13.2	—
benzodiazepines	**3.3**	3.0	—	10.6	12.6	**2.3**	2.8	—	—	—	—
Possibly Associated With Both Groups											
progesterones	**2.5**	3.0	—	—	—	—	—	—	—	—	—
solvents	**2.4**	3.2	—	—	—	—	—	—	4.4	4.5	7.3
marijuana	**1.6**	1.7	—	—	—	—	—	—	3.6	2.8	—

Multivarate adjusted odds ratios are shown in **bold** type (from column B of Tables 5.13 and 5.15); univariate odds ratios are shown in standard type (from Tables 5.7 and 5.12).

[1]Odds ratio for exposure both influenza and ibuprofen relatives to absences of both (additive effect);

[2]Odds ratio for exposure to influenza;

(p) = previous preterm births;

IVS = intact ventricular septum; VSD = ventricular septal defect; DORV = double-outlet right ventricle; TOF = Tetralogy of Fallot; PS = pulmonary valve stenosis; PAtr = pulmonary atresia; Sc = supracristal.

Proband

Among cases, multiple noncardiac anomalies included a variety of chromosome abnormalities and syndromic associations, several of probable single-gene origin.

These associations were rare in the transposition group, but common among those with normally related (spiral) great arteries. Micro-deletion of chromosomal region 22q11, now believed to be a common defect,[69,118,385] was not recognized in infants during the 1980s, as it was reported only once[186] in a patient with DiGeorge syndrome. No case of conotruncal-face anomaly, a designation probably rarely made in infancy.

Family History

Familial congenital heart disease did not appear as a risk factor in any of the statistical analyses; however the familial conotruncal tendency described by Corone[91] was apparent in spite of the very low sibling occurrence rate of congenital heart disease. Review of the literature of reported sibships of probands with outflow tract defects reveals a remarkable degree of concordance, especially if one includes pulmonic stenosis and ventricular septal defect as related (forme fruste) anomalies (Table 5.19).

To examine this relationship further, we also reviewed the Baltimore-Washington Infant Study cases in whom the index diagnoses was either pulmonic stenosis or ventricular septal defect, and found two parents with tetralogy of Fallot, one in each diagnostic group. There were eight siblings (1 of a pulmonic stenosis case and 7 of ventricular septal defect cases), with a conotruncal defect; transposition of great arteries in 3, tetralogy of Fallot in 2, and persistent arterial trunk in 2. There was a striking male predominance in the siblings with outflow tract defects (7 of 8) compared to the index cases with pulmonic stenosis or ventricular septal defect (2 of 8).

Monogenic inheritance of conotruncal malformations has been suggested by several authors,[294,360] a possibility of special interest in view of the recently discovered chromosomal micro-deletions in this malformation group.[69] Furthermore, the studies on the hereditary outflow tract defects in the keeshond dog after generations of inbreeding demonstrated the presence of a single major gene.[337] Nevertheless the rarity of familial cases indicates that an hereditary origin cannot account for the majority of cases.

Heritable Blood Disorders

The third indication of a possible genetic susceptibility is in the association of outflow tract defects with familial heritable blood disorders. The association of parental hemophilia in cases, but never in controls, was reported previously,[154] and led to the hypothesis that inborn errors in endothelial cell functions may be relevant to developmental disturbances.[142] As endothelial cell functions contribute importantly to cardiovascular homeostasis,[180,378] this remains a promising hypothesis.

Table 15.19
Outflow Tract Defects: Sibships Reported in The Literature

Reported Diagnosis of Proband	Outflow Tract Defects in Affected Sibling				Other Congenital Heart Disease (CHD) in Affected Sibling				
	Transposition	Tetralogy	Double-Outlet Right Ventricle	Truncus	Pulmonic Stenosis	Ventricular Septal Defect	Atrial Septal Defect	Patent Ductus Arteriosus	Severe Lesions
Transposition (n=60)	7[1,3,4,8,10]	20[1,2,3,7,8,12]	2[1,3]	2[7,12]	4[3,7,8]	12[3,4,7,8]	5[3,4,7,8]	5[7,10]	1-Endocardial cushion defect[17] 1-Aortic atresia[5] 1-Aortic stenosis[1]
Tetralogy (n=122)	3[1,2,3,5,8,18]	—	0	4[3,16]	39[2,3,5,7,8,14,18]	32[2,3,5,7,8,17]	5[3,7,14]	6[3,7]	1-Single Ventricle + Pulmonary Atresia[4] 4-complex[14]
Double-Outlet Right Ventricle (n=6)	—	—	2[17]	2[3,16]	0	1[3]	0	0	1-Endocardial cushion defect[17]
Truncus (n=14)	—	—	—	6[3,15,16,17]	4[5,7,16,17]	3[3,4,7]	0	1[7]	—

Each cell represents the number of sibling pairs reported (references in superscript): [1]Briard, et al. 1984; [2]Campbell 1965; [3]Corone, et al. 1983; [4]Davidson 1967; [5]Ehlers, Engle 1966; [7]Fraser, Hunter 1975; [8]Fuhrmann 1968; [10]McKeown 1953; [12]Sorenson 1951; [14]Lamy, et al. 1957; [15]Zavala, et al. 1992; [16]Pierpont, et al. 1988; [17]Allen, et al. 1986; [18]Sanchez-Cascos 1989.

Maternal Diabetes

Our results confirm earlier reports of an association between maternal diabetes and abnormalities of the cardiac outflow tract. Specifically, our results indicate that the risk of outflow tract defects among infants of mothers with diabetes is four to five times greater than in infants of mothers without diabetes. Furthermore, the results of stratified and adjusted analyses highlighted the fact that the risk varies with the type of diabetes, the strongest association with overt diabetes and insulin users. In addition, the associations varied for specific diagnoses, and were strongest with severe outcomes.

An analysis was also done of all cases of double-outlet right ventricle (ie, with and without transposition), and the odds ratio for the combined group was 9.1 (confidence interval 2.6 to 31.2) for overt diabetes and 13.1 (confidence interval 3.7 to 46.3) for mothers taking insulin. The high frequency of multiple noncardiac malformations among infants of mothers with diabetes was previously described[153] and suggests either that diabetes is a strong nonspecific teratogen or that mothers with diabetes are likely to have other exposures with teratogenic potential. However, among the mothers with diabetes in the Baltimore-Washington Infant Study none took benzodiazepines, ibuprofen, progesterone, or clomiphene citrate, or were exposed to solvents or ionizing radiation.

Environmental Factors

The two groups of outflow tract defects, those with and those without transposition, showed some differences as well as some similarities in possible risk factors. Candidate risk factors associated with transposition only (especially with the intact ventricular septum subgroup) were influenza and ibuprofen, whereas factors associated with the group with normally related great arteries alone were reproductive problems and exposure to metronidazole.

As duration of exposure to factors such as influenza and ibuprofen is usually short-term, the opportunity of an exposure during a critical period of embryogenesis is probably limited, as is the likelihood of teratogenesis. This limited opportunity for exposure and the possibility that different morphogenetic mechanisms may have nonoverlapping critical exposure windows might explain why the associations with influenza and ibuprofen could not be found for the group with normally related great arteries.

Associations between prenatal use of benzodiazepines and adverse pregnancy outcome have been suggested before.[58,59,238] In the Baltimore-Washington Infant Study, this association was also found between maternal benzodiazepine use and Ebstein's anomaly of the tricuspid valve.[93] Adams et al[10] report an association of benzodiazepine use with conotruncal anomalies in univariate analyses. Our results using multivariate method are consistent with this latter observation. Although various types of benzodiazepines were taken by case and control mothers, diazepam was the most frequent one reported. This may reflect benzodiazepine prescription practices in the study region rather than an association unique to diazepam.

The association of defects with normally related arteries with previous pregnancy loss suggests susceptibility for embryo lethality, either from genetic factors or from recurrent or permanent effects of exposures to xenobiotics. The persistence of an association with previous miscarriages in the multiple anomaly/multiplex, but not the isolated/simplex data subset tends to support the former possibility, especially as the stronger associations were seen when the chromosome subset was included. Further support for possible genetic factors is the observation that the association with fertility medicines was limited to the subgroup with multiple malformations and affected families.

Nevertheless, recurrent environmental exposures or chronic effects from such factors may also be important, as suggested by the association with poverty. Socioeconomic disadvantage, expressed as a low socioeconomic score, which had been identified as a risk factor for cardiac situs and looping defects, was also found among outflow tract defects in both the isolated/simplex and multiple/multiplex subsets of the normally related great artery group. None of the usually suspected coexisting associations of alcohol or other abuse substances were found. This observation suggests that additional factors connected with poverty, such as nutritional deficiencies, may be of importance. An evaluation of this possibility will be carried out in the subsets of cases whose mothers completed the Harvard Food Frequency questionnaire.

Conclusions

The heterogeneity of outflow tract defects by great artery relationships is confirmed in these findings. However there is also evidence of homogeneity in the association of both the parallel (transposed) and spiral (normally related) great artery groups with diabetes and with maternal benzodiazepine exposure. Exposure to these factors during the critical developmental period is likely to be regular and of long duration. Diabetes and benzodiazepine are known to be risk factors for abnormal development associated also with various noncardiac malformations.[386] We highlight the importance of these findings as both factors provide opportunities for clinical preventive interventions and for renewed research efforts.

Chapter 6

Atrioventricular Septal Defects With and Without Down Syndrome

From: Ferencz C, Loffredo CA, Correa-Villaseñor A, Wilson PD, eds: *Genetic & Environmental Risk Factors of Major Cardiovascular Malformations: The Baltimore-Washington Infant Study 1981–1989.* Armonk, NY: Futura Publishing Co., Inc; ©1997.

Atrioventricular septal defects include a spectrum of anomalies involving the atrioventricular junction, ie, the atrioventricular valves and the leading edges of the atrial and ventricular septums. In a *complete* atrioventricular septal defect there is a single, common valve with large anterior and posterior bridging leaflets and defects of the atrial (ostium primum) and ventricular septums, allowing communication across all four cardiac chambers.[17] The defect is labeled partial in the presence of separate right and left atrioventricular valves, frequently with a cleft (tri-leafed) left-sided valve with a predominant atrial communication.

Speculation on the possible morphogenetic mechanisms of these defects is longstanding. Abbott[5] described the many anatomic variations observed by previous authors and concurred in their opinion of *a primary arrest of the endocardial cushions.* More recently, Markwald and colleagues[280] describe the abnormal transformation of endothelial lining cells to matrix cells in the process of valve maturation. Building on this work, Clark,[82] in his mechanistic classification of cardiac defects, categorizes atrioventricular septal defects as *defects of extracellular matrix formation.* Anderson's recent historical and anatomic evaluation of atrioventricular septal defects[16] highlights details of valvar anatomy that may be responsible for the clinical and hemodynamic variations in manifestations. The term "atrioventricular septal defect" emphasizes the abnormality of the common atrioventricular junction as the hallmark of these malformations, and is preferred in the European literature in contrast to the term "endocardial cushion defect," used in the American literature.[16,442]

Among cardiac anomalies, atrioventricular septal defect is unique in its characteristic association with Down syndrome. Reviews of morphological and morphogenetic variations in atrioventricular septal defects[16,17,442] have not suggested any distinction of the hearts of patients with Down syndrome from the hearts of patients with normal chromosomes. However, differences in associated valvar and vascular anomalies for syndromic and nonchromosomal cases of atrioventricular septal defects have been described,[111,277] suggesting different morphogenetic mechanisms. In a comparison of isolated and syndromic endocardial cushion defects, Carmi et al[72] found fewer complete defects and more associated cardiovascular malformations, particularly left-sided obstructive lesions, in the isolated form, providing further support for the hypothesis of different morphogenetic mechanisms for the Down syndrome and nonsyndromic forms of atrioventricular septal defects.

No previous population-based study has described the frequency with which Down syndrome occurs among all cases of atrioventricular septal defects, or the extent to which cases with Down syndrome and those without chromosomal or Mendelian disorders are similar in epidemiologic characteristics. This chapter evaluates various epidemiologic characteristics of atrioventricular septal defects in infants with normal cardio-visceral relationship in a population-based case-control study of congenital cardiovascular malformations. To elucidate possible differences in associated anomalies and familial and environmental factors, infants with atrioventricular septal defects and Down syndrome and infants with nonchromosomal and non-Mendelian atrioventricular septal defects are evaluated separately.

Study Population

Between January 1, 1981 and December 31, 1989 the Baltimore-Washington Infant Study enrolled 4390 infants with cardiovascular malformations from a regional cohort of 906,626 livebirths. Cases of atrioventricular septal defects with laterality and looping defects (N = 38) and with outflow tract defects (N = 5) are not included in this chapter. Atrioventricular septal defects as a primary cardiac diagnosis (N = 320) represent 7.3% of all cardiovascular malformations and a regional overall prevalence of 3.5 per 10,000 livebirths. Family interviews were conducted in 266 families of infants with atrioventricular septal defects (90% of these with Down syndrome and 86% of the nonchromosomal cases) (Figure 6.1).

Cardiac Abnormalities

Cardiovascular defects additional to atrioventricular septal defects were evaluated for the two major subgroups of atrioventricular septal defects, those with Down syndrome and those with non-Mendelian, nonchromosomal atrioventricular

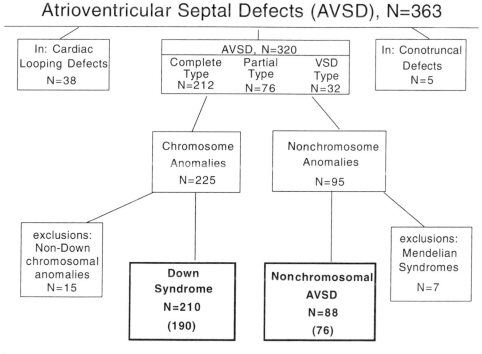

Numbers in parentheses indicate interviewed families

Figure 6.1: Cases with a primary diagnosis of atrioventricular septal defect (N = 320) were divided into two major subgroups for analysis: cases with Down syndrome and cases with nonchromosomal atrioventricular septal defects (AVSD). Cases with AVSD secondary to cardiac looping or conotruncal defects were included in Chapters 4 and 5.

Table 6.1
Atrioventricular Septal Defects (AVSD):
Associated Cardiac Anomalies
Baltimore-Washington Infant Study (1981–1989)

Associated Cardiac Anomaly	AVSD with Down Syndrome N=210	Nonchromosomal AVSD N=88
Pulmonic Stenosis (PS), total	12 (5.7%)	6 (6.8%)
PS only	11	4
PS with anomalous systemic veins	0	1
PS with PAPVR	0	1
PS with right aortic arch	1	0
Left-Sided Obstructive Lesions, total	11 (5.2%)	20 (22.7%)
hypoplastic left ventricle (HLV) only	1	7
coarctation of aorta (CoA) only	9	7
aortic stenosis only	0	1
HLV with CoA	0	3
CoA with aortic stenosis	1	1
CoA with anomalous systemic veins	0	1
Aortic Arch Anomalies (only), total	3 (1.4%)	0
right aortic arch	1	0
double aortic arch	2	0
Anomalous Systemic Veins (only), total	2 (1.0%)	3 (3.4%)
Total cases with associated cardiac anomalies:	28 (13.3%)	29 (32.9%)

PAPVR=partial anomalous pulmonary venous return.

septal defects (Table 6.1). Pulmonic stenosis occurred with similar frequencies in both subgroups. Left-sided obstructive lesions were rare among the Down syndrome cases (5.2%) but frequent among the nonsyndromic cases (23%), the latter encompassing various combinations of left ventricular, aortic valve, and coarctation components. The Down syndrome group included three cases with aortic arch anomalies, among which one had right aortic arch and pulmonic stenosis; no aortic arch anomalies were among the left-sided anomalies in the nonchromosomal group. Rare cases of anomalous systemic veins occurred in both subgroups of atrioventricular septal defects.

A separation of cases by the severity of the atrioventricular malformation revealed that about two thirds of atrioventricular septal defects were classified as complete, and among these, cases with Down syndrome predominated (Figure 6.2). Among infants with partial atrioventricular defects, cases with Down syndrome were less frequent but still made up more than one third of this subgroup.

Noncardiac Anomalies

Among the 320 cases 70% were associated with chromosomal anomalies (N = 225) and 2% with Mendelian syndromes (N = 7). The remaining cases—that is, cases without chromosomal or monogenic conditions (ie, nonsyndromic atri-

Atrioventricular Septal Defects (N=320)
Distribution of noncardiac anomalies by type of defect

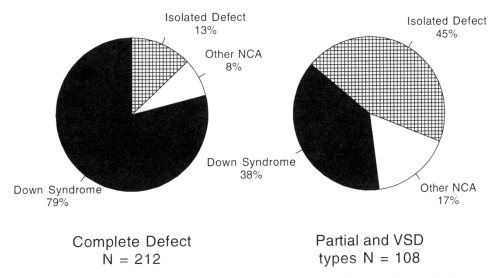

Complete Defect
N = 212

Partial and VSD
types N = 108

Figure 6.2: Down syndrome occurred in the majority of cases with *complete* AVSD. In contrast, Down syndrome was only half as prevalent, among cases with *partial* AVSD.

oventricular septal defects)—included infants with isolated atrioventricular septal defects (N = 73), infants with other associated organ defects (N = 5), and infants with non-Mendelian associations (N = 10).

Chromosomal and Mendelian Disorders

The predominant anomaly among all cases of atrioventricular septal defects was Down syndrome (N = 210), accounting for 66% of all atrioventricular septal defects and for 93% of the 225 cases with chromosome disorders (Table 6.2).

Additional chromosome disorders among cases of atrioventricular septal defect included nine cases of trisomy 18, and single cases of trisomy 4q, ring chromosome 22, trisomy 1 p/deletion 3q, deletion 10p, deletion 4q, and iso-chromosome 12p. Mendelian syndromes were diagnosed in 7 cases, including 2 cases of Ellis-Van Creveld syndrome (siblings), 2 of Smith Lemli-Opitz syndrome, and single cases of Schinzel-Gideon, Holt-Oram, and Van de Woude syndromes.

Non-Mendelian Associations and Other Noncardiac Defects

Among the 10 cases with non-Mendelian associations, 3 had asplenia or polysplenia, and were diagnosed as Ivemark complex, even in the absence of other laterality defects; there were 2 cases of Cornelia de Lange and 2 of Gold-

Table 6.2
Atrioventricular Septal Defects (N=320):
Noncardiac Anomalies
Baltimore-Washington Infant Study (1981–1989)

Chromosomal Abnormalities:	225	**(70.3%)**
Down Syndrome	210	
Trisomy 18	9	[#1567, 2378, 3678, 10173, 10508, 10782, 10827, 11276, 11517]
Trisomy 4q	1	[#3637]
Ring chromosome 22	1	[#3799]
Trisomy 1q/Deletion 3q	1	[#11054]
Deletion 10p	1	[#3225]
Deletion 4q	1	[#10642]
Iso-chromosome 12p	1	[#3516]
Syndromes	17	**(5.3%)**
a. **Mendelian Syndromes:**	7	
Ellis-Van Creveld	2	[##2548, 11553 - sibs]
Smith-Lemli-Optiz	2	[##3253, 11212]
Schinzel-Giedion	1	[#11370]
Holt-Oram	1	[#1380 - inherited]
Van der Woude	1	[#11060]
b. **Non-Mendelian Associations:**	10	
Ivemark	3	{##3280, 3330, 10992]
Cornelia de Lange	2	[##3407, 11460]
Goldenhar	2	[#3744, 10390]
VACTERL	1	[#10188]
CHARGE	1	[#1674]
Hydantoin embryopathy	1	[#2944]
Organ defects:	5	**(1.6%)**
Tracheoesophageal fistula	2	[#3816, 3460]
Polycystic kidney	1	[#3401]
Duplicate collecting system (kidney)	1	[#2173]
Hypospadias	1	[#11403]
No Associated Anomalies	73	**(22.8%)**

enhar syndrome. Other multiorgan anomalies including VACTERL and CHARGE associations, and hydantoin embryopathy, occurred as single cases. Among the five single-organ defects associated with atrioventricular septal defects there were two cases of tracheoesophageal fistula, and single cases of polycystic kidney disease, duplicate collecting system in the kidney, and hypospadias.

Descriptive Analyses

Prevalence by Time, Season, and Area of Residence

The overall prevalence of atrioventricular septal defects was 3 to 5 per 10,000 regional livebirths. For both of the large subgroups of atrioventricular defects (infants with atrioventricular septal defects and Down syndrome and infants with nonchromosomal and non-Mendelian atrioventricular septal defects), prevalence

Table 6.3
Atrioventricular Septal Defects (AVSD): Prevalence
Baltimore-Washington Infant Study (1981–1989)

	AVSD with Down Syndrome	Nonchro-mosomal AVSD	Area Births	prevalence per 10,000	
				AVSD with Down Syndrome	Nonchro-mosomal AVSD
Total Subjects	210	88	906,626	2.32	0.97
Year of Birth:					
1981–1983	63	21	274,558	2.29	0.76
1984–1986	79	37	295,845	2.67	1.25
1987–1989	63	28	336,223	1.87	0.83
Birth Quarter:					
1st (Jan–Mar)	45	21	219,145	2.05	0.96
2nd (Apr–Jun)	56	23	231,777	2.42	0.99
3rd (Jul–Sep)	57	27	233,626	2.44	1.16
4th (Oct–Dec)	52	17	222,078	2.34	0.77
Area of Residence:					
urban	41	21	208,568	1.96	1.00
suburban	122	59	584,022	2.09	1.01
rural	27	8	112,318	2.40	0.71

(per 10,000) was slightly higher from 1984 to 1986 (2.7 and 1.2, respectively) than in other years of the study (2.3 and 0.8 from 1981 to 1983 and 1.9 and 0.8 from 1987 to 1989, respectively) but showed no significant variation with season of birth (Table 6.3). For Down syndrome cases prevalence was slightly higher among rural compared to urban residents (2.4 and 2.0, respectively) whereas among nonchromosomal cases the reverse was true (0.7 and 1.0, respectively).

Diagnosis and Course

Compared to nonsyndromic cases, Down syndrome cases were more likely to be diagnosed earlier, to require cardiac surgery in the first year of life, and to have a slightly better survival to 1 year of age (Table 6.4). Nonchromosomal heart defects were more likely to be diagnosed solely by echocardiography.

Gender, Race, and Twinning

The proportion of females among infants with Down syndrome (58.1 percent) was higher than among nonsyndromic cases (55.7 percent) or controls (49.1 percent). The two subgroups of cases were similar to controls in race (Table 6.5). Among the cases, 6 were twins: 2 occurred in the Down syndrome group; 1 had Goldenhar syndrome; 3 occurred with isolated cardiac defects. The proportion of twin cases in the nonchromosomal group was three-fold that of controls. All co-twins were normal. There were no twins among excluded cases.

Table 6.4
Atrioventricular Septal Defects (AVSD): Clinical Data
Baltimore-Washington Infant Study (1981–1989)

	AVSD with Down Syndrome N=210		Nonchromosmal AVSD N=88	
	no.	(%)	no.	(%)
Age at Diagnosis:				
<1 week	49	(23.3)	20	(22.7)
1–4 weeks	99	(47.1)	26	(29.5)
5–24 weeks	51	(24.3)	29	(32.9)
25–52 weeks	11	(5.2)	13	(14.9)
Method of Diagnosis:				
echocardiography only	38	(18.1)	30	(34.1)
echocardiography and cardiac catheterization	31	(14.8)	15	(17.0)
echocardiography and surgery	2	(1.0)	1	(1.1)
echocardiography, cardiac catheterization, and surgery	117*	(55.7)	33**	(37.5)
autopsy following other methods	20	(9.5)	7	(8.0)
autopsy only	2	(1.0)	2	(2.3)
Surgery During First Year of Life	143	(68.1)	41	(46.6)
Follow-Up:				
not followed	3	(1.4)	4	(4.5)
followed:	207		84	
died in the first year of life	50	(24.2)	28	(33.3)
alive	157	(75.8)	56	(66.7)
median age at last visit	11months		12 months	

*Includes 1 cardiac catherization only, 1 cardiac catheterization and surgery.
**All were echocardiography, cardiac catheterization and surgery.

Table 6.5
Atrioventricular Septal Defect (AVSD): Infant Characteristics
Baltimore-Washington Infant Study (1981–1989)

Characteristic	AVSD with Down Syndrome N=210		Nonsyndromic AVSD N=88		Controls N=3572	
	no.	(%)	no.	(%)	no.	(%)
Gender:						
male	88	(41.9)	39	(44.3)	1,817	(50.9)
female	122	(58.1)	49	(55.7)	1,755	(49.1)
Race:*						
white	142	(69.3)	57	(64.8)	2,362	(66.1)
black	57	(27.8)	29	(32.9)	1,109	(31.0)
other	6	(2.9)	2	(2.3)	98	(2.7)
Twin births	2	(1.0)	4	(4.5)	53	(1.5)

*Missing race in 5 cases with Down Syndrome and 3 controls.

Table 6.6
Atrioventricular Septal Defect (AVSD):
Fetal Growth of Characteristics
Baltimore-Washington Infant Study (1981–1989)

Characteristic	AVSD with Down Syndrome no. (%)	Nonchromosomal AVSD no. (%)	Controls no. (%)
Number of Families Interviewed	190	76	3572
Birthweight (grams)			
<2500	26 (13.7)	18 (23.7)	252 (7.1)
2500–3500	134 (70.5)	36 (47.4)	1853 (51.9)
>3500	30 (15.8)	22 (28.9)	1467 (41.0)
mean ±standard error	2998±36	2977±95	3351±10
range	1446–4500	640–5120	340–5273
Gestational Age (weeks)			
<38	36 (19.0)	11 (14.5)	339 (9.5)
38+	154 (81.0)	65 (85.5)	3233 (90.5)
mean	38.7	38.9	39.6
range	27–44	28–42	20–47
Size for Gestational Age			
small (SGA)	32 (16.8)	15 (19.7)	211 (5.9)
normal	146 (76.8)	54 (71.1)	2712 (75.9)
large	12 (6.4)	7 (9.2)	649 (18.2)
Odds Ratio for SGA (95% CI)	3.2 (2.2–4.8)	3.9 (2.2–7.0)	1.0 (reference)

Birthweight and Gestational Age

Compared to controls, both subgroups of cases of atrioventricular septal defects were born after shorter gestation and had lower birthweights (Table 6.6). Also compared to controls, cases with Down syndrome had a three-fold greater risk, and those with nonchromosomal heart defects had a nearly four-fold greater risk of being small for gestational age, as defined by the population standard of Brenner and colleagues[61] (infants below the 10th percentile of gestational age-specific birthweights). The mean birthweights and gestational ages were similar in the two subgroups of cases.

Sociodemographic Characteristics

Sociodemographic factors, including parental education and occupation, family income, and maternal marital status, did not differ significantly among cases and controls (Table 6.7). Furthermore, when compared on a socioeconomic index based on the above variables, the proportions of families with high, medium, and low scores were similar for the atrioventricular septal defect (AVSD) subgroups and controls (not shown).

Table 6.7

Atrioventricular Septal Defects (AVSD):
Analysis of Sociodemiographic Characteristics
Baltimore-Washington Infant Study (1981–1989)

Interviewed Families	Controls N=3572 no.	(%)	AVSD with Down Syndrome N=190 no.	(%)	Odds Ratio (95% CI)	Nonchromosomal AVSD N=76 no.	(%)	Odds Ratio (95% CI)
Maternal Marital Status:								
not married	990	(27.7)	41	(21.6)	0.7 (0.5–1.0)	19	(25.0)	0.9 (0.5–1.5)
married	2582	(72.3)	149	(78.4)	1.0 (reference)	57	(75.0)	1.0 (reference)
Maternal Education:								
<high school	659	(18.4)	21	(11.1)	0.5 (0.3–0.9)	8	(10.5)	0.6 (0.3–1.3)
high school	1265	(35.4)	71	(37.4)	0.9 (0.7–1.3)	34	(44.7)	1.3 (0.8–2.1)
college	1648	(46.1)	98	(51.5)	1.0 (reference)	34	(44.7)	1.0 (reference)
Paternal Education:								
<high school	650	(18.2)	25	(13.2)	0.6 (0.4–0.9)	15	(19.7)	1.2 (0.6–2.2)
high school	1298	(36.3)	62	(32.6)	0.8 (0.5–1.0)	29	(38.2)	1.1 (0.7–1.9)
college	1624	(45.5)	103	(54.2)	1.0 (reference)	32	(42.1)	1.0 (reference)
Annual Household Income:								
<$10,000	686	(19.2)	28	(14.9)	0.7 (0.4–1.0)	12	(16.0)	0.7 (0.4–1.3)
$10,000–$19,999	699	(19.6)	40	(21.3)	0.9 (0.6–1.4)	16	(21.3)	0.9 (0.5–1.6)
$20,000–$29,999	737	(20.6)	36	(19.1)	0.8 (0.5–1.2)	12	(16.0)	0.6 (0.3–1.2)
$30,000+	1373	(38.4)	84	(44.7)	1.0 (reference)	35	(46.7)	1.0 (reference)

Maternal Occupation:					
not working	1142 (32.0)	59 (31.1)	0.9 (0.6–1.3)	23 (30.3)	0.8 (0.4–1.5)
clerical/sales	1119 (31.3)	61 (32.1)	0.9 (0.6–1.4)	24 (31.6)	0.9 (0.5–1.6)
service	444 (12.4)	21 (11.0)	0.8 (0.5–1.4)	7 (9.2)	0.6 (0.3–1.5)
factory	137 (3.8)	7 (3.7)	0.9 (0.4–2.0)	4 (5.3)	1.2 (0.4–3.6)
professional	730 (20.4)	42 (22.1)	1.0 (reference)	18 (23.7)	1.0 (reference)
Paternal Occupation:					
not working	246 (6.9)	12 (6.3)	0.7 (0.4–1.4)	5 (6.7)	0.9 (0.3–2.5)
clerical/sales	618 (17.3)	34 (17.9)	0.8 (0.5–1.3)	14 (18.7)	1.0 (0.5–2.0)
service	480 (13.4)	18 (9.5)	0.6 (0.3–1.0)	6 (8.0)	0.6 (0.2–1.4)
factory	1279 (35.8)	64 (33.7)	0.8 (0.5–1.1)	29 (38.7)	1.0 (0.6–1.8)
professional	947 (26.5)	62 (32.6)	1.0 (reference)	21 (28.0)	1.0 (reference)
Month of Pregnancy Confirmation:					
1st month	511 (14.3)	28 (14.7)	1.0 (reference)	13 (17.3)	1.0 (reference)
2nd month	1955 (54.7)	107 (56.3)	1.0 (0.7–1.5)	34 (45.3)	0.7 (0.4–1.3)
3rd month	663 (18.6)	35 (18.4)	1.0 (0.6–1.6)	17 (22.7)	1.0 (0.5–2.1)
4th month or later	416 (11.6)	20 (10.5)	0.9 (0.5–1.6)	11 (14.7)	1.0 (0.5–2.3)

Data are missing for income (2 cases with Down syndrome, 1 nonchromosomal case, 77 controls) paternal occupation (1 nonchromosomal case, 2 controls), and pregnancy confirmation date (1 nonchromosomal case, 27 controls).

Potential Risk Factors

Familial Cardiac and Noncardiac Anomalies

Details of the familial abnormalities in first-degree relatives are shown in Table 6.8. The patterns of malformations in families are shown for the Down and nonchromosomal subgroups as well as for the remaining 22 infants with non-Down syndromic AVSD, and were strikingly similar in the two subgroups, with a total recurrence of an atrioventricular septal defect in 5 instances and of a ventricular septal defect in 2. Recurrence of an atrioventricular septal defect in euploid and aneuploid members within the same family was noted for three infants in the chromosome groups. It is notable that each of the three affected mothers had a partial defect in association with a cleft (tri-leafed) mitral valve, shown at cardiac surgery. Also of interest is an infant with AVSD and Holt-Oram syndrome whose mother had the same syndrome associated with an atrial septal defect of the ostium secundum type. Concordance of familial heart disease was high: 8 of 11 families affected were concordant for atrioventricular septal defect or ventricular septal defect; pulmonic stenosis in the mother of case #1625 was concordant with pulmonic stenosis in the infant with AVSD. Pedigrees of these families were previously published.[143]

Among familial noncardiac disorders, there was a wide range of anomalies in first-degree relatives of Down patients (13/190: 7%) and of nonchromosomal cases, (6/76: 8%) among which the only notable finding may be the presence of five heritable blood disorders in the Down subgroup. Case #2614 had a presumably non-hemophilia bleeding tendency, while his brother and two male maternal cousins had hemophilia. The pedigree of this family has been published.[143,152]

Genetic and Environmental Factors

Univariate Analyses

Familial cardiovascular malformations occurred more frequently among the first-degree relatives of infants in both groups than among first-degree relatives of control infants. Among the Down syndrome cases, the frequency was twice as great as in the control families (Table 6.9). The case excess was even greater among nonsyndromic cases.

Maternal Age and Reproductive History

Univariate analysis revealed that increasing maternal age was associated with increasing risk of atrioventricular septal defect only in the presence of Down syndrome, with a stepwise increase in the odds ratio among mothers over 30 years of age (Figure 6.3). Paternal age was also associated only with the Down syndrome defects, but this association was limited to fathers 40 years of age or older. In contrast with controls and with the nonsyndromic cases where one third of infants were first-born, of the Down syndrome cases only 1 in 5 were first-born (Table 6.9). The proportion of multiparous mothers reporting three or more previous miscarriages was greater among Down syndrome cases than among controls, with an

Table 6.8
Atrioventricular Septal Defects (AVSD):
Congenital Anomalies Among First-Degree Relatives
Baltimore-Washington Infant Study (1981–1989)

	Proband		Family	
ID	Sex	Noncardiac anomalies	Relative	Diagnosis

I. Probands With AVSD and Down Syndrome (N=190)
A. Congenital Heart Defects:

1458	F	Down	Mother	AVSD
1994	M	Down	Mother	AVSD
1234	F	Down	Father	Bicuspid aortic valve
2860	F	Down	Sister	Ventricular septal defect, Dandy-Walker
3274	F	Down	Half-Sister	AVSD

B. Congenital Defects of Other Organs:

10826	M	Down	Mother	Rh incompatible
1495	M	Down	Mother	Hemivertebrae
1434	F	Down	Father	Pilonidal cyst
1897	M	Down	Father	Thalassemia
2581	M	Down	Father	Sickle cell
2923	F	Down	Father	Rh incompatible
3617	F	Down	Father	Ichthyosis
10823	M	Down	Father	Dysplastic kidney
1303	F	Down	Brother	Polydactyly
10111	F	Down	Brother	Charcot-Marie
2614	M	Down	Half-Brother	Hemophilia
2851	F	Down	Sister	Hypoplasia of 3 fingers
10978	F	Down	Sister	Hydronephrosis

II. Probands With Nonchromosomal AVSD (N=76)
A. Congenital Heart Defects:

10122	F	—	Mother	AVSD
1625	F	—	Mother	Pulmonary stenosis
10672	F	—	Father	Cardiomyopathy
		—	Brother	Ventricular septal defect
2380	M	—	Brother	AVSD
11173	F	—	Half-Brother	Ventricular septal defect

B. Congenital Defects of Other Organs:

1640	M	—	Mother	Duplicated right kidney
2887	F	—	Father	Lung cysts
3485	M	—	Father	Polydactyly
11403	M	Hypospadias	Father	Hypospadias
			Sister	Congenital deafness
10078	M	—	Sister	Anencephaly

III. Probands With AVSD and Other Chromosomal and Mendelian Syndromes (N=18)
A. Congenital Heart Defects:

1380	F	Holt-Oram	Mother	Atrial septal defect, Holt-Oram

B. Congenital Defects of Other Organs:

11054	F	Deletion 3q/ Trisomy 1p	Mother	Dysplastic right kidney

Table 6.9

Atrioventricular Septal Defect (AVSD): Univariate Analysis of Risk Factors • Baltimore-Washington Infant Study (1981–1989)

Variable	Controls N=3572		AVSD + Down Syndrome (N=190)				Nonchromosomal AVSD (N=76)			
	no.	(%)	no.	(%)	O.R.	(95% CI)	no.	(%)	O.R.	(95% CI)
Family History (1st degree relatives)										
Cardiac defects	43	(1.2)	5	(2.6)	2.21	(0.87–5.67)	5	(6.6)	**5.78**	**(2.22–15.03)**
noncardiac anomalies	165	(4.6)	12	(6.3)	1.39	(0.76–2.55)	5	(6.6)	1.45	(0.58–3.65)
Parental Age										
maternal age:										
<20	535	(15.0)	18	(9.5)	1.00	(reference)	9	(11.8)	1.00	(reference)
20–24	696	(19.5)	30	(15.8)	1.28	(0.71–2.32)	10	(13.2)	0.85	(0.34–2.11)
25–29	1120	(31.4)	47	(24.7)	1.24	(0.72–2.17)	29	(38.2)	1.53	(0.71–3.24)
30–34	845	(23.7)	54	(28.4)	**1.90**	**(1.10–3.27)**	22	(29.0)	1.55	(0.70–3.39)
35–39	336	(9.4)	30	(15.8)	**2.65**	**(1.46–4.84)**	5	(6.6)	0.88	(0.29–2.66)
40+	40	(1.1)	11	(5.8)	**8.17**	**(3.61–18.48)**	1	(1.3)	1.49	(0.18–12.02)
paternal age:										
<20	289	(8.1)	11	(5.8)	1.00	(reference)	6	(8.0)	1.00	(reference)
20–24	557	(15.7)	27	(14.2)	1.27	(0.62–2.60)	10	(13.3)	0.86	(0.31–2.40)
25–29	1008	(30.6)	53	(27.9)	1.28	(0.66–2.48)	20	(26.7)	0.89	(0.35–2.22)
30–34	917	(25.8)	48	(25.3)	1.38	(0.70–2.68)	22	(29.3)	1.15	(0.36–2.87)
35–39	491	(13.8)	33	(17.4)	1.77	(0.88–3.55)	15	(20.0)	1.47	(0.56–3.83)
40+	214	(6.0)	18	(9.5)	**2.21**	**(1.02–4.77)**	2	(2.6)	0.45	(0.09–2.25)
Maternal Reproductive History										
No. previous pregnancies										
0	1159	(32.4)	41	(21.6)	1.00	(reference)	25	(32.9)	1.00	(reference)
1	1097	(30.7)	63	(33.2)	**1.62**	**(1.09–2.43)**	27	(35.6)	1.24	(0.71–2.18)
2	709	(19.8)	40	(21.0)	**1.59**	**(1.02–2.49)**	10	(13.1)	0.70	(0.34–1.50)
3+	607	(17.0)	46	(24.0)	**2.14**	**(1.39–3.30)**	14	(18.1)	1.33	(0.70–2.53)
No. previous miscarriages:										
0	2891	(80.9)	140	(73.7)	1.00	(reference)	57	(75.0)	1.00	(reference)
1	550	(15.4)	36	(19.0)	1.35	(0.93–1.97)	15	(19.7)	1.38	(0.77–2.46)
2	104	(2.9)	8	(4.2)	1.59	(0.76–3.32)	4	(5.3)	1.95	(0.69–5.48)
3+	27	(0.8)	6	(3.1)	**4.59**	**(1.86–11.30)**	0		—	
No. previous stillbirths:										
0	3537	(99.0)	188	(98.9)	1.00	(reference)	72	(94.7)	1.00	(reference)
1+	35	(1.0)	2	(1.1)	1.07	(0.26–4.50)	4	(5.3)	**5.61**	**(1.94–16.21)**

Illnesses and Medications

	N	(%)	OR	(CI)	N	(%)	OR	(CI)
diabetes:								
overt	23	(0.6)	0.82	(0.11–6.08)	5	(6.6)	10.87	(4.02–29.40)
insulin-treated	16	(0.5)	—	—	5	(6.6)	15.65	(5.58–43.90)
gestational	115	(3.2)	0.98	(0.43–2.26)	3	(4.0)	1.24	(0.38–4.00)
influenza	278	(7.8)	1.63	(1.04–2.57)	1	(1.3)	0.16	(0.02–1.14)
antitussives	23	(0.6)	2.48	(0.74–8.32)	3	(4.0)	6.34	(1.86–21.59)
diuretics*	23	(0.6)	0.82	(0.11–6.08)	4	(5.3)	8.57	(2.89–25.42)
ibuprofen**	119	(3.3)	2.49	(1.42–4.34)	2	(2.6)	0.78	(0.28–2.32)
benzodiazepines	35	(1.0)	1.07	(0.26–4.50)	3	(4.0)	4.15	(1.24–13.81)

Lifestyle Exposures

	N	(%)	OR	(CI)	N	(%)	OR	(CI)
alcohol (any amount)	2101	(58.8)	1.15	(0.85–1.55)	49	(64.5)	1.27	(0.79–2.03)
smoking (cigarettes/day)								
0	2302	(61.8)	1.00	(reference)	47	(61.8)	1.00	(reference)
1–10	565	(15.8)	0.99	(0.65–1.48)	6	(7.9)	0.52	(0.22–1.33)
11–20	529	(14.8)	1.05	(0.70–1.58)	14	(18.4)	1.29	(0.70–2.41)
>20	176	(4.9)	0.63	(0.27–1.45)	9	(11.8)	2.50	(1.21–5.19)
cocaine	42	(1.2)	1.35	(0.41–4.39)	3	(4.0)	3.45	(1.05–11.40)

Home and Occupational Exposures

	N	(%)	OR	(CI)	N	(%)	OR	(CI)
freq. fireplace heating	419	(11.7)	1.76	(1.21–2.56)	12	(15.8)	1.41	(0.76–2.64)
painting:								
mother	367	(10.3)	1.77	(1.19–2.63)	8	(10.5)	1.03	(0.49–2.15)
father	757	(21.2)	1.36	(0.98–1.90)	18	(23.7)	1.15	(0.68–2.19)
both	158	(4.4)	1.72	(0.89–3.03)	3	(4.0)	0.89	(0.28–2.85)
varnishing:								
mother	106	(3.0)	1.25	(0.57–2.72)	5	(6.6)	2.30	(0.91–5.82)
father	268	(7.5)	1.06	(0.62–1.82)	9	(11.8)	1.65	(0.82–3.36)
both	32	(0.9)	2.38	(0.83–6.80)	3	(4.0)	4.54	(1.36–15.18)
welding (father)	228	(6.4)	1.82	(1.14–2.92)	2	(2.6)	0.40	(0.10–1.68)
ionizing radiation:								
mother	48	(1.3)	1.98	(0.78–5.40)	1	(1.3)	0.98	(0.13–7.19)
father	32	(0.9)	2.38	(0.83–6.80)	3	(4.0)	4.54	(1.36–15.18)

Race of Infant

	N	(%)	OR	(CI)	N	(%)	OR	(CI)
white	2362	(66.2)	1.2	(0.9–1.6)	51	(67.1)	1.0	(0.6–1.7)
nonwhite	1207	(33.8)	1.0	(reference)	25	(32.9)	1.0	(reference)

*diuretics = triamterine +/- hydrochlorthiazide; **ibuprofen = taken for menstrual pain (LMP) in 13/15 mothers of cases (AVSD+Down Syndrome); risk factors are maternal unless labelled as paternal.
O.R. = odds ratio; CI = confidence interval.

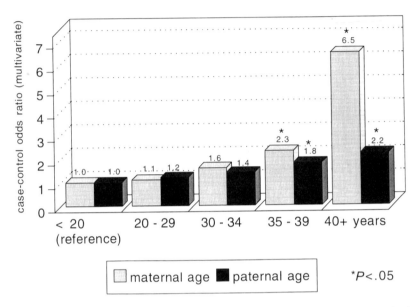

**Atrioventricular Septal Defect with Down Syndrome
Odds Ratios for Parental Age**

Baltimore-Washington Infant Study (1981 - 1989)

Figure 6.3: The multivariate case-control odds ratio for each parental age stratum is shown for mothers (lighter shading) and fathers (darker shading) of infants with AVSD and Down syndrome (the odds ratio is shown on top of the bar). Relative to parents who were younger than 20 years old, the risk of having an affected infant increased with increasing parental age, especially among mothers who were 40 years or older at the time of conception.

odds ratio of 4.6; no significant difference was observed between nonsyndromic cases and controls. On the other hand, the proportion of mothers reporting previous stillbirths was greater for the nonsyndromic group than for Down syndrome cases and controls.

Maternal Illnesses and Medications

Questions about maternal health and medications revealed contrasting findings in the two subgroups (Table 6.9). Overt diabetes and insulin-treated diabetes were strongly associated with nonsyndromic cases only; there was no case-control difference for gestational diabetes. There were no other maternal illness associated with cases of AVSD.

Intake of antitussives, diuretics, and benzodiazepines was reported more frequently by mothers of the nonsyndromic cases than by mothers of controls. Among mothers of the Down syndrome cases, the frequencies of these exposures were similar to those of controls; the only medication with a significant association was ibuprofen, taken by mothers for relief of menstrual pain in the last menstrual period prior to conception. Among reported benzodiazepines, the predominant medication was diazepam in both nonsyndromic cases (3/3 users) and controls

(18/35 users). Diuretic medication, specifically triameterene-hydrochlorothiazide combination therapy for hypertension, was reported by four nonsyndromic case mothers and 23 controls.

Lifestyle Exposures

Case-control differences in maternal cigarette smoking were found only within nonsyndromic cases, who showed a 2.5-fold odds ratio of smoking over 20 cigarettes (greater than 1 pack) per day compared to controls (Table 6.9). Maternal use of cocaine was also associated with this subgroup. Because heavy smokers tend to have other exposures, we examined the possible coteratogenic influences of alcohol and caffeine intake with smoking and cocaine use, but found no interaction nor an independent effect of either alcohol or cocaine.

Parental Home and Occupational Exposures

The Down syndrome cases were associated with frequent use of fireplaces for home heating, maternal paint exposures at home and at the work place, and paternal occupational exposures involving welding. In contrast, the nonsyndromic case group was associated with varnishing by both parents and by paternal occupational exposure to ionizing radiation. Further analyses revealed no linear trends of risk with exposure frequency for painting, varnishing, welding or ionizing radiation with either subgroup of cases.

Multivariate Analysis

Associations identified by univariate analysis were examined further with several logistic regression models. For each major subgroup, the Down syndrome and the nonsyndromic atrioventricular septal defect, associations with genetic and environmental factors were examined alone and in the presence of maternal reproductive factors.

For analysis of Down syndrome cases without maternal reproductive factors, the final regression model included maternal age, ibuprofen, frequent fireplace heating, maternal painting, and paternal welding (Table 6.10). Increasing maternal age was associated with an increasing risk of atrioventricular septal defect with Down syndrome; among mothers, ages 40 years and older, the risk was six-fold compared to mothers younger than 20 years. This association, as well as that of maternal use of ibuprofen, was significant at the 99% confidence level, unlike the three environmental exposures (fireplace, painting, and welding), which had smaller and barely significant associations at the 95% confidence level. Paternal age was not a significant factor even when maternal age was excluded from the regression. Inclusion in the regression of maternal reproductive factors revealed no significant associations of reproductive variables nor any evidence of effect modification or of confounding.

For analysis of nonsyndromic cases without maternal reproductive factors, the final regression model included family history of cardiac defects in first degree-relatives, overt diabetes, antitussives, diuretic medication, paternal expo-

Table 6.10

Atrioventricular Septal Defects: Results of Multivariate Logistic Regression
Baltimore-Washington Infant Study (1981–1989)

Variable	Adjusted O.R.	95% CI	99% CI
Atrioventricular Septal Defect with Down Syndrome (N=190)			
maternal age (years):			
<20	1.0	(reference)	(reference)
20–29	1.1	0.6–1.9	0.5–2.3
30–34	1.6	0.9–2.9	0.7–3.5
35–39	2.3	1.2–4.4	1.0–5.4
40+	6.5	2.7–15.6	2.1–20.4
ibuprofen	2.4	1.3–4.2	1.1–5.0
fireplace heating	1.5	1.0–2.2	0.9–2.5
painting	1.6	1.0–2.3	0.9–2.6
welding (paternal)	1.7	1.1–2.8	0.9–3.2
Nonchromosomal Atrioventricular Septal Defect (N=76)			
family history of CVM	7.0	2.6–18.9	1.9–25.8
overt diabetes	9.3	3.1–28.2	2.2–39.6
antitussives	8.9	2.6–30.6	1.7–45.2
diuretic medication[1]	7.3	2.1–25.0	1.4–37.0
ionizing radiation (paternal)	5.7	1.7–19.3	1.1–28.5
varnishing (both parents)	5.6	1.7–18.9	1.1–28.2
previous stillbirth[2]	7.5	2.1–26.6	1.4–39.6

[1]medication = triamterene +/- hydrochlorothiazide; [2]one or more stillbirths among women with multiple previous pregnancies.
CVM = cardiovascular malformations; CI = confidence interval.

sure to ionizing radiation, exposure of both parents to varnishing, and previous stillbirths among mothers with multiple previous pregnancies. All of these factors exhibited strong associations that were significant at the 99% confidence level. Inclusion in the regression of maternal reproductive factors revealed no significant associations of reproductive variables nor any evidence of effect modification or of confounding.

Discussion

Atrioventricular septal defects are known to encompass two major subgroups of lesions, distinguishable primarily by the presence or absence of Down syndrome and secondarily by a variation in the frequency of associated anomalies.[72,111,113,277] In the present evaluation, these two major phenotypes of atrioventricular septal defects were evaluated. The subgroup associated with Down syndrome comprised two thirds of all the cases. The second, less-prevalent subgroup, classified as nonsyndromic, consisted of infants with isolated cardiac lesions, associated cardiac lesions, and non-Mendelian disorders. In both subgroups, associated cardiac anomalies included pulmonic stenosis and left-sided obstructive lesions; the latter were four times more frequent in nonsyndromic than in the Down syndrome cases. Furthermore, these two subgroups exhibited different epidemiologic features. Cases with Down syndrome were associated (significant at 99% confidence level) only with maternal age and pain medications, whereas nonsyndromic cases were associated with family history of cardiac anomalies, maternal diabetes, history of stillbirths in previous pregnancies, and several environmental factors, including maternal medications and occupational or avocational exposures.

Differences in prevalence between the Down syndrome and nonsyndromic atrioventricular septal defects might arise from differential ascertainment if affected infants were generally asymptomatic or if the cardiac defects were difficult to detect. However, in this study neither of these is a likely possibility. Independent of the presence of Down syndrome, AVSD results in a variety of clinical manifestations in infancy, including murmurs and congestive heart failure, which can be detected by clinical examination and confirmed by echocardiography.

Also, differential ascertainment between the two subgroups is not likely to account for the lower prevalence of associated anomalies among the Down syndrome cases, since presence of this syndrome may actually elicit a more intensive search for other anomalies.[221] Thus, the differences observed between Down syndrome and nonsyndromic cases are probably real, and further support the hypothesis of heterogenous etiologies and morphogenetic mechanisms.

In the Baltimore-Washington Infant Study, the high prevalence of Down syndrome in patients with atrioventricular syndrome was well within the range of 52 percent to 74 percent, noted in previous studies.[111,166,351,402] This observation and the previously noted high prevalence of atrioventricular septal defects among patients with Down syndrome[176,221,402,437] are consistent with a genetic basis. Cytogenetic studies have allowed the construction of a phenotypic map for Down syndrome, according to which the minimal region likely to contain the genes determining the common congenital cardiac anomalies in Down syndrome is 21q22.→qter.[230] However, recent studies show that a gene of a familial form of isolated atrioventricular septal defect is not linked with chromosome 21.[98,177] Therefore, either chromosome 21 has another putative gene related to atrioventricular septal defects or some other mechanisms (but not triplication of a specific gene) responsible for these cardiac defects in infants with Down syndrome.

Atrioventricular septal defect was also common in trisomy 18 (9 of 44 cases in the Baltimore-Washington Infant Study) but the spectrum of types of cardiovascular malformations in this trisomy is very wide.[143,152,431] Not evident in our study was a chromosomal defect strongly associated with atrioventricular septal defects, 8p deletion.[113,156]

Atrioventricular septal defect is a characteristic cardiac manifestation for a few Mendelian syndromes of multiple congenital abnormalities. It is a common type of cardiac defect for the Ellis-van Creveld syndrome, where it can be found in approximately 20% of all cases.[24,168,353,457] These defects are also frequent in the patients with the Smith-Lemli-Opitz syndrome: out of 65 reported cases with cardiovascular malformations, 14 had different variants of atrioventricular septal defects.*

Atrioventricular septal defects are not characteristic for the Holt-Oram syndrome; only two out of 100 reported patients had this form of cardiac defect.[398,460] We could not find reports of atrioventricular septal defects in the Schinzel-Gideon or the Van der Woude syndromes.

References 36, 42, 80, 101, 197, 212, 233, 237, 381.

Atrioventricular septal defects occur frequently as component cardiovascular malformations in the Ivemark and polysplenia complex.[202] In the Baltimore-Washington Infant Study, in the non-Mendelian subgroup, three cases had asplenia or polysplenia without evidence of other laterality defects. Atrioventricular septal defects were mentioned in the literature also in the Cornelia de Lange[208] and Goldenhar[396] syndromes, VACTERL,[104] and CHARGE[102,327] associations. Examples of those were also observed among the cases in this study. Among single-organ noncardiac defects in this study, renal and tracheoesophageal abnormalities suggest possible forme fruste manifestations of these associations.

The familial recurrence of endocardial cushion defect has provided evidence of genetic transmission of this defect.[105,114,133,352,368] Nora[324] notes that the highest intrafamilial concordance among individual cardiac malformations was in endocardial cushion defect (ie, atrioventricular septal defect), and he considers endocardial cushion maldevelopment as the most discriminating defect.

The Baltimore-Washington Infant Study has added further important information on familial recurrence of the malformation in three mothers with atrioventricular septal defect (all operated and of the partial type) who had infants with atrioventricular septal defect, two with Down syndrome. Additionally, an atrial septal defect and ventricular septal defect that might be considered to be a partially concordant malformation was present in four additional case families. Less concordance, or a partial concordance, was noted in nine of 11 families with cardiovascular malformations in first-degree relatives. In addition, there was partial concordance of pulmonary stenosis in a mother whose infant had an AVSD with pulmonary stenosis.

Although in Down syndrome, AVSD is the predominant, but not the only form of cardiac disease, it is necessary to consider that other genetic, maternal, and environmental factors may contribute to the occurrence of associated defects.[221]

Risk factor analysis of the selective subset of Down syndrome and atrioventricular septal defects revealed only two associations with a reasonable degree of reliability: one with maternal age and one with ibuprofen. The association with older maternal age probably reflects the subgroup of cases in whom nondisjunction was the main basis for aneuploidy. The association with ibuprofen has not been noted before, and warrants further study since it might reflect a chance occurrence.

Nonchromosomal nonsyndromic cases of atrioventricular septal defects presented an entirely different risk profile: our results revealed multiple possible risk factors including family history of cardiac anomalies, maternal diabetes, history of prior pregnancy loss, and environmental factors. Since these same risk factors are associated with looping and outflow tract anomalies it is clear that they are responsible for errors in primary cardiogenesis. It appears likely that some alterations associated with advancing maternal age explain the occurrence of the very same cardiac defect in Down syndrome and other chromosomal anomalies.

Chapter 7

Ventricular Septal Defects

Section A:
Overview

Section B:
Membranous Type

Section C:
Muscular Type

From: Ferencz C, Loffredo CA, Correa-Villaseñor A, Wilson PD, eds: *Genetic & Environmental Risk Factors of Major Cardiovascular Malformations: The Baltimore-Washington Infant Study 1981–1989.* Armonk, NY: Futura Publishing Co., Inc; ©1997.

Section A

Overview

Ventricular septal defect may appear to be a "simple" anomaly of the heart, but depending on its location, size, and associated pathologic conditions, it produces a wide range of clinical entities with the varying medical and surgical needs of affected patients.

Ventricular septal defect occurring as a single cardiac anomaly is now the most frequent form of congenital heart disease; however, in the early writings of the century, ventricular septal defect was not considered common. Among the 1000 cases of congenital heart disease studied by Maude Abbott, only 55 had a defect of the ventricular septum as the primary lesion: one was observed in a fetus, the others in living patients up to the age of 79 years.[2,6] The usual patient was one without symptoms or cardiac enlargement but with a long, drawn out systolic murmur accompanied by a pronounced thrill above the left sternal border. This clinical picture described by Roger in 1879[364] constitutes a definite clinical pathologic entity, and is still referred to as the "maladie de Roger." In these patients a limited amount of blood crosses from the high-pressure left ventricle into the low-pressure right ventricle with a small increase in pulmonary bloodflow.

A greater complexity of defects and of multiple clinical manifestations became evident when attention turned to younger patients, especially infants, and the surgical approach demanded detailed information on anatomic and physiologic features as well as on the natural course of the disease. In a classic paper from the Mayo Clinic, Becu and Fontana together with DuShane, Kirklin, Burchell, and Edwards[34] report a detailed necropsy study of 50 hearts in which a ventricular septal defect was the principal malformation; the defects were described by their location, size, associated cardiovascular malformations, associated noncardiac anomalies, and by the causes of death.

An unfavorable course of patients with ventricular septal defect was also evaluated by studies of the pulmonary arteries, which in infancy can control the physiologic consequences of a large communication between the ventricles. Prenatally, when the lungs are not functioning, the pulmonary circulation maintains a high resistance by thick-walled, constricted small pulmonary arteries. With the first breath of the neonate the lungs expand and the pulmonary vascular resistance falls dramatically and then more slowly with the regression of the thick musculature of the arterial walls. In the presence of a large ventricular septal defect the high pulmonary arterial resistance initially controls the shunt of blood from the left to the right ventricle, but with the regression of the arterial musculature, the shunting to the lungs may increase excessively, leading to congestive heart failure.[106] Some infants will develop a gradually increasing pulmonary vascular resistance with hypertensive changes in the arterial walls, which at first favorably controls the shunt but can become progressive and constitute a contraindication to surgery.

The studies of Hoffman and Rudolph[198] present a more favorable aspect of the natural history of ventricular septal defects in infancy. Their longitudinal observations on hemodynamic alterations demonstrated that improvement in clinical course was frequent and principally due to a decreasing size and even spontaneous closure of the ventricular septal defect in infancy or in early childhood. This phenomenon explains, at least in part, the rarity of ventricular septal defects in adults.

Kaplan et al[215] studied 400 patients by clinical and physiological studies, and separated the natural course of the disease, by age groups, into six components: 1) death from heart failure and pulmonary hypertension in infancy, 2) morbidity in infancy and childhood due to pulmonary infections related to a high blood flow, 3) development of right ventricular outflow stenosis, 4) spontaneous closure of the defect, 5) development of subacute bacterial endocarditis, 6) death from pulmonary hypertension in adult life. There is also a large group of patients who remain asymptomatic and undetected in infancy. These authors emphasized the rarity of death from ventricular septal defect between the ages of 2 and 15, and the rarity of ventricular septal defects in adults, having found only a single case of "maladie de Roger" in the autopsy population of 6983 in the University of Cincinnati College of Medicine.

These early investigations determined the risk factors for mortality and for complications of ventricular septal defects. Emphasis on the various anatomic types came with the increasing interest in morphogenesis, which separated ventricular septal defects according to their relationship to the outflow tract and to the atrioventricular junction.[425] The significance of these variations was already described by Rokitansky[367] in 1875 in a now rarely available classic publication. A classification of congenital heart disease by embryonic mechanism was proposed by Clark,[82] who allocated the anterior supracristal (subpulmonary) defect to the outflow group, and the posterior inflow type of defect to the atrioventricular septal defect group; defects of the membranous septum, considered to result from embryonic hemodynamic disturbances, were separated from those of the muscular part of the septum, which may be due to excessive cell death. This mechanistic approach is of help to the epidemiologists who seek to identify specific disease categories.

The development of left ventricular angiography and later cineangiography continued to refine the distinctions of the anatomic types of ventricular septal defect. Within the last decade, two-dimensional and Doppler flow echocardiography has permitted clinicians to clearly see the precise structural characteristics of abnormal septation.

This capability of two-dimensional echocardiography resulted in an increased interest in the diagnosis of ventricular septal defects and a rapid rise in the population prevalence among infants. Numerous studies, including our own by Martin,[283] Wilson,[453] Ferencz and Neill,[149] and others[321] demonstrate that defects requiring cardiac catheterization and surgery remained at stable levels and that the temporal increase was confined to small defects, which were identified by the noninvasive evaluations.

At the time of the Baltimore-Washington Infant Study design (1979–1980) it was assumed that cardiac catheterization would be performed in infants with abnormal cardiac manifestations (enlargement of the heart, congestive heart failure, or failure to thrive) to quantify the physiologic changes, and that such defects were likely to be of moderate size; defects diagnosed by *echocardiography only* would represent small-sized defects. This assumption was no longer valid after the two-dimensional echocardiography and Doppler flow studies identified defects more precisely. The precise defect size was not reported to the study, but a code for "small" defects was introduced based on the echocardiography results. The designation of "moderate/large" defects was based on the diagnostic procedures of cardiac catheterization and/or surgery.

Our etiologic study included a total of 1466 infants with ventricular septal defects (Figure 7A.1). Cases were separated into the different anatomic types of ventricular septal defects according to their presumed mechanistic origins.[82] Thus, supracristal defects (ie, defects in the anterior septum above the cristal muscle of the right ventricle) were evaluated with outflow tract defects in Chapter 5, while defects in the posterior (inflow) portion of the septum were considered with atrioventricular septal defects in Chapter 6. Cases with multiple types of ventricular septal defects and those with an unspecified type of defect were excluded from analysis. According to the hierarchical order of primary diagnoses, obstructive defects such as coarctation of the aorta and pulmonic stenosis took precedence over septal defects.

Two types of ventricular septal defects, membranous and muscular, remained as primary diagnoses; these are considered as two separate diagnostic groups. The number of enrolled cases and their division by size and by associated noncardiac defects is shown in Figure 7A.2. For each group, the number of cases with interviewed families is shown in parentheses. Since it was not possible to interview all of the families of the rapidly increasing case population, the proportion of small defects was selected for interviews by random sampling strategy.[342]

Risk factor analyses included only those cases with interviewed families. The final risk factor analyses did not distinguish defects by size, but subdivided the groups by the potential association with genetic factors (see Chapter 3).

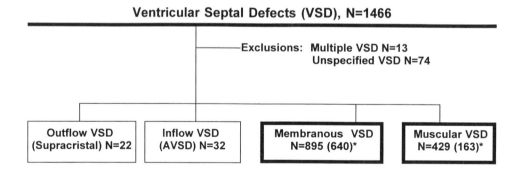

*Numbers in parenthesis represent interviewed cases.

Figure 7A.1: 1466 infants had ventricular septal defects (VSD) as primary diagnoses. Excluding multiple and unspecified defects, there were four main types: outflow defects, inflow defects (Chapters 5 and 6), membranous ventricular septal defects, and muscular ventricular septal defects (described in this chapter).

Ventricular Septal Defects:
Distribution by Size of Defect and Associated Noncardiac Anomalies

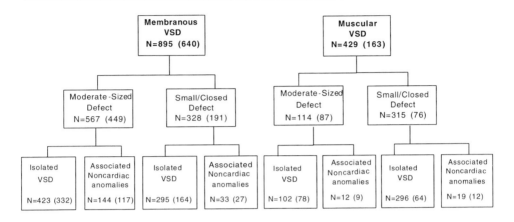

( )= Interviewed families

Figure 7A.2: Membranous and muscular ventricular septal defects (VSD) were each subdivided according to the size of the defect and the presence of associated noncardiac anomalies.

Section B

Membranous Type

Closure of the membranous part of the ventricular septum constitutes the final critical conjunction that completes the separation of the right and left side of the heart (see Chapter 2). Defects of the membranous ventricular septum represent failure of the intended fusion of tissues from three sources: the atrioventricular cushions, the right and left conal swellings, and the muscular ventricular septum.[382]

Although in this report a "pure" defect of the membranous ventricular septum is considered as an anatomic entity, it was recognized that a small defect may simply represent normal delay of closure, while a larger defect may be a forme fruste manifestation of major inflow or outflow anomalies. An objective preliminary to the risk factor analysis was the evaluation of possible heterogeneity by size of defect.

Study Population

There were 895 infants with membranous ventricular septal defects identified during the 9-year study period, a prevalence of almost 1 per 1000 livebirths. The descriptive analyses represented in Table 7B.1 through Table 7B.5 include all enrolled cases, while the information obtained from interviews is shown in Table 7B.6 through Table 7B.12.

Due to a rapid increase in enrollment during the last 4 years of the study, it was not possible to interview every family, so a random sampling strategy was used to select families for interview (Figure 7B.1) Of the 567 families of infants with moderate-sized defects, we sampled 513 for interview. Of these, 449 (87.5%) were interviewed. Of the 328 families of infants with small-sized defects, we sampled 235 for interview, of which 191 (81.3%) participated.

Cardiac Abnormalities

Coexisting cardiac defects of a lower diagnostic hierarchy included atrial septal defect and patent arterial duct alone or in combination in a total of 141 (15.8%) cases (Table 7B.1). The proportion with these associated diagnoses was almost three times as great in patients with moderate/large defects (20.6%) as in those with small defects (7.3%). These combined left-to-right shunts undoubtedly aggravated the clinical picture and contributed to the designation of greater severity.

Right aortic arch, associated with a small proportion of cases (1.3%) may indicate a forme fruste conotruncal anomaly. Cardiomyopathy also coexisted with a small proportion of membranous ventricular septal defect.

Noncardiac Anomalies

Twenty percent of the cases of membranous ventricular septal defects had associated noncardiac anomalies (Table 7B.2). Noncardiac anomalies were more fre-

128

Membranous Ventricular Septal Defects (VSD), N=895

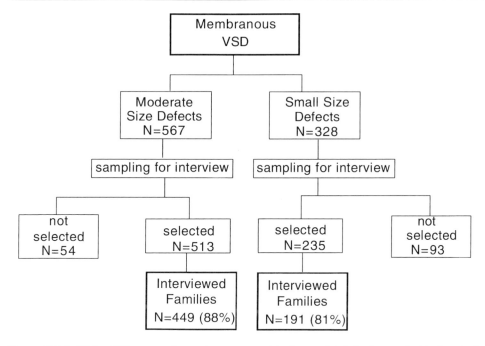

Figure 7B.1: Of the 895 cases of membranous ventricular septal defects enrolled in the study, 567 had moderate-sized defects and 328 had small defects (some of which closed during the first year of life). Families were randomly selected for interview: parental interview rates were 88% for moderate defects and 81% for small defects.

Table 7B.1
Membranous Ventricular Septal Defects:
Associated Cardiac Anomalies
Baltimore-Washington Infant Study (1981–1989)

Associated Cardiac Anomalies	All cases N=895 no. (%)	Moderate Defects N=567 no. (%)	Small Defects N=328 no. (%)
atrial septal defect only	88 (9.8)	70 (12.3)	18 (5.5)
atrial septal defect and patent arterial duct	19 (2.1)	19 (3.4)	0
patent arterial duct only	34 (3.8)	28 (4.9)	6 (1.8)
right aortic arch	12 (1.3)	9* (1.6)	3 (0.9)
cardiomyopathy	5 (0.6)	4 (0.7)	1 (0.3)
total with associated cardiac anomalies	158 (17.7)	130 (22.9)	18 (5.5)

*One patient had a double aortic arch.

Table 7B.2
Membranous Ventricular Septal Defect:
Noncardiac Anomalies
Baltimore-Washington Infant Study (1981–1989)

Type of Noncardiac Anomaly	Total Cases N=895	Moderate Size Defects N=567	Small Defects N=328
Chromosomal Abnormalities:	87 (9.7%)	78 (13.8%)	9 (2.7%)
Down syndrome	64*	59	5 [#3748, 10054, 10055, 10479, 3222]
Trisomy 18	12	11**	1 [#11356]
Trisomy 13	4	4 [#1286, 1705, 10151, 10749]	0
Tetrasomy X	1	0	1 [#11688]
Trisomy 3q	1	1 [#2435]	0
Trisomy 7q	1	1 [#10041]	0
Trisomy 9q+13q	1	1 [#3373]	0
Trisomy 10p	1	0	1 [#3331]
Unspecified	2	1 [#10958]	1 [#10575]
Syndromes	36 (4.0%)	17 (3.0%)	9 (2.7%)
a. Mendelian Syndromes:	19	15	4
Holt-Oram	2	2 [#1376, 2798]	0
Noonan	2	1 [#1996]	1 [#10545]
3c (Ritscher-Schinzel)	2	2 [#1777, 10082]	0
Achondroplasia	1	1 [#1547]	0
Apert	1	1 [#11192]	0
Townes-Brocks	1	0	1 [#10081]
Williams	1	1 [#3699]	0
Campomelic dysplasia	1	1 [#1473]	0
Fancomi (lethal variant)	1	1 [#3396]	0
Meckel	1	1 [#3080]	0
Pompe	1	1 [#1405]	0
Albinism	1	1 [#3509]	0
Bartter	1	1 [#1954]	0
Thrombocytopenia	3	1 [#2856]	2 [#3431, 10744]
b. Non-Mendelian Syndromes:	17	12	5
Fetal alcohol syndrome	5	4 [#1452, 2045, 10372, 11069]	1 [#11014]
VACTERL	4	4 [#1269, 1328, 2959, 11194]	0
Goldenhar	3	1 [#1576]	2 [#3731, 2781]
De Lange	3	1 [#11561]	2 [#10475, 11519]
CHARGE	1	1 [#2910]	0
Di George	1	1 [#10167]	0
Multiple, Nonclassified Anomalies:	5 (0.6%)	4 (0.7%)	1 (0.3%)
Cleft palate, chalasia	1	1 [#1399]	0
Cleft palate, anal atresia	1	0	1 [#11748]
Tracheo-esophageal fistula, intestinal atresia	2	2 [#10905]	0
Polycystic kidney, hypospadias	1	1 [#10159]	0
Organ Defects	49 (5.5%)	35 (6.2%)	14 (4.3%)
Central Nervous System:			
Holoprosencephaly	1	1 [#1968]	0
Macrocephaly	1	1 [#2225]	0

Table 7B.2—Continued
Membranous Ventricular Septal Defect:
Noncardiac Anomalies
Baltimore-Washington Infant Study (1981–1989)

Type of Noncardiac Anomaly	Total Cases N=895	Moderate Size Defects N=567	Small Defects N=328
Spina bifida	1	1 [#10386]	0
Dandy-Walker	2	2 [#1638, 11030]	0
Orofacial Defects:			
Facial dysmorphisms	2	2 [#3508, 10501]	0
Cleft palate	1	0	1 [#3598]
Gastrointestinal Defects:			
Omphalocele	6	3 [#1657, 2758, 10193]	3 [#10035, 10483, 11260]
Intestinal atresia	4	1 [#1503]	3 [#2155, 3451, 11744]
Anal atresia	4	2 [#1935, 3110]	2 [#3659, 11249]
Tracheoesophageal fistula	2	2 [#1550, 10755]	0
Diaphragmatic hernia	2	1 [#11304]	1 [#1764]
Pyloric stenosis	1	1 [#2770]	0
Urogenital Anomalies:			
Hypospadias	7	5 [#1315, 2258, 3700, 10556, 10622]	2 [#2829, 10283]
Agenesis of kidney	2	1 [#3025]	1 [#10203]
Polycystic kidney	2	2 [#10968, 11096]	0
Hydronephrosis	2	2 [#2285, 10933]	0
Ectopic kidney	1	1 [#2356]	0
Duplicate collecting system	1	0	1 [#11122]
Sexual ambiguity	1	1 [#3494]	0
Skeletal Defects:			
Polydactyly	2	2 [#1220, 2229]	0
Limb reduction	1	1 [#10843]	0
Pectus excavatum	1	1 [#10896]	0
Hemivertebrae	1	1 [#10578]	0
Miscellaneous Anomalies:			
Hypothyroidism	1	1 [#1716]	0
No Associated Anomalies	723 (80.8%)	437 (77.1%)	296 (90.2%)

*Cases with Down Syndrome are listed in Appendix C; ** cases with Trisomy 18 (moderate size): #1039, 1936, 2136, 3060, 3395, 3554, 3685, 10079, 10114, 10695, 11523.

quent among moderate-sized than among small-sized defects (23% versus 9.8%). The difference was greatest for chromosome defects (13.8% and 2.7% respectively) with an almost ten-fold difference in Down syndrome (10.4% and 1.5% respectively).

Mendelian syndromes were rare in both groups (2.6% and 1.2%), and no particular type predominated. However among defects of individual organ systems, central nervous system defects occurred only in the moderate group. Small ventricular septal defect was not reported with skeletal anomalies of the spine and chest, and there were fewer small defects with urogenital anomalies.

Descriptive Analyses

Prevalence by Time, Season, and Area of Residence

The overall prevalence of membranous ventricular septal defects was 9.9 cases per 10,000 livebirths, almost twice as high for moderate-sized as for small-sized defects (Table 7B.3). Over time, there was a slight decline in prevalence of defects of moderate size. There was, however, a very pronounced increase in the prevalence of small defects: from 1 per 10,000 from 1981 to 1983 to 6 per 10,000 from 1987 to 1989. There was no variation in prevalence with quarter of birth. There was a higher prevalence in urban than in rural areas; this urban-rural difference was more apparent for moderate than for small defects.

Diagnosis and Course

Approximately half of the cases of membranous ventricular septal defects were diagnosed by 4 weeks of age, and over 90% by 24 weeks of age (Table 7B.4). The proportion of defects diagnosed under 4 weeks of age was slightly greater for small (58.5%) than for moderate (49.4%) defects. Moderate and small defects differed in the methods by which they were classified. Almost all small defects were classified as such by echocardiography, except for 8 (2.4%) which were classified as small based in the findings of catheterization (Figure 7B.2). Seven infants with moderate-sized defects were diagnosed only at autopsy: 6 of the 7 died on the first day of life from multiple congenital anomalies; the seventh infant died suddenly at home on the 4th day of life and had only a ventricular septal defect.

Table 7B.3
Membranous Ventricular Septal Defect (VSD): Prevalence
Baltimore-Washington Infant Study (1981–1989)

	Membranous VSD			Prevalence (per 10,000)			
	All Cases	Moderate Defects	Small Defects	Area Births	All Cases	Moderate Defects	Small Defects
Total Subjects	895	567	328	906,626	9.87	6.25	3.62
Year of Birth:							
1981–1983	222	193	29	274,558	8.09	7.03	1.06
1984–1986	288	194	44	295,845	9.73	6.56	3.17
1987–1989	377	172	205	336,223	11.21	5.11	6.10
Birth Quarter:							
1st (Jan–Mar)	224	141	83	219,145	10.22	6.43	3.79
2nd (Apr–Jun)	230	152	78	231,777	9.92	6.56	3.36
3rd (Jul–Sep)	211	133	78	233,626	9.03	5.69	3.34
4th (Oct–Dec)	230	141	89	222,078	10.36	6.35	4.01
Area of Residence*							
urban	195	141	54	208,568	9.35	6.76	2.59
suburban	374	263	111	584,022	6.4	4.5	1.90
rural	71	45	26	112,318	6.32	4.01	2.31

*Interviewed families.

Table 7B.4
Membranous Ventricular Septal Defect (VSD): Clinical Data
Baltimore-Washington Infant Study (1981–1989)

	All Membranous VSD N=895		Moderate Defects N=567		Small Defects N=328	
	no.	(%)	no.	(%)	no.	(%)
Age at Diagnosis:						
<1 week	124	(13.9)	74	(13.1)	50	(15.2)
1–4 weeks	348	(38.9)	206	(36.3)	142	(43.3)
5–24 weeks	350	(39.1)	233	(41.1)	117	(35.7)
25–52 weeks	73	(8.1)	54	(9.5)	19	(5.8)
Method of Diagnosis:						
echocardiography only	630	(70.4)	318	(56.1)	317	(96.6)
echocardiography and cardiac catheterization	90	(10.1)	77	(13.6)	8	(2.4)
echocardiography and surgery	10	(1.1)	10	(1.8)	0	
echocardiography, cardiac catherization, and surgery	141	(15.8)	141*	(24.9)	0	
autopsy following other methods	17	(1.9)	14	(2.5)	3	(1.0)
autopsy only	7	(0.8)	7	(1.2)	0	
Surgery During First Year of Life	144	(16.1)	144	(25.4)	0	
Follow-Up:						
not followed	75	(8.4)	30	(5.3)	45	(13.7)
followed:	820		537		283	
died in the first year of life	56	(6.8)	50	(8.8)	6	(2.1)
alive	764	(93.2)	487	(91.2)	277	(97.9)
median age at last visit	10 months		10 months		9 months	

*Includes 7 cardiac catheterization only, 6 cardiac catheterization and surgery (without echocardiography).

Surgery during the first year of life was performed in one fourth of the cases with moderate defects, but in none of the cases with small defects (Table 7B.4). The proportion alive at the time of the diagnostic update was nearly 98% for cases with small defects and was 91% for cases with moderate defects. However in this diagnostic group, the time of diagnostic update was frequently short and a number of defects, both moderate and small, were reported as having closed during the observation period.[253] Overall, 8% of the infants with a membranous ventricular septal defect were not followed after their initial registration into the study (5% of moderate-sized defects and 14% of small defects). The median age of the infant at the last visit to a cardiologist was 10 months.

Gender, Race, and Twinning

Compared to controls, cases of ventricular septal defect had a slight deficiency of males, irrespective of defect size (Table 7B.5). Also, compared to controls, cases had a slight excess of black infants, particularly among the moderate-sized defects. Furthermore, the frequency of twins was slightly greater among infants with moderate-sized defects than among controls. The frequency of twins among infants with small defects was similar to that of controls.

Membranous Ventricular Septal Defects
Diagnosis by Echocardigraphy
Baltimore-Washington Infant Study (1981-1989)

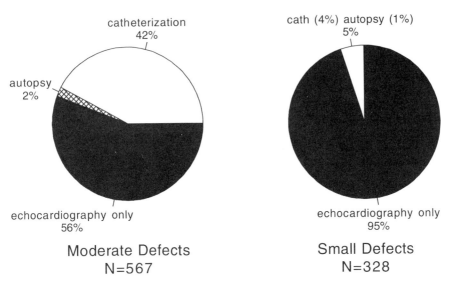

Moderate Defects
N=567

Small Defects
N=328

Figure 7B.2: Echocardiography predominated as the diagnostic method even among infants with moderate-sized defects. Among infants with small defects, almost all were diagnosed solely by echocardiography.

Table 7B.5

Membranous Ventricular Septal Defect (VSD): Infant Characteristics
Baltimore-Washington Infant Study (1981–1989)

Characteristic	All Cases N=895 no. (%)		Moderate Defects N=567 no. (%)		Small Defects N=328 no. (%)		Controls N=3572 no. (%)	
Gender:								
male	423	(47.3)	270	(47.6)	153	(46.6)	1,817	(50.9)
female	472	(52.7)	297	(53.4)	175	(53.4)	1,755	(49.1)
Race*:								
white	496	(55.4)	316	(55.7)	180	(54.9)	2,362	(66.1)
black	319	(35.6)	208	(36.7)	111	(33.8)	1,109	(31.0)
other	25	(2.8)	16	(2.8)	9	(2.7)	98	(2.7)
Twin births:	19	(2.1)	15	(2.6)	4	(1.2)	53	(1.5)

*Missing data in 27 moderate and 28 small defects.

Birthweight and Gestational Age

Information on fetal growth parameters obtained in the family interviews is the last comparison by defect size. Compared to controls, the mean birthweight of cases with moderate-sized defects was about 300 g lighter, and that of cases with small-sized/closed defects was 200 g lighter (Table 7B.6). Compared to controls, cases had a higher frequency of low birthweight (ie, <2500 g); this frequency was nearly three-fold greater among cases with moderate-sized defects, and two-fold greater among cases with small/closed defects. The mean gestational age was similar for cases and controls. However, the frequency of prematurity (ie, gestational age <38 weeks) was higher for cases than for controls, particularly for cases with moderate-sized defects. Furthermore, the frequency of small for gestational age (SGA) at birth was greater for cases than for controls. This excess in risk of intrauterine growth retardation was more prominent for infants with moderate-sized defects.

Similarity of Small and Moderate/Large Defects

Review of the comparisons between moderate/large defects and those designated as small did not suggest notable differences in risk factors. It was clear that small defects differed from controls; ie, they represented mild, but definite evidence of congenital heart disease. In the risk factor analyses we combined defects of all sizes and considered the variations in associated noncardiac anomalies as a probable indicator of etiologic heterogeneity.

Table 7B.6
Membranous Ventricular Septal Defect (VSD):
Fetal Growth Characteristics
Baltimore-Washington Infant Study (1981–1989)

Characteristic	All Cases no. (%)	Moderate Defects no. (%)	Small Defects no. (%)	Controls no. (%)
Number of Families Interviewed	640	449	191	3572
Birthweight (grams)				
<2500	116 (18.1)	86 (19.2)	30 (15.9)	252 (7.1)
2500–3500	350 (54.7)	253 (56.5)	97 (51.3)	1853 (51.9)
>3500	171 (26.7)	109 (24.3)	62 (32.8)	1467 (41.0)
mean±standard error	3083±29	3053±34	3155±53	3351±10
range	528–5358	528–5358	539–4608	340–5273
Gestational age (weeks)				
<38	102 (15.9)	75 (16.7)	27 (14.2)	339 (9.5)
38+	536 (83.8)	373 (83.3)	163 (85.8)	3233 (90.5)
mean	39.0±0.1	38.9±0.1	39.1±0.2	39.6
range	20–46	20–46	26–43	20–47
Size for Gestational Age				
small (SGA)	112 (17.5)	82 (18.3)	30 (15.7)	211 (5.9)
normal	444 (69.4)	315 (83.3)	129 (67.5)	2712 (75.9)
large	84 (13.1)	52 (11.5)	32 (16.8)	549 (18.2)
Odds Ratio for SGA (95% CI)	3.4 (2.7–4.3)	3.6 (2.7–4.7)	3.0 (2.0–4.5)	1.0 (reference)

Sociodemographic Characteristics

Compared to control mothers, case mothers were more likely to be unmarried and to not have had the pregnancy confirmed until late in the pregnancy (Table 7B.7). Also, compared to control fathers, case fathers were more likely to be unemployed. Cases and controls were similar with respect to all other sociodemographic characteristics such as parental education, annual household income, and maternal occupation.

Table 7B.7
Membranous Ventricular Septal Defect:
Analysis of Sociodemographic Characteristics
Baltimore-Washington Infant Study (1981–1989)

Interviewed Families	Cases N=640 no.	(%)	Controls N=3572 no.	(%)	Odds Radio (95% CI)	
Maternal Marital Status:						
not married	227	(35.5)	990	(27.7)	**1.4**	**(1.2–1.7)**
married	413	(64.5)	2582	(72.3)	1.0	(reference)
Maternal Education:						
<high school	128	(20.0)	659	(18.4)	1.1	(0.9–1.4)
high school	227	(35.5)	1265	(35.4)	1.0	(0.8–1.3)
college	285	(44.5)	1648	(46.1)	1.0	(reference)
Paternal Education:						
<high school	121	(18.9)	650	(18.2)	1.1	(0.9–1.4)
high school	244	(38.1)	1298	(36.3)	1.1	(0.9–1.3)
college	275	(43.0)	1624	(45.5)	1.0	(reference)
Annual Household Income:						
<$10,000	137	(21.9)	686	(19.2)	1.1	(0.9–1.4)
$10,000–$19,999	117	(18.7)	699	(19.6)	0.9	(0.7–1.2)
$20,000–$29,999	121	(19.4)	737	(20.6)	0.9	(0.7–1.1)
$30,000+	250	(40.0)	1373	(38.4)	1.0	(reference)
Maternal Occupation:						
not working	187	(29.2)	1142	(32.0)	0.9	(0.7–1.1)
clerical/sales	202	(31.6)	1119	(31.3)	0.9	(0.7–1.2)
service	88	(13.8)	444	(12.4)	1.0	(0.8–1.4)
factory	24	(3.8)	137	(3.8)	0.9	(0.6–1.5)
professional	139	(21.7)	730	(20.4)	1.0	(reference)
Paternal Occupation:						
not working	57	(8.9)	246	(6.9)	**1.4**	**(1.0–1.9)**
clerical/sales	86	(13.4)	618	(17.3)	0.8	(0.6–1.1)
service	80	(12.5)	480	(13.4)	1.0	(0.7–1.3)
factory	255	(39.8)	1279	(35.8)	1.2	(0.9–1.4)
professional	162	(25.3)	947	(26.5)	1.0	(reference)
Month of Pregnancy Confirmation:						
1st month	80	(12.6)	511	(14.3)	1.0	(reference)
2nd month	361	(56.8)	1955	(54.7)	1.2	(0.9–1.5)
3rd month	100	(15.7)	663	(18.6)	1.0	(0.7–1.3)
4th month or later	95	(14.9)	416	(11.6)	**1.5**	**(1.1–2.0)**

Data are missing for income (15 cases, 77 controls), paternal occupation (2 controls), and pregnancy confirmation month (4 cases, 27 controls).

Potential Risk Factors

Familial Cardiac and Noncardiac Anomalies

Among first-degree relatives of cases of membranous ventricular septal defects, cardiac anomalies occurred in 2.5% of the families (Table 7B.8). Five mothers and 4 fathers had congenital heart disease: a concordant ventricular septal defect was present in 1 maternal-infant pair and in 2 paternal-infant pairs. Among 13 affected sibling pairs, 6 were concordant for ventricular septal defect, but in 1 of those (#1239) a second affected sibling had transposition of the great arteries. Thus, that family and six other families (2 others with transposition, 3 with tetralogy of Fallot and 1 with common arterial trunk) showed evidence of a "conotruncal tendency,"[90] a finding of special interest to geneticists who search for families with chromosome 22q11 deletion.[70,118]

Also of special interest is the occurrence of Noonan syndrome with ventricular septal defect (case #1996) in the offspring of a mother with pulmonic stenosis who also had cleft palate, and also case #2435, whose ventricular septal defect occurred with trisomy 3q, but whose brother had an isolated ventricular septal defect.

Noncardiac anomalies among first-degree relatives of cases of membranous ventricular septal defect included major abnormalities in parents and in siblings (Table 7B.9). Worth noting was the presence of parental blood disorders in a parent in nine families and in a sibling in a 10th family: sickle cell disease in 5, thalassemia in 2, thrombocytopenia in 2, and von Willebrand's disease in 1. The association of blood disorders in case families, but not in controls, was noted early in the study,[138,142,154] but the occurrence remained rare and the information remains too meager to draw any conclusions.

Other families of interest include that of case #11192, with Apert syndrome in mother and child; and case #1405, with familial Pompe's disease and the combined presence of cardiomyopathy with a ventricular septal defect in the proband.[150]

Siblings also had severe disorders, notably sickle cell disease in 1 (noted above), anencephaly in 3, and hydrocephaly, microcephaly, and spina bifida each in 1 sibling. Four of the 6 sibs with central nervous system anomalies occurred in siblings of probands with trisomy: Down syndrome (N = 2), trisomy 18 (N = 1), and trisomy 13 (N = 1).

Genetic and Environmental Factors

The analytic strategy used in the evaluation of potential risk factors is shown in Figure 7B.3. Univariate and multivariate analysis was performed for the total group, followed by evaluations of two subgroups: families classified as isolated/simplex (N = 459) and those defined as multiple anomalies/multiplex (N = 181), according to the classification method described in Chapter 3. The latter group was further subdivided by the presence and absence of a chromosome anomaly in the proband.

Univariate Analysis

Univariate analysis of cases of membranous ventricular septal defects as a group identified case excesses in the history of cardiac and noncardiac defects in

Table 7B.8
Membranous Ventricular Septal Defect:
Congenital Heart Defects in First-Degree Relatives
The Baltimore-Washington Infant Study (1981–1989)

	Proband			Family	
ID	Size of Defect	Sex	Noncardiac Anomalies	Relative	Diagnosis
			A. Parents		
1996	moderate	M	Noonan	Mother	Pulmonic stenosis, cleft palate
1241	moderate	M		Mother	Tetralogy of Fallot
2884	moderate	F	Tracheoesophageal fistula, intestinal atresia	Mother	Coronary artery anomaly
3008	moderate	M		Mother	Ventricular septal defect
1317	moderate	F		Mother	Patent arterial duct
1700	moderate	F		Father	Ventricular septal defect
2715	moderate	M		Father	Ventricular septal defect
10044	small	F		Father	Patent arterial duct
11118	small	F		Father	Suspect CVM
			B. Siblings		
1239	moderate	M		Brother	Complete transposition
				Sister	Ventricular septal defect
3261	moderate	F		Brother	Tetralogy of Fallot
1538	moderate	M		Maternal Half-Brother	Tetralogy of Fallot
2309	moderate	F		Sister	Common arterial trunk
3628	moderate	F		Brother	Situs/looping defect
1569	moderate	M		Brother	Ventricular septal defect
2435	moderate	M	Trisomy 3q	Brother	Ventricular septal defect
2159	moderate	F		Sister	Ventricular septal defect
10312	small	F		Brother	Complete transposition
10989	small	F		Brother	Complete transposition
2957	small	M		Sister	Ventricular septal defect
10726	small	F		Sister	Ventricular septal defect
11758	small	F		Sister	Possible CVM
				Brother	Pectus excavatum
				Maternal Half-Brother	Pectus excavatum

CVM-cardiovascular malformation.

Table 7B.9
Membranous Ventricular Septal Defect:
NonCardiac Anomalies Among First-Degree Relatives
The Baltimore-Washington Infant Study (1981–1989)

	Proband			Family	
ID	Size of Defect	Sex	Noncardiac Anomalies	Relative	Diagnosis
A. Parents					
2258	moderate	M	Hypospadias	Mother	Sickle cell
2306	moderate	M		Mother	Sickle cell
2337	moderate	M	Down syndrome	Mother	Thalassemia
11358	small	M		Mother	Thalassemia
10223	moderate	M		Mother	Thrombocytopenia
1345	moderate	M		Mother	RH incompatibility
11192	moderate	M	Apert	Mother, Grandmother	Apert
11219	moderate	M		Mother	Ehlers-Danlos
3123	small	M		Mother	Cleft lip
1725	moderate	M	Down syndrome	Mother	Absent ear
2219	small	F		Father	Sickle cell
2939	moderate	M		Father	Thrombocytopenia
3494	moderate	F		Father	Von Willebrand
3588	moderate	F		Father	Hydrocephalus
3046	small	F		Father	Hydrocephalus
1405	moderate	M	Pompe	Father's twin	Pompe
1555	moderate	M	Tracheoesophageal fistula	Father, Paternal Aunt	Thyroglossal cyst
				Paternal Uncle	Down syndrome Pseudoxanthoma
B. Siblings					
1841	moderate	F		Maternal Half-Sister	Anencephaly
2136	moderate	M	Trisomy 18	Sister	Anencephaly
11252	small	F		Maternal Half-Sister	Anencephaly
1846	moderate	M	Down syndrome	Brother	Hydrocephalus
10749	moderate	M	Trisomy 13	Maternal Half-Sister	Microcephaly
11189	moderate	F	Down syndrome	Sister	Spina bifida
1321	moderate	F		Maternal Half-Brother	Renal agenesis
1583	moderate	M		Brother	Pyloric stenosis
10536	moderate	F		Maternal Half-Brother	Pyloric stenosis
10267	moderate	F		Maternal Half-Brother	Down syndrome
2229	moderate	M	Polydactyly	Brother	Polydactyly
11360	moderate	F		Paternal Half-Sister	Sickle Cell
2890	moderate	M		Sister	Cataract

Membranous Ventricular Septal Defects:
Subsets of Interviewed Families for Risk Factors Analysis

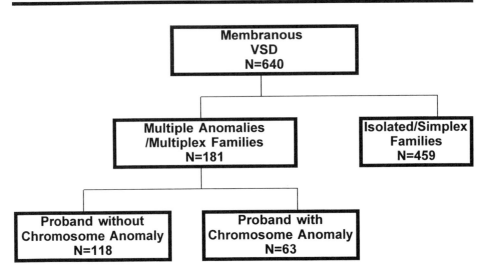

Figure 7B.3: This flow chart depicts the analytic strategy used in the evaluation of risk factors. All interviewed case families were analyzed first, followed by separate evaluations of two subgroups: the isolated/simplex subset and the multiple anomalies/multiplex subset. In addition, the latter group was subdivided according to the presence or absence of a chromosomal anomaly in the proband.

first-degree relatives, maternal reproductive history, selected maternal illnesses, use of medications, and lifestyle and home exposures (Table 7B.10). A family history of cardiac anomalies was almost three times as great in cases as in controls. For affected siblings it was five times as great. Also, compared to control mothers, case mothers reported more previous pregnancies, more previous spontaneous abortions, and more frequent bleeding during pregnancy. Furthermore, the frequency of overt diabetes in case mothers was nearly three-fold that of control mothers. Use of ibuprofen and metronidazole, exposure to general anesthesia, use of cocaine, and home exposure to pesticides were all more frequent among case mothers. Use of cocaine and marijuana was more frequent among case fathers as compared to control fathers.

Multivariate Analysis

Multiple logistic regression analysis was performed first on all cases of membranous ventricular septal defect as a group. The variables in the resulting model were family history of cardiac anomalies; maternal overt diabetes, exposure to general anesthesia, use of cocaine, and exposure to pesticides; paternal use of cocaine; and race of the infant (Table 7B.11).

Additional regression analyses were conducted on the isolated/simplex and multiple anomalies/multiplex subgroups to elucidate these associations and to take into account possible differences in risks due to associated anomalies and family history of birth defects. In the analysis of isolated/simplex cases, the largest subgroup, potential risk factors that were significant at the 5% level included: ma-

Table 7B.10
Membranous Ventricular Septal Defect:
Univariate Analysis of Risk Factors
Baltimore-Washington Infant Study (1981–1989)

Variable	Cases N=640 no. (%)		Controls N=3572 no. (%)		Odds Ratio (95% CI)	
Family History (first-degree relatives)						
Congenital Heart Disease:						
parents	9	(1.4)	28	(0.8)	1.8	(0.9–3.9)
siblings	13	(2.0)	15	(0.4)	**4.9**	**(2.3–10.5)**
total	22	(3.4)	43	(1.2)	**2.9**	**(1.8–4.9)**
Noncardiac Anomalies:						
parents	17	(2.7)	115	(3.2)	0.8	(0.5–1.4)
siblings	13	(2.0)	50	(1.5)	1.5	(0.8–2.7)
total	30	(4.7)	165	(4.6)	1.0	(0.7–1.5)
Parental Age						
maternal age:						
<20	91	(14.2)	507	(14.2)	1.0	(0.8–1.3)
20–39	353	(55.1)	2009	(56.2)	1.0	(reference)
30+	196	(30.6)	1056	(29.6)	1.1	(0.9–1.3)
paternal age:						
<20	46	(7.2)	226	(6.3)	1.2	(0.8–1.7)
20–29	297	(46.4)	1724	(48.3)	1.0	(reference)
30+	297	(46.4)	1622	(45.4)	1.1	(0.9–1.3)
Maternal Reproductive History						
number of previous pregnancie(s):						
none	204	(31.9)	1159	(32.4)	1.0	(reference)
one	168	(26.2)	1097	(30.7)	0.9	(0.7–1.1)
two	120	(18.8)	709	(19.9)	1.0	(0.7–1.2)
three or more	148	(23.1)	607	(17.0)	**1.4**	**(1.1–1.8)**
previous miscarriage(s)	135	(21.1)	681	(19.1)	1.1	(0.9–1.4)
previous induced abortions	161	(25.2)	691	(19.3)	**1.4**	**(1.2–1.7)**
bleeding during pregnancy	124	(19.4)	534	(15.0)	**1.3**	**(1.1–1.7)**
Illnesses and Medications						
diabetes:						
overt	11	(1.7)	23	(0.6)	**2.7**	**(1.3–5.6)**
gestational	24	(3.8)	115	(3.2)	1.2	(0.7–1.8)
influenza	53	(8.3)	278	(7.8)	1.1	(0.8–1.5)
ibuprofen	32	(5.0)	119	(3.3)	**1.5**	**(1.0–2.3)**
metronidazole	5	(0.8)	8	(0.2)	**3.5**	**(1.1–10.8)**
general anesthesia	32	(5.0)	105	(2.9)	**1.7**	**(1.2–2.6)**
Lifestyle Exposures						
cocaine:						
mother	21	(3.3)	42	(1.2)	**2.9**	**(1.7–4.8)**
father	44	(6.9)	111	(3.1)	**2.3**	**(1.6–3.3)**
marijuana (father)	137	(21.4)	555	(15.5)	**1.5**	**(1.2–1.8)**
alcohol (any amount)	378	(59.1)	2101	(58.8)	1.0	(0.9–1.2)
smoking (cigarettes/day):						
none	400	(62.5)	2302	(64.5)	1.0	(reference)
1–10	115	(18.0)	565	(15.8)	1.2	(0.9–1.5)
11–20	94	(14.7)	529	(14.8)	1.0	(0.8–1.3)
>20	31	(4.8)	176	(4.9)	1.0	(0.7–1.5)

Table 7B.10—Continued

Variable	Cases N=640 no. (%)		Controls N=3572 no. (%)		Odds Ratio (95% CI)	
Home and Occupational Exposures						
ionizing radiation (occupational)	12	(1.9)	48	(1.3)	1.4	(0.7–2.7)
pesticides	195	(30.5)	926	(25.9)	**1.3**	**(1.0–1.5)**
Race of Infant						
white	380	(59.5)	2362	(66.2)	1.0	(reference)
nonwhite	259	(40.5)	1207	(33.8)	**1.3**	**(1.1–1.6)**

Risk factors are *maternal* unless labelled as paternal.

Table 7B.11

All Cases with Membranous Ventricular Septal Defect (N=640): Multivariate Model of Genetic and Environmental Risk Factors Baltimore-Washington Infant Study (1981–1989)

Variable	O.R.	95% CI	99% CI
Family History:			
family history of CVM	3.1	1.9–5.3	1.6–6.2
Maternal Exposures:			
overt diabetes	2.9	1.4–6.1	1.1–7.6
general anesthesia	1.8	1.1–2.6	1.0–2.9
cocaine	1.9	1.0–2.7	0.8–4.3
pesticides	1.3	1.0–1.5	0.9–1.6
Paternal Exposures:			
cocaine	1.9	1.3–2.9	1.1–3.3
Sociodemographic Characteristics:			
race (nonwhite)	1.3	1.1–1.6	1.1–1.7

SES score and reproductive history variables were not significant and were neither effect modifiers nor confounders of the variables shown above.
O.R. = odds ratio; CVM = cardiovascular malformations.

ternal exposures to general anesthesia, cocaine, pesticides, and heat in occupational settings; paternal exposure to marijuana, and occupational ionizing radiation; and nonwhite race of the infant (Table 7B.12). Of these, the only variable that was significant at the 1% level was maternal cocaine use.

For purposes of analysis, the total group of multiple anomalies/multiplex cases (Set A, N = 181) was subdivided into those that excluded probands with chromosome defects (Set B, N = 118) and those that included only probands with chromosome defects (Set C, N = 63).

For the total subset of multiple anomalies/multiplex cases (set A), potential risk factors that were significant at the 5% level included: maternal overt diabetes, use of ibuprofen, use of metronidazole, work in auto body repair, and use of hair dyes, as well as paternal use of cocaine, white race, and increasing maternal age (Table 7B.13, set A). Of these, overt diabetes, metronidazole use, and maternal age were significant at the 1% level. The variables that were significant at the 5% level for set A, and only these variables, were used for analysis in sets B and C.

Multivariate models of the multiple anomalies/multiplex subset excluding probands with chromosome defects (Set B) revealed a strengthening of some of the associations noted in the total group and an attenuation of others. According to this model, the risk of multiple anomalies/multiplex ventricular septal defects exclusive of probands with chromosome disorders was increased 6.6-fold if there was overt diabetes in the mother, 2.4-fold by maternal use of ibuprofen, 12.2-fold by maternal use of metronidazole, 8-fold by maternal work in auto body repair, and 1.5-fold by nonwhite race. Of these associations, all but race were significant at the 1% level. For this subset of defects there was no maternal age effect.

In contrast, multivariate analysis of multiple anomalies/multiplex subset of cases with chromosome defects (Set C) identified only three significant risk fac-

Table 7B.12
"Isolated/Simplex" Subset (N=459) of Membranous
Ventricular Septal Defect Cases: Multivariate Model of Risk Factors
Baltimore-Washington Infant Study (1981–1989)

Variable	Cases	Controls	Odds Ratio	95% CI	99% CI
Maternal Exposures:					
general anesthesia	24	101	1.8	1.1–2.8	0.9–3.2
cocaine	15	39	2.4	1.3–4.4	1.1–5.4
pesticides	141	857	1.3	1.0–1.6	0.9–1.7
occupational heat exposure	3	3	7.9	1.5–40.3	0.9–67.0
Paternal Exposures:					
marijuana	100	521	1.4	1.1–1.8	1.0–1.9
ionizing radiation	9	29	2.4	1.1–5.2	0.9–6.6
Sociodemographic Characteristics:					
race (nonwhite)	187	1148	1.3	1.1–1.6	1.0–1.7

SES score and reproductive history variables were not significant and were neither effect modifiers nor confounders.

Table 7B.13

"Multiple Anomalies/Multiplex" Subsets of Membranous Ventricular Septal Defect: Multivariate Models of Risk Factors
Baltimore-Washington Infant Study (1981–1989)

Variable	Cases	Controls	Set A. All Families (N=181)			Set B. Families Excluding Chromosomal Cases (N=118)			
			Odds Ratio	95% CI	99% CI	Cases	Odds Ratio	95% CI	99% CI
Maternal Exposures:									
diabetes	5	22	3.9	1.4–10.7	1.1–14.6	5	6.6	2.4–18.2	1.8–25.0
ibuprofen	14	112	1.9	1.0–3.5	0.9–4.2	11	2.4	1.2–4.8	1.0–5.9
metronidazole	3	7	7.5	1.9–30.3	1.2–46.9	3	12.2	3.0–50.2	1.9–78.0
auto body repair	3	10	4.7	1.2–19.3	0.7–29.9	3	8.0	2.0–32.4	1.3–49.9
hair dyes	21	214	1.7	1.1–2.9	0.9–3.3	10	1.2	0.6–2.5	0.5–3.1
Paternal Exposures:									
cocaine	15	106	2.1	1.0–4.1	0.8–5.1	9	1.6	0.7–3.9	0.5–5.1
Sociodemographic Characteristics:									
race (nonwhite)	72	1148	1.4	1.0–2.0	0.9–2.2	50	1.5	1.0–2.2	0.9–2.5
maternal age:									
<20	22	482	1.0	(reference)		17	1.0	(reference)	
20–29	85	1872	1.3	1.1–1.6	1.0–1.7	63	1.1	0.8–1.3	0.7–1.4
30–34	54	704	1.7	1.2–2.5	1.1–2.8	29	1.1	0.7–1.8	0.6–2.1
35–39	16	232	2.3	1.3–3.9	1.1–4.6	7	1.2	0.6–2.4	0.5–2.9
40+	4	28	3.0	1.5–6.1	1.2–7.6	2	1.3	0.5–3.2	0.4–4.3

SES score and reproductive history variables were not significant and were neither confounders nor effect modifiers of the variables shown above; there were no significant interactions involving time period.

tors (Table 7B.14). These variables were: maternal use of hair dyes, paternal use of cocaine, and increasing maternal age. According to this model, the prevalence of membranous ventricular septal defect associated with chromosomal disorders increased 2.6-fold with maternal use of hair dyes, 3.3-fold with paternal use of cocaine, and increased monotonically with maternal age from age 20 to age 40. The prevalence of an offspring with a membranous ventricular septal defect and a chromosomal disorder was 11 times greater among mothers 40 years of age and older than among mothers younger than 20 years of age. Of these variables, all but paternal cocaine use were significant at the 1% level.

Discussion

Certain methodological issues related to the size of the defects warrant further comment. First, there was no information on the actual size of the defects, therefore, we were not able to determine the accuracy of the classification of lesions by size in order to conduct a more reliable assessment by defect size. Second, the study period encompassed the advent of two-dimensional echocardiography in the study area and this resulted in increased referrals and an increase in prevalence of small defects. However, the stability of the prevalence of moderate defects, and the above similarities of small- and moderate-sized defects suggest that changes in referral patterns probably did not result in selection of a sample of cases that was biased with respect to the exposures examined in this study.

Table 7B.14
Chromosome Subset of Membranous Ventricular Septal Defect:
Multivariate Model of Risk Factors
Baltimore-Washington Infant Study (1981–1989)

Variable	Cases	Odds Ratio	95% CI	99% CI
			Set C. Proband with Chromosome Defect (N=63)	
Maternal Exposures:				
diabetes	9	– –		
ibuprofen	3	1.2	0.3–3.8	0.2–5.6
metronidazole	0	– –		
auto body repair	0	– –		
hair dyes	11	**2.6**	**1.3–5.2**	**1.0–6.4**
Paternal Exposures:				
cocaine	6	**3.3**	**1.2–9.1**	0.9–12.5
Sociodemographic Characteristics:				
race (nonwhite)	22	1.3	0.7–2.2	0.6–2.6
maternal age:				
<20	5	1.0	(reference)	
20–29	22	**1.8**	**1.4–2.4**	**1.3–2.6**
30–34	25	**3.3**	**2.4–4.4**	**2.3–4.7**
35–39	9	**6.1**	**4.6–8.0**	**4.3–8.7**
40+	2	**11.1**	**8.5–14.5**	**7.8–15.8**

SES score and reproductive history variables were not significant and were neither confounders nor effect modifiers of the variables shown above; there were not significant interactions involving time period.

Third, it is possible that the presence of associated anomalies or of a family history of malformations might lead to increased surveillance for and detection of small defects, and therefore to a greater case-control difference in associated anomalies. However, our findings of a greater frequency of associated anomalies among moderate-sized than among small-sized defects suggest that such a selection bias was not a major issue in this study. Finally, many of the small membranous ventricular septal defects may close within the first year of life and infant mortality associated with them may be negligible. Nevertheless, we believe that small defects are real abnormalities and should be considered as such in epidemiologic studies.

In our case-control analysis of all cases of membranous ventricular septal defects as a group, we found associations with family history of cardiac defects as well as with maternal diabetes and other environmental factors. We considered a positive family history of birth defects and the presence of associated anomalies in the proband as possible indicators of different morphogenetic mechanisms and etiologies.

The subgroup of membranous ventricular septal defects designated as isolated/simplex comprised 70% of all cases of membranous ventricular septal defects and had no strong risk factors. This observation is consistent with the hypothesis that membranous ventricular septal defects are due to an imbalance of intracardiac blood flow prior to completion of septation.[82] The possible origins of the imbalance in blood flow remain unclear; it is unknown whether parental factors or exposures would play a role in such hemodynamic disturbances. Altered intracardiac blood flow could also explain some of the dynamic aspects of the defects such as decrease in defect size and delayed closure, while the relative sizes of the heart and the defect change during the growth of the infant.

The analysis by subsets of cases of membranous ventricular septal defects highlights two other important points. One is that the greater number of risk factors for the multiple anomalies/multiplex group compared to the isolated/simplex suggests that a familial history of defects enhances the susceptibility of an embryo to teratogenesis by various kinds of risk factors, and that the associated anomalies in the proband may be either an indicator of such susceptibility or a manifestation of the effects of a given exposure. Another point is that there is a substantial degree of etiologic heterogeneity within the multiple anomalies/multiplex subgroup. Thus, analysis of this total subgroup could lead to attenuation or failure to find a number of subset-specific associations with potential prevention implications.

Paradoxically, this evaluation of the possible etiologic heterogeneity of membranous ventricular septal defects allowed us to identify an overlap in risk factors with other morphogenetic types of defects. The associations that we found for the subset of multiple anomalies/multiplex, excluding cases with chromosome disorders, were strikingly similar to those obtained in the evaluations of outflow tract defects with normally related great arteries (Chapter 5) and atrioventricular septal defects without chromosomal disorders (Chapter 6). This illustrates that membranous ventricular septal defects belong to the spectrum of outflow and inflow tract anomalies. For the subset of multiple anomalies/multiplex cases limited to those with chromosomal disorders, the only association of moderate magnitude and precision was maternal age. This finding is similar to the findings for cases of atrioventricular septal defects with chromosomal disorders [infants with atrioventricular septal defects and Down syndrome (Chapter 6)], and suggests that membranous ventricular septal defects are also part of the spectrum of those atrioventricular septal defects.

Noteworthy is the fact that Down syndrome and other trisomies that are predominantly associated with atrioventricular septal defects are also associated with some outflow tract defects and with ventricular septal defects. In the Baltimore-Washington Infant Study, not all of the infants with Down syndrome had an atrioventricular septal defect: 22 (6.2%) had outflow tract defects (tetralogy of Fallot, N = 20; double-outlet right ventricle, N = 1; supracristal ventricular septal defect, N = 1), and 64 (18.1%) had a membranous ventricular septal defect.

These observations suggest that in the etiologically related groups of outflow tract defects with normal great arteries, atrioventricular septal defects and membranous ventricular septal defects, respectively, the cardiac abnormality represents only one of the multiorgan disturbances of the embryonic systems.

Summary

This study of membranous ventricular septal defect has revealed findings of clinical, morphogenetic, and etiologic significance.

First we were able to address the longstanding clinical question regarding the clinical identification of ever smaller defects of the membranous septum in terms of their significance in family counseling. In other words, are they really defects as opposed to late closures? For the interviewed families of the case group of small defects the answer has to be affirmative: there were similar though rarer and milder associations with other cardiac and noncardiac defects and with fetal growth retardation, which certainly set aside the group of small defects from the controls. There are however a number of caveats:

- there was no information in this study on the *actual* size of the defects, although in the final 4 years, the cardiologists reported small defects on the basis of size criteria.
- the Baltimore-Washington Infant Study encompassed the *initial* time of use of two-dimensional echocardiography and the nature of referrals changed with the years, a change possibly related to defect size.
- many of the small defects were reported to have closed within the defined 1-year time period. However the 1-year time limit was not always met. It had not been the intention of the Baltimore-Washington Infant Study to evaluate the issue of spontaneous closure, but according to the data compiled by Lewis and Loffredo,[253] the rate of closure was probably quite high.
- Small ventricular septal defects may be discovered by preferential screening in families with coexisting congenital abnormalities, a possibly important departure from the recognition of larger defects on the basis of clinical manifestations in the proband.

Since the evidence suggested that small defects represent real congenital heart disease, we analyzed the *total* case group of membranous ventricular septal defects in one single analysis and then performed further analyses in the more informative subgroups that were determined by genetic factors. These subgroups, defined by the presence of noncardiac anomalies in the proband and cardiac and noncardiac anomalies in the family, were associated with different risk factors. This fact illuminated the etiologic heterogeneity of ventricular septal defects.

The isolated/simplex membranous ventricular septal defect group (those in whom the proband had no associated anomalies nor did any first-degree relative have a cardiac or noncardiac abnormality) represented 70% of all cases. No strong risk factors were evident for this group.

A far more serious developmental deviation is indicated in ventricular septal defect of the membranous type associated with noncardiac abnormalities and/or a positive family history. In this multiple/multiplex group we demonstrated the presence of several risk factors: maternal diabetes, the use of various medications, and exposure to environmental xenobiotics. Dichotomizing this subgroup into families including chromosome cases and those without chromosome defects allocated some risk factors to one group and other risk factors to the second group *without any overlap*.

In probands with chromosome defects, the major risk factor for the chromosomal subset was the increased risk of advancing maternal age, a finding even more striking than that found in atrioventricular septal defect with Down syndrome, which is an 11-fold (versus a 6.5-fold) difference in risk of an infant with membranous ventricular septal defect in mothers over 40 years of age. Only two other risk factors were significantly associated: hair dyes and paternal cocaine exposure.

Section C

Muscular Type

Defects of the muscular ventricular septum were recognized in early descriptions of cardiovascular malformations,[367] but were deserving of only the briefest mention as a rare anomaly of no notable clinical significance. Peacock's statement[338] on ventricular septal defects, *in some cases, also, openings occur near the apex,* was expanded only slightly by Abbott:[6] *more rarely still, defects occur at the lower part of the septum and then not uncommonly they are multiple.* Taussig[407] noted the surgical implications of multiple ventricular septal defects, which *may be small and difficult to find, but together may place an intolerable burden on the heart.* Beyond this caution muscular ventricular septal defects remained to be fully discovered only by the introduction of echocardiography. The Baltimore-Washington Infant Study documented the rising predominance of this defect, which, in the final year of case registration (1989), constituted the second largest diagnostic subset, even though multiple ventricular defects were not included in the count.

Formation of the muscular ventricular septum occurs by coalescence of the primitive trabeculation and by cellular proliferation.[84] A defect can arise by abnormal trabecular organization or by foci of cellular death during the rapid growth and remodeling of the ventricles. In his developmental classification, Clark[82] separates muscular ventricular septal defects from the other forms of septal defects by noting that they have a *unique* developmental mechanism.

The rising clinical importance of muscular ventricular septal defect was previously described by Martin et al.[283] It is noted that the prevalence of isolated ventricular septal defect confirmed by catheterization, surgery, or autopsy did not change between 1981 and 1984; however, there was a 400% increase in the diagnosis of muscular ventricular septal defects between the first two and the second two years (1981 to 1982 and 1983 to 1984) of the Baltimore-Washington Infant Study. This change in prevalence was parallel with the increased use of high-resolution two-dimensional echocardiography. Lewis et al (1996)[253] report on the occurrence of membranous and muscular ventricular septal defects in the Baltimore-Washington Infant Study region and on the frequency of spontaneous closure during the first year of life. Of muscular defects, 33.5% were reported to have closed spontaneously, compared to 12.7% of membranous defects. Our coding system permits us to classify cases by size of the defect and by closure, providing an opportunity to examine potential etiologic differences among these subpopulations of infants with muscular ventricular septal defects.

Study Population

Cardiac Abnormalities

In this evaluation, cases with muscular ventricular septal defects constitute a "pure" anatomic group described separately from all other types of ventricular

149

septal defects. Between 1981 and 1989, 429 cases were enrolled in the study (Figure 7C.1), of whom 114 (26.6%) were moderate to large in size and 315 (73.4%) were coded as small or having closed during the first year of life. In both categories the majority of infants had isolated heart defects (89% of moderate and 94% of small defects).

Due to the rapidly rising case enrollment of small defects in the last 4 years of the study, it was not possible to interview every family, and a random sampling strategy was used to select affected families for interview. Of the 114 families of infants with moderate defects, we sampled 93 to interview. Of these, 87 families (94%) agreed to participate. Of the 315 families of infants with small defects, we sampled 82 to interview. Of these, 76 families (93%) agreed to participate.

Descriptive information on infant and clinical characteristics are presented for the entire group of 429 enrolled cases by defect size. Case-control comparisons on potential risk factors are restricted to the subset of interviewed families.

Associated Cardiac Abnormalities

Of the 429 cases with muscular ventricular septal defects, 3.5% had an atrial septal defect (secundum type) and 5.1% had an associated patent arterial duct (Table 7C.1). Both of these associated defects are two to three times more likely

Muscular Ventricular Septal Defects (VSD), N=429

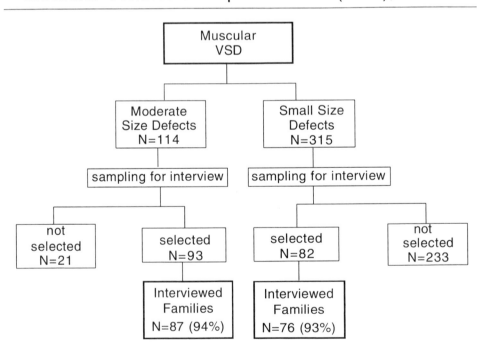

Figure 7C.1: Of the 429 cases of muscular ventricular septal defects enrolled in the study, 114 were classified as having moderate- to large-sized defects, and 315 had small defects, some of which closed in the first year of life. Families were randomly selected for interview, and over 90% agreed to participate.

Table 7C.1
Muscular Ventricular Septal Defects:
Associated Cardiac Anomalies
Baltimore-Washington Infant Study (1981–1989)

Associated Anomaly	Total Cases N=429	Moderate Defects N=114	Small Defects N=315
atrial septal defect	15 (3.5%)	7 (6.1%)	8 (2.5%)
patent arterial duct	22 (5.1%)	13 (11.4%)	9 (2.9%)
partial anomalous pulmonary venous return	1 (0.2%)	1 (0.9%)	0
anomalous origin of right subclavian artery	1 (0.2%)	0	1 (0.3%)
right-sided aortic arch	1 (0.2%)	1 (0.9%)	0
hypertrophic, obstructive cardiomyopathy	1 (0.2%)	1 (0.9%)	0

only 4 patients above had >1 associated anomaly (all had moderate size VSD):
   case #1413 - patent arterial duct, cardiomyopathy;
      #3817 - patent arterial duct, atrial septal defect;
      #11482 - patent arterial duct, atrial septal defect;
      #11422 - right aortic arch, atrial septal defect.

to be diagnosed in infants with moderate-sized ventricular septal defects than among those with small defects. Other cardiac abnormalities, each occurring in single patients, included partial anomalous pulmonary venous return, anomalous origin of the right subclavian artery, right-sided aortic arch, and hypertrophic, obstructive cardiomyopathy. All occurred in infants with moderate-sized defects except for the patient with the subclavian artery anomaly (Table 7C.1)

Noncardiac Anomalies

The great majority (92.8%) of infants with muscular ventricular defect was unaffected by noncardiac anomalies. Chromosome anomalies were diagnosed in 11 patients: 8 had Down syndrome, 2 had trisomy 18, and 1 had a 13q deletion (Table 7C.2). Mendelian syndromes were diagnosed in 4 infants, including 2 cases with Holt-Oram. Among the three cases with non-Mendelian associations one infant was diagnosed with fetal alcohol syndrome. Non-syndromic organ defects were present in 13 patients: organ malformations affecting the central nervous system, gastrointestinal tract, urogenital tract were diagnosed in 10 of those infants.

Descriptive Analyses

Prevalence by Time, Season, and Area of Residence

The 429 enrolled cases of muscular ventricular septal defect represent a regional prevalence of 4.7 cases per 10,000 livebirths (1.3 for moderate defects and 3.5 for small defects). Prevalence by year of birth (Figure 7C.2) shows a fairly

Table 7C.2
Muscular Ventricular Septal Defect:
Noncardiac Anomalies
Baltimore-Washington Infant Study (1981–1989)

Type of Noncardiac Anomaly	Total Cases N=429	Moderate Defects N=114	Small Defects N=328
Chromosomal Abnormalities:	N=11 (2.6%)	N=4 (3.5%)	N=7 (2.2%)
Down syndrome	8*	1 [#2194]	7 [#2776, 3277, 10127, 10625, 10777, 10828, 10830]
Trisomy 18	2	2 [#11482, 3717]	0
Deletion 13q	1	1 [#11422]	0
Syndromes	N=7 (1.6%)	N=5 (4.4%)	N=2 (0.6%)
a. Mendelian Syndromes:	4	2	2
Holt-Oram	2	0	2 [#10460, 11244]
Pena-Shokeir	1	1 [#1413]	0
Thomas	1	1 [#3817]	0
b. Non-Mendelian Associations:	3	3	0
Prader-Willi	1	1 [#10621]	0
Goldenhar	1	1 [#1323]	0
Fetal alcohol syndrome	1	1 [#1066]	0
Organ and Other Defects:	N=13 (3.0%)	N=3 (2.6%)	N=10 (3.2%)
thalassemia	1	0	1 [#10243]
growth hormone deficiency	1	0	1 [#11411]
encephalocele	1	1 [#3085]	0
hydrocephalus	1	0	1 [#11476]
tracheoesophageal fistula	1	0	1 [#2478]
gastroschisis	1	0	1 [#10324]
pyloric stenosis	1	0	1 [#11104]
jejunal atresia	1	1 [#11475]	0
agenesis of kidney	1	0	1 [#11603]
hydronephrosis	1	0	1 [#3808]
sex ambiguity	1	0	1 [#10730]
polydactyly	2	1 [#11010]	1 [#3733]
No associated anomalies	N=398 (92.8%)	N=102 (89.5%)	N=296 (94.0%)

*Cases with Down syndrome are listed in Appendix 6.

steady occurrence rate of moderate size defects, in contrast to the large increase in the occurrence rate of small defects from 0.11 per 10,000 in the first 3 years to 7.70 per 10,000 in the last 3 years; and the increase over the 8 years was about 90-fold. Moderate defects showed little variation in prevalence by birth quarter, but small defects occurred more frequently during July through September compared to other birth quarters (Table 7C.3). Moderate-sized defects were more likely to occur in urban and suburban areas than in rural areas of the region; in contrast the prevalence of small defects was greatest in the rural areas.

PREVALENCE OF MUSCULAR VENTRICULAR SEPTAL DEFECTS BALTIMORE-WASHINGTON INFANT STUDY (1981-1989)

Figure 7C.2: Over the 9 years of the study, the annual prevalence of moderate-sized muscular ventricular septal defects remained fairly constant among the regional livebirths. In contrast, there was an enormous increase in the prevalence of small defects.

Table 7C.3
Muscular Ventricular Septal Defect (VSD): Prevalence
Baltimore-Washington Infant Study (1981–1989)

	Muscular VSD				Prevalence (per 10,000)		
	All Cases	Moderate Defects	Small Defects	Area Births	**All Cases**	Moderate Defects	Small Defects
Total Subjects	**429**	114	315	906,626	**4.73**	1.26	3.47
Year of Birth:							
1981–1983	**23**	20	3	274,558	**0.84**	0.73	0.11
1984–1986	**105**	52	53	295,845	**3.54**	1.76	1.79
1987–1989	**301**	42	259	336,223	**8.95**	1.25	7.70
Birth Quarter:							
1st (Jan–Mar)	**78**	25	53	219,145	**3.56**	1.14	2.42
2nd (Apr–Jun)	**111**	30	81	231,777	**4.79**	1.29	3.49
3rd (Jul–Sep)	**129**	29	100	233,626	**5.52**	1.24	4.28
4th (Oct–Dec)	**111**	30	81	222,078	**5.00**	1.35	3.65
Area of Residence*:							
urban	**40**	24	16	208,568	**1.92**	1.15	0.76
suburban	**102**	57	45	584,022	**1.74**	0.98	0.77
rural	**21**	6	15	112,318	**1.87**	0.53	1.33

*Interviewed families.

Diagnosis and Course

Almost two thirds of the infants with muscular ventricular septal defects were diagnosed in the first 4 weeks of life and over 90% presented by the 24th week (Table 7C.4). There were no notable differences in the age at diagnosis of moderate defects compared to small defects. Cardiac catheterization to establish the diagnosis, performed in only one fifth of the infants with moderate-sized defects, was rare among infants with small defects (Figure 7C.3), among whom echocardiography was the sole diagnostic method in 98%. Two of the cases were discovered only at autopsy.

Surgery for the ventricular septal defect was performed during the first year of life in eight patients (7%) with moderate size defects and no patients with a small defects. In four additional cases with moderate size defects and three with small defects, ligation of an associated patent arterial duct was performed during the first year. Almost all infants with muscular ventricular septal defect survived the first year of life, but three cases with moderate defects died: 1 due to cardiomyopathy (congestive heart failure), 1 because of an encephalocele, and 1 with multiple anomalies characteristic of chromosome 13q deletion. The single case with a small defect who died had a massive Group B strep infection at 3 months of age. Nineteen percent of the infants with muscular ventricular septal defects were

Table 7C.4
Muscular Ventricular Septal Defect (VSD): Clinical Data
Baltimore-Washington Infant Study (1981–1989)

	All Muscular VSD N=429		Moderate Defects N=114		Small Defects N=315	
	no.	(%)	no.	(%)	no.	(%)
Age at Diagnosis:						
<1 week	69	(16.1)	18	(15.8)	51	(16.2)
1–4 weeks	200	(46.6)	50	(43.9)	150	(47.6)
5–24 weeks	134	(31.2)	40	(35.1)	94	(29.8)
25–52 weeks	26	(6.1)	6	(5.2)	20	(6.4)
Method of Diagnosis:						
echocardiography only	405	(94.4)	98	(86.0)	307	(97.5)
echocardiography and cardiac catheterization	12	(2.8)	7	(6.1)	5	(1.6)
echocardiography and surgery	1	(0.2)	1	(0.9)	0	
echocardiography, cardiac catheterization and surgery	7	(1.6)	6	(5.3)	0	
autopsy following other methods	2	(0.5)	1	(0.9)	1	(0.3)
autopsy only	2	(0.5)	1	(0.9)	1	(0.3)
Surgery for VSD During First Year of Life	8	(1.9)	8	(7.0)	0	
Follow-Up:						
not followed	81	(18.9)	19	(16.7)	62	(19.7)
followed:	348		95		253	
died in the first year of life	4	(1.1)	3	(3.2)	1	(0.4)
alive	344	(98.9)	92	(96.8)	252	(99.6)
median age at last visit	8 months		8 months		7 months	

Diagnosis of Muscular Ventricular Septal Defects
Baltimore-Washington Infant Study (1981-1989)

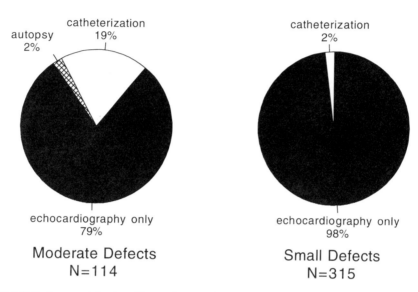

Figure 7C.3: Among infants with moderate-sized muscular ventricular septal defects (left-hand pie diagram) the predominant mode of diagnosis was echocardiography alone, but nearly one fifth were diagnosed by cardiac catheterization, and the remaining 2% by autopsy. Among infants with small defects, echocardiography was the sole method of diagnosis in 98% of the infants.

not followed after registration in the study. Among those who were followed, the median age of the infant at the last visit was 8 months (7 months among those with small defects).

Gender, Race, and Twinning

There was an excess in female infants with muscular ventricular septal defects relative to controls. The excess was more pronounced among those with moderate-sized defects (Table 7C.5). Racial distributions in infants with muscular ventricular septal defects were similar to controls. Twin births occurred in nearly equal proportions of cases and controls.

Birthweight and Gestational Age

Overall, cases with muscular ventricular septal defect were more likely than controls to have low birthweight and be premature (Table 7C.6). When subdivided by size of the defect, infants with moderate-sized defects were more than twice as likely than controls to be of low birthweight and they were also more likely to be premature. Birthweight and gestational age in infants with small defects were similar to that in controls. Infants with moderate defects, but not those with small defects, were significantly growth retarded relative to controls.

Table 7C.5

Muscular Ventricular Septal Defect (VSD): Infant Characteristics
Baltimore-Washington Infant Study (1981–1989)

Characteristic	All Cases N=429 no. (%)		Moderate Defects N=114 no. (%)		Small Defects N=315 no. (%)		Controls N=3572 no. (%)	
Gender:								
male	183	(42.7)	45	(39.5)	138	(43.8)	1,817	(50.9)
female	246	(57.3)	69	(60.5)	177	(56.2)	1,755	(49.1)
Race*:								
white	283	(70.3)	73	(67.0)	210	(71.7)	2,362	(66.1)
black	99	(24.6)	33	(30.3)	66	(22.5)	1,109	(31.0)
other	20	(5.0)	3	(2.7)	17	(5.8)	98	(2.7)
Twin births	5	(1.2)	2	(1.8)	3	(1.0)	53	(1.5)

*Missing data in 5 moderate and 22 small defects.

Table 7C.6

Muscular Ventricular Septal Defect (VSD): Fetal Growth Characteristics
Baltimore-Washington Infant Study (1981–1989)

Characteristic	All Cases no. (%)		Moderate Defects no. (%)		Small Defects no. (%)		Controls no. (%)	
Number of Families Interviewed	163		87		76		3572	
Birthweight (grams)								
<2500	22	(13.5)	17	(19.5)	5	(6.6)	252	(7.1)
2500–3500	71	(43.6)	39	(44.8)	32	(42.1)	1853	(51.9)
>3500	70	(42.9)	31	(35.7)	39	(51.3)	1467	(41.0)
mean±standard error	3328±52		3234±71		3435±73		3351±10	
range	960–4649		1408–4451		960–4649		340–5273	
Gestational Age (weeks)								
<38	20	(12.3)	12	(16.7)	8	(10.5)	339	(9.5)
38+	143	(87.7)	75	(83.3)	68	(89.5)	3233	(90.5)
mean±standard error	39.4±0.2		39.3±0.2		39.5±0.2		39.6	
range	26–43		29–43		26–43		20–47	
Size for Gestational Age								
small (SGA)	17	(10.4)	12	(13.8)	5	(6.6)	211	(5.9)
normal	112	(68.7)	63	(72.4)	49	(64.5)	2712	(75.9)
large	34	(20.9)	12	(13.8)	22	(28.9)	649	(18.2)
Odds Ratio for SGA (95% CI)	1.9 (1.1–3.1)		2.5 (1.4–4.8)		1.1 (0.4–2.8)		1.0 (reference)	

Sociodemographic Characteristics

An analysis of sociodemographic characteristics among all cases with muscular ventricular septal defects indicates that case families were more likely to have household incomes of at least $30,000 than were controls, and fathers of cases were more likely than controls to hold professional occupations (Table 7C.7). In addition, significantly fewer case mothers than control mothers had late confirmation of their pregnancies.

Table 7C.7
Muscular Ventricular Septal Defect (all cases):
Analysis of Sociodemographic Characteristics
Baltimore-Washington Infant Study (1981–1989)

Interviewed Families	Cases N=163		Controls N=3572		Odds Ratio (95% CI)	
	no.	(%)	no.	(%)		
Maternal Marital Status:						
not married	37	(22.7)	990	(27.7)	0.8	(0.5–1.1)
married	126	(77.3)	2582	(72.3)	1.0	(reference)
Maternal Education:						
<high school	27	(16.6)	659	(18.4)	0.9	(0.6–1.4)
high school	62	(38.0)	1265	(35.4)	1.1	(0.8–1.5)
college	74	(45.4)	1648	(46.1)	1.0	(reference)
Paternal Education:						
<high school	28	(17.2)	650	(18.2)	0.8	(0.5–1.2)
high school	48	(29.4)	1298	(36.3)	0.7	(0.5–1.0)
college	87	(53.4)	1624	(45.5)	1.0	(reference)
Annual Household Income:						
<$10,000	22	(13.8)	686	(19.2)	**0.5**	**(0.3–0.9)**
$10,000–$19,999	33	(20.8)	699	(19.6)	0.8	(0.5–1.2)
$20,000–$29,999	23	(14.5)	737	(20.6)	**0.5**	**(0.3–0.8)**
$30,000+	81	(50.9)	1373	(38.4)	1.0	(reference)
Maternal Occupation:						
not working	43	(26.4)	1142	(32.0)	0.7	(0.5–1.1)
clerical/sales	54	(33.1)	1119	(31.3)	0.9	(0.6–1.4)
service	22	(13.5)	444	(12.4)	0.9	(0.5–1.6)
factory	5	(3.1)	137	(3.8)	0.7	(0.3–1.8)
professional	39	(23.9)	730	(20.4)	1.0	(reference)
Paternal Occupation:						
not working	11	(6.7)	246	(6.9)	0.7	(0.4–1.4)
clerical/sales	22	(13.5)	618	(17.3)	**0.6**	**(0.3–0.9)**
service	19	(11.7)	480	(13.4)	0.6	(0.4–1.1)
factory	51	(31.3)	1279	(35.8)	**0.6**	**(0.4–0.9)**
professional	60	(36.8)	947	(26.5)	1.0	(reference)
Month of Pregnancy Confirmation:						
1st month	30	(18.5)	511	(14.3)	1.0	(reference)
2nd month	90	(55.6)	1955	(54.7)	0.8	(0.5–1.2)
3rd month	29	(17.9)	663	(18.6)	0.7	(0.4–1.3)
4th month of later	13	(8.0)	416	(11.6)	**0.5**	**(0.8–1.0)**

Data are missing for income (4 cases, 77 controls), paternal occupation (2 controls), and pregnancy confirmation month (1 case, 27 controls).

Potential Risk Factors

Familial Cardiac and Noncardiac Anomalies

Three cases had family members with congenital heart disease and in each of three additional cases, the mother had mitral valve prolapse (Table 7C.8). There was concordance of the defect in two families, but the type of ventricular septal defect was not known for either the affected father or sister. Of interest is the occurrence of a male sibling with common arterial trunk in the family of a female proband with a small muscular ventricular septal defect.

No structural noncardiac anomalies were reported in these families. Only the mother of a female proband was reported to have had a "cyst on chest" removed in infancy; none of the available hospital records confirmed this.

Genetic and Environmental Factors

The case groupings created for the case-control evaluation of potential risk factors included the total group of muscular ventricular septal defects (N = 163) and separate analyses of the moderate (N = 87) and small (N = 76) subsets. The multivariate analysis of small defects was restricted to cases and controls born after 1983, since only 3 cases were diagnosed in the 1981 to 1983 enrollment period.

Table 7C.8
Muscular Ventricular Septal Defect:
Congenital Defects and Hereditary Disorders in First-Degree Relatives
The Baltimore-Washington Infant Study (1981–1989)

	Proband			Family	
ID	Size of Defect	Sex	Noncardiac Anomalies	Relative	Diagnosis
A. Congenital Heart Defects:					
10243	small	M	Thalassemia	Sister	Ventricular septal defect
				Mother	Thalassemia
2158	moderate	M		Father	Ventricular septal defect
2853	small	F		Brother	Common arterial trunk
2760	moderate	F		Mother	Mitral valve prolapse
10374	moderate	M		Mother	Mitral valve prolapse, kyphoscoliosis
10570	small	F		Mother	Mitral valve prolapse
B. Congenital Defects of Other Organs:					
10770	small	F		Mother	Cyst on chest (infancy)
2776	small	F	Down syndrome	Brother	Cysteinuria
				2 Sisters	

Univariate Analysis

Family history of congenital heart disease and noncardiac anomalies was not significantly different between cases and controls, regardless of defect size (Table 7C.9). There were no case-control differences in either parental age distributions or maternal reproductive history variables. Three maternal medication variables were reported in excess by case mothers: gastrointestinal medications taken for the relief of acute symptoms, antitussives, and acetaminophen. Only gastrointestinal medications were significant in the small defect group. Maternal alcohol consumption and cocaine use was weakly associated with small muscular ventricular defects. Among home and occupational exposures, only hair dyes showed a significant case excess only in the group with small defects. There were no case-control differences in any paternal exposures.

Multivariate Analysis

Separate models for all muscular ventricular septal defects, moderate-sized defects, and small-sized defects are shown in Table 7C.10. Three factors were associated with the total case group: gastrointestinal medications, acetaminophen, and hair dyes. The latter two associations failed to reach significance at the 99% confidence level.

Analyses by subgroup revealed that gastrointestinal medications were associated with moderate-sized defects. The association with antitussive medications, with an adjusted odds ratio of 4.7, was not significant at the 1% level. The risk associated with maternal use of acetaminophen depended on the time period: relative to nonusers, mothers who took the drug and gave birth between 1984 and 1987 were at significantly increased risk, but there was no significant increase in risk to drug-using mothers who delivered earlier or later in the decade.

The analysis of small defects also revealed an effect modification of exposure risks by time period. In this group limited to infants born after 1983, there was a significant risk of alcohol consumption by mothers whose infants were born from 1984 to 1986 (odds ratio = 1.8), but this risk increased even further (odds ratio = 2.6) in the period from 1987 to 1989. This was the only significant case-control difference to emerge from the multivariate analysis of small defects. There was no evidence of confounding by race, socioeconomic status, or coexposures.

Discussion

In contrast to previous epidemiologic studies[329,415] that grouped together all types of ventricular septal defects, we analyzed cases by anatomic type and by the size of the defect. In the present chapter we focused on *muscular* ventricular septal defect.

Cases with muscular ventricular septal defects differed from those with membranous defects and from all other types of congenital heart disease in several ways. Among the unique features of cases with muscular ventricular septal defects are the dramatic increase in prevalence by echocardiographic diagnosis, the rarity of associated congenital anomalies in the proband and of a family history of malformations, and the indications of a somewhat favorable socioeconomic status.

Table 7C.9

Muscular Ventricular Septal Defect (VSDmu):
Univariate Analysis of Potential Risk Factors by Size of Defect
Baltimore-Washington Infant Study (1981–1989)

Variable	Controls N=3572		All VSDmu N=163				Moderate Defects N=87				Small Defects N=76			
	no.	(%)	no.	(%)	Odds Ratio	(95% CI)	no.	(%)	Odds Ratio	(95% CI)	no.	(%)	Odds Ratio	(95% CI)
Family History (first degree relatives)														
congenital heart disease	43	(1.2)	3	(1.8)	1.5	0.5–5.0	1	(1.1)	1.0	0.1–6.9	2	(2.6)	2.2	0.5–9.3
noncardiac anomalies	165	(4.6)	2	(1.2)	0.3	0.1–1.0	0	—	—	—	2	(2.6)	0.6	0.1–2.3
Parental Age														
maternal age:														
<20	507	(14.2)	23	(14.1)	1.1	0.7–1.8	13	(14.9)	1.2	0.6–2.2	10	(13.2)	1.0	0.5–2.0
20–29	2009	(56.2)	83	(50.9)	1.0	(reference)	44	(50.6)	1.0	(reference)	39	(51.3)	1.0	(reference)
30+	1056	(29.6)	57	(35.0)	1.3	0.9–1.8	30	(34.5)	1.3	0.8–2.1	27	(35.5)	1.3	0.8–2.2
paternal age:														
<20	226	(6.3)	9	(5.5)	1.0	0.5–2.0	6	(6.9)	1.3	0.5–3.1	3	(3.9)	0.7	0.2–2.2
20–29	1724	(48.3)	70	(42.9)	1.0	(reference)	36	(41.4)	1.0	(reference)	34	(44.7)	1.0	(reference)
30+	1622	(45.4)	84	(51.5)	1.3	0.9–1.8	45	(51.7)	1.3	0.9–2.1	39	(51.3)	1.2	0.8–1.9
Maternal Reproductive History														
number of previous pregnancie(s):														
none	1159	(32.4)	58	(35.6)	1.0	(reference)	30	(34.5)	1.0	(reference)	28	(36.8)	1.0	(reference)
one	1097	(30.7)	46	(28.2)	0.8	0.6–1.2	23	(26.4)	0.8	0.5–1.4	23	(30.3)	0.9	0.5–1.5

	N	(%)	OR	CI	N	(%)	OR	CI	N	(%)	OR	CI
two	709	(19.9)	1.0	0.7–1.6	22	(25.3)	1.2	0.7–2.1	14	(18.4)	0.8	0.4–1.6
three or more	607	(17.0)	0.8	0.5–1.2	12	(13.8)	0.8	0.4–1.5	11	(14.5)	0.8	0.4–1.5
previous miscarriage(s)	681	(19.1)	0.8	0.6–1.3	18	(20.7)	1.1	0.7–1.9	9	(11.8)	0.6	0.3–1.1

Illnesses and Medications

diabetes:

	N	(%)	OR	CI	N	(%)	OR	CI	N	(%)	OR	CI
overt	23	(0.6)	1.9	0.4–8.2	1	(1.1)	1.8	0.2–13.4	1	(1.3)	2.1	0.3–15.4
gestational	115	(3.2)	1.3	0.6–2.9	4	(4.6)	1.4	0.5–4.0	3	(4.0)	1.2	0.4–4.0
influenza	278	(7.8)	1.3	0.8–2.2	8	(9.2)	1.2	0.6–2.5	8	(10.5)	1.4	0.7–2.9
gastrointestinal medications	115	(3.2)	**2.8**	**1.6–5.0**	9	(10.3)	**3.0**	**1.4–6.4**	6	(7.9)	**2.6**	**1.1–6.1**
antitussives	23	(0.6)	2.9	0.9–9.7	3	(3.5)	**5.5**	**1.6–18.7**	0	—	—	—
acetaminophen	1485	(41.6)	**1.5**	**1.1–2.0**	48	(55.2)	**1.7**	**1.1–2.7**	36	(47.4)	1.3	0.8–2.0

Lifestyle Exposures

	N	(%)	OR	CI	N	(%)	OR	CI	N	(%)	OR	CI
alcohol, any amount	2101	(58.8)	**1.4**	**1.0–1.9**	54	(62.1)	1.1	0.7–1.8	57	(71.0)	**1.7**	**1.0–2.8**
cocaine	42	(1.2)	1.6	0.5–5.1	0	—	—	—	3	(4.0)	**3.5**	**1.0–11.4**

smoking (cigarettes/day):

	N	(%)	OR	CI	N	(%)	OR	CI	N	(%)	OR	CI
none	2302	(64.5)	1.0	(reference)	59	(67.8)	1.0	(reference)	58	(76.3)	1.0	(reference)
1–10	565	(15.8)	0.8	0.5–1.2	15	(17.2)	1.0	0.6–1.8	7	(9.2)	0.5	0.2–1.1
11–20	529	(14.8)	0.7	0.4–1.2	11	(12.6)	0.8	0.4–1.6	8	(10.5)	0.6	0.3–1.2
>20	176	(4.9)	0.6	0.2–1.4	2	(2.3)	0.4	0.1–1.8	3	(4.0)	0.7	0.2–2.2

Home and Occupational Exposures

	N	(%)	OR	CI	N	(%)	OR	CI	N	(%)	OR	CI
ionizing radiation (occupational)	48	(1.3)	2.3	0.9–5.9	2	(2.3)	1.7	0.4–7.2	3	(4.0)	3.0	0.9–9.9
hair dyes	238	(6.7)	**1.7**	**1.0–2.9**	8	(9.1)	1.4	0.8–2.1	10	(13.2)	**2.1**	**1.1–4.2**

Race of Infant

	N	(%)	OR	CI	N	(%)	OR	CI	N	(%)	OR	CI
white	2362	(66.2)	1.5	1.0–2.1	61	(70.1)	1.2	0.8–1.9	60	(79.0)	**1.9**	**1.1–3.3**
nonwhite	1207	(33.8)	1.0	(reference)	26	(29.9)	1.0	(reference)	16	(21.0)	1.0	(reference)

Risk factors are *maternal* unless labelled as paternal. VSDmu=muscular ventricular septal defect.

Table 7C.10
Muscular Ventricular Septal Defect (VSDmu):
Multivariant Models
Baltimore-Washington Infant Study (1981–1989)

	Cases	Controls	Adjusted Odds Ratio	95% CI	99%CI
All VSDmu (N=163)					
gastrointestinal medications	14	115	2.7	1.5–4.9	1.3–5.8
acetaminophen	84	1485	1.4	1.0–2.0	0.9–2.2
hair dyes	18	238	1.7	1.0–2.8	0.8–3.3
Moderate-Sized VSDmu (N=87)					
gastrointestinal medications	9	115	2.8	1.3–5.9	1.0–7.7
antitussives	3	23	4.7	1.3–16.7	0.7–19.6
Acetaminophen Use by Time Period:					
no acetaminophen use	40	2099	1.0	(reference)	
acetaminophen users:					
1981–1983 births	8	519	0.7	0.3–1.5	0.2–2.1
1984–1987 births	34	808	2.0	1.2–3.3	1.1–4.4
1988–1989 births	5	146	1.5	0.6–3.9	0.5–4.0
Small VSDmu (N=73)*					
Alcohol Use by Time Period:					
no alcohol consumption	19	968	1.0	(reference)	
alcohol use (any):					
1984–1986 births	31	859	1.8	1.0–3.3	0.9–3.9
1987–1989 births	23	459	2.6	1.4–4.8	1.1–5.8

*Analysis restricted to cases and controls born after 1983.

Possible Genetic Risk Factors

Associated congenital anomalies were relatively rare in probands with muscular defects (7%) compared to probands with membranous defects (20%), and no specific syndrome or organ defect predominated. Down syndrome occurred in 8 cases, 7 of whom had a small defect (2.2%). The array of reported conditions including single cases of severe anomalies such as central nervous system defects, tracheoesophageal fistula, and pyloric stenosis, suggests the possibility of selection bias in that diagnosis of the heart defect was more likely to occur among sick infants being treated for other conditions. However the vast majority of cases (93%) had no congenital anomalies.

Although these rates of familial anomalies are lower than those reported by families of infants with membranous ventricular septal defects, the reported familial cases are of interest in that there is concordance of the cardiac defect in two instances, and partial concordance with a conotruncal anomaly in one. In addition, two heritable disorders, thalassemia and cystinuria, affected multiple members in the respective families.

Changes in Prevalence and Risk Factors

The huge increase in prevalence of cases diagnosed solely by echocardiography in contrast to the stable prevalence of cases diagnosed by catheterization represents another possible type of selection bias. As previous investigations[283,453] have noted, this increase is attributable to the increasing use of high-resolution two-dimensional echocardiography that occurred in pediatric cardiology during the decade of the 1980s, and which has now become standard practice.

A recent report by Roguin et al,[365] which employed uniform follow-up, reports that 89% of 45 infants diagnosed solely by echocardiography experienced spontaneous closure of the defect by 10 months of age. The epidemiologic implication of these observations is that the large number of cases with clinically trivial heart defects that spontaneously close may "dilute" the characteristics of cases with true defects, among whom there may be potential etiologic clues such as family history and parental exposures to toxicants. Indeed, some authors[301,365] interpret the closure data as being consistent with a view of muscular ventricular septal defect as "delayed physiologic development" rather than as a "disease."

In this chapter we attempted to address the issue of potential etiologic differences by separating the cases based on the size of the defect. Several observations in our data suggest that this subdivision is valid. In terms of fetal growth measurements, infants with small muscular ventricular septal defects were similar to controls on all measures. In addition, the statistical associations with maternal medications and lifestyle exposures were nonoverlapping in the two subgroups. All associations in both groups involved over-the-counter medications and legally available substances (ie, alcohol). Of interest is the interaction with the time period in which they were used; for two of these variables (acetaminophen and alcohol), this possibly reflects changes in available pharmaceutical formulations as well as changing societal concerns and attitudes towards the use of alcohol by pregnant women.

Reported Epidemiologic Studies

Several previous studies that examined potential risk factors for ventricular septal defects as a group failed to separate cases by anatomic phenotype, and to our knowledge only one published case-control study[365] has focused specifically on muscular ventricular septal defects.

Roguin[365] (1995) achieved an analysis of risk factors specifically for muscular defects in a study that performed color Doppler echocardiography on 1053 consecutive neonates at the Western Galilee Hospital in Israel. In the analysis of 56 cases and 975 controls, the prevalence of the defect was, as it was in our study, somewhat higher among females than among males, but no case-control differences were found in birthweight or gestational age. Similar to our results, there was no association with either family history, parental ages, or maternal chronic diseases, but unlike our study there were no data collected on maternal use of alcohol or over-the-counter medications.

Our findings clearly demonstrate the importance of differentiating ventricular septal defects by anatomic types and associated noncardiac defects.

Chapter 8

Left-Sided Obstructive Lesions

Section A:
Overview

Section B:
Hypoplastic Left Heart Syndrome

Section C:
Coarctation of Aorta

Section D:
Aortic Stenosis

Section E:
Bicuspid Aortic Valve

Section F:
Summary and Conclusions

From: Ferencz C, Loffredo CA, Correa-Villaseñor A, Wilson PD, eds: *Genetic & Environmental Risk Factors of Major Cardiovascular Malformations: The Baltimore-Washington Infant Study 1981–1989.* Armonk, NY: Futura Publishing Co., Inc; ©1997.

Section A

Overview

Obstructive defects of the left heart and aorta represent a major group of cardiovascular malformations. In the Baltimore-Washington Infant Study the use of a hierarchial order of primary diagnoses allocated 632 cases (14% of all cardiovascular malformations) to the left-sided obstructive defects group, which excludes those cases in which the obstructive defects occurred in association with major structural anomalies of the heart considered in previous chapters. This malformation group is morphogenetically heterogenous, as some anomalies of the left atrium and mitral valve may represent variants of primary abnormalities of the atrioventricular region, while abnormalities of the aortic arch system (such as double aortic arch and interrupted arch) are defects of the branchial arch system. The remaining "core" of left-sided obstructive anomalies consists of the anatomic phenotypes of hypoplastic left heart syndrome, coarctation of the aorta, aortic stenosis, and bicuspid aortic valve. These latter defects are thought to represent abnormal *secondary* morphogenesis of the heart, which occurs during the long period of development and remolding as the fetus grows several hundred-fold into the newborn infant.[84] The factors that influence this secondary developmental phase are controlled by somatic genes and include various hemodynamic control mechanisms, the completion of autonomic innervation, and the expansion of the vascular beds including that of the coronary arteries.[83,85]

Left-sided obstructive lesions have been of special interest to the Baltimore-Washington Infant Study research group, as the co-occurrence within families of different anatomic phenotypes indicated a prominent role of genetic factors.[52-56,271,272] A common genetic origin is also suggested by the association of a chromosome anomaly, Turner syndrome (monosomy X), with each of the four anatomic phenotypes. These potentially common genetic and morphogenetic origins impart the impression of homogeneity of these cases, described as "flow lesions" in some of our previous publications, but they are in fact morphogenetically heterogenous, as noted below.

Historical and Diagnostic Considerations

Coarctation of the aorta is described in historic case reports and pathologic studies in children and adults. Those patients were clinically recognized either by the dramatic adaptive manifestations such as hypertension in the arms and the tortuous collateral circulation with visible pulsations of the intercostal arteries, or by complications such as bacterial endocarditis, which damaged cusps of the associated bicuspid aortic valve. Infants were only rarely diagnosed with coarctation until the critical manifestations of congestive heart failure brought them to the attention of pediatric cardiologists. Now, even mild coarctation of the aorta is recognized in infants by ultrasonic evaluation.

166

Aortic stenosis was also a common diagnosis among adult patients who experienced shortness of breath, cardiac failure, or bacterial endocarditis. However, the history of this abnormality is quite different, as it was considered to be a consequence of rheumatic inflammation of the aortic valve followed by fibrosis of the valve. The differential diagnosis of acquired versus congenital stenosis of the aortic valve was only recognized in the past 25 to 30 years[362] when it was noted that some of the valves were abnormal in structure (usually bicuspid) prior to the development of fibrosis and calcification. At the same time it was recognized that the typical finding of a systolic murmur over the aortic area developed gradually over the childhood years. The recognition of critical aortic stenosis in infancy is relatively recent; the clinical manifestations are very similar to those of the much more severe combined anomalies known as the hypoplastic left heart syndrome, a complex of serial obstructive defects with a hypoplastic (poorly developed) left ventricle, and a ductus-dependent systemic circulation.

The morphogenetic and pathologic features of the diagnostic phenotypes of left-sided obstructive lesions have been examined by detailed anatomic and histological studies in the laboratory of Dr. Adriana Gittenberger-de Groot in Leiden, Holland. These uniquely thorough doctoral thesis studies were reported by Elzenga[127-129] and Kappetein[216] and have provided some hypotheses regarding the developmental heterogeneity of left-sided obstructive defects.

Analysis Plan

In the evaluation of the four presumably related forms of left-sided obstructive lesions, we first combined all patients for the analysis of potential risk factors and then we conducted separate univariate and multivariate analyses for each anatomic phenotype in the search of distinct epidemiologic features for each. As mentioned in Chapter 3, separate multivariate analyses were conducted for the isolated/simplex and multiple anomalies/multiplex subsets.

All cases reported as hypoplastic left heart syndrome were studied. For coarctation of the aorta, the analyzed subgroup is limited to "pure" coarctation, with which only bicuspid aortic valve and/or patent arterial duct may coexist. Aortic stenosis as a "pure" group includes only valvar stenosis. Bicuspid aortic valve without obstruction—now diagnosed in infants with two-dimensional echocardiography—is included as a diagnostic entity, although a progression to obstruction (ie, aortic stenosis) is known to occur. Thus no ultimate delineation of bicuspid aortic valve from aortic stenosis can be made, as each undergoes temporal and quantitative changes with age.

Risk factor analyses for the combined group of left-sided defects are presented in Section A of this chapter. Risk factor analyses for each malformation subgroup are presented in Section B through Section E. In each section we describe a brief historical perspective, the clinical characteristics of the cases, and the results of the case-control analyses of potential risk factors. A final comparative overview of the findings is presented in Section F.

Risk Factor Analysis of the Total Group

Diagnostic Definition

This section evaluated the case group of left-sided obstructive lesions composed of those with a diagnosis of hypoplastic left heart syndrome, coarctation of the aorta, valvar aortic stenosis, and bicuspid aortic valve, after excluding rare diagnoses, multiple diagnoses, and anatomic variants (Figure 8A.1). This core group of "pure" diagnoses is composed of a total of 429 cases; 383 (89%) families were interviewed.

Genetic Relationships Among Left-Sided Heart Defects

Two lines of evidence suggested a genetic relationship among the four types of left-sided defects: familial recurrence and the diagnostic associations with the Turner syndrome (monosomy X) phenotype. The familial aggregations of left-sided obstructive lesions of different types are illustrated in the pedigrees of 13 families of case probands with hypoplastic left heart syndrome, coarctation of the aorta, aortic stenosis, and bicuspid aortic valve, respectively (Figure 8A.2).

The diagnosis of the relative was concordant with that of the proband in five

Left-Sided Obstructive Heart Defects, N=632

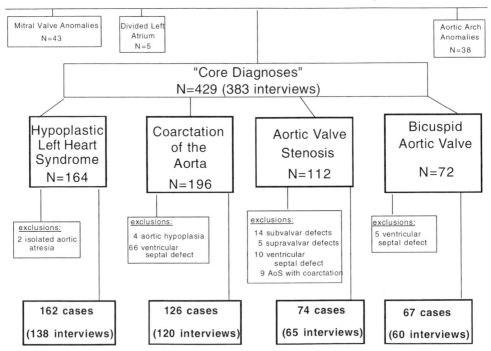

Figure 8A.1: Left-sided obstructive heart defects represent a wide array of developmental anomalies. In this chapter we focus on the core diagnoses of hypoplastic left heart syndrome, coarctation of the aorta, aortic valve stenosis, and bicuspid aortic valve. Anatomic variants and cases with associated heart defects were excluded from analysis.

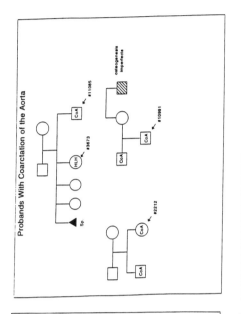

Probands With Coarctation of the Aorta

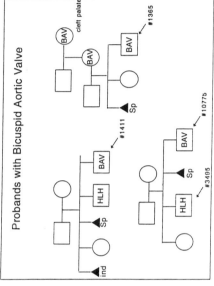

Probands with Bicuspid Aortic Valve

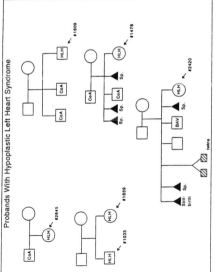

Probands With Hypoplastic Left Heart Syndrome

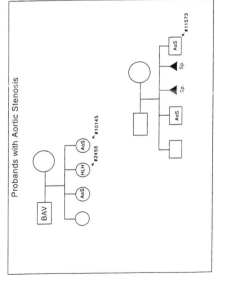

Probands with Aortic Stenosis

Figure 8A.2: Familial Left-Sided Heart Defects: Baltimore-Washington Infant Study (1981–1989). These pedigrees illustrate the occurrence of left-sided obstructive heart defects among the first-degree relatives of affected case infants; identification numbers and arrows indicate the probands in our study. In some families there were two affected siblings who were independently ascertained during the 9 years of the study.

of these families and the anatomic phenotypes were discordant in eight families. Except for hypoplastic left heart syndrome (a lethal anomaly) and for aortic stenosis, concordant and discordant diagnoses appeared in parent-child pairs as well as sibling pairs. Three families were identified by reascertainment of the family through subsequently affected infants (cases #1033/1889, #2458/10145, and #3495/10775).

The frequency of occurrence of congenital heart disease in first-degree relatives is shown in Table 8A.1, separately for both parents and for full siblings in the total group and in the diagnostic subgroups, the details of which are described in the appropriate diagnostic sections.

Congenital heart defects occurred in 4 mothers, only 1 of whom had a left-sided obstructive lesion (her infant had bicuspid aortic valve), and in 4 fathers, 3

Table 8A.1

Left-Sided Obstructive Defects:

Familial Aggregation of Congenital Heart Defects

Baltimore-Washington Infant Study (1981–1989)

Diagnosis of the Proband	Total Interviewed Cases	Relatives with Congenital Heart Defects		
		Mothers n (%)	Fathers n (%)	Full Sibs n (%)
Total group of Left-Sided Obstructive Defects	383	4 (1.0) 1=left sided defects 3=other defects	4 (1.0) 3=left-sided defects 1=other defects	18/266 (6.8) 12=left-sided defects 6=other defects
Hypoplastic Left Heart Syndrome (HLH)	138	0	1 (0.7) 1=CoA	7/101 (6.9) 1=HLH 1=VSD 3=CoA 1=BAV 1-Tetralogy
Coarctation of the Aorta (CoA)	120	2 (1.7) 1=VSD 1=PDA	2 (1.7) 1=CoA 1=SI totalis	4/80 (5.0) 1=CoA 1=AVSD+Down 1=HLH 1=ASD
Aortic Valve Stenosis (AoS)	65	0	1 (1.5) 1=BAV	4/46 (8.7) 2=AoS 1=VSD 1=HLH
Bicuspid Aortic Valve (BAV)	60	2 (3.3) 1=BAV 1=CM	0	3/39 (7.7) 2=HLH 1=VSD
Controls	3572	9 (0.3) 1=left-sided defects 8=other defects	19 (0.5) 2=left-sided defects 17=other defects	10/2202 (0.5) 3=left-sided defects 7=other defects

ASD=atrial septal defect; AVSD=atrioventricular septal defect; CM=cardiomyopathy; PDA=patent arterial duct; SI=situs inversus; VSD=ventricular septal defect.

of whom had left-sided obstructive lesions. A cardiovascular malformation was present in 18 of 266 (6.8%) of previously born full siblings, but was concordant only in 12 of 266 (4.5%). These proportions of affected siblings are markedly greater than in the control population.

All 18 infants with Turner syndrome (monosomy X) identified in the Baltimore-Washington Infant Study were found to have left-sided heart lesions (Table 8A.2), but not all cases were among those categorized as "pure" defects. Therefore, only 13 of the Turner syndrome cases will be considered in the sections below.

These findings support the earlier reports from the Baltimore-Washington Infant Study of an important role of genetic factors: all four anatomic phenotypes occurred in various combinations within familial aggregations, and all except bicuspid aortic valve occurred in association with Turner syndrome.

Familial aggregation introduced some methodological issues concerning the analysis. In order to maintain independence of the risk factor data we had to eliminate the three families with multiply ascertained infants in the risk factor analyses of the total group. However, we included each of those families in the respective diagnostic subgroup when the sibling's diagnosis differed from that of the proband.

Infant Characteristics

Analysis of infant gender, race, and size-for-gestational-age (Table 8A.3) revealed an overall case-control difference in these characteristics in the combined group of left-sided heart defects, with a male excess (odds ratio 1.5, 95% confidence interval 1.2–1.8) and an excess of white infants (odds ratio 1.5, 95% confidence interval 1.2–1.9). There was also a significant elevation in the proportion of infants classified as being small-for-gestational age (odds ratio 2.5, 95% confidence interval 1.8–3.5).

Table 8A.2
Left-Sided Obstructive Defects: Turner Syndrome
Baltimore-Washington Infant Study (1981–1989)

Cardiac Anomaly	Total Number of Cases	Cases with Turner	
hypoplastic left heart	164	3	(1.8%)
aortic valve stenosis	74	3	(4.0%)
*subvalvar aortic stenosis	14	1	(7.1%)
coarctation of the aorta	126	7	(5.6%)
*coarctation of the aorta with ventricular septal defect	66	4	(6.1%)
All cases with left-sided defects	632	18	(2.8%)
Total cases in BWIS	4390	18	(0.4%)

*Not included in the statistical analysis.

Table 8A.3

Left-sided Obstructive Defects and Controls:
Gender, Race, and Size for Gestational Age Distributions
Baltimore-Washington Infant Study (1981–1989)

Infant Characteristic	All Defects Combined N=423[1]		Controls N=3572	
Gender:				
male	255	(60.3)	1,817	(50.9)
female	168	(39.7)	1,755	(49.1)
Race:				
white	309	(74.6)	2,362	(66.1)
black	98	(23.7)	1,109	(31.0)
other	7	(1.7)	98	(1.5)
Size for Gestational Age[2]:				
small	52	(13.8)	211	(5.9)
normal	264	(70.0)	2712	(75.9)
large	61	(16.2)	649	(18.2)

[1]all registered cases, regardless of interview participation; excluding families with multiple probands; [2]analysis restricted to interviewed families.

Potential Risk Factors

Univariate Analysis

The univariate case-control comparisons are shown in Table 8A.4. These infants were more than four times more likely than controls to have a reported family history of congenital heart defects, with a greater likelihood of having affected siblings (odds ratio 7.8) than affected parents (odds ratio 2.4). There was no excess of familial noncardiac anomalies. Neither the parental age distributions nor the maternal reproductive history variables differed significantly between cases and controls. Overt and gestational diabetes were associated with case status (odds ratio 1.7 and 1.6 respectively), but only those with gestational diabetes reached statistical significance at the 5% level. Maternal epilepsy was associated with case status (odds ratio 3.4): only 2 of the 5 case mothers and 11 of the 14 control mothers who had epilepsy reported taking antiepileptic medications. Maternal use of macrodantin, benzodiazepines, antidepressants, and sympathomimetic medications appeared as potential risk factors.

Lifestyle exposures (cigarette smoking, alcohol consumption, and use of recreational drugs) did not differ significantly between cases and controls. Parental exposures to organic solvents at home and in the workplace were strongly associated with left-sided heart defects, particularly maternal exposures to degreasing solvents (odds ratio 3.5) and to miscellaneous solvents (eg, paint thinners). In several case families, *both* parents reported solvent exposures, with odds ratios of 3.2 and 4.2 respectively for the two solvent categories. Analysis by the summary score of solvent exposure weighted by frequency of use showed a dose-response effect with a 2.5-fold risk for those who had frequent exposures.

Table 8A.4
All Left-Sided Obstructive Defects Univariate Analysis of Risk Factors
Baltimore-Washington Infant Study (1981–1989)

Variable	Cases N=377 no. (%)		Controls N=3572 no. (%)		Odds Ratio (95% CI)	
Family History (first-degree relatives)						
congenital heart disease:	19	(5.0)	43	(1.2)	**4.4**	**(2.5–7.6)**
parents	7	(1.9)	28	(0.8)	**2.4**	**(1.0–5.5)**
siblings	12	(3.2)	15	(0.4)	**7.8**	**(3.6–16.8)**
noncardiac anomalies:	25	(6.6)	165	(4.6)	1.5	(0.9–2.3)
parents	15	(4.0)	111	(3.1)	1.3	(0.7–2.2)
siblings	10	(2.7)	54	(1.5)	1.8	(0.9–3.5)
Parental Age						
maternal age:						
<20	45	(11.9)	507	(14.2)	0.8	(0.6–1.7)
20–29	211	(56.0)	2009	(56.2)	1.0	(reference)
30+	121	(32.1)	1056	(29.6)	1.1	(0.9–1.4)
paternal age:						
<20	22	(5.8)	226	(6.3)	0.9	(0.5–1.5)
20–29	184	(48.8)	1724	(48.3)	1.0	(reference)
30+	171	(45.4)	1622	(45.4)	1.0	(0.8–1.2)
Maternal Reproductive History						
number of previous pregnancies						
none	112	(29.7)	1159	(32.4)	1.0	(reference)
one	124	(32.9)	1097	(30.7)	1.2	(0.9–1.5)
two	69	(18.3)	709	(19.9)	1.0	(0.7–1.4)
three or more	72	(19.1)	607	(17.0)	1.2	(0.9–1.7)
previous miscarriage(s)	84	(22.3)	681	(19.1)	1.2	(0.9–1.6)
Illnesses and Medications						
diabetes:						
overt	4	(1.1)	23	(0.6)	1.7	(0.6–4.8)
gestational	19	(5.0)	115	(3.2)	**1.6**	**(1.0–2.6)**
influenza	35	(9.3)	278	(7.8)	1.2	(0.8–1.8)
epilepsy	5	(1.3)	14	(0.4)	**3.4**	**(1.2–9.5)**
macrodantin	5	(1.3)	13	(0.4)	**3.7**	**(1.3–10.4)**
benzodiazepines	8	(2.1)	35	(1.0)	**2.2**	**(1.0–4.8)**
antidepressants	3	(0.8)	6	(0.2)	**4.8**	**(1.2–19.1)**
sympathomimetics	47	(12.5)	259	(7.3)	**1.5**	**(1.1–2.0)**
Lifestyle Exposures						
alcohol (any amount)	22	(60.7)	2101	(58.8)	1.1	(0.9–1.3)
smoking (cigarettes/day):						
none	24	(63.9)	2302	(64.5)	1.0	(reference)
1–10	59	(15.7)	565	(15.8)	1.0	(0.7–1.3)
11–20	58	(15.4)	529	(14.8)	1.0	(0.8–1.4)
>20	19	(5.0)	176	(4.9)	1.0	(0.6–1.7)
Home and Occupational Exposures						
ionizing radiation (occupational)	5	(1.3)	48	(1.3)	1.0	(0.4–2.5)
degreasing solvents:						
mother	7	(1.9)	19	(0.5)	**3.5**	**(1.5–8.5)**
father	71	(18.8)	553	(15.5)	**1.3**	**(1.0–1.7)**
both parents	4	(1.1)	12	(0.3)	**3.2**	**(1.0–9.9)**

Table 8A.4—Continued
All Left-Sided Obstructive Defects Univariate Analysis of Risk Factors
Baltimore-Washington Infant Study (1981–1989)

Variable	Cases N=377 no. (%)		Controls N=3572 no. (%)		Odds Ratio (95% CI)	
miscellaneous solvents:						
mother	15	(4.0)	64	(1.8)	**2.3**	**(1.3–4.0)**
father	35	(9.3)	267	(7.5)	1.3	(0.9–1.8)
both parents	7	(1.9)	16	(0.5)	**4.2**	**(1.7–10.3)**
maternal solvent score:						
0 (no exposures)	314	(83.3)	2932	(82.1)	1.0	(reference)
1–11 (1 or more agents, each <once a week)	25	(6.6)	358	(10.0)	1.0*	(1.0–1.1)
12–89 (1 or more agents, each 1–3 times a week)	27	(7.1)	214	(6.0)	1.2*	(1.0–1.3)
90–540 (1 or more agents, each daily)	11	(3.0)	68	(1.9)	**2.5***	**(1.3–4.8)**
occupational exposure to cold temperature (father)	4	(1.1)	9	(0.3)	**4.2**	**(1.3–13.9)**
Race of Infant						
white	284	(75.5)	2362	(66.2)	**1.6**	**(1.2–2.0)**
nonwhite	92	(24.5)	1207	(33.8)	1.0	(reference)

*Odds ratio and confidence intervals are from logistic regression for the scores of 5, 39, and 225 (midpoints of the score ranges 1–11, 12–89, and 90–540, respectively); risk factors are *maternal* unless labelled as paternal.

Multivariate Analysis

In the multivariate analysis of all cases with left-sided obstructive defects, family history of congenital heart disease continued to show a strong association (adjusted odds ratio 4.4), which was significant at the 1% level (Table 8A.5). None of the associations involving maternal illnesses or medications remained significant at the 1% level. Maternal exposures to degreasing and/or miscellaneous solvents emerged as the most stable of the solvent associations, and paternal occupational exposure to extremely cold temperature was also significantly associated. A racial difference was noted: whites were 50% more likely than nonwhites to have an affected infant, an association that was not confounded or modified by socioeconomic status.

Isolated/Simplex Subset

A total of 288 cases with left-sided heart defects met the criteria for inclusion in this subset. These infants had isolated congenital heart disease (ie, no noncardiac anomalies), and no first-degree relative had any type of birth defect. The results of the multivariate evaluation of this subset in comparison to 3318 controls free of birth defects and free of affected relatives (Table 8A.6) are similar to those obtained for the total group, except for the exclusion of the family history variable (by definition of the case subset). A different solvent variable emerged in the

Table 8A.5
All Left-Sided Obstructive Defects (N=377):
Multivariate Analysis of Risk Factors
Baltimore-Washington Infant Study (1981–1989)

Risk Factor	Adjusted Odds Ratio	95% CI	99% CI
Family History			
family history of congenital heart disease	4.4	2.5–7.6	2.1–9.0
Maternal Illness and Medications			
gestational diabetes	1.7	1.0–2.8	0.9–3.3
epilepsy	3.5	1.2–9.9	0.9–13.6
sympathomimetics	1.5	1.1–2.2	0.9–2.4
Home and Occupational Exposures			
degreasing or miscellaneous solvents (mother)	2.5	1.5–4.1	1.3–4.8
occupational cold temperature (father)	4.9	1.5–16.5	1.0–24.0
Sociodemographic Factors			
white race	1.5	1.2–1.9	1.1–2.1

Table 8A.6
Left-Sided Obstructive Defects: Isolated/Simplex Subset (N=288)
Multivariate Analysis of Risk Factors
Baltimore-Washington Infant Study (1981–1989)

Risk Factors	Cases	Controls	Adjusted Odds Ratio	95% CI	99% CI
Maternal Illnesses and Medications					
gestational diabetes	16	104	1.8	1.0–3.1	0.9–3.7
sympathomimetics	33	225	1.7	1.1–2.5	0.9–2.8
macrodantin	4	10	3.9	1.2–12.8	0.8–18.5
Home and Occupational Exposures					
degreasing or miscellaneous solvents (both parents)	8	24	4.0	1.8–9.2	1.4–11.9
occupational extreme cold temperature (father)	3	9	4.7	1.3–17.7	0.8–26.7
Sociodemographic Factors					
white race	214	2170	1.5	1.1–1.9	1.0–2.1

analysis of the isolated/simplex subset: exposure of *both parents* to degreasing or miscellaneous solvents at home or at work was the only strong solvent-related association. Only this solvent variable and race maintained significance at the 1% level.

Multiple Anomalies/Multiplex Subset

There were 89 cases in which either the infant had associated noncardiac anomalies or there were affected family members (cardiac or noncardiac anomalies). The multivariate analysis (Table 8A.7, part A) revealed two new associations: maternal use of benzodiazepines (odds ratio 3.8) and maternal history of one or more previous premature births (odds ratio 2.8). The associations of solvent-related exposures were stronger than for isolated/simplex: both of the solvent-related associations (maternal exposure to degreasing solvents, odds ratio 6.0, and paternal exposure to varnishing, odds ratio 2.4) were significant at the 1% level. In addition, the association with white race was slightly stronger in the multiple/multiplex than in the isolated/simplex subset.

In an additional analysis (Table 8A.7, part B) we took account of the presence of chromosome anomalies among the probands by restricting the analysis to cases free of chromosome anomalies. The results for this small subset (N = 63) revealed odds ratios for the associations with benzodiazapines, previous premature births, and white race that are similar to those in Part 7A. However, the odds ratio for maternal degreasing solvent exposures was enhanced to 8.6 in this subset of cases, which contained all three of the exposed mothers of infants in the multiple anomalies/multiplex subset.

Summary

Evaluation of the patients with left-sided obstructive defects supports the generally held view that these affect predominantly males and white infants. Univariate evaluation of environmental risk factors revealed maternal illness, medicinal drugs, and most notably solvent exposure as potential risk factors. In the multivariate analyses, solvent exposure was the most prominent risk factor for cases in the isolated/simplex subset, and in the multiple anomaly/multiplex subset it was strongest after exclusion of cases with chromosome defects.

Table 8A.7

Left-Sided Obstructive Defects: Multiple Anomalies/Multiplex Subset Multivariate Analysis of Risk Factors

Baltimore-Washington Infant Study (1981–1989)

Risk Factor	A. All Families (N=89)					B. Excluding Chromosome Defects (N=63)			
	Cases	Controls	Adjusted Odds Ratio	95% CI	99% CI	Cases	Adjusted Odds Ratio	95% CI	99% CI
Maternal Illnesses and Medications									
benzodiazepines	4	33	3.8	1.3–11.1	0.9–15.5	3	4.1	1.2–13.9	0.8–20.5
Maternal Reproductive History									
previous preterm birth	10	156	2.8	1.4–5.5	1.1–6.8	7	2.8	1.2–6.3	0.9–8.1
Home and Occupational Exposures									
degreasing solvents (mother)	3	17	6.0	1.7–21.3	1.2–31.6	3	8.6	2.4–30.7	1.6–45.7
varnishing (father)	16	240	2.4	1.4–4.3	1.2–5.1	13	2.8	1.5–5.3	1.2–6.5
Sociodemographic Factors									
white race	71	2170	2.0	1.2–3.4	1.0–4.1	51	2.1	1.1–4.1	0.9–5.0

Section B

Hypoplastic Left Heart Syndrome

The term hypoplastic left heart syndrome (HLHS) designates a complex of severe obstructive defects encompassing atresia (complete obstruction) or severe stenosis (partial obstruction) of the aortic and/or the mitral valve associated with hypoplasia (underdevelopment) of the left ventricle and of the ascending aorta.

Unlike the other diagnoses considered in this study, this is a clinically and empirically defined "syndrome," manifested by the onset of cardiac failure in the first few days of life, and with early death in the vast majority of cases. Lev[246] first described this complex of anomalies as the "hypoplasia of the aortic tract complex," and emphasized as the salient feature the very small aorta, which is supplied by retrograde perfusion from the pulmonary artery through a widely patent arterial duct. An intact ventricular septum was a criterion for this diagnosis.[40]

When Noonan and Nadas[322] used the term "hypoplastic left heart syndrome" in a clinical/pathologic description of 101 cases, their case definition was different. A large proportion of their cases (77 of 101) had pathologic features such as hypoplasia of the aortic arch or atresia of the transverse aortic arch, usually with associated septal defects. These features today are not considered part of the typical complex. The feature that they had in common with today's definition was underdevelopment of the left ventricle due to either valvar obstructions or to unbalanced bloodflow through intracardiac defects.

The term "hypoplastic left heart syndrome" is currently used to describe a diminutive left ventricle with underdevelopment of the mitral and aortic valves,[242] the malformation that is the target of the staged operative approach developed by Norwood.[326] It is also similarly defined by Allan[12] for the fetal echocardiographic diagnosis by the reduced dimensions of the left ventricle and aorta below the 5th percentile for gestational age.

Rowe and Mehrizi[375] use the term "hypoplastic left ventricular syndrome" as a "functional clinical generalization," but prefer to consider the malformation under the group title of "aortic atresia." This, in turn, allows for the possibility of a normally developed left ventricle in the presence of a ventricular septal defect.[343]

Clinicians recognize the hypoplastic left heart syndrome by manifestations of congestive cardiac failure in early infancy, with shock and acidosis and a characteristic echocardiogram of a hypoplastic left ventricle, stenosis of the mitral and aortic valves, and a small aorta,[158] but with extensive morphological variations.[160]

Infants with hypoplastic left heart syndrome may appear normal at birth, but evidence of intrauterine growth retardation and of a proportionately diminished head volume[372,373] have been noted. Associated noncardiac malformations including chromosomal and syndromic anomalies vary in frequency depending on the cardiac case definitions.[313] A recent report[184] emphasizes the occurrence of cranial and central nervous system abnormalities as well as minor dysmorphic anomalies, further suggesting heterogeneity of the case group.

A geographically based study reported from Oregon[307] describes 98 cases (from 1971 to 1986) of hypoplastic left heart syndrome. Some of the cases are associated with major intracardiac anomalies. In that epidemiologic study, noncardiac anomalies were present in 11%. Twenty of the 98 infants underwent the Norwood procedure, with improved survival to 1 month of age, but without apparent long-term improvement. Cardiac transplantation has been performed in some infants, but in general the outlook remains poor.[192] These considerations heighten the need for etiologic studies that may lead to primary prevention strategies.

Study Population

Hypoplastic left heart syndrome, as reported in this study, was defined by the reporting pediatric cardiologists using the ISC codes provided for hypoplastic left ventricle with obstruction of the mitral and/or aortic valves and a ductus-dependent systemic circulation. We considered this a single diagnostic group of left ventricular hypoplasia. The case group consists of 162 infants with hypoplastic left heart syndrome (3.7% of all cardiovascular malformations) with a 9-year regional prevalence of 1.78 per 10,000 livebirths.

Cardiac Abnormalities

The Baltimore-Washington Infant Study cardiac diagnostic coding system did not capture the anatomic variations and defects that are part of the syndrome, but not separately recorded. There were, however, 20 infants (12%) in whom an additional ventricular septal defect was diagnosed, and three who were reported to have partial anomalous pulmonary venous return.

Hypoplastic left heart syndrome as a secondary diagnosis in the hierarchical system occurred in 40 instances: the primary diagnoses in these 40 infants are listed on the diagnostic flow sheet (Figure 8B.1).

Hypoplastic Left Heart Syndrome
Baltimore-Washington Infant Study (1981-1989)

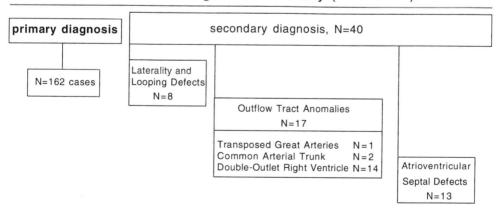

Figure 8B.1: Hypoplastic left heart syndrome occurred as a primary diagnosis in 162 cases. In another 40 infants, this diagnosis was secondary to the primary cardiogenesis defects of laterality and looping, outflow tract formation, and atrioventricular septation.

Noncardiac Anomalies

Noncardiac anomalies occurred in 24 cases (15%). Chromosome abnormalities, syndromes, and organ defects each accounted for roughly a third of these (Table 8B.1). Of the 9 infants with chromosome anomalies, 3 had Turner syndrome (monosomy X), 5 had trisomies, and 1 had a deletion. Four different Mendelian syndromes were diagnosed, each in a single case, and four infants were described as having non-Mendelian associations of multiple defects, including two cases of the VACTERL association. These anomalies are similar to those described in the literature.

Descriptive Analyses

Prevalence by Time, Season, and Area of Residence

The overall prevalence of hypoplastic left heart syndrome was 1.78 per 10,000 livebirths, but the prevalence declined throughout the study years, from 2.37 in the time from 1981 to 1983, to 1.79 from 1984 to 1986, and to 1.31 from 1987 to 1989 (Table 8B.2). At the same time the total prevalence of all congenital heart disease increased primarily due to increased detection of mild defects by echocardiogra-

Table 8B.1

Hypoplastic Left Heart Syndrome (N=162):

Noncardiac Anomalies

Baltimore-Washington Infant Study (1981–1989)

Chromosomal Abnormalities:	N=9	(5.6%)
Down syndrome	1	[# 10220]
Trisomy 18	2	[# 2802, 2883]
Trisomy 13	2	[# 2318, 3716]
Turner syndrome	3	[#1497, 2979, 3366]
Deletion 1q	1	[# 2360]
Syndromes	N=8	(4.9%)
a. Mendelian Syndromes:	4	
Smith-Lemli-Opitz	1	[# 1338]
3C (Ritscher-Schinzel)	1	[# 10576]
Lazjuk	1	[# 1268]
Renal-hepatic-pancreatic dysplasia	1	[# 2420]
b. Non-Mendelian Associations:	4	
Fetal alcohol syndrome	1	[# 10569]
VACTERL	2	[# 1065, 1067]
Holzgreve	1	[# 10841]
Multiple, Nonclassified Anomalies:	N=1	(0.6%)
Microcephaly, unspecified "hand defect"	1	[# 1256]
Organ Defects:	N=6	(3.7%)
Urethral stenosis	2	[# 1823, 11274]
Eye defects	2	[# 3276, 10752]
Sexual ambiguity	1	[# 10023]
Hydrocephalus	1	[# 11551]
No Associated Anomalies	N=138	(85.2%)

Table 8B.2

Hypoplastic Left Heart Syndrome: Prevalence-Characteristics

Baltimore-Washington Infant Study (1981–1989)

	Cases	Area Birth	Prevalence per 10,000
Total Subjects	162	906,626	1.78
Year of Birth:			
1981–1983	65	274,558	2.37
1984–1986	53	295,845	1.79
1987–1989	44	336,223	1.31
Birth Quarter:			
1st (Jan–Mar)	39	219,145	1.77
2nd (Apr–Jun)	46	231,777	1.98
3rd (Jul–Sep)	48	233,626	2.05
4th (Oct–Dec)	29	222,078	1.30
Area of Residence:			
urban	48	208,568	2.30
suburban	88	584,022	1.51
rural	26	112,318	2.31

phy. Thus the proportion of total cases represented by hypoplastic left heart syndrome declined from 6% in the time from 1981 to 1983, to 2.5% from 1987 to 1989.

Prevalence was highest among spring and summer births and lowest among fall births. The prevalence among residents of suburban areas was approximately two thirds of the prevalence in urban and rural area residents.

Diagnosis and Course

The diagnosis was made very early. Seventy-six percent of the cases were diagnosed at less than 1 week of age, and 96%, by 4 weeks of age (Table 8B.3); only one case was over 6 months of age at diagnosis.

The diagnosis of hypoplastic left heart syndrome was made by echocardiography alone in 27.8% of cases, 39 cases (24%) underwent diagnostic cardiac catheterization. Diagnosis was confirmed at autopsy in 70 cases (43.2%). Twelve of these 70 patients were diagnosed only at autopsy, of these, four weighed less than 2000 grams, and a fifth case had a chromosome anomaly. Fifty-three patients (32.7%) had surgery, and 22 infants were alive at the final update, representing 13.6% of the entire group.

Gender, Race, and Twinning

The proportion of males in cases (58.6%) was somewhat greater than in controls (50.9%) (Table 8B.4). The proportion of white cases (65%) was comparable to the proportion of white infants in the region (66%). Two cases (1.2%) were from multiple births: one was a twin of unknown zygosity and the other was one of triplets. All co-twins were normal.

Table 8B.3

Hypoplastic Left Heart Syndrome: Clinical Data

Baltimore-Washington Infant Study (1981–1989)

	Cases N=162 no.	(%)
Age at Diagnosis:		
<1 week	123	(75.9)
1–4 weeks	33	(20.4)
5–24 weeks	5	(3.1)
25–52 weeks	1	(0.6)
Method of Diagnosis:		
echocardiography only	45	(27.8)
echocardiography and cardiac catheterization	16	(9.9)
echocardiography and surgery	8	(4.9)
echocardiography, cardiac catherization, and surgery	23*	(14.2)
autopsy following other methods	58	(35.8)
autopsy only	12	(7.4)
Surgery During First Year of Life	53	(32.7)
Follow-Up:		
not followed	1	(0.6)
followed:	161	
died in the first year of life	140	(87.0)
alive	21	(13.0)
median age at last visit	11 months	

*Includes 1 with cardiac catherization only.

Table 8B.4

Hypoplastic Left Heart Syndrome: Infant Characteristics

Baltimore-Washington Infant Study (1981–1989)

Characteristic	Cases N=162 no.	(%)	Controls N=3572 no.	(%)
Gender:				
male	95	(58.6)	1,817	(50.9)
female	67	(41.4)	1,755	(49.1)
Race:				
white	102	(64.6)	2,362	(66.1)
black	53	(33.5)	1,109	(31.0)
other	3	(1.9)	98	(2.7)
Twin births	2	(1.2)	53	(1.5)

Birthweight and Gestational Age

This information, provided by the mothers who were interviewed, was available for 138 cases (85.2%). The mean birthweight was 3071 grams (Table 8B.5), 280 grams less than the mean birthweight of controls. Low birthweight (less than 2500 g) was twice as frequent as that of controls, although the mean gestational age of the cases was not reduced. Additionally there was a four-fold odds ratio for being small for gestational age.

Sociodemographic Factors

There were no case-control differences in any of the sociodemographic variables. The distributions of parental education and occupation, family income, maternal marital status, and the time of pregnancy confirmation were very similar in cases and controls (Table 8B.6).

Potential Risk Factors

Familial Cardiac and Noncardiac Anomalies

Cardiac anomalies (not counting maternal mitral valve prolapse) were reported in first-degree relatives of seven infants with the hypoplastic left heart syndrome (Table 8B.7). In five of these seven families (which are shown in pedigrees in Figure 8A.2) the cardiac anomaly was a left-sided obstructive lesion: 1 infant was born to a father with coarctation of the aorta; 3 brothers (2 in one family) had

Table 8B.5
Hypoplastic Left Heart Syndrome: Fetal Growth Characteristics
Baltimore-Washington Infant Study (1981–1989)

Characteristic	Cases no.	(%)	Controls no.	(%)
Number of Families Interviewed	138		3572	
Birthweight (grams)				
<2500	20	(14.5)	252	(7.1)
2500–3500	84	(60.9)	1853	(51.9)
>3500	34	(24.6)	1467	(41.0)
mean±standard error	3071±59		3351±10	
range	680–4621		340–5273	
Gestational Age (weeks)				
<38	24	(17.4)	339	(9.5)
38+	114	(82.6)	3233	(90.5)
mean	39.1±0.2		39.6±0.1	
range	28–43		20–47	
Size for Gestational Age				
small (SGA)	30	(21.7)	211	(5.9)
normal	91	(65.9)	2712	(75.9)
large	17	(12.3)	649	(18.2)
Odds Ratio for SGA (95% CI)	**4.4**	**(2.9–6.8)**	1.0	(reference)

Table 8B.6

Hypoplastic Left Heart Syndrome: Sociodemiographic Characteristics

Baltimore-Washington Infant Study (1981–1989)

Interviewed Families	Cases N=138 no.	(%)	Controls N=3572 no.	(%)	Odds Ratio (95% CI)	
Maternal Marital Status:						
not married	36	(26.1)	990	(27.7)	0.9	(0.6–1.4)
married	102	(73.9)	2582	(72.3)	1.0	(reference)
Maternal Education:						
<high school	21	(15.2)	659	(18.4)	0.8	(0.5–1.3)
high school	50	(36.2)	1265	(35.4)	1.0	(0.7–1.4)
college	67	(48.6)	1648	(46.1)	1.0	(reference)
Paternal Education:						
<high school	25	(18.1)	650	(18.2)	1.0	(0.6–1.6)
high school	50	(36.2)	1298	(36.3)	1.0	(0.7–1.5)
college	63	(45.7)	1624	(45.5)	1.0	(reference)
Annual Household Income:						
<$10,000	26	(19.1)	686	(19.2)	1.0	(0.6–1.7)
$10,000–$19,999	31	(22.8)	699	(19.6)	1.2	(0.8–1.9)
$20,000–$29,999	28	(20.6)	737	(20.6)	1.0	(0.6–1.6)
$30,000+	51	(37.5)	1373	(38.4)	1.0	(reference)
Maternal Occupation:						
not working	48	(34.8)	1142	(32.0)	1.1	(0.7–1.7)
clerical/sales	43	(31.2)	1119	(31.3)	1.0	(0.6–1.6)
service	14	(10.1)	444	(12.4)	0.8	(0.4–1.5)
factory	4	(2.9)	137	(3.8)	0.7	(0.3–2.1)
professional	29	(21.0)	730	(20.4)	1.0	(reference)
Paternal Occupation:						
not working	10	(7.2)	246	(6.9)	0.9	(0.4–1.7)
clerical/sales	22	(15.9)	618	(17.3)	0.7	(0.4–1.3)
service	20	(14.5)	480	(13.4)	0.9	(0.5–1.5)
factory	41	(29.7)	1279	(35.8)	0.7	(0.4–1.0)
professional	45	(32.6)	947	(26.5)	1.0	(reference)
Month of Pregnancy Confirmation:						
1st month	20	(14.6)	511	(14.3)	1.0	(reference)
2nd month	69	(50.4)	1955	(54.7)	0.9	(0.5–1.5)
3rd month	31	(22.6)	663	(18.6)	1.2	(0.7–2.1)
4th month or later	17	(12.4)	416	(11.6)	1.0	(0.5–2.0)

Data are missing for income (2 cases, 77 controls), paternal occupation (2 controls) and pregnancy confirmation month (1 case, 27 controls).

coarctation of the aorta, and 1 had a bicuspid aortic valve; only 1 infant had a brother with a fully concordant abnormality. A sister with a ventricular septal defect and a sister with tetralogy of Fallot were reported in two separate families. Noncardiac malformations included thanatophoric dwarfism (a possible dominant mutation of a chondrodystrophy) in a sister of the sibling pair with hypoplastic left heart syndrome in the case (#10912) and tetralogy of Fallot, in the sibling. Noncardiac defects were reported in eight families, encompassing a range of anomalies including two mothers with congenital deafness.

Table 8B.7
Hypoplastic Left Heart Syndrome
Congenital Defects and Hereditary Disorders in First-Degree Relatives
The Baltimore-Washington Infant Study (1981–1989)

Proband	Sex	Diagnosis	Relative	Diagnosis
A. Congenital Heart Defects:				
1889*	F		Brother	Hypoplastic left heart
2645*	F		Father	Coarctation of aorta
1609*	M		2 Brothers	Coarctation of aorta
1478*	F		Brother	Coarctation of aorta
2420*	F		Brother	Bicuspid aorta valve
1252	F		Sister	Ventricular septal defect
10912	M		Sister	Tetralogy
			Sister	Thanatophoric dysplasia
3276	M	Cataract	Mother	Mitral valve prolapse
B. Congenital Defects of Other Organs:				
1352	F		Mother	Deafness
1386	F		Mother	Deafness
2259	M		Mother	Syndactyly
1512	F	Holt-Oram	Father	Limb reduction[1]
10931	M		Father	Pectus exacavatum
10841	F	Cleft Palate	Father	Pilonidal cyst
2648	M		Brother	Down syndrome
11166	M		Brother	Sickle Cell

[1]Two missing fingers on right hand, 1 missing on left hand, arms shortened.
*See pedigrees in Section A, Figure 2.

Genetic and Environmental Factors

Univariate Analysis

Compared to controls, cases were much more likely to report a family history of congenital heart disease (odds ratio 4.4). There were no significant case-control differences in parental age distributions, maternal reproductive history variables, or maternal lifestyle exposure (Table 8B.8). Maternal diabetes mellitus was reported more frequently among cases than controls. Three mothers of cases reported overt diabetes (odds ratio 3.4), all were treated with insulin.

One paternal medical factor, exposure to general anesthesia during the 6 months prior to the study pregnancy, was significantly more frequent among cases than among controls. General anesthesia was reported in association with dental work (N = 3), orthopedic surgery (N = 2), ear, nose, and throat surgery (N = 1), appendectomy (N = 1), and in an unknown procedure (N = 1).

Among home and occupational agents, exposure to degreasing solvents was significantly more frequent among case mothers. In each of the three instances, the exposure occurred during cleaning of weapons; two exposures were at home and one was occupational (military). Miscellaneous parental solvent exposures were also reported more frequently among cases, affecting both the mother and the father (odds ratio 6.6) by use of photographic solvents, xylene, printing inks and paint thinners.

Table 8B.8
Hypoplastic Left Heart Syndrome: Univariate Analysis of Risk Factors
Baltimore-Washington Infant Study (1981–1989)

Variable	Cases N=138 no.	(%)	Controls N=3572 no.	(%)	Odds Ratio (95% CI)
Family History **(first-degree relatives)**					
congenital heart disease	7	(5.1)	43	(1.2)	**4.4 (1.9–9.9)**
noncardiac anomalies	8	(5.8)	165	(4.6)	1.3 (0.6–2.6)
Parental Age					
maternal age:					
<20	20	(14.5)	507	(14.2)	1.0 (0.6–1.7)
20–29	77	(55.8)	2009	(56.2)	1.0 (reference)
30+	41	(29.7)	1056	(29.6)	1.0 (0.7–1.5)
paternal age:					
<20	9	(6.5)	226	(6.3)	1.1 (0.5–2.2)
20–29	65	(47.1)	1724	(48.3)	1.0 (reference)
30+	64	(46.4)	1622	(45.4)	1.0 (0.7–1.5)
Maternal Reproductive **History**					
number of previous pregnancies:					
none	42	(30.4)	1159	(32.4)	1.0 (reference)
one	47	(34.1)	1097	(30.7)	1.2 (0.8–1.8)
two	21	(15.2)	709	(19.9)	0.8 (0.5–1.4)
three or more	28	(20.3)	607	(17.0)	1.3 (0.8–2.1)
previous miscarriage(s)	30	(21.7)	681	(19.1)	1.2 (0.8–1.8)
Illnesses and Medications					
diabetes:					
overt	3	(2.2)	23	(0.6)	**3.4 (1.0–11.5)**
gestational	6	(4.4)	115	(3.2)	1.4 (0.6–3.2)
influenza	16	(11.6)	278	(7.8)	1.5 (0.9–2.7)
general anesthesia (paternal)	8	(5.8)	98	(2.7)	**2.2 (1.0–4.6)**
Lifestyle Exposures					
alcohol (any amount)	80	(58.0)	2101	(58.8)	1.0 (0.7–1.4)
smoking (cigarettes/day):					
none	88	(63.8)	2302	(64.5)	1.0 (reference)
1–10	23	(16.7)	565	(15.8)	1.1 (0.7–1.7)
11–20	21	(15.2)	529	(14.8)	1.0 (0.6–1.7)
>20	6	(4.4)	176	(4.9)	0.9 (0.4–2.1)
Home and **Occupational Exposures**					
ionizing radiation (occupational)	4	(2.9)	48	(1.3)	2.2 (0.8–6.2)
degreasing solvents	3	(2.2)	19	(0.5)	**4.2 (1.2–14.2)**
miscellaneous solvents:					
mother	6	(4.4)	64	(1.8)	**2.5 (1.1–5.9)**
father	15	(10.9)	267	(7.5)	1.5 (0.9–2.6)
both parents	4	(2.9)	16	(0.5)	**6.6 (2.2–20.1)**
Race of Infant					
white	91	(65.9)	2362	(66.2)	1.0 (0.7–1.4)
nonwhite	47	(34.1)	1207	(33.8)	1.0 (reference)

Risk factors are *maternal* unless labelled as paternal.

Multivariate Analysis

The multivariate model of genetic and environmental risk factors is shown in Table 8B.9. Family history was strongly associated with this defect (odds ratio 4.8). Maternal overt diabetes was associated with hypoplastic left heart syndrome (odds ratio 3.9). The confidence interval was wide and not significant at the 1% level. This association was based on reported maternal diabetes in three cases, all of whom were born prior to 1985. Maternal exposures to degreasing or miscellaneous solvents showed a more than three-fold increase in risk, and this association remained statistically significant at the 1% level. In this adjusted analysis there was no increased risk for the combined exposure of both parents to solvents. Paternal exposure to general anesthesia (odds ratio 2.4) was significantly associated at the 5% level.

Finally, we separately analyzed the subset of cases and controls classified as isolated/simplex, in which the proband was free of noncardiac anomalies and the family members had no congenital anomalies (data not shown). In this set of 105 cases, the multivariate model of potential risk factors revealed the same associations detected in the total group of hypoplastic left heart syndromes: overt diabetes (odds ratio = 3.2), maternal solvent exposures (odds ratio = 3.4), and paternal exposure to general anesthesia (odds ratio 2.7). Only the solvent association reached the 1% significance level. There were too few cases classified as multiple anomalies/multiplex (N = 33) to analyze separately.

Discussion

Letters and phone calls from parents of infants with hypoplastic left heart syndrome indicated that among all case groups, the emotional trauma and anxiety was greatest in these families and they feared the possibility of a similar abnormality in a subsequent pregnancy. The findings of this study provide information on the association of a probable genetic risk and also suggest a possible role of environmental risk factors, avoidance of which could constitute a preventive action.

Table 8B.9
Hypoplastic Left Heart Syndrome (N=138)
Multivariate Analysis of Potential Risk Factors
Baltimore-Washington Infant Study (1981–1989)

Variable	Cases	Controls	Adjusted Odds Ratio	95% CI	99% CI
family history of congenital heart defects	7	43	4.8	2.1–10.8	1.6–14.0
overt diabetes	3	23	3.9	1.2–13.2	0.8–19.3
solvent exposures*	9	80	3.4	1.6–6.9	1.3–8.6
paternal general anesthesia	8	98	2.4	1.1–5.0	0.9–6.3

*Degreasing or miscellaneous solvents.

Family History

This diagnostic case group is clearly heterogenous.

A genetic component is evident in the association with Turner syndrome, which links this defect with the other left-sided obstructions (Section A, Table 2), however the affected proportion of hypoplastic left heart cases is low (1.8% versus 4.0% to 7.1% of other left-sided obstructions).

The occurrence of hypoplastic left heart syndrome in autosomal trisomies has been frequently described but it is rare. Only one case with hypoplastic left heart syndrome and Down syndrome was enrolled over the 9-year period of the Baltimore-Washington Infant Study; there were 2 cases with trisomy 13 and 2 cases with trisomy 18. Among the syndromic cases, the occurrence of Mendelian and non-Mendelian associations and the VACTERL association has been noted.[313] The Smith-Lemli-Opitz syndrome with its renal, gastrointestinal, and limb defects, and coexistent urethral stenosis in hypoplastic left heart syndrome may suggest the occurrence of this cardiac defect as a midline field defect.

The family history of cardiac defects is notable, with an odds ratio of 4.8. Six of the seven infants with a family history of cardiac defects were born in the first 5 years of the study (1981 to 1985). In the early study years fetal echocardiography was still uncommon in the community, but as a positive family history became a common indication for fetal cardiac ultrasound studies, a declining frequency of cases with a positive family history of congenital heart disease occurred in the second half of the Baltimore-Washington Infant Study, presumably because of interrupted pregnancies following diagnosis of hypoplastic left heart syndrome by fetal ultrasound.

Maternal Illnesses

The true association between maternal diabetes and hypoplastic left heart syndrome may also be underestimated, as all of the cases with reported maternal diabetes were born prior to 1985; again, in the later years, fetal echocardiography was performed because of maternal diabetes and could also have resulted in interrupted pregnancies for hypoplastic left heart syndrome. Thus the strong association with overt diabetes in the total case group (adjusted odds ratio 3.9) represents an important area for future investigations into the categorization and causation of hypoplastic left heart syndrome. The variability of case definition and the association of left ventricular hypoplasia with major intracardiac defects as a secondary diagnosis raises the possibility that among the primary (analyzed) case group, some infants might etiologically belong to the other group, ie, the heart may not have been normal at any stage of development.

Environmental Risk Factors

In the univariate analysis, maternal exposure to various solvents appeared as the strongest risk factor, and the association with degreasing and miscellaneous solvents remained significant in the adjusted analysis (odds ratio 3.4). This association may represent an environmental origin for some of the cases. This association also has a potential for elucidating the mechanism of the fetal disturbance, which may originate from altered early-fetal hemodynamics.

Variability of Anatomic Features

In the introduction to this chapter we noted the variability of anatomic definitions of this syndrome, and in the diagnostic flow sheet (Figure 8B.1) we showed the occurrence of a hypoplastic left ventricle as a secondary diagnosis. However, a ventricular septal defect remained as a coexisting diagnosis in 20 of the analyzed cases and was described as small in 5 cases. To evaluate the possibility of an effect of this "misclassification" on the risk factor analyses, we reviewed the individual records of each of these cases. Sixteen families were interviewed.

Notably, one infant (#1899) with hypoplastic left heart syndrome with a ventricular septal defect had a brother with hypoplastic left heart syndrome without a ventricular septum. None of the mothers had diabetes, but in several families solvent exposure was reported for one or both parents.

These findings do not suggest a difference of this small subgroup of cases from the remaining group.

Regrettably, very little has been done so far to investigate the epidemiology of hypoplastic left heart syndrome or its developmental origin. The recent study by Tikkanen and Heinonen[414] evaluated potential risk factors for a small case group with hypoplastic left heart syndrome (N = 34), but there were no significant associations with either solvent exposures or with diabetes. New methods of molecular studies may offer promising possibilities for the elucidation of possible genetic susceptibilities to environmental agents. Certainly the avoidance of exposures to chemicals represents a reasonable precaution consistent with current prenatal recommendations to avoid alcohol, medicines, and drug intake before and during pregnancy.

Section C

Coarctation of Aorta

Coarctation is a narrowing of the aorta in the region of the junction of the embryonic fourth and sixth embryonic arches, represented by the arch of the aorta and the arterial duct, respectively. Proximal to the arterial duct is a segment of aorta known as the isthmus, which in fetal life carries a reduced blood flow as left ventricular blood is directed to the head and upper extremities above this segment, and right ventricular blood is directed into the pulmonary artery and through the patent arterial duct into the lower extremities. The relationship of the coarcted (narrowed) segment of the aorta to the arterial duct has long been of interest because of clinical and etiologic considerations.[211]

Morgagni's remarkable "state-of-the-art" five-volume work[306] describes clinical-pathologic correlations of many cardiac defects (mitral stenosis, aortic stenosis, coronary sclerosis, and aortic aneurysm among them),[78] but not coarctation of the aorta in the usual anatomic position, as has been erroneously reported. In Letter XVIII, Articles 5 and 6, Morgagni debates the causes of dilatation of one or other ventricle and mentions the case of a colleague, Marc Antonio Laurentio, of a man who died after having had a slight fever for some time and was found to have *so great a dilatation of the heart, especially the right ventricle as had never been seen before and the aorta was contracted to an amazing narrowing near to the heart.*

Abbott[3] (1928) credits *Monsieur* Paris, prosector of the Hôtel Dieu Hospital in Paris, with the first description of a typical case. The collateral circulation was dramatically described by *Monsieur* Paris in that first documented case of coarctation of the thoracic aorta.[335] Among the great number of cadavers injected for dissection was that of a very thin woman of about 50 years of age in whom the resin-tallow mixture tainted with lamp black introduced at the base of the aorta penetrated easily and in an increased amount and made visible many tortuous thoracic arteries, the "zig-zag" (sic) pattern of which resembled the beads of a rosary. Proceeding therefore with special care, the prosector identified a marked constriction of the aorta between the arterial ligament and the first intercostal artery; the lumen was no bigger than the quill of a feather. The prosector correctly described the course of the arterial circulation as it bypassed the narrowed aortic segment. Alas, we are told nothing more. However, as Monsieur Desault (the professor) *conserved the anatomic specimen in his cabinet,* additional recorded information might still possibly be discovered.

Hamilton and Abbott[194] describe the history of a boy "of high intelligence" who had a "huge collateral circulation" and died at age 14 in a coma, probably due to cerebral hemorrhage. Stimulated by this occurrence Abbott performed a "historical retrospect" of 200 autopsied cases in patients over 2 years of age.[3] The late complications of coarctation often appeared dramatically in those "historic" cases, some of which, such as the "Bernoise peasant who died 6 days after carrying a heavy bag of corn to market" or "an Austrian officer who fought in the French Revolution" who also died suddenly, are illustrated in Abbott's Atlas.[6]

In the large series of patients reported by Reifenstein et al,[359] in whom coarctation was found at autopsy, deaths from aortic rupture, endocarditis, and central nervous system hemorrhage occurred mostly during the first 3 decades of life; while deaths from congestive heart failure occurred later. The association of hypertension in the upper extremities and evidence of renal failure was described as noteworthy by Paul Dudley White,[445] in a patient with "Bright's disease" (chronic renal failure) who was eventually cured by surgery of the aorta.

The classic clinical picture of coarctation in the older child and adult consisted of hypertension in the upper extremities, reduced or absent femoral pulses, and a continuous murmur over the back that results from the compensatory flow through dilated intercostal collateral arteries. Radiologists noted erosion of the rib margins by the dilated collateral arteries.

The infant form of coarctation and the importance of coarctation as a cause of death in infancy was not recognized until the 1950s.[26] The classic presentation of coarctation of the aorta in infancy is that of a critical reduction in left ventricular outflow that develops acutely at the time of ductal closure, or from the neonatal fall in pulmonary vascular resistance, which diminishes the ductal blood flow into the descending aorta. In either case, systemic blood flow is acutely compromised and the infant may present in extremis with shock and acidosis. This clinical picture is more frequent when coarctation is associated with intracardiac defects, and may be confused with sepsis or metabolic disorders.

The first reported surgical repair of coarctation was performed by Crafoord[99] (Sweden) in 1944. Crafoord used a technique of turning down the left subclavian artery to bypass the stenosed segment of the aorta. Gross and Hufnagel[191] developed an alternate operation with resection of the stenosed segment and end-to-end anastomosis, and reported their first patient treated by this procedure in 1945. Somewhat later coarctation repairs in infancy were reported.[270,310] Although the expectations for a surgical cure were high, persistent elevation of blood pressure and recurrence of the coarctation still remain important problems.[217]

Etiologic considerations were first raised by Bonnet[49] who divided coarctation into preductal and postductal types designated as "infantile" (with ductal patency) and "adult" (ductus closed). This grouping failed to attract support for the theory that ductal tissue extending into the aorta accounted for the sharp constriction of the adult type.

The detailed histological studies of Elzenga et al,[128,130] however, demonstrate the presence of ductal tissue in the aorta, but cannot conclude that it has an etiologic role in the aortic constriction. The possibility that maldevelopment of the neural crest may play a role in the origin of this malformation has also been considered.[216]

Since the epidemiologic study is unable to subdivide the case group, as pathologists can, a clarification of these morphogenetic issues can not be derived from our data.

Study Population

In the Baltimore-Washington Infant Study all infant patients were observed over a time period during which noninvasive diagnostic methods assumed a prominent role in the evaluation of suspected cardiovascular anomalies. Thus the infants diagnosed with coarctation of the aorta represent a broad range of severity of the anomaly, as the current ability to obtain excellent quality ultrasound images and Doppler flow measurements can establish the presence of mild forms of coarctation. In addition, the restriction for analysis of the case group to "pure" coarctation has excluded infants with associated noncardiac anomalies (Figure 8A.1).

Cardiac Abnormalities

Coarctation of the aorta as a principal cardiac diagnosis was present in 196 infants. This discussion is limited to 126 cases of "pure" coarctation of the aorta which, according to the hierarchical assignment of the principal diagnosis, takes precedence over associated intracardiac defects, irrespective of the severity of the hemodynamic effects of each diagnostic component. Only bicuspid aortic valve and patent arterial duct remain as associated diagnoses. Bicuspid aortic valve was present in one third of the patients with coarctation of the aorta; the arterial duct is often intermittently or persistently patent in early infancy.

Noncardiac Anomalies

Among the 126 infants with "pure" coarctation, 112 (89%) had no associated noncardiac anomalies and all of the remaining cases had multiple noncardiac defects (Table 8C.1). Chromosome anomalies were present in 10 patients (7.9%), of whom 7 had Turner syndrome. Four patients had malformation syndromes: 2 were Mendelian and 2 were non-Mendelian associations.

Table 8C.1
Coarctation of Aorta (N=126): Noncardiac Anomalies
Baltimore-Washington Infant Study (1981–1989)

Chromosomal Abnormalities:	N=10	(7.9%)
Turner syndrome	7	[# 1724, 1887, 2462, 3696, 10808, 11046, 11567]
Down syndrome	1	[# 3058]
Trisomy 13	1	[# 1266]
Partial trisomy, unspecified	1	[# 1255]
Syndromes	N=4	(3.2%)
a. Mendelian Syndromes:	2	
Roberts	1	[# 2808]
Peters-plus	1	[# 11390]
b. Non-Mendelian Associations:	2	
VACTERL	1	[# 2398]
Goldenhar	1	[# 10433]
Organ Defects	N=0	
No Associated Anomalies	N=112	(88.9%)

Descriptive Analysis

Prevalence by Time, Season, and Area of Residence

During the years 1981 to 1989, the 126 cases of "pure" coarctation of the aorta enrolled in the Baltimore-Washington Infant Study represented a regional prevalence of 1.39 cases per 10,000 livebirths. Prevalence declined slightly during the 9-year time period from 1.78 per 10,000 births in the first 3 years to 1.14 in the last 3 years (Table 8C.2). There was little variation in prevalence by birth quarter. Rates for this malformation were twice as high in rural and suburban areas of the region than in urban areas.

Diagnosis and Course

Most infants with coarctation were diagnosed early in life: 23% were diagnosed in the first week of life, two thirds were diagnosed by the end of the first month, and 90% were diagnosed by 6 months of age (Table 8C.3). All patients (except 4 diagnosed only at autopsy) underwent echocardiography with Doppler determination of a pressure gradient at the coarctation site. Cardiac catheterization followed in about two thirds of the cases, surgery followed without catheterization in about one fifth. Of the four patients whose diagnosis was made only at autopsy, one had a chromosome abnormality and one had Robert's syndrome. Surgical intervention during the follow-up period was performed in 92 cases (73%); the all-cause mortality was 7%.

Gender, Race, and Twinning

The proportion of males in this diagnostic group did not differ significantly from controls (58% versus 51%; Table 8C.4). There was an increased proportion of

Table 8C.2
Coarctation of Aorta: Prevalence
Baltimore-Washington Infant Study (1981–1989)

	Cases	Area Births	Prevalence per 10,000
Total Subjects	126	906,626	1.39
Year of Birth:			
1981–1983	49	274,558	1.78
1984–1986	36	295,845	1.22
1987–1989	39	336,223	1.14
Birth Quarter:			
1st (Jan–Mar)	34	219,145	1.55
2nd (Apr–Jun)	30	231,777	1.29
3rd (Jul–Sep)	29	233,626	1.24
4th (Oct–Dec)	33	222,078	1.49
Area of Residence:			
urban	16	208,568	0.77
suburban	88	584,022	1.51
rural	22	112,318	1.96

Table 8C.3

Coarctation of Aorta: Clinical Data

Baltimore-Washington Infant Study (1981–1989)

	Cases N=126	
	no.	(%)
Age at Diagnosis:		
<1 week	29	(23.0)
1–4 weeks	54	(42.9)
5–24 weeks	30	(23.8)
25–52 weeks	13	(10.3)
Method of Diagnosis:		
echocardiography only	23	(18.3)
echocardiography and cardiac catheterization	8	(6.3)
echocardiography and surgery	25	(19.8)
echocardiography, cardiac catheterization, and surgery	64*	(50.8)
autopsy following other methods	2	(1.6)
autopsy only	4	(3.2)
Surgery During First Year of Life	92	(73.0)
Follow-Up:		
not followed	4	(3.2)
followed:	122	
died in the first year of life	9	(7.4)
alive	113	(92.6)
median age at last visit	11 months	

*Includes 4 cardiac catheterization and surgery.

Table 8C.4

Coarctation of Aorta: Infant Characteristics

Baltimore-Washington Infant Study (1981–1989)

Characteristic	Cases N=126		Controls N=3572	
	no.	(%)	no.	(%)
Gender				
male	73	(57.9)	1,817	(50.9)
female	53	(42.1)	1,755	(49.1)
Race:				
white	98	(79.0)	2,362	(66.1)
black	24	(19.4)	1,109	(31.0)
other	2	(1.6)	98	(2.7)
Twin births	4	(3.2)	53	(1.5)

*Missing race in 2 cases and 3 controls.

white infants among the cases (79% versus 66%). Coarctation of the aorta occurred in four twin cases (3.2%), twice the rate of twinning as among the controls
(1.5%); all co-twins were unaffected.

Birthweight and Gestational Age

Data from the family interview on all 120 infants (95.2%) with coarctation
showed that they were twice as likely as controls to have low birthweight
(≤2500 grams) (Table 8C.5) and were also more likely to be born prematurely;
in addition, there was a two-fold odds ratio for small-for-gestational-age. The
mean birthweight of cases was 198g less than the average birthweight of control
infants.

Sociodemographic Characteristics

There were few case-control differences in the distributions of sociodemographic characteristics (Table 8C.6). Case mothers were less likely to be unmarried than were controls, but the groups were not different in terms of parental education, income, and occupation. Fathers working in clerical or sales-related
occupations were under-represented among the cases relative to controls, but did
not differ from controls in education level.

Table 8C.5
Coarctation of Aorta: Fetal Growth Characteristics
Baltimore-Washington Infant Study (1981–1989)

Characteristic	Cases no.	(%)	Controls no.	(%)
Number of Families Interviewed	120		3572	
Birthweight (grams)				
<2500	17	(14.2)	252	(7.1)
2500–3500	65	(54.2)	1853	(51.9)
>3500	38	(31.7)	1467	(41.0)
mean±standard error	3153±64		3351±10	
range	992–4479		340–5273	
Gestational Age (weeks)				
<38	18	(15.1)	339	(9.5)
38+	101	(84.9)	3233	(90.5)
mean±standard error	39.2±0.2		39.6±0.1	
range	26–44		20–47	
Size for Gestational Age				
small (SGA)	14	(11.7)	211	(5.9)
normal	88	(73.3)	2712	(75.9)
large	18	(15.0)	649	(18.2)
Odds Ratio for SGA (95% CI)	**2.1**	**(1.2–3.7)**	1.0	(reference)

Table 8C.6

Coarctation of Aorta: Sociodemographic Characteristics

Baltimore-Washington Infant Study (1981–1989)

Interviewed Families	Cases N=120 no.	(%)	Controls N=3572 no.	(%)	Odds Ratio (95% CI)	
Maternal Marital Status:						
not married	24	(20.0)	990	(27.7)	**0.7**	**(0.4–1.0)**
married	96	(80.0)	2582	(72.3)	1.0	(reference)
Maternal Education:						
<high school	19	(15.8)	659	(18.4)	1.0	(0.6–1.7)
high school	53	(44.2)	1265	(35.4)	1.4	(0.9–2.1)
college	48	(40.0)	1648	(46.1)	1.0	(reference)
Paternal Education:						
<high school	25	(20.8)	650	(18.2)	1.2	(0.8–2.0)
high school	44	(36.7)	1298	(36.3)	1.1	(0.7–1.6)
college	51	(42.5)	1624	(45.5)	1.0	(reference)
Annual Household Income:						
<$10,000	18	(15.2)	686	(19.2)	0.8	(0.4–1.3)
$10,000–$19,999	23	(19.5)	699	(19.6)	0.9	(0.6–1.6)
$20,000–$29,999	29	(24.6)	737	(20.6)	1.1	(0.7–1.8)
$30,000+	48	(40.7)	1373	(38.4)	1.0	(reference)
Maternal Occupation:						
not working	37	(30.8)	1142	(32.0)	0.9	(0.5–1.5)
clerical/sales	35	(29.2)	1119	(31.3)	0.8	(0.5–1.4)
service	16	(13.3)	444	(12.4)	1.0	(0.5–1.8)
factory	5	(4.2)	137	(3.8)	1.0	(0.4–2.6)
professional	27	(22.5)	730	(20.4)	1.0	(reference)
Paternal Occupation:						
not working	6	(5.0)	246	(6.9)	0.6	(0.2–1.3)
clerical/sales	13	(10.8)	618	(17.3)	**0.5**	**(0.3–0.9)**
service	14	(11.7)	480	(13.4)	0.7	(0.4–1.2)
factory	45	(37.5)	1279	(35.8)	0.8	(0.5–1.2)
professional	42	(35.0)	947	(26.5)	1.0	(reference)
Month of Pregnancy Confirmation:						
1st month	18	(15.2)	511	(14.3)	1.0	(reference)
2nd month	75	(63.6)	1955	(54.7)	1.1	(0.6–1.8)
3rd month	16	(13.5)	663	(18.6)	0.7	(0.3–1.2)
4th month or later	9	(7.6)	416	(11.6)	0.6	(0.3–1.4)

Data are missing for income (2 cases, 77 controls), paternal occupation (2 controls) and pregnancy confirmation month (2 cases, 27 controls).

Potential Risk Factors

Familial Cardiac and Noncardiac Anomalies

Congenital heart defects were reported among first-degree relatives in eight families (Table 8C.7). Among these were three with left-sided heart defects: two with coarctation of the aorta, and one with the hypoplastic left heart syndrome in a sibling. The pedigrees of these three families are shown in Figure 8A.2. Five other families reported the occurrence of congenital heart defects in first-degree rela-

Table 8C.7

Coarctation of Aorta: Congenital Defects and

Hereditary Disorders in First-Degree Relatives

The Baltimore-Washington Infant Study (1981–1989)

Proband	Sex	Noncardiac Anomalies	Relative	Diagnosis
A. Congenital Heart Defects:				
2212*	F	—	Brother	Coarctation of aorta
10991*	M	—	Father	Coarctation of aorta
11085*	M	—	Sister	Hypoplastic left heart
2611	M	—	Brother	Atrioventricular septal defect, Down syndrome
1952	F	—	Mother	Ventricular septal defect
10914	F	—	Brother	Atrial septal defect
2072	M	—	Mother	Patent arterial duct
1544	M	—	Father	Situs inversus totalis, normal heart
2240	M	—	Mother	Mitral valve prolapse
10068	F	—	Mother	Mitral valve prolapse
11165	M	—	Mother	Mitral valve prolapse
10808	F	Turner	Father	Mitral valve prolapse
11046	F	Turner	Father	Mitral valve prolapse
B. Congenital Defects of Other Organs:				
1362	M	—	Brother	Cataract
2683	M	—	Mother	Cleft palate
1379	F	—	Mother	Uterus bicornis
1420	F	—	Mother	Butterfly vertebrae
C. Hereditary Disorders				
2251	M	—	Father, Paternal Grandmother	Thalassemia
2274	F	—	Father, Paternal Aunt	Wilson's disease
10492	F	—	Father, Paternal Grandmother	Polydactyly

*See pedigrees in Section A, Figure 2.

tives who had various nonconcordant diagnoses. Mitral valve prolapse, though not counted as a congenital malformation in the risk factor analysis, was reported by five parents, two of whom were fathers of probands affected with Turner syndrome.

Single congenital defects of other organ systems were reported in four families and in three families the defects were hereditary: thalassemia, Wilson's disease, and polydactyly. Osteogenesis imperfecta, a rare heritable disorder, occurred in the maternal uncle of case #10991, whose father had coarctation.

Genetic and Environmental Factors

Univariate Analysis

Family history of congenital heart disease among first-degree relatives was significantly associated with coarctation of the aorta, with a nearly six-fold odds ratio (Table 8C.8). Parental age distributions did not differ significantly among

Table 8C.8
Coarctation of Aorta Univariate Analysis of Risk Factors
Baltimore-Washington Infant Study (1981–1989)

Variable	Cases N=120 no.	(%)	Controls N=3572 no	(%)	Odds Ratio (95% CI)	
Family History (first-degree relatives)						
congenital heart disease	8	(6.7)	43	(1.2)	**5.9**	**(2.7–12.8)**
noncardiac anomalies	7	(5.8)	165	(4.6)	1.3	(0.6–2.8)
Parental Age						
maternal age:						
<20	16	(13.3)	507	(14.2)	1.0	(0.6–1.7)
20–29	65	(54.2)	2009	(56.2)	1.0	(reference)
30+	39	(32.5)	1056	(29.6)	1.1	(0.7–1.7)
paternal age:						
<20	9	(7.5)	226	(6.3)	1.1	(0.5–2.2)
20–29	63	(52.5)	1724	(48.3)	1.0	(reference)
30+	48	(40.0)	1622	(45.4)	0.8	(0.6–1.2)
Maternal Reproductive History						
number of previous pregnancies:						
none	40	(33.3)	1159	(32.4)	1.0	(reference)
one	37	(30.8)	1097	(30.7)	1.0	(0.6–1.5)
two	22	(18.3)	709	(19.9)	0.9	(0.5–1.5)
three or more	21	(17.5)	607	(17.0)	1.0	(0.6–1.7)
previous miscarriage(s)	27	(22.5)	681	(19.1)	1.2	(0.8–1.9)
use of clomiphene citrate	4	(3.3)	20	(0.5)	**6.1**	**(2.1–18.2)**
Illnesses and Medications						
diabetes:						
overt	1	(0.8)	23	(0.6)	1.3	(0.2–9.7)
gestational	4	(3.3)	115	(3.2)	1.0	(0.3–2.9)
influenza	5	(4.2)	278	(7.8)	0.5	(0.2–1.3)
epilepsy	3	(2.5)	14	(0.4	**6.5**	**(1.8–23.0)**
macrodantin	4	(3.3)	13	(0.4)	**9.4**	**(3.0–29.4)**
guaifenesin	5	(4.2)	59	(1.7)	**2.6**	**(1.0–6.6)**
sympathomimetics	16	(13.3)	259	(7.3)	**2.0**	**(1.1–3.4)**
Lifestyle Exposures						
alcohol (any amount)	79	(65.8)	2101	(58.8)	1.3	(0.9–2.0)
smoking (cigarettes/day):						
none	71	(59.2)	2302	(64.5)	1.0	(reference)
1–10	21	(17.5)	565	(15.8)	1.2	(0.7–2.0)
11–20	19	(15.8)	529	(14.8)	1.2	(0.7–1.9)
>20	9	(7.5)	176	(4.9)	1.7	(0.8–3.4)
Home and Occupational Exposures						
ionizing radiation (occupational)	0		48	(1.3)	—	—
painting	11	(9.2)	367	(10.3)	0.9	(0.5–1.7)
varnishing	4	(3.3)	106	(3.0)	1.1	(0.4–3.1)
degreasing solvents	2	(1.7)	19	(0.5)	3.2	(0.7–13.8)
miscellaneous solvents	5	(4.2)	64	(1.8)	2.4	(0.9–6.0)

Table 8C.8—Continued
(Univariate Analysis of Risk Factors)

Variable	Cases N=120 no.	(%)	Controls N=3572 no	(%)	Odds Ratio (95% CI)	
Home and Occupational Exposures (cont'd)						
maternal solvent score:						
0 (no exposures)	103	(85.8)	2932	(82.1)	1.0	(reference)
1–11 (1 or more agents, each <once a week)	3	(2.5)	348	(10.0)	1.0*	(1.0–1.1)
12–89 (1 or more agents, each 1–3 times a week)	9	(7.5)	214	(6.0)	1.3*	(1.1–1.4)
90–540 (1 or more agents, each daily)	5	(4.2)	68	(1.9)	4.1*	(1.8–9.3)
Race of Infant						
white	94	(79.0)	2362	(66.2)	1.9	(1.2–3.0)
nonwhite	25	(21.0)	1207	(33.8)	1.0	(reference)

*Odds ratio and confidence intervals are from logistic regression for the scores of 5, 39, and 225 (midpoints of the score ranges 1–11, 12–89, and 90–540, respectively; risk factors are *maternal* unless labelled as paternal.

cases and controls, nor did the number of previous pregnancies and miscarriages. The fertility medication clomiphene citrate taken by four case mothers and 20 control mothers during the critical period, yielded a six-fold odds ratio.

Epilepsy was reported by 3 case mothers (2.5%) and only 14 controls (0.4%), with an odds ratio of 6.5. There was no case-control difference in the anticonvulsant medications taken; 1 case mother and 3 control mothers took no anticonvulsants during pregnancy.

Several medications reported by mothers were significantly associated with coarctation. Macrodantin (odds ratio 9.4) was used in the treatment of urinary tract infections in four cases and 13 controls. Use of expectorant medications containing guaifenesin and sympathomimetic drugs, predominantly pseudoephedrine hydrochloride, was also more prevalent among the case mothers than the controls.

Among lifestyle exposures and home and occupational exposures, only maternal solvent exposures were more prevalent among the cases. Increasing values of the summary solvent score were associated with increasing odds ratios: for a mother with infrequent exposures (score = 1 to 11) the odds ratio was the same as that for unexposed mothers, but those with moderate (score = 12 to 89) and heavy exposures (score = 90+) the odds ratios increased to 1.3 and 4.1 respectively.

Multivariate Analysis

In the multivariate analysis family history of congenital heart disease remained a strong and highly significant risk factor (Table 8C.9). All of the maternal illness and medication variables detected in the univariate analysis (clomiphene citrate, epilepsy, macrodantin, and sympathomimetics) were significant risk factors in the multivariate model, except for guaifenesin exposure which lost significance after adjustment for race and coexposures. Only the exposures to clomiphene citrate and macrodantin, however, were significant at the 1% level. The maternal solvent score continued to be associated with coarctation, with an odds ratio of 3.19 for mothers with daily exposures at home and/or at work, but the 99%

Table 8C.9
Coarctation of Aorta: Multivariate Model
Baltimore-Washington Infant Study (1981–1989)

Variable	Adjusted Odds Ratio	95% CI	99% CI
family history of congenital heart disease	4.6	2.0–10.8	1.5–13.9
epilepsy	5.3	1.4–20.2	0.9–30.6
clomiphene citrate	4.5	1.5–14.0	1.0–19.9
macrodantin	6.7	2.0–22.0	1.4–31.8
sympahtomimetics	1.8	1.1–3.2	0.9–3.8
solvent score:			
0 (not exposed)	1.00	(reference)	
1–11 (one or more agents, <once per week)	1.03	1.01–1.05	0.99–1.05
12–89 (one or more agents, 1–3 times per week)	1.22	1.04–1.43	0.99–1.50
90–540 (one or more agents, daily)	3.19	1.29–7.90	0.97–10.17
race (white)	1.7	1.1–2.7	0.9–3.1

*Odds ratios and confidence intervals for the scores 5, 39, and 225.

confidence intervals were just below unity. White race (odds ratio 1.7) was associated with this defect, but this association was not significant at the 1% level.

Eighty-nine cases were classified as isolated/simplex, and 31 as multiple anomalies/multiplex: the former group was large enough to permit multivariate analysis of potential risk factors, as shown on Table 8C.10. Five of the associations detected in the isolated/simplex subset were also found in the total group of coarctation cases and had similar odds ratios: maternal epilepsy, clomiphene citrate, macrodantin, sympathomimetics, and white race. However only the macrodantin exposure was significant at the 1% level. One new exposure, maternal smoking of two or more packs of cigarettes per day (4 case mothers and 35 controls), was associated with this subgroup, with an odds ratio of 3.5 (significant at the 5% level). The maternal solvent score, which was associated with the total coarctation group but not at the 1% level, was not associated with the isolated/simplex subset.

Discussion

This case-control evaluation of 126 infants with "pure" coarctation of the aorta resulted in several findings of importance. The malformation was not equally distributed within the population. It showed a predominance among infants of white race, with an almost two-fold risk in the adjusted analysis. Such a predominance of coarctation in white infants has been noted and discussed by several authors.[95,404,420]

Major noncardiac malformations were present in 11% of the cases, among which the most prominent was the Turner syndrome. Among other associated malformations, the VACTERL and Goldenhar syndromes are consistent with a hypothesis of an abnormality of neural crest derived ectomesenchyme.[129,216]

The well-known familial aggregation of left-sided obstructive defects was observed in this case population, however five of the eight familial cases had other

Table 8C.10
Coarctation of the Aorta:
Isolated/Simplex Subset (N=89) Multivariate Analysis of Risk Factors
Baltimore-Washington Infant Study (1981–1989)

Risk Factors	Cases	Controls	Adjusted Odds Ratio	95% CI	99% CI
Maternal Illnesses and Medications					
epilepsy	3	14	5.7	1.5–22.1	0.9–33.6
clomiphene citrate	3	17	4.9	1.3–18.3	0.9–27.5
macrodantin	3	10	6.7	1.7–27.1	1.1–41.7
sympathomimetics	14	225	2.2	1.2–4.0	0.9–4.8
Lifestyle Exposures					
maternal smoking (2 or more packs per day)	4	35	3.5	1.2–10.4	0.8–14.6
Sociodemographic Factors					
white race	70	2170	1.8	1.0–2.9	0.9–3.5

types of cardiovascular malformations, including septation defects. Also of note is the presence of heritable disorders in some of the families.

This malformation group was associated with maternal epilepsy and with several medical therapies, as well as with a dose-dependent exposure of the mother to industrial solvents. Solvent exposures were also associated with coarctation in a study of 50 cases and 756 controls in Finland,[419] but the odds ratio of 1.6 was not significant (95% confidence interval 0.6 to 4.2). However, in contrast to our analysis of "pure" coarctation, the Finnish study included 20 cases (40%) with ventricular septal defects, as well as six cases who had outflow anomalies such as transposed great arteries or common arterial trunk, and their case group was therefore quite different from ours.

Section D

Aortic Stenosis

Throughout the first half of this century aortic stenosis was considered a late outcome of acute rheumatic fever with fibrotic constriction of the aortic valve and fusion of its valve cusps. The classic clinical findings were those observed in adulthood when symptomatic patients presented with a heart murmur and congestive heart failure or bacterial endocarditis. Some patients suffered from chest pain and some died suddenly during exertion. The differentiation of *rheumatic* aortic stenosis from *congenital* aortic stenosis is a relatively recent phenomenon, with the development of pediatric cardiology as a specialty and the decline of rheumatic fever as a cause of aortic stenosis. Pediatricians who cared for children with active rheumatic carditis did not observe the development of aortic stenosis, even over years of follow-up, while internists saw cases with severe aortic stenosis and presumed it was of rheumatic origin.[446] Calcified stenosed valves in older patients could not easily be defined regarding their original anatomic structure, but the likelihood of a congenital anomaly in most cases became increasingly evident.[164,282]

Paul Dudley White described the characteristic case history of congenital aortic stenosis as related by a 40 year old patient: a heart murmur heard at the age of 12 years as a boy scout, again at 18 years as a sophomore in college (allowed to play football) and at 30, when he was turned down for military service. He was still asymptomatic at 40 years of age but over the next 6 years developed a "feeling of oppression" over the left chest on vigorous physical activity. He underwent a surgical valvotomy, which improved the function of a heavily calcified bicuspid aortic valve.[443]

A child with aortic stenosis is usually identified by the presence of a murmur in the absence of symptoms. This is usually a white male child. By myth if not data, these boys have a predilection for participation in competitive athletics, or perhaps the necessary medical examination brings them to attention. The murmur is maximal at the second right intracostal space, with a thrill in the suprasternal notch and radiation of the murmur to the carotid vessels. Diminished pulse volume and other abnormal characteristics are less helpful in children than in adults in assessing the degree of aortic obstruction. Thus echocardiographic examination and close follow-up is essential as the severity of obstruction increases with time.[227]

For infants the clinical picture is quite different. Critical aortic obstruction may present in the first week of life with low cardiac output and respiratory distress. The physical findings of these infants are usually those of poor perfusion with a soft cardiac murmur and soft heart sounds.[374] These infants often have myocardial dysfunction, endocardial fibroelastosis, and varying degrees of left ventricular hypoplasia. Mitral insufficiency may be a consequence of infarction of the left ventricular papillary muscles. Intervention is urgent and the mortality is high.

Surgical intervention for valvar aortic stenosis has undergone gradual changes during the time period of the Baltimore-Washington Infant Study. Valvo-

202

tomy with cardiopulmonary bypass (open valvotomy) has been replaced by balloon aortic valvuloplasty, which has become a feasible technique for the relief of severe aortic stenosis in infants. Many studies attempt to determine predictors of mortality by reviewing echocardiographic and cardiac catheterization data[195,361] as well morphometric characteristics in autopsy specimens[245] in relation to the types of operation contemplated. The long-term outlook may not be favorable, especially if the aortic valve anomaly is associated with abnormalities of left ventricular size and structure.

The Baltimore-Washington Infant Study evaluated congenital aortic stenosis for associations that might define differences and similarities to the closely related malformations of bicuspid aortic valve on the mild end of the anatomic spectrum of anomalies and of hypoplastic left heart syndrome on the severe end.

Study Population

Within the group of left-sided cardiac defects, 112 infants had a principal diagnosis of aortic stenosis. Aortic stenosis was associated with a ventricular septal defect in 10 infants and with aortic coarctation in nine. Excluding these cases as well as those with obstructions located above or below the aortic valve ring left 74 infants diagnosed with "pure" aortic valve stenosis.

Noncardiac Anomalies

Thirteen infants (17.6%) had associated noncardiac anomalies, a proportion similar to that found in the other left heart lesions. A diagnosis of Turner syndrome was confirmed in three infants with aortic stenosis. Other chromosome anomalies included 2 cases with trisomy 18, 1 with Down syndrome, and 1 with trisomy 1q (Table 8D.1). No infant with aortic stenosis had a reported Mendelian syndrome or a recognized association of multiple congenital anomalies. Among six patients with an additional noncardiac anomaly, two had cleft palate.

Table 8D.1
Aortic Stenosis (=74): Noncardiac Anomalies
Baltimore-Washington Infant Study (1981–1989)

Chromosomal Abnormalities:	N=7	(9.5%)
Turner syndrome	3	[# 1753, 10431, 10434]
Trisomy 18	2	[# 2561, 10271]
Down syndrome	1	[# 10040]
Trisomy 1q	1	[# 10783]
Syndromes	N=0	
Organ Defects:	N=6	(8.1%)
Cleft palate	2	[# 2287, 2374]
Cataract	1	[# 10398]
Hypospadias	1	[# 2361]
Pyloric stenosis	1	[# 11318]
Deficiency Thyroxin Binding Globulin	1	[# 2733]
No Associated Anomalies	N=61	(82.4%)

Descriptive Analyses

Prevalence by Time, Season, and Area of Residence

The total prevalence of pure aortic stenosis over the 9-year study period was 0.81 cases per 10,000 livebirths. Prevalence of the defect increased from 0.44 in the first 3 years of the study to 1.11 in the time from 1984 to 1986, (a period of increased use of echocardiography) and then declined somewhat (to 0.83) during 1987 to 1989 (Table 8D.2). Affected births during the fourth quarter (October through December) were slightly more common than in earlier quarters. Urban areas showed a low prevalence of infant aortic stenosis compared to suburban and rural areas.

Diagnosis and Course

Diagnosis of aortic stenosis was accomplished within the first week of life in 41% of the infants, and in the first month of life—the neonatal period—in approximately two thirds of the cases; only 15% of cases presented in the second half of the first year (Table 8D.3). Nearly half of the cases were diagnosed solely by echocardiography; the remainder also had cardiac catheterization except for one case that was discovered at autopsy.

Not all infants showed the clinical picture of critical aortic stenosis; only 38% underwent intervention (surgery or balloon aortic valvotomy) prior to 1 year of age. The all-cause mortality during the follow-up period was 14%.

Gender, Race, and Twinning

Compared to controls, there was an excess of males in this case group (Table 8D.4), and the racial distributions were strikingly different: 94.4% of the case in-

Table 8D.2
Aortic Stenosis: Prevalence
Baltimore-Washington Infant Study (1981–1989)

	Cases	Area Births	Prevalence per 10,000
Total Subjects	74	906,626	0.81
Year of Birth:			
1981–1983	12	274,558	0.44
1984–1986	33	295,845	1.11
1987–1989	28	336,223	0.83
Birth Quarter:			
1st (Jan–Mar)	17	219,145	0.78
2nd (Apr–June)	16	231,777	0.69
3rd (Jul–Sep)	17	233,626	0.73
4th (Oct–Dec)	24	222,078	1.08
Area of Residence:			
urban	8	208,568	0.38
suburban	56	584,022	0.96
rural	10	112,318	0.89

Table 8D.3
Aortic Stenosis: Clinical Data of Infants
Baltimore-Washington Infant Study (1981–1989)

	Cases N=74 no.	(%)
Age at Diagnosis:		
<1 week	30	(40.5)
1–4 weeks	17	(23.0)
5–24 weeks	16	(21.6)
25–52 weeks	11	(14.9)
Method of Diagnosis:		
echocardiography only	35	(47.3)
echocardiography and cardiac catheterization	10	(13.5)
echocardiography and surgery	1	(1.4)
echocardiography, cardiac catheterization, and surgery	24*	(32.4)
autopsy following other methods	3	(4.1)
autopsy only	1	(1.4)
Surgery During First Year of Life	28	(37.8)
Follow-Up:		
not followed	2	(2.7)
followed:	72	
died in first year of life	10	(13.9)
alive	62	(86.1)
median age at last visit	10 months	

*Includes 1 cardiac catherization and surgery (without echocardiography).

Table 8D.4
Aortic Stenosis: Infant Characteristics
Baltimore-Washington Infant Study (1981–1989)

Characteristic	Cases N=74 no.	(%)	Controls N=3572 no.	(%)
Gender:				
male	53	(71.6)	1,817	(50.9)
female	21	(28.4)	1,755	(49.1)
Race:*				
white	68	(94.4)	2,362	(66.1)
black	3	(4.2)	1,109	(31.0)
other	1	(1.4)	98	(2.7)
Twin births	1	(1.4)	53	(1.5)

*Missing race in 2 cases and 3 controls.

fants were white versus 66.1% of controls. A single case of aortic stenosis occurred in a twin birth.

Family interviews obtained for 65 cases (88%) permit case-control comparisons on additional characteristics as follows.

Birthweight and Gestational Age

Fetal growth characteristics of cases and controls are displayed in Table 8D.5. Elevated rates of low birthweight and prematurity characterized the cases, and mean birthweights and gestational ages of cases were reduced below those of controls. The proportion of small-for-gestational-age infants among the cases was not significantly different from that of controls. As the unadjusted data reflect differences in the gender and racial distributions, a comparison was made of fetal growth data in the white infants with aortic stenosis and in white controls. This comparison shows a greater excess of low birthweight and prematurity among the cases, but the case-control odds ratio for small-for-gestational age was not significantly elevated.

Sociodemographic Characteristics

Comparisons of sociodemographic data of cases and controls are shown in Table 8D.6. In addition to being white, case mothers were more likely to be mar-

Table 8D.5
Aortic Stenosis: Fetal Growth Characteristics
Baltimore-Washington Infant Study (1981–1989)

Characteristic	Cases		Controls	
	All	Whites	All	Whites
Number of Families				
Interviewed	65	62	3572	2362
Birthweight (grams)				
<2500	8 (12.3)	7 (11.3)	252 (7.1)	106 (4.5)
2500–3500	30 (46.2)	29 (46.8)	1853 (51.9)	1130 (47.9)
>3500	27 (41.5)	26 (41.9)	1467 (41.0)	1125 (47.6)
mean±standard error	3296±76	3293±76	3351±10	3454±11
range	1616–4352	1616–4312	340–5273	425–5273
Gestational Age (weeks)				
<38	10 (15.1)	10 (16.1)	339 (9.5)	164 (6.9)
38+	55 (84.9)	52 (83.9)	3233 (90.5)	2196 (93.1)
mean±standard error	38.4±0.8	38.4±0.8	39.6±0.1	39.8±0.04
range	30–44	30–44	20–47	20–47
Size for Gestational Age				
small (SGA)	5 (7.7)	4 (6.5)	211 (5.9)	91 (3.9)
normal	46 (70.8)	45 (72.6)	2712 (75.9)	1769 (74.9)
large	14 (21.5)	13 (21.0)	649 (18.2)	502 (21.3)
Odds Ratio for SGA				
(95% CI)	1.3 (0.5–3.3)	1.7 (0.6–4.8)	1.0 (reference)	1.0 (reference)

Table 8D.6

Aortic Stenosis: Sociodemiographic Characteristics

Baltimore-Washington Infant Study (1981–1989)

Interviewed Families	Cases N=65 no.	(%)	Controls N=3572 no.	(%)	Odds Ratio (95% CI)	
Race of Infant:						
nonwhite	3	(4.6)	1207	(33.8)	1.0	(reference)
white	62	(95.4)	2362	(66.2)	**10.6**	**(3.3–33.7)**
Maternal Marital Status:						
not married	7	(10.8)	990	(27.7)	1.0	(reference)
married	58	(89.2)	2582	(72.3)	**3.2**	**(1.4–7.0)**
Maternal Education:						
<high school	7	(10.8)	659	(18.4)	1.0	(reference)
high school	16	(24.6)	1265	(35.4)	1.2	(0.5–2.9)
college	42	(64.6)	1648	(46.1)	**2.4**	**(1.1–5.4)**
Paternal Education:						
<high school	8	(12.3)	650	(18.2)	1.0	(reference)
high school	19	(29.2)	1298	(36.3)	1.2	(0.5–2.7)
college	38	(58.5)	1624	(45.5)	1.9	(0.9–4.1)
Annual Household Income:						
<$10,000	9	(13.8)	686	(19.2)	1.0	(reference)
$10,000–$19,999	11	(16.9)	699	(19.6)	1.2	(0.5–2.9)
$20,000–$29,999	12	(18.5)	737	(20.6)	1.2	(0.5–3.0)
$30,000+	33	(50.8)	1373	(38.4)	1.8	(0.9–3.9)
Maternal Occupation:						
not working	15	(23.1)	1142	(32.0)	1.0	(reference)
clerical/sales	16	(24.6)	1119	(31.3)	1.1	(0.5–2.2)
service	7	(10.8)	444	(12.4)	1.2	(0.5–3.0)
factory	5	(7.7)	137	(3.8)	**2.7**	**(1.0–7.7)**
professional	22	(33.8)	730	(20.4)	**2.3**	**(1.2–4.5)**
Paternal Occupation:						
not working	1	(1.5)	246	(6.9)	0.2	(0.1–1.7)
clerical/sales	18	(27.7)	618	(17.3)	1.6	(0.9–3.0)
service	4	(6.2)	480	(13.4)	0.5	(0.2–1.3)
factory	23	(35.4)	1279	(35.8)	1.0	(reference)
professional	19	(29.2)	947	(26.5)	1.1	(0.6–2.1)
Month of Pregnancy Confirmation:						
1st month	15	(23.1)	511	(14.3)	**3.1**	**(1.0–9.3)**
2nd month	38	(58.5)	1955	(54.7)	2.0	(0.7–5.7)
3rd month	8	(12.3)	663	(18.6)	1.3	(0.4–4.2)
4th month or later	4	(6.2)	416	(11.6)	1.0	(reference)

Data are missing for income (77 controls), paternal occupation (2 controls), pregnancy confirmation month (27 controls).

ried and well educated, to hold professional occupations as well as factory occupations, and to seek prenatal care in the first month of pregnancy. Mean household incomes appeared higher among the cases, with an odds ratio of 1.8 for incomes over $30,000, but this was not statistically significant.

Potential Risk Factors

Familial Cardiac and Noncardiac Anomalies

Congenital heart defects in first-degree relatives of the proband were reported in three families, and mitral valve prolapse in a parent was reported in two additional families (Table 8D.7). In two families a sibling of the male proband was concordant for aortic stenosis, a brother in one family, and a sister in another. Case #10145 had three affected relatives, all with left-sided obstructive lesions of varying severity: the father had bicuspid aortic valve, one sister had hypoplastic left heart syndrome and one sister had aortic stenosis (see Figure 8A.2). The only case of aortic stenosis in a twin pair in our series was a female proband with Turner syndrome whose female co-twin was diagnosed with a ventricular septal defect.

In three additional case families, noncardiac anomalies were reported among relatives. Of interest is the family (#10398) of a female proband who had cataracts. Her father, a paternal uncle, and the paternal grandfather were each affected by a rare heritable disorder, osteogenesis imperfecta. In another family, a sister of a male proband was diagnosed with Sturge-Webber syndrome.

Genetic and Environmental Factors

Univariate Analysis

Compared to the total control group, a significantly elevated case-control odds ratio of 4.0 was detected for family history of congenital heart disease (Table 8D.8). There was an elevated odds ratio for maternal age over 30 years. A

Table 8D.7
Aortic Stenosis: Congenital Defects and
Hereditary Disorders in First-Degree Relatives
The Baltimore-Washington Infant Study (1981–1989)

Proband	Sex	Diagnosis	Relative	Diagnosis
A. Congenital Heart Defects:				
11573*	M		Brother	Aortic stenosis
10145*	M		Father	Bicuspid aortic valve
			Sister	Hypoplastic left heart
			Sister	Aortic stenosis
10434	F	Turner	Female Twin	Ventricular septal defect
3383	M		Mother	Mitral valve prolapse
10797	M		Father	Mitral Valve prolapse
B. Congenital Defects of Other Organs:				
10398	F	cataract	Father	Osteogenesis imperfecta
			Paternal Grandmother	
			Paternal Uncle	
10420	M		Sister	Sturge-Webber
10839	M		Maternal Half-Sister	Hemivertebrae

*See pedigree in Section A, Figure 2.

Table 8D.8
Aortic Stenosis: Univariate Analysis of Risk Factors
Baltimore-Washington Infant Study (1981–1989)

Variable	Cases N=65 no.	(%)	Controls N=3572 no.	(%)	Odds Ratio (95%CI)	
Family History						
(first-degree relatives)						
congenital heart disease	3	(4.6)	43	(1.2)	**4.0**	**(1.2–13.1)**
noncardiac anomalies	3	(4.6)	165	(4.6)	1.0	(0.3–3.2)
Parental Age						
maternal age:						
<20	3	(4.6)	507	(14.2)	0.4	(0.1–1.2)
20–29	33	(50.8)	2009	(56.2)	1.0	(reference)
30+	29	(44.6)	1056	(29.6)	**1.7**	**(1.0–2.8)**
paternal age:						
<20	0		226	(6.3)	—	—
20–29	30	(46.2)	1724	(48.3)	1.0	(reference)
30+	35	(53.8)	1622	(45.4)	1.2	(0.8–2.0)
Maternal Reproductive						
History						
number of previous pregnancies:						
none	15	(23.1)	1159	(32.4)	1.0	(reference)
one	19	(29.2)	1097	(30.7)	1.3	(0.7–2.6)
two	18	(27.7)	709	(19.9)	2.0	(0.9–3.9)
three or more	13	(20.0)	607	(17.0)	1.7	(0.8–3.5)
previous miscarriage(s)	15	(23.1)	681	(19.1)	1.3	(0.7–2.3)
irregular periods:						
total	19	(29.9)	574	(16.1)	**2.2**	**(1.3–3.7)**
treated with progesterone	3	(4.6)	48	(1.3)	**3.6**	**(1.1–11.7)**
Illnesses and Medications						
diabetes:						
overt	1	(1.5)	23	(0.6)	2.4	(0.3–18.1)
gestational	7	(10.8)	115	(3.2)	**3.6**	**(1.6–8.1)**
influenza	9	(13.9)	278	(7.8)	1.9	(0.9–3.9)
tranquilizers*	3	(4.6)	48	(1.3)	**3.6**	**(1.1–11.7)**
Lifestyle Exposures						
alcohol (any amount)	39	(60.0)	2101	(58.8)	1.1	(0.6–1.7)
smoking (cigarettes/day):						
none	47	(72.3)	2302	(64.5)	1.0	(reference)
1–10	6	(9.2)	565	(15.8)	0.5	(0.2–1.2)
11–20	10	(15.4)	529	(14.8)	0.9	(0.5–1.8)
>20	2	(3.1)	176	(4.9)	0.6	(0.1–2.3)
Home and						
Occupational Exposures						
ionizing radiation (occupational)	1	(1.5)	48	(1.3)	1.1	(0.2–8.9)
Race of Infant						
white	62	(95.4)	2362	(66.2)	**10.6**	**(3.3–33.7)**
nonwhite	3	(4.6)	1207	(33.8)	1.0	(reference)

*2 cases and 35 controls used benzodiazepines; 1 case used buspirone; risk factors are *maternal* unless labelled as paternal.

greater proportion of case mothers reported having irregular periods in the 6 months prior to the proband's conception date, than did control mothers, and a greater proportion of case mothers took progesterone during the critical period. Maternal diabetes was also associated with aortic stenosis, but unlike other diagnostic groups described in previous chapters the case excess was in self-reported gestational diabetes, not overt (pregestational) diabetes. None of the affected case mothers reported taking insulin during pregnancy. Maternal use of tranquilizers was associated with aortic stenosis, with two case mothers reporting benzodiazepine exposure and one reporting exposure to buspirone. There were no case-control differences in parental lifestyle exposures or in home and occupational exposures.

Multivariate Analysis

Since the majority of cases (all but 3) were white, multivariate analysis included two models: a) an analysis of all subjects including race as a factor in the regression, and b) a separate analysis of white cases and white controls. These results are shown in Table 8D.9.

In the analysis of all subjects (Part A of Table 8D.9), gestational diabetes and family history of congenital heart disease were each strongly associated with case status, each showing a nearly four-fold odds ratio. White race was very strongly associated, with a high odds ratio of 10.5. When nonwhite subjects were excluded from the analysis (Part B of Table 8D.9), the same associations of gestational diabetes and family history were detected, each with a somewhat higher odds ratio than in the analysis of part A.

Discussion

Examination of all of this relatively small group of infant cases with aortic stenosis reveals two findings of overriding importance: the great predominance of infants of white race and the obvious heterogeneity within the group as shown by the presence of noncardiac anomalies in almost one fifth of the cases. Half of these

Table 8D.9
Aortic Stenosis: Multivariate Models
Baltimore-Washington Infant Study (1981–1989)

Variable	Adjusted Odds Ratio	95% CI	99% CI
Part A-All subjects: N=65 cases and 3,572 controls			
family history of congenital heart disease	3.7	1.1–12.4	0.8–14.1
gestational diabetes	3.9	1.7–8.8	1.6–13.8
race (white)	10.5	3.3–33.4	2.9–43.0
Part B-White subjects only: N=62 cases and 2,362 controls			
family history of congenital heart disease	3.8	1.1–12.9	0.8–18.7
gestational diabetes	4.2	1.8–9.5	1.4–12.2

cases had chromosome defects, including 3 cases with Turner syndrome, 3 with trisomies, and 1 with a partial trisomy. No Mendelian disorders or non-Mendelian associations of malformations were present.

Within the total group of enrolled cases, a comparison was made of the analyzed case group to the nonanalyzed group of cases (ie, those excluded because of associated cardiac malformations or nonrespondents). As this selection bias was also possible for coarctation and bicuspid aortic valve, the comparison of analyzed and excluded patients is presented in Section F of this chapter.

The univariate analyses revealed an odds ratio of 10.6 for white race and other associations typically characteristic of the white regional population, including maternal college education, married status, professional workers, and early confirmation of pregnancy, as well as somewhat greater paternal age. Other than the positive family history, the multivariate model revealed only gestational diabetes as a risk factor, significant at the 1% level and based on seven affected mothers. In some other groups we considered the possibility that gestational diabetes might represent an indicator of some medical and social disadvantage, but for this case group this contrasts with the findings of socioeconomic advantages.

In a previous report on white-black differences in cardiovascular malformations[95] we examined 2087 Baltimore-Washington Study cases registered during the first 6 years of the study (1981 to 1986) in comparison to controls (N = 2721); there were 67 cases of aortic stenosis enrolled and none of the cases were excluded in the analyses. We found an unadjusted case-control odds ratio for white race of 3.6 (95% confidence interval 1.7 to 7.6). Race was then evaluated by socioeconomic strata. In higher socioeconomic strata the white-black difference diminished compared to that in the lower strata, although the excess risk for white race remained. This finding was not duplicated in the current analysis of the "pure" cases of aortic stenosis.

Section E

Bicuspid Aortic Valve

Bicuspid aortic valve diagnosed as a cardiovascular malformation in infants is recent, and entirely due to echocardiographic studies. Pediatric cardiology textbooks, if they mention this diagnosis at all, will refer to it as a trivial diagnosis associated with other diagnoses of importance, or mention its possible evolution into aortic stenosis and the potential for bacterial endocarditis in later life.

In the Baltimore-Washington Infant Study bicuspid aortic valve came to special attention because of its coincidence in families affected by various left-sided obstructive cardiac anomalies. For this reason we have created this separate section of the chapter to examine this anomaly and its possible similarities and differences from the other left-sided defects.

Study Population

Cardiac Abnormalities

Seventy-two infants with a primary diagnosis of bicuspid aortic valve were enrolled, five of whom also had a ventricular septal defect. Their exclusion from the analysis resulted in a group of 67 infants with "pure" bicuspid aortic valve.

Noncardiac Anomalies

Eleven infants (16.4%) had associated noncardiac anomalies (Table 8E.1), among whom there were 2 cases of Turner syndrome and 1 case of each Down syndrome, trisomy 13, and deletion 4p. No Mendelian syndromes were diagnosed, but one infant had characteristics of the VACTERL association and another was reported with the diagnosis of dihydantoin embryopathy.

Descriptive Analyses

Prevalence by Time, Season, and Area of Residence

The total prevalence of bicuspid aortic valve during the 9 years of the study was 0.74 cases per 10,000 livebirths. No temporal trend in prevalence was observed, but the prevalence was much higher from 1984 to 1986 than in the other years. The prevalence of first-quarter (January to March) births was lower than the prevalence in the other birth quarters (Table 8E.2). Urban areas showed a higher birth prevalence than suburban areas, while rural areas of the region had the lowest prevalence.

Table 8E.1
Bicuspid Aortic Valve (N=67) Noncardiac Anomalies
Baltimore-Washington Infant Study (1981–1989)

Chromosomal Abnormalities:	**N=5**	**(7.4%)**
Turner's syndrome	2	[#2056, 10070]
Down syndrome	1	[#3357]
Trisomy 13	1	[#3740]
Deletion 4p	1	[#10256]
Syndromes	**N=2**	**(3.0%)**
a. Mendelian Syndromes	**0**	
b. Non-Mendelian Associations:	**2**	
VACTERL	1	[#3660]
Dihydantoin embryopathy	1	[#10547]
Organ Defects:	**N=4**	**(6.0%)**
Cleft palate	1	[#2998]
Diaphragmatic hernia	1	[#11015]
Hydronephrosis	1	[#1902]
Pancytopenia	1	[#2180]
No Associated Anomalies	**N=56**	**(83.6%)**

Diagnosis and Course

In contrast to other forms of left-sided defects, the diagnosis of bicuspid aortic valve was accomplished relatively late: 67% of cases were diagnosed after the first month of life and nearly a quarter after 24 weeks (Table 8E.3). Echocardiography was the sole method of diagnosis in 97% of the cases, and no case was diagnosed by autopsy alone or had surgical intervention in the first year of life. Of 2 infants who died early, 1 had hydrops fetalis (pancytopenia) and 1 had a diaphragmatic hernia.

Table 8E.2
Bicuspid Aortic Valve: Prevalence
Baltimore-Washington Infant Study (1981–1989)

	Cases	Area Births	Prevalence per 10,000
Total Subjects	67	906,626	0.74
Year of Birth:			
1981–1983	18	274,558	0.66
1984–1986	29	295,845	0.98
1987–1989	20	336,223	0.59
Birth Quarter:			
1st (Jan–Mar)	11	219,145	0.50
2nd (Apr–Jun)	19	231,777	0.82
3rd (Jul–Sep)	17	233,626	0.73
4th (Oct–Dec)	20	222,078	0.90
Area of Residence:			
urban	18	208,568	0.86
suburban	43	584,022	0.74
rural	6	112,318	0.53

Table 8E.3
Bicuspid Aortic Valve: Clinical Data
Baltimore-Washington Infant Study (1981–1989)

	Cases N=67 no.	(%)
Age at Diagnosis:		
<1 week	14	(20.9)
1–4 weeks	8	(11.9)
5–24 weeks	30	(44.8)
25–52 weeks	15	(22.4)
Method of Diagnosis:		
echocardiography only	65	(97.0)
cardiac catheterization only	1	(1.5)
cardiac catheterization and echocardiography	1	(1.5)
autopsy	0	
Surgery During First Year of Life	0	
Follow-Up:		
not followed	12	(17.9)
followed:	55	
died in the first year of life	2	(3.6)
alive	53	(96.4)
median age at last visit	9 months	

Gender, Race, and Twinning

There was a slight preponderance of males among the cases (57%) compared to controls (51%). Racial distributions were similar in the two groups (Table 8E.4). Only a single case occurred in a twin birth (1.5%), the same proportion of twins as in the controls.

Information obtained by interview on 60 cases (90%) provided information on other characteristics.

Birthweight and Gestational Age

Compared to controls, infants with bicuspid aortic valve had only a slightly elevated rate of low birthweight and somewhat more elevated rate of prematurity (Table 8E.5). Cases had a mean birthweight deficit of only 39 grams, and there was no significant elevation in the odds ratio of being small for gestational age.

Sociodemographic Characteristics

There were no significant case-control differences in the distributions of sociodemographic variables (Table 8E.6). Case families did not differ significantly from controls in maternal marital status, parental education, family income, parental occupations, or in the month of pregnancy confirmation.

Table 8E.4
Bicuspid Aortic Valve: Infant Characteristics
Baltimore-Washington Infant Study (1981–1989)

Characteristic	Cases N=67 no.	(%)	Controls N=3572 no.	(%)
Gender:				
male	38	(56.7)	1,817	(50.9)
female	29	(43.3)	1,755	(49.1)
Race*:				
white	45	(68.2)	2,362	(66.1)
black	20	(30.3)	1,109	(31.0)
other	1	(1.5)	98	(2.7)
Twin births	1	(1.5)	53	(1.5)

*Missing race in 1 case and 3 controls.

Table 8E.5
Bicuspid Aortic Valve: Fetal Growth Characteristics
Baltimore-Washington Infant Study (1981–1989)

Characteristic	Cases no.	(%)	Controls no.	(%)
Number of Families Interviewed	60		3572	
Birthweight (grams)				
<2500	5	(8.5)	252	(7.1)
2500–3500	28	(47.5)	1853	(51.9)
>3500	26	(44.0)	1467	(41.0)
mean±standard error	3312±99		3351±10	
range	928–4933		340–5273	
Gestational Age (weeks)				
<38	9	(15.0)	339	(9.5)
38+	51	(85.0)	3233	(90.5)
mean ±standard error	39.2±0.3		39.6±0.1	
range	29–42		20–47	
Size for Gestational Age				
small (SGA)	4	(6.7)	211	(5.9)
normal	42	(70.0)	2712	(75.9)
large	14	(23.3)	649	(18.2)
Odds Ratio for SGA (95% CI)	1.1	(0.4–3.0)	1.0	(reference)

Table 8E.6

Bicuspid Aortic Valve: Sociodemiographic Characteristics

Baltimore-Washington Infant Study (1981–1989)

Interviewed Families	Cases N=60 no. (%)		Controls N=3572 no. (%)		Odds Ratio (95% CI)	
Maternal Marital Status:						
not married	17	(28.3)	990	(27.7)	1.0	(0.6–1.8)
married	43	(71.7)	2582	(72.3)	1.0	(reference)
Maternal Education:						
<high school	5	(8.3)	659	(18.4)	0.5	(0.2–1.2)
high school	28	(46.7)	1265	(35.4)	1.4	(0.8–2.3)
college	27	(45.0)	1648	(46.1)	1.0	(reference)
Paternal Education:						
<high school	10	(16.6)	650	(18.2)	1.0	(0.5–2.1)
high school	25	(41.7)	1298	(36.3)	1.3	(0.7–2.2)
college	25	(41.7)	1624	(45.5)	1.0	(reference)
Annual Household Income:						
<$10,000	7	(11.7)	686	(19.2)	0.5	(0.2–1.1)
$10,000–$19,999	8	(13.3)	699	(19.6)	0.5	(0.2–1.1)
$20,000–$29,999	14	(23.3)	737	(20.6)	0.9	(0.5–1.6)
$30,000+	30	(50.0)	1373	(38.4)	1.0	(reference)
Maternal Occupation:						
not working	12	(20.0)	1142	(32.0)	0.5	(0.3–1.1)
clerical/sales	22	(36.7)	1119	(31.3)	1.0	(0.5–2.0)
service	9	(15.0)	444	(12.4)	1.1	(0.4–2.5)
factory	3	(5.0)	137	(3.8)	1.1	(0.3–4.0)
professional	14	(23.3)	730	(20.4)	1.0	(reference)
Paternal Occupation:						
not working	2	(3.3)	246	(6.9)	0.4	(0.1–1.9)
clerical/sales	11	(18.3)	618	(17.3)	0.9	(0.4–2.0)
service	8	(13.3)	480	(13.4)	0.9	(0.4–2.0)
factory	20	(33.3)	1279	(35.8)	0.8	(0.4–1.6)
professional	18	(30.0)	947	(26.5)	1.0	(reference)
Month of Pregnancy Confirmation:						
1st month	7	(11.7)	511	(14.3)	1.0	(reference)
2nd month	33	(55.0)	1955	(54.7)	1.2	(0.5–2.8)
3rd month	12	(20.0)	663	(18.6)	1.3	(0.5–3.4)
4th month or later	7	(11.7)	416	(11.6)	1.2	(0.4–3.5)

Data are missing for income (1 case, 77 controls), paternal occupation (1 case, 2 controls) and pregnancy confirmation month (1 case, 27 controls).

Potential Risk Factors

Familial Cardiac and Noncardiac Anomalies

Familial aggregation of left-sided cardiac defects is evident in three families of probands with bicuspid aortic valve (see Figure 8A.2). In one of these families, both the mother and maternal grandmother had a diagnosis of bicuspid aortic valve. In each of two additional families, the proband's brother had hypoplastic left heart syndrome. These and other families with congenital anomalies are described in Table 8E.7. In addition to the familial cases with left-sided defects,

Table 8E.7
Bicuspid Aortic Valve Congenital Defects and
Hereditary Disorders in First Degree Relatives
The Baltimore-Washington Infant Study (1981–1989)

Proband	Sex	Diagnosis	Relative	Diagnosis
A. Congenital Heart Defects:				
1365*	M	—	Mother	Bicuspid aortic valve
			Maternal Grandmother	Bicuspid aortic valve, cleft palate
1411*	M	—	Brother	Hypoplastic left heart
10775*	M	—	Brother	Hypoplastic left heart
11313	F	—	Sister	Ventricular septal defect
10502	F	—	Mother	Cardiomyopathy
3016	M	—	Father and Paternal Grandfather	Mitral valve prolapse
B. Congenital Defects of Other Organs:				
1755	M	—	Brother	Polydactyly
2007	F	—	Father	Hydrocephaly
10547	M	dihydantoin embryopathy	Mother	Cavernous angioma, familial

*Pedigrees in Section A, Figure 2.

there was one family in which the proband's sister had a ventricular septal defect, and another family in which the mother had cardiomyopathy. Mitral valve prolapse (not included in the statistical analysis of family history) was reported in both the father and paternal grandfather of a male proband. Noncardiac anomalies were reported only in three families: in one of these, a female proband with dihydantoin embryopathy was born to a mother who had a familial form of cavernous angioma.

Genetic and Environmental Factors

Univariate Analysis

Univariate case-control comparisons are shown in Table 8E.8. There was a significant excess of cases with family history of cardiac defects (odds ratio 7.5), but cases and controls did not differ significantly in the proportion with familial noncardiac anomalies. Parental ages and maternal reproductive history variables were distributed similarly among cases and controls. No case mother reported overt diabetes and there was no case-control difference in reported gestational diabetes. Significantly more case mothers reported taking ibuprofen compared to controls (odds ratio 3.8). Neither alcohol nor smoking was associated with bicuspid aortic valve.

Among home and occupational exposures, the only association was with maternal exposures to degreasing solvents (odds ratio 6.4) based on the reports of two case mothers, both of whom cleaned guns at home (once) during the critical period.

Table 8E.8
Bicuspid Aortic Valve Univariate Analysis of Risk Factors
Baltimore-Washington Infant Study (1981–1989)

Variable	Cases N=60 no.	Cases N=60 (%)	Controls N=3572 no.	Controls N=3572 (%)	Odds Ratio (95% CI)	
Family History (first-degree relatives)						
congenital heart disease	5	(8.3)	43	(1.2)	**7.5**	**(2.8–19.6)**
noncardiac anomalies	3	(5.0)	165	(4.6)	1.1	(0.2–2.0)
Parental Age						
maternal age:						
<20	6	(10.0)	507	(14.2)	0.7	(0.3–1.6)
20–29	36	(60.0)	2009	(56.2)	1.0	(reference)
30+	18	(30.0)	1056	(29.6)	0.9	(0.5–1.7)
paternal age:						
<20	4	(6.7)	226	(6.3)	1.1	(0.4–3.3)
20–29	27	(45.0)	1724	(48.3)	1.0	(reference)
30+	29	(48.3)	1622	(45.4)	1.1	(0.7–1.9)
Maternal Reproductive History						
number of previous pregnancies:						
none	16	(26.7)	1159	(32.4)	1.0	(reference)
one	22	(36.7)	1097	(30.7)	1.5	(0.8–2.8)
two	9	(15.0)	709	(19.9)	0.9	(0.4–2.1)
three or more	13	(21.7)	607	(17.0)	1.6	(0.7–3.2)
previous miscarriage(s)	14	(23.3)	681	(19.1)	1.3	(0.7–2.4)
Illnesses and Medications						
diabetes:						
overt	0	—	23	(0.6)	—	
gestational	2	(3.3)	115	(3.2)	1.0	(0.3–4.3)
influenza	7	(11.7)	278	(7.8)	1.6	(0.7–3.5)
ibuprofen	7	(11.7)	119	(3.3)	**3.8**	**(1.7–8.6)**
Lifestyle Exposures						
alcohol (any amount)	35	(58.3)	2101	(58.8)	1.0	(0.6–1.6)
smoking (cigarettes/day):						
none	40	(66.7)	2302	(64.5)	1.0	(reference)
1–10	10	(16.7)	565	(15.8)	1.0	(0.5–2.0)
11–20	8	(13.3)	529	(14.8)	0.9	(0.4–1.9)
>20	2	(3.3)	176	(4.9)	0.7	(0.1–2.7)
Home and Occupational Exposures						
ionizing radiation (occupational)	0	—	48	(1.3)	—	
degreasing solvents	2	(3.3)	19	(0.5)	**6.4**	**(1.5–28.3)**
Race of Infant						
white	41	(68.3)	2362	(66.2)	1.1	(0.6–1.9)
nonwhite	19	(31.7)	1207	(33.8)	1.0	(reference)

Risk factors are *maternal* unless labelled as paternal.

Multivariate Analysis

Three factors were associated with bicuspid aortic valve in the multivariate analysis of potential risk factors (Table 8E.9). Family history of congenital heart

Table 8E.9
Bicuspid Aortic Valve (N=60):
Multivariate Analysis of Risk Factors
Baltimore-Washington Infant Study (1981–1989)

Risk Factors	Odds Ratio	95% CI	99% CI
family history of congenital heart disease	7.2	2.7–19.1	2.0–25.8
ibuprofen	4.1	1.8–9.3	1.4–11.9
maternal age >34 years	2.5	1.3–4.8	1.0–5.9

disease was the strongest of these associations, with an adjusted odds ratio of 7.2. Maternal use of ibuprofen was also a strong association (odds ratio 4.1). Both associations were significant at the 1% level, as was the association with maternal age over 34 years. The case group was too small to conduct separate analyses of isolated versus multiple affected cases.

Discussion

The epidemiologic evaluation of infants with bicuspid aortic valve complements the evaluations of the other pure left-sided obstructive lesions. The benign clinical findings are consistent with a trivial deviation from normal, evident in the similarity of the case group to the controls, with two exceptions: an associated risk of family history of cardiac malformations and an environmental risk factor of ibuprofen exposure during the critical period (significant at the 1% level).

In this case group two other findings are of interest: the absence of a difference in race and of gender from the control population is in sharp contrast with the finding for aortic stenosis, in which most cases were white, and in which there was a male preponderance. This suggests that our infants with bicuspid aortic valve do not represent the same or even a mild form of aortic obstructive disease.

Section F

Summary and Conclusions

The comparative evaluations of the four major anatomic phenotypes of left-sided obstructive heart defects resulted in surprising and striking evidence of heterogeneity of these diagnostic subgroups and also of unexpected similarities among them. In this section of the chapter we compare and contrast the results from the diagnosis-specific sections of this chapter, focusing first on the descriptive analyses (Table 8F.1 and Table 8F.2) that shed light on important characteristics of the infants themselves, and focusing secondly on the risk factor analyses (Table 8F.3 and Table 8F.4) that may provide etiologic clues for future investigations.

Descriptive Analyses

Comparison of the characteristics of the infants in each phenotype group to those of controls is shown on Table 8F.1.

Infant Characteristics

Noncardiac anomalies occurred in a similar proportion of infants across the four diagnostic groups (11% to 18%), and were greatly in excess compared to controls (1.7% affected). A similar pattern was observed for chromosomal anomalies, but Mendelian syndromes were completely absent in infants with aortic stenosis and in those with bicuspid aortic valve.

In gender distributions, there was an excess of males with only aortic stenosis and hypoplastic left heart syndrome, while the slightly elevated proportion of males with coarctation of aorta and bicuspid aortic valve was not significantly different from controls. More striking is the racial distribution: a great excess of white infants was present among the cases of aortic stenosis and to a lesser degree in coarctation, but there was no racial difference among the infants with hypoplastic left heart syndrome and those with bicuspid aortic valve compared to controls.

The mean birthweight was lower than that of controls with hypoplastic left heart syndrome and with coarctation, for both of which the mean gestational age was similar to controls. There was an elevated odds ratio for being small for gestational age (Table 8F.1), with a four-fold excess for the former diagnosis and two-fold excess for the latter. Infants with aortic valve stenosis and those with bicuspid aortic valve were not more likely than controls to be small for gestational age.

Race and Gender: Epidemiologic Considerations

The prominence of white race among infants with coarctation and aortic stenosis has previously been suggested in the literature.[284,404,420] It is also ex-

Table 8F.1

Left-Sided Obstructive Defects and Controls: Comparison of Infant Characteristics

Baltimore–Washington Infant Study (1981–1989)

Infant Characteristic	Hypoplastic Left Heart Syndrome N=162 n (%)	Coarction of the Aorta N=126 n (%)	Aortic Valve Stenosis N=74 n (%)	Bicuspid Aortic Valve N=67 n (%)	Controls N=3572 n (%)
Noncardiac anomalies:					
any	24 (14.8)	14 (11.1)	13 (17.6)	11 (16.4)	61 (1.7)
chromosome abnormalities	9 (5.6)	10 (7.9)	7 (9.5)	5 (7.4)	3 (0.1)
Mendelian syndromes	4 (2.5)	2 (1.6)	0	0	4 (0.1)
Gender					
male	95 (58.6)	73 (57.9)	53 (71.6)	38 (56.7)	1817 (50.9)
female	67 (41.4)	53 (42.1)	21 (28.4)	29 (43.3)	1755 (49.1)
odds ratio (95% CI) for males	**1.4 (1.0–1.9)**	1.3 (0.9–1.9)	**2.4 (1.5–4.1)**	1.3 (0.8–2.1)	1.0 (reference)
Race					
white	102 (64.6)	98 (79.0)	68 (94.4)	45 (68.2)	2362 (66.1)
black	53 (33.5)	24 (19.4)	3 (4.2)	20 (30.3)	1109 (31.0)
other	3 (1.9)	2 (1.6)	1 (1.4)	1 (1.5)	98 (1.5)
odds ratio (95% CI) for whites	0.9 (0.7–1.3)	**1.9 (1.2–3.0)**	**8.7 (3.2–23.9)**	1.1 (0.6–1.8)	1.0 (reference)
Mean Birth Weight (g)	3071	3153	3296	3312	3351
Mean Gestational Age (weeks)	39.1	39.2	38.4	39.2	39.6
Size for Gestational Age					
small	30 (21.7)	14 (11.7)	5 (7.7)	4 (6.7)	211 (5.9)
normal	91 (65.9)	88 (73.3)	46 (70.8)	42 (70.0)	2712 (75.9)
large	17 (12.3)	18 (15.0)	14 (21.5)	14 (23.3)	649 (18.2)
odds ratio (95% CI) for small infants	**4.4 (2.9–6.8)**	**2.1 (1.2–3.7)**	1.3 (0.5–3.3)	1.1 (0.4–3.0)	1.0 (reference)

[1]Analysis restricted to interviewed families.

amined in an earlier report from the Baltimore-Washington Infant Study.[96] In that study, among all 112 infants with aortic valve stenosis only 12% were non-white, and among the group of 74 with "pure" aortic stenosis, only 5% were non-white.

We evaluated the effect of restricting the analysis to "pure" diagnoses by examining the racial distribution for the stepwise diminishing subsets. We looked at all enrolled cases, the excluded ("non-pure") cases, the cases with "pure" diagnoses, and the subset in which the families were interviewed. The results are shown in Table 8F.2. For the group of aortic valve stenosis there was a difference between the original total group of enrolled cases and the "pure" diagnostic group, as the exclusion of 38 subjects with associated cardiac anomalies also excluded eight of the 11 black infants.

One might question the interpretation of the almost completely white group of infants with "pure" aortic stenosis: is this due to an actual high risk for white infants or is there a selection bias for the early recognition of white infants, perhaps by preferential selection for echocardiographic study? Such a selection bias seems unlikely, since the same diagnostic considerations would likely be present for cases with hypoplastic left heart syndrome, which did not show significant variation by race. In contrast, for coarctation of the aorta the slight excess of white infants was not altered by the exclusion of 70 of the 196 total enrolled cases, which included 15 of the 39 black infants.

An analysis of gender distributions before and after diagnostic exclusions revealed a similar pattern to that of the racial distributions (data not shown); only in the group of infants with aortic stenosis did the gender distributions change significantly as a result of diagnostic exclusions. In that group, the preponderance of males among all cases (64%; odds ratio 1.7; 95% confidence interval 1.2 to 2.6) became somewhat greater (72%; odds ratio 2.5; 95% confidence interval 1.5 to 4.4) after excluding cases that did not fit the definition of "pure" aortic valve stenosis.

Storch and Mannick[404] report findings similar to ours. In their study in Louisiana there was little variation in the distribution of hypoplastic left heart syndrome by sex or race, but the prevalence of aortic stenosis and coarctation (analyzed as a single group in their study) was markedly higher among whites and among male infants. Similar results were reported by Torfs et al[420] in California.

Regarding the possible origins of fetal growth retardation, the neonatal anthropometry of infants with isolated hypoplastic left heart syndrome and with coarctation was studied in detail by Rosenthal[372] who used additional information on infant length and head circumference from the birth records of the Baltimore-Washington Infant Study cases. He found evidence that suggested that different mechanisms of growth impairment may be operating. Most notably, infants with hypoplastic left heart syndrome had an especially small head volume for birthweight, even after accounting for their global growth retardation, and infants with coarctation of the aorta had diminished birthweight and birthlength but a normal head circumference (ie, greater head volume for birthweight). That study suggests that fetal circulatory abnormalities may influence the abnormal pattern of fetal growth, a hypothesis consistent with the presumed hemodynamic origin of these malformations.

Table 8F.2
Left-Sided Obstructive Defects: Race Distribution Baltimore-Washington Infant Study (1981–1989)

Diagnostic Group	N	White n (%)	Black n (%)	Other n (%)	comments
Hypoplastic Left Heart					
All	164	104 (65.6)	53 (32.5)	3 (1.8)	Distributions are
Exclusions	2	2 —	0 —	0 —	not
After Exclusions (pure)	162	102 (64.6)	53 (33.5)	3 (1.9)	different from controls
Interviewed	138	91 (65.9)	45 (32.6)	2 (1.4)	(O.R.=0.9)
Coarctation of the Aorta					
All	196	150 (77.3)	39 (20.1)	5 (2.6)	Moderate
Exclusions	70	52 —	15 —	3 —	excess of whites
After Exclusions (pure)	126	98 (79.0)	24 (19.4)	2 (1.6)	(O.R.=1.9)
Interviewed	120	94 (79.0)	24 (20.0)	1 (1.0)	
Aortic Valve Stenosis					
All	112	97 (88.2)	11 (10.0)	2 (1.8)	Great excess of
Exclusions	38	29 —	8 —	1 —	whites
After Exclusions (pure)	74	68 (94.4)	3 (4.2)	1 (1.4)	
Interviewed	65	62 (95.4)	3 (4.6)	0 —	(O.R.=1.1)
Bicuspid Aortic Valve					
All	72	47 (67.1)	21 (30.0)	2 (2.9)	Distributions are
Exclusions	5	2 —	1 —	1 —	not
After Exclusions (pure)	67	45 (68.2)	20 (30.3)	1 (1.5)	different from controls
Interviewed	60	41 (68.3)	19 (31.7)	0 —	(O.R.=1.1)

RACE*

Data are missing in 4 cases with hypoplastic left heart, 2 with coarctation, 2 with aortic stenosis, 1 with bicuspid aortic valve; O.R.=odds ratio.

Potential Risk Factors

All four phenotypes had a strong association with familial congenital heart disease, but striking differences among them were apparent in the other risk factors identified, including maternal illnesses and a range of xenobiotic exposures (Table 8F.3). Each phenotype had a specific risk profile with the largest number of factors in coarctation of the aorta. The association of maternal solvent exposure was striking for hypoplastic left heart syndrome and for coarctation, but it was not apparent for aortic stenosis or for bicuspid aortic valve.

Evaluation of the Effect of Doubly Ascertained Families

A special concern in the interpretation of these associations was the potential for artifactual odds ratio inflation due to the inclusion of two affected siblings in a family in some of the analyses, which violates the assumption of independence of

Table 8F.3
Left-Sided Obstructive Defects: Summary of Risk Factors
Baltimore-Washington Infant Study (1981–1989)

Risk Factors	Total Case Group N=377	Hypoplastic Left Heart N=138	Coarctation of the Aorta N=120	Aortic Stenosis N=65	Bicuspid Aortic Valve N=60
Family History					
congenital heart disease	4.4	4.8	4.6	3.7	7.2
Maternal Illnesses and Medications					
diabetes	1.7 gest.	3.9 overt	—	3.9 gest.	—
epilepsy	3.5	—	5.3	—	—
clomiphene citrate	—	—	4.5	—	—
sympathomimetics	1.5	—	1.8	—	—
benzodiazepines	3.8 multiple/ multiplex	—	—	—	—
macrodantin	3.9 isolated/ simplex	—	6.7	—	—
ibuprofen	—	—	—	—	4.1
Home and Occupational Exposures					
solvents	2.5	3.4 any exposure[1]	3.2 daily exposure[2]	—	—
general anesthesia	—	2.4	—	—	—
occupational cold temperature (paternal)	4.9	—	—	—	—
Sociodemographic Characteristics					
maternal age >34	—	—	—	—	2.5
white race	1.5	—	1.7	10.5	—

Numbers shown in the table are adjusted odds ratios from multivariate analysis—all are statistically significant at the 5% level.
[1]degreasing or miscellaneous solvents;
[2]solvent score.

Table 8F.4
Left-Sided Obstructive Defects: Evaluation of Case Exclusions on Identified Risk Factors
Baltimore-Washington Infant Study (1981–1989)

	Odds Ratio (95% CI)		Odds Ratio (95% CI)	
Aortic Stenosis (N=65)	Analysis with Case # 10145 Included		Analysis with Case Excluded	
family history of CVM	3.7	(1.1–12.4)	2.5	(0.6–10.5)
gestational diabetes	3.9	(1.7–8.8)	3.9	(1.7–8.8)
white race	10.5	(3.3–33.4)	10.4	(3.2–33.2)
Bicuspid Aortic Valve (N=60)	Analysis with Case # 10775 Included		Analysis with Case Excluded	
family history of CVM	5.2	(2.7–19.1)	5.9	(2.0–17.2)
ibuprofen	4.1	(1.8–9.3)	4.1	(1.8–9.4)
maternal age >34 years	2.5	(1.3–4.8)	2.3	(1.1–4.6)
Coarctation of Aorta (N=120)	Analysis with Case # 11085 Included		Analysis with Case Excluded	
family history of CVM	4.6	(2.0–10.8)	3.8	(1.5–9.5)
epilepsy	5.3	(1.4–20.2)	5.3	(1.4–20.2)
clomiphene citrate	4.5	(1.5–14.0)	4.6	(1.5–14.2)
macrodantin	6.7	(2.0–22.0)	6.7	(2.0–21.9)
sympathomimetics	1.8	(1.8–3.2)	1.8	(1.1–3.2)
white race	1.7	(1.1–2.7)	1.8	(1.1–2.9)
solvent score (225)	3.2	(1.3–7.9)	3.3	(1.3–7.9)

CVM = cardiovascular malformation.

observations in the logistic regression analyses. In three case families (#10145, #10775, and #11085) the first ascertained sibling had hypoplastic left heart syndrome and a subsequent sibling was ascertained with either coarctation, aortic stenosis, or bicuspid aortic valve. Removing these specific cases from the analysis evaluated their impact on the risk factor associations, as shown in Table 8F.4. As expected, the odds ratios for family history of cardiovascular malformations were diminished when these cases were removed from the multivariate analysis, but it is noteworthy that the environmental risk factor associations remained essentially unchanged.

Conclusions

There can be no doubt about the heterogeneity of these four anatomic phenotypes of left-sided obstructive lesions. As noted in the introduction, no sharp distinctions can be made between them by anatomic criteria alone. Among the epidemiologic factors studied, fetal growth retardation and solvent exposure link hypoplastic left heart syndrome and coarctation, while white race may link hypoplastic left heart syndrome and aortic stenosis. Obviously the anatomic case definitions, even though limited to pure diagnoses, are still not fully appropriate.

Chapter 9

Right-Sided Obstructive Defects

Section A:
Overview

Section B:
Pulmonary Valve Stenosis

Section C:
Pulmonary Valve Atresia with Intact
Ventricular Septum

Section D:
Tricuspid Atresia with Normally Related
Great Arteries

From: Ferencz C, Loffredo CA, Correa-Villaseñor A, Wilson PD, eds: *Genetic & Environmental Risk Factors of Major Cardiovascular Malformations: The Baltimore-Washington Infant Study 1981–1989.* Armonk, NY: Futura Publishing Co., Inc; ©1997.

Section A

Overview

Obstructive cardiovascular malformations of the right side of the heart and pulmonary artery differ from the varied forms of left-sided obstructive defects as they cannot be grouped together either by a shared family history or by associations of common genetic defects.

The malformations in this group include pulmonary valve stenosis (Section B in this chapter), pulmonary atresia with intact ventricular septum (Section C in this chapter), and tricuspid atresia with normally related great arteries (Section D in this chapter). Pulmonary valve stenosis is the predominant diagnosis (66% of total group). The two smaller groups, pulmonary atresia with intact ventricular septum, and tricuspid atresia, are considered unique and unrelated entities; the former with a late and the latter with a very early embryonic origin. Cases of tricuspid atresia were classified in our study into those with transposition and those without transposition. As transposed great arteries constituted a principal cardiac diagnosis, the cases with transposition were allocated to the outflow tract defect group (Chapter 5) as complex transpositions; only cases of tricuspid atresia with normally related great arteries are discussed here. Peripheral pulmonary artery stenosis was not evaluated as a primary diagnostic group because of the uncertainty of diagnosis in premature infants where it may be present as a functional abnormality.

In pulmonary valve stenosis the obstruction is limited to the valvar level—the pulmonary valve is thickened and dome shaped. The right ventricle is usually normally formed, although often markedly hypertrophied. Pulmonic stenosis is considered the typical cardiac defect in the Noonan syndrome. Cases of severe pulmonary stenosis require cardiac catheterization and surgical intervention in the first year of life. Their life course is in marked contrast to that of infants with mild pulmonic stenosis, a diagnosis now frequently made by the widespread use of Doppler and color echocardiography as diagnostic tools.

Pulmonary atresia with intact ventricular septum is a complex cardiovascular malformation. The atresia (complete obstruction) may involve the valve alone or the main pulmonary artery and the right ventricular outflow tract as well. The right ventricle is often abnormal with variable degrees of hypoplasia and endocardial fibroelastosis and with sinusoidal communications that connect the high-pressure right ventricle to the coronary arteries. These connections may contain multiple stenoses, and surgical relief of the high right ventricular pressure may result in myocardial infarction and death. In both pulmonary stenosis and pulmonary atresia, the presence of a normally formed pulmonary arterial tree has given rise to the opinion that these obstructions may have been acquired in later embryonic life.

Tricuspid atresia is a failure of communication from the right atrium to the right ventricle. Morphological studies have distinguished two forms of tricuspid atresia: 1) that which occurs when the tricuspid valve is replaced by a diaphrag-

Right-Sided Obstructive Defects, N=549

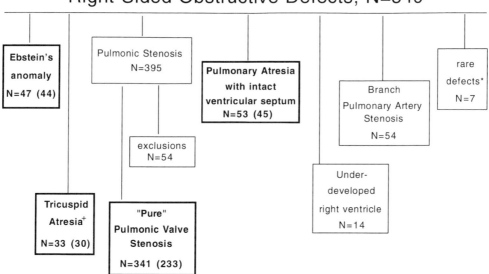

+with normally related great arteries
*Rare defects= pulmonary artery sling (n=1); hypoplastic artery tree (=4); Uhl's anomaly (n=2)
numbers in parentheses indicate interviewed families

Figure 9A.1: Flow chart illustrating right-sided obstructive heart defects, among which pulmonic stenosis is the most prevalent. Analyses of "pure" pulmonic stenosis, pulmonary atresia with intact ventricular septum, tricuspid atresia, and Ebstein's anomaly are each described in this book. Numbers in parentheses indicate the number of families interviewed.
Findings on Ebstein's anomaly were previously reported and are summarized in Chapter 10. Tricuspid atresia with transposition of the great arteries was discussed in Chapter 5.

matic structure at the site of the external atrioventricular connection and 2) when there is a complete failure of the atrium to appose the right ventricle.[20] While this distinction between the two forms would be valuable in an etiologic evaluation, it was not made in the clinical evaluation of our cases. In tricuspid atresia with normally related great arteries the pulmonary flow is usually reduced, and the affected babies present as newborns with cyanosis and a murmur. In contrast, when babies have tricuspid atresia and transposed great arteries, it may not be recognized until a few weeks of age. These babies also have a murmur but they are less cyanotic, as they have an increased pulmonary blood flow.

One searches for a parallel of the right heart obstructions with the left-sided obstructions, specifically the complex called hypoplastic left heart syndrome, which is a global underdevelopment of the valves, chambers, and the artery arising from the left side of the heart. One might speculate that a comparable malformation of the right heart is a lethal embryonic malformation, or that it does not occur, or that the pulmonary arterial obstructions were acquired only after the ventricles had developed.

The various forms of right-sided obstructions are shown on the flowchart (Figure 9A.1), which indicates the malformations selected for detailed analyses.

Section B

Pulmonary Valve Stenosis

Obstruction to the flow of right ventricular blood into the lungs may occur in association with major cardiovascular malformations due to narrowing of the outflow tract (infundibulum), abnormalities of the pulmonic valve, or constrictive defects of the main pulmonary artery or its branches. However, pulmonary valve stenosis also occurs as a single defect with distinct anatomic, physiologic, and perhaps, etiologic characteristics. Described by Paul Wood as *pulmonary valve stenosis with normal aortic root,*[455] this anomaly has not attracted the same level of inquiry that has been given to other severe cardiovascular malformations. This relative disinterest may reflect the perception that the defect has a favorable clinical outcome, either because it is often mild or because of successful relief of the obstruction by surgical or balloon catheter interventions.

There are now fewer and fewer pediatric cardiologists who can still remember the severe clinical manifestations of this unique anomaly of the pulmonary valve. Indelible in the author's mind (CF) is her patient seen in the 1940s: a little girl who had the round rosy face later considered to be typical, ascites, and a large pulsating liver. She cooperated gently as we unremittingly pursued every known medical therapy for her severe congestive heart failure; her weakness was progressive and death came quietly. No striking auscultatory findings indicated congenital heart disease, only a huge heart and a markedly abnormal electrocardiogram. It is likely that in general medical practice of that time the true cause of her heart failure would not have been recognized. However, without surgery, even the expert diagnosis would have been in vain.

Fifteen years later the picture was quite different: another patient was a deeply cyanotic 1-year-old in 1960 when he underwent cardiac catheterization, became moribund immediately thereafter, and required emergency surgery with a transventricular pulmonary valvotomy in the middle of the night. He is now in his thirties, an active blue collar worker supporting a family with two healthy boys. The contrast in the life courses of these two patients with identical cardiac defects illustrates the hallmark importance of definitive cardiac surgery.

The major five-volume work of Morgagni was credited by Maude Abbott[6] as including the first description of "pure pulmonary stenosis." Indeed, Letter XVII, Articles 12 and 13 contain a description of the heart of a 15-year-old girl in whom the opening of the fused pulmonary valve cusps was "like a barley corn." The characteristic pathologic features of a thick right ventricle, a large right atrium, and a widely patent foramen ovale were also described. In addition, Morgagni clearly described the consequent circulatory alterations and their reflection in the clinical manifestations of the patient.[306] (Figure 9B.1).

In 1858 Peacock[338] presented a precise illustration of such a fused dome-shaped pulmonary valve with the dilated pulmonary artery, the marked right ventricular hypertrophy proximal to the obstruction, and the patent foramen ovale,

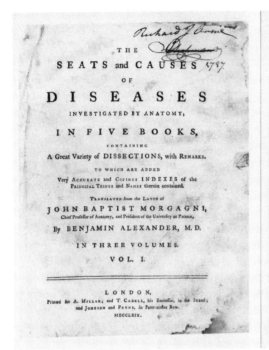

THE

SEATS and CAUSES

OF

DISEASES

INVESTIGATED BY ANATOMY;

IN FIVE BOOKS,

CONTAINING

A Great Variety of DISSECTIONS, with REMARKS.

TO WHICH ARE ADDED

Very ACCURATE and COPIOUS INDEXES of the
PRINCIPAL THINGS and NAMES therein contained.

TRANSLATED from the LATIN of

JOHN BAPTIST MORGAGNI,
Chief Professor of Anatomy, and President of the University at PADUA,

By BENJAMIN ALEXANDER, M.D.

IN THREE VOLUMES.

VOL. I.

LONDON,

Printed for A. MILLAR; and T. CADELL, his Successor, in the Strand;
and JOHNSON and PAYNE, in Pater-noster Row.

MDCCLXIX.

Letter XVII.   Articles 12, 13.

12. A virgin who, from the very time of her birth, had always lain sick, especially on account of her very great debility, breath'd short, and had her skin ting'd all over with a kind of livid colour: at length, when she came to be about sixteen years of age, she died. She had a heart that was very small, and, towards the apex, in a manner roundish. The right ventricle was of the form that the left generally us'd to be, and the left of the same form that the right us'd to be; and although the right was wider than the left, yet it had thicker parietes. The right auricle, in like manner, was universally enlarg'd, and twice as big as the left, and twice as thick; betwixt the two, even then, the foramen ovale was open, so as to admit the little finger. Of the three triangular valves, one had a proper bigness, the two others were less than usual. The sigmoid, which lie at the mouth of the pulmonary artery, were at their basis indeed natural, but in their upper part seem'd cartilaginous, nay indeed they had already a small ossification; and were so connected together in this part, that they did but just leave a little foramen, not bigger than a barley-corn, through which the blood was sent out. And at this foramen also were some small, fleshy, and membranous productions, plac'd in such a manner, that they might supply the places of valves, by yielding to the blood that was going out, and by resisting that which was about to return.

13. I should suppose that this virgin had the beginnings of that disorder at the mouth of the pulmonary artery, from her original formation; to which disorder, being gradually more and more encreas'd, every thing she suffer'd when living, and what were found in the dead body, are without doubt to be referr'd: that is to say, the less quick, and less ready entrance of the blood into that artery, from this cause, was a reason why, on the one hand, a less quantity of it should be transmitted through that artery and its corresponding vein, to the left auricle and left ventricle, and from this should be sent to the whole body; and, on the other hand, that a greater quantity of blood than is natural should remain in the right ventricle, right auricle, and all the veins. From whence the colour of the whole skin, in a manner livid, and the dilatation of the right ventricle and right auricle, and the continu'd communication by the foramen ovale, by reason of its valve being urg'd towards the left side, by the great quantity of blood from the right, whereas but little urg'd it on the left side, and applied it to the edge of the foramen. But, for contrary reasons, the left auricle and left ventricle were neither sufficiently open'd out and dilated, nor sufficiently strong; and the proper influx of blood to the brain, and to all other parts, being deficient, that very great debility, and difficult respiration were the consequences; and these even for that very reason, because from the small portion of blood entering into so large and firm a vessel as the pulmonary artery, it could neither be sufficiently expanded and dilated, nor consequently contract and restore itself, as it ought to do, in order to carry the blood properly through the lungs.

Figure 9B.1: Title page and Letter XVII, Articles 12 and 13 from the first English translation by Benjamin Alexander (1769) of Giovanni Baptista Morgagni's masterwork "De Sedibus et Causis Morborum per Anatomen Indagatis," (1761) in which a case of isolated pulmonic stenosis was first described.

which permitted right-to-left shunting and relieved the progressive distention of the right atrium (See Chapter 2, Figure 2.7).

Probably in no other type of congenital cardiac malformation is there a greater similarity between one specimen and the next, wrote Edwards[121] in the first modern textbook on the pathology of the heart, referring to the fused dome-shaped valve.

The malformation was thought to be rare. Among 1000 cases of congenital heart disease presented in her Atlas, Abbott[6] found only 25 cases of pulmonic stenosis with intact ventricular septum ranging in age from 4 to 57 years. Taussig,[406] who had directed the children's cardiac clinic at the Johns Hopkins Hospital since 1930, reported in the first edition of her book *Congenital Malformations of the Heart* (1947) that she had not seen such a case in her Baltimore practice during those 17 years. When, together with Engle in 1950[135] she established clinical criteria for the diagnosis of this malformation, these criteria were based on observations of only three cases. Wood,[455] (1950) in London, England studied 2000 cases of congenital heart disease in children over 2 years of age and among them found 170 with pulmonic valve stenosis, indicating to him a frequency *second only to atrial septal defect.* He also described the inexorable course of "pure" right heart failure in his patients, and in some he noted a cardiofacial syndrome, a "moonface" similar to that induced by steroid administration.[455] This report came after the introduction of trans-ventricular pulmonary valvotomy in 1948 by Russell Brock,[62] a major turning point in the clinical fate of affected patients who were referred to medical centers in increasing numbers and at all ages to seek such dramatic postoperative relief. Brock also traveled abroad to demonstrate his operation to surgeons in various countries. This author (CF) saw the first operation performed by Brock at the Johns Hopkins Hospital in Baltimore in 1949, and participated in the evaluation of physiologic studies on 34 patients from the Laboratory of Richard Bing.[276]

Clinical perceptions of pulmonic valve stenosis underwent further changes once attention was turned to infants. Luke[262] studied the infant population among operated patients at Johns Hopkins from 1949 to 1962 and noted the atypical manifestations in sick infants with advanced congestive failure in whom the malformation was "extremely dangerous" because of vague symptoms such as restlessness and irritability and few clinical signs except for poor heart sounds.

Twenty years later, the opposite, the mild-end-of-severity spectrum, came to light by the diagnostic application of ultrasound and two-dimensional and Doppler flow echocardiography. Now there was another great increase in diagnosed cases, this time of seemingly well infants with minor cardiovascular findings. "Dystrophic" valves were seen in patients with the characteristics of the Noonan syndrome, a heritable anomaly in which familial cases may also display pulmonary valve stenosis and other types of cardiac defects, some with and others without the associated syndromic features of the anomaly.[56,143]

Etiologic hypotheses were stated in the earliest works. Fetal endocarditis was the cause presumed by Peacock and other earlier writers.[338] Abbott also believed that this malformation *has to be due to an inflammation after septation of the heart is completed.*[7] Becu and Sommerville[33] described pathologic changes in the myocardium presumably due to healed fetal endomyocarditis, but their hypothesis was not well accepted. A genetic origin for this malformation was suggested on the basis of familial aggregations of pulmonic stenosis and its association with other heritable disorders.

Recognition: Changes By Time

The Baltimore-Washington Infant Study case enrollment years (1981 to 1989) encompassed a time period of major changes in the diagnosis and management of pulmonic valve stenosis.

The introduction of two-dimensional echocardiography in all participating centers by 1983, and later the use of color Doppler echocardiographs, made it possible to visualize slight abnormalities of the pulmonic valve and to estimate hemodynamic alterations and pressure gradients across the valve. Abnormalities described as "turbulent" blood flow, "valvar thickening" and "dystrophic valve" provided explanations for slight murmurs.

Coincident with these diagnostic advances, transcatheter balloon valvuloplasty was introduced and applied in the treatment of children and then of infants.[214] The simplicity and safety of this procedure altered the indications for cardiac catheterization: cases with moderately severe pulmonic obstruction could be evaluated and treated, if necessary, within a single procedure, although some infants with critical pulmonic stenosis still required a surgical valvotomy.

These alterations in diagnostic and treatment methods also altered the nature of the case population that was referred to the regional Pediatric Cardiology Centers. There was an increase in the referrals of infants with mild abnormalities of the heart who could be studied once by echocardiography and then did not require further evaluations in infancy. The magnitude of this new case population increased rapidly. Their registration into the Baltimore-Washington Infant Study overwhelmed the team's ability to carry out home visits, and interviews had to be allocated by sampling. It also became important to subdivide the case population and separate those with mild "dystrophic" abnormalities of the pulmonary valve from those with severe obstructive disease.

Categorization of Severity

In the Baltimore-Washington Infant Study the anatomic cardiac diagnoses were coded without note of their hemodynamic consequences. Thus initially no provision was made to recognize mild pulmonic stenosis. In the early 1980s it could be assumed that cases diagnosed only by echocardiography had no clinically significant obstructions of the pulmonic valve. However as the number of referred cases increased, the cardiologists introduced a special code (#449) in 1985 to indicate a mild abnormality. This code was applied to new cases at registration as well as to cases seen for reevaluation, so that the use of this code extended back to cases born in 1983 and 1984. In 1987 the cardiologists established a uniform definition of mild defects as a right ventricular-pulmonary artery pressure gradient, by Doppler echocardiogram, of less than 25 mm.

Reported studies indicate that the course of neonates with mild forms of obstruction do not worsen in severity because the valve area increases with age.[37] Patients with moderate pulmonary stenosis (gradient of 40 to 50 mm) require closer follow-up in case they need balloon dilation of the pulmonary valve.[169]

Prior to the analysis of this case group, a review of all study forms took into account the cardiologists' notes, recommendations for follow-up, and the mother's report in the interviews. Three case categories were created for analysis. Pulmonic stenosis was defined as:

- *mild* if coded ISC 449 and accompanied by notes such as "turbulent flow," "slightly dystrophic," and recorded gradients of less than 25 mm, and if no close follow-up was advised;
- *moderate* if coded ISC 448, diagnosis was made only by echocardiogram, and notes similar to the above or recorded gradients of less than 40 mm, and only late follow-up recommended;
- *severe* if coded ISC 448 and the patient underwent cardiac catheterization with or without balloon valvuloplasty or surgery.

Study Population

Cardiac Abnormalities

During the years from 1981 through 1989, 395 cases of pulmonic stenosis suitable for risk factor analyses were enrolled in the Baltimore-Washington Infant Study.[152] This current evaluation is based on 341 cases of pulmonary valve stenosis with intact ventricular septum, having excluded from this group pulmonary stenosis in association with a ventricular septal defect, obstructions at both valvar and right ventricular or infundibular levels, and cases with both pulmonary and aortic valve stenosis (Figure 9B.2). The remaining group of 341 cases represents 7.8% of all cardiovascular malformations registered in the Baltimore-Washington Infant Study.

Noncardiac Anomalies

In this large case group 31 cases (9%) had associated noncardiac anomalies, the distribution of which differed markedly in the severity stratified subgroups (Table 9B.1). No chromosomal abnormalities occurred in infants with severe pulmonary stenosis, but eight of these cases (12.5%) had a malformation syndrome. Noonan syndrome was the predominant associated anomaly. Among the 10 cases of Noonan syndrome in the entire pulmonic stenosis group, six were diagnosed in the severe subgroup (9.4%), three in the mild/moderate group (2.4%) and only a single case among the 152 infants with very mild defects (0.7%). In contrast, major organ defects were more frequent in the mild groups and may have served to attract attention to possible cardiovascular anomalies.

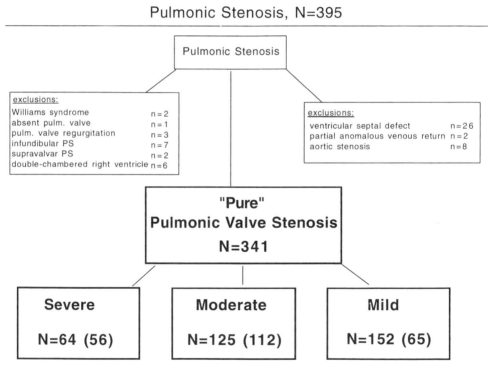

Pulmonic Stenosis, N=395

Pulmonic Stenosis

exclusions:
Williams syndrome	n = 2
absent pulm. valve	n = 1
pulm. valve regurgitation	n = 3
infundibular PS	n = 7
supravalvar PS	n = 2
double-chambered right ventricle	n = 6

exclusions:
ventricular septal defect	n = 26
partial anomalous venous return	n = 2
aortic stenosis	n = 8

"Pure"
Pulmonic Valve Stenosis
N=341

Severe	Moderate	Mild
N=64 (56)	N=125 (112)	N=152 (65)

Numbers in parentheses indicate interviewed families.

Figure 9B.2: Flow chart of the total number of cases in the Baltimore-Washington Infant Study who were registered with a diagnosis of pulmonic stenosis (395 cases), the diagnostic exclusions that resulted in the analytic subset of "pure" cases of pulmonic valve stenosis, and its subsets based on clinical severity. Numbers in parenthesis indicate the number of families interviewed.

Table 9B.1
Pulmonic Valve Stenosis (PS):
Noncardiac Anomalies
Baltimore-Washington Infant Study (1981–1989)

	Total Group N=341	Severe PS N=64	Moderate PS N=125	Very Mild PS N=152
Chromosomal Abnormalities:	4 (1.2%)	0%	3 (2.4%)	1 (0.7%)
Down syndrome	2	0	2 [# 10057, 10229]	0
Trisomy 1q	1	0	1 [# 3765]	0
47XXX/46XX	1	0	0	1 [#10681]
Syndromes	14 (4.1%)	8 (12.5%)	3 (2.4%)	3 (2.0%)
a. Mendelian Syndrome				
Noonan	10	6 [#1959, 2316, 2825,10526, 10834, 11443]	3 [#10050, 10316, 11510]	1 [#10481]
Ehlers-Danlos	1	1 [# 1851]	0	0
Prune belly	1	0	0	1 [#2511]
Opitz-Frias	1	0	0	1 [#10092]
b. Non-Mendelian Associations:				
CHARGE	1	1 [# 3251]	0	0
Multiple, Nonclassified Anomalies:	2 (0.6%)	1 (1.6%)	0	1 (0.7%)
Absent right kidney, dysmorphic facies	1	1 [# 1537]	0	0
Micropenis, inguinal hernia, deafness	1	0	0	1 [# 3693]
Organ Defects:	8 (2.3%)	1 (1.6%)	3 (2.4%)	4 (2.6%)
Cataract	1	1 [# 1442]	0	0
Microcephaly	2	0	0	2 [# 2962, 10467]
Hydrocephalus	1	0	0	1 [# 3627]
Glaucoma	1	0	0	1 [# 3653]
Omphalocele	1	0	1 [# 3722]	0
Adreno-genital association	1	0	1 [# 3384]	0
Polydactyly	1	0	1 [# 2161]	0
No Associated Anomalies	313 (91.8%)	55 (84.4%)	116 (92.8%)	143 (94.1%)

Descriptive Analyses

Prevalence by Time, Season, and Area of Residence

The total regional prevalence of pulmonic valve stenosis was 3.78 per 10,000 livebirths. The prevalence of pulmonic valve stenosis increased sharply during the 9-year study period, from 1.46 per 10,000 livebirths in the first year, to a maximum of 5.4 during the next-to-last year (1988) (Figure 9B.3). This dramatic change rep-

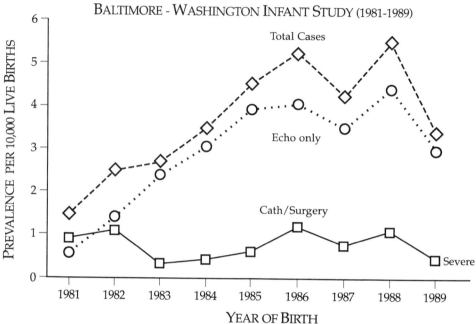

**PREVALENCE OF PULMONARY VALVE STENOSIS
BY METHOD OF DIAGNOSIS**
BALTIMORE - WASHINGTON INFANT STUDY (1981-1989)

Figure 9B.3: Prevalence of pulmonary valve stenosis per 10,000 livebirths rose during the decade, indicated on the graph as diamond-shaped markers. This trend in prevalence was mirrored closely by the subset of cases diagnosed solely by echocardiography (echo), indicated by circles, but the prevalence of cases diagnosed by cardiac catheterization and/or surgery (squares) remained fairly constant during the study.

resented an increase in the prevalence of defects diagnosed by echocardiography and there was little change in the prevalence of cases who underwent cardiac catheterization and/or surgery.

Further evaluation of the prevalence of cases diagnosed by echocardiography by study year illustrates the effect of the new code for mild defects introduced in 1985 (Figure 9B.4).

Categorized by three levels of severity, Table 9B.2 shows the stability of the distribution of the small number of severe cases over time, season, and geographic region of residence. In contrast the very mild cases increased over time and were concentrated in the urban geographic area.

Diagnosis and Course

Overall the diagnosis of pulmonary valve stenosis was rarely established in the first week of life (8.2% for all cases, 15.6% for severe cases) but about one third of cases were recognized in the first months of life even in the very mild category (Table 9B.3). About one fifth of the cases in all severity groups were diagnosed after 6 months of age.

All cases who underwent cardiac catheterization were considered to have se-

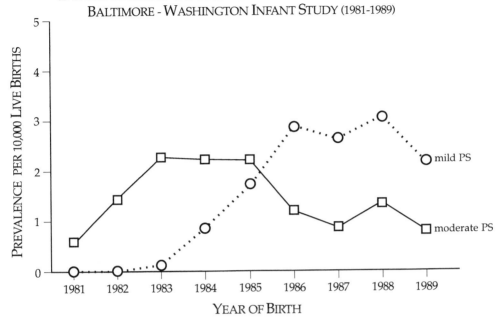

PREVALENCE OF PULMONARY VALVE STENOSIS
DIAGNOSISED BY ECHOCARDIOGRAPHY ONLY
BALTIMORE - WASHINGTON INFANT STUDY (1981-1989)

Figure 9B.4: Temporal changes in the prevalence of pulmonary valve stenosis cases who were diagnosed solely by echocardiography. After 1983, increasingly large numbers of clinically mild cases (circles) were diagnosed.

vere obstructions. Half of the severe cases underwent surgery in the first year of life, and balloon valvuloplasty was not consistently recorded in the research file. The diagnosis of pulmonic valve stenosis was never established by autopsy alone, ie, no case was found in the community search. During the first year of life only a single patient died as a complication of cardiac catheterization.

The length of follow-up varied by the severity of the lesion. Overall, 11% of these infants were not followed after registration into the study, but 18% of those with very mild defects were not followed in contrast to complete follow-up of all those with severe lesions. The median age of infants at their last visit during the first year of life was 11 months for infants with severe pulmonic stenosis but only 9 months for those with moderate or mild lesions.

Gender, Race, and Twinning

A predominance of female infants was noted among the very mild and moderate cases, but the male to female proportion among severe cases was equal to controls (Table 9B.4). Similarly, the racial distribution of severe cases equaled that of controls, while among the very mild and moderate cases there was an excess of black infants. Multiple births were more common in cases than in controls (2.6% total versus 1.5%). In one set of male dizygotic twins mild pulmonic stenosis was suspected in the co-twin, but was not proven.

Table 9B.2

Pulmonic Valve Stenosis:Prevalence

Baltimore-Washington Infant Study (1981–1989)

| | Number of Cases | | | | | Prevalence per 10,000 Livebirths | | | |
	All Cases	Severe Defects	Moderate Defects	Very Mild Defects	Area Births	All Cases	Severe Defects	Moderate Defects	Very Mild Defects
Total Subjects	341	64	125	152	906,626	3.78	0.71	1.38	1.68
Year of Birth:									
1981–1983	60	20	38	2	274,558	2.22	0.73	1.38	0.07
1984–1986	131	18	55	58	295,845	4.43	0.60	1.86	1.96
1987–1989	144	22	31	91	336,223	4.31	0.65	0.92	2.71
Birth Quarter:									
1st (Jan–Mar)	77	20	24	33	219,145	3.56	0.91	1.10	1.51
2nd (Apr–Jun)	78	16	29	33	231,777	3.41	0.69	1.25	1.42
3rd (Jul–Sep)	103	17	39	47	233,626	4.41	0.73	1.67	2.01
4th (Oct–Dec)	83	11	33	39	222,078	3.74	0.50	1.49	1.76
Area of Residence:									
urban	140	21	37	82	208,568	6.71	1.01	1.77	3.93
suburban	167	34	78	55	584,022	2.86	0.58	1.33	0.94
rural	34	9	10	15	112,318	3.03	0.80	0.89	1.34

Table 9B.3
Pulmonic Stenosis: Clinical Data
Baltimore-Washington Infant Study (1981–1989)

	All Cases N=341		Severe Defects N=64		Moderate Defects N=125		Very Mild Defects N=152	
	no.	(%)	no.	(%)	no.	(%)	no.	(%)
Age at Diagnosis:								
<1 week	28	(8.2)	10	(15.6)	8	(8.4)	10	(6.6)
1–4 weeks	76	(22.3)	15	(23.4)	29	(23.2)	32	(21.1)
5–24 weeks	172	(50.4)	24	(37.5)	68	(54.4)	80	(52.6)
25–52 weeks	65	(19.1)	15	(23.4)	20	(16.0)	30	(19.7)
Method of Diagnosis:								
echocardiography only	270	(79.2)	3	(4.7)	124	(99.2)	143	(94.1)
echocardiography and cardiac catheterization	41	(12.0)	32	(50.0)	0		9	(5.9)
echocardiography and surgery	3	(0.9)	2	(3.1)	1	(0.8)	0	
echocardiography, cardiac catheterization, and surgery	27	(7.9)	27*	(42.2)	0		0	
autopsy following other methods	0		0		0		0	
autopsy only	0		0		0		0	
Surgery During First Year of Life	31	(9.1)	31	(48.4)	0		0	
Follow-Up:								
not followed	38	(11.1)	0		10	(8.0)	28	(18.4)
followed:	303		64		115		124	
died in the first year of life	1	(0.3)	1	(1.5)	0	(0)	0	
alive	302	(99.7)	63	(98.5)	115	(100)	124	(100)
median age at last visit	10 months		11 months		9 months		9 months	

*Includes 3 cardiac catheterization only and 2 cardiac catheterization and surgery (without echocardiography).

Table 9B4
Pulmonic Valve Stenosis: Infant Characteristics
Baltimore-Washington Infant Study (1981–1989)

Characteristic	All Cases N=341 no. (%)	Severe Defects N=64 no. (%)	Moderate Defects N=125 no. (%)	Very Mild Defects N=152 no. (%)	Controls N=3572 no. (%)
Gender:					
male	138 (40.5)	33 (51.6)	49 (39.2)	56 (36.8)	1817 (50.9)
female	203 (59.5)	31 (48.4)	76 (60.8)	96 (63.2)	1755 (49.1)
Race*:					
white	171 (50.1)	44 (68.8)	65 (53.3)	62 (42.5)	2362 (66.1)
black	150 (44.0)	19 (29.7)	52 (42.6)	79 (54.1)	1109 (31.0)
other	11 (3.2)	1 (1.5)	5 (4.1)	5 (3.4)	98 (2.7)
Twin births	9 (2.6)	2 (3.1)	2 (1.6)	5 (3.3)	53 (1.5)

*Missing race in 9 cases and 3 controls.

Birthweight and Gestational Age

Information on birthweight and gestational age was obtained in the maternal interview (233 cases). The mean birthweight of the case infants was low (Table 9B.5), a finding shared with almost all other forms of congenital heart disease. The deficit was least in the severe group, but infants in all three groups showed a greater proportion of low birthweight and of prematurity than controls. Mean birthweights by severity of pulmonic stenosis, gender, and race are shown in Table 9B.5a for cases and controls. Among white infants, females were significantly smaller than males in the combined moderate/mild case group as well as among controls, but there was no significant male/female disparity in the birthweights of severely affected cases. There were no small-for-gestational-age females among cases with severe pulmonic stenosis, in contrast to moderate/mild lesions, which affected 4 of 56 white females (7%) and 10 of 48 black females (21%) classified as small for gestational age.

These descriptive data suggest that the diagnosis of very mild and moderate pulmonic valve stenosis may have been subject to a selection bias favoring the urban, black and small-for-gestational-age subpopulation, ie, high-risk infants on whom echocardiography may have been performed on the slightest indication of possible congenital heart disease.

Sociodemographic Characteristics

Comparing all cases to controls there were few sociodemographic characteristics that were associated with pulmonic stenosis (Table 9B.6). Mothers of cases were more likely to be unmarried, but did not differ from controls mothers in education, income, or occupation. Fathers of cases and controls did not differ in educational level, but case fathers were somewhat more likely than controls to hold

Table 9B.5

Pulmonic Valve Stenosis: Fetal Growth Characteristics
Baltimore-Washington Infant Study (1981–1989)

Characteristics	All Cases no. (%)	Severe Defects no. (%)	Moderate Defects no. (%)	Very Mild Defects no. (%)	Controls no. (%)
Number of Families Interviewed	233	56	112	65	3572
Birthweight (grams)					
<2500	52 (22.2)	13 (23.2)	24 (21.4)	15 (23.1)	252 (7.1)
2500–3500	126 (54.3)	28 (50.0)	59 (52.7)	39 (60.0)	1853 (51.9)
>3500	55 (23.5)	15 (26.8)	29 (25.9)	11 (16.9)	1467 (41.0)
mean±standard error	2991±47	3049±85	3005±74	2920±88	3351±10
range	822–5046	1328–4080	822±5046	1219–4352	340–5273
Gestational Age (weeks)					
<38	56 (24.4)	15 (26.8)	22 (19.6)	19 (29.2)	339 (9.5)
38+	177 (75.6)	41 (73.2)	90 (80.4)	46 (70.8)	3233 (90.5)
mean±standard error	38.4±0.2	38.4±0.4	38.6±0.3	38.4±0.4	39.6±0.1
range	24–45	30–42	24–45	30–43	20–47
Size for Gestational Age*					
small (SGA)	31 (13.2)	5 (8.9)	16 (14.2)	10 (15.4)	211 (5.9)
normal	178 (76.5)	44 (78.6)	85 (75.9)	49 (75.4)	2712 (75.9)
large	24 (10.3)	7 (12.5)	11 (9.8)	6 (9.2)	649 (18.2)
Odds Ratio for SGA (95% CI)	2.4 (1.6–3.6)	1.6 (0.6–4.0)	2.7 (1.5–4.6)	2.9 (1.5–5.8)	1.0 (reference)

*Using as reference.

Table 9B.5a
Pulmonic Valve Stenosis (PS):
Mean Birth Weight by Race, Gender, and Severity
Baltimore-Washington Infant Study (1981–1989)

	Whites			Blacks		
Gender	Severe PS	Moderate and Mild PS	Controls	Severe PS	Moderate and Mild PS	Controls
males	3168*	3425	3508	2935	2752	3198
	(n=16)	(n=35)	(n=1202)	(n=11)	(n=32)	(n=560)
females	3154	3026	3398	2556	2672	3087
	(n=22)	(n=56)	(n=1159)	(n=6)	(n=48)	(n=548)
P-valve**	0.947	0.010	0.0001	0.287	0.629	0.004

*mean birthweight in grams; ** P-value from t-test of gender difference in mean birthweight within diagnostic group and race group.

service-related occupations. There was no significant case-control difference in the months of pregnancy confirmation.

Potential Risk Factors

Familial Cardiac and Noncardiac Anomalies

Families of infants with pulmonic valve stenosis are presented in Table 9B.7. The findings are in order by concordance of the defect. Pulmonic stenosis occurred in a mother, a father, a sister, a maternal half-sister, and a brother, while pulmonary atresia occurred in another brother—a remarkable concordance of the anomaly in six (35%) of the 17 affected families. Pulmonic stenosis in the infant also occurred in association with tetralogy of Fallot and with ventricular septal defects in five families, associations previously described[56] as possibly concordant in terms of abnormal outflow tract development. Furthermore, among the three families in which the type of congenital heart disease could not be ascertained, at least one may be concordant in that both the father and the infant had Noonan syndrome. In three instances familial cardiac diagnoses were discordant from that of the patient.

Among the concordant defects case #1959 is of special interest, as this was an infant with severe pulmonic stenosis and the Noonan syndrome whose mother had typical severe pulmonic stenosis and was operated on as a child in one of our pediatric cardiology centers. No question of a diagnosis of Noonan syndrome had been raised in her case.[56,143] This case and four others were infants with severe pulmonic stenosis, as was the infant with Noonan syndrome whose brother also had the syndrome but presumably had a normal heart (case #2316).

Table 9B.6
Pulmonic Valve Stenosis: Analysis of Sociodemiographic Characteristics
Baltimore-Washington Infant Study (1981–1989)

Interviewed Families	Cases N=234 no.	(%)	Controls N=3572 no.	(%)	Odds Ratio (95% CI)	
Maternal Marital Status:						
not married	84	(35.9)	990	(27.7)	**1.5**	**(1.1–1.9)**
married	150	(64.1)	2582	(72.3)	1.0	(reference)
Maternal Education:						
<high school	43	(18.4)	659	(18.4)	1.0	(0.7–1.4)
high school	82	(35.0)	1265	(35.4)	1.0	(0.7–1.3)
college	109	(46.6)	1648	(46.1)	1.0	(reference)
Paternal Education:						
<high school	42	(17.9)	650	(18.2)	1.0	(0.7–1.4)
high school	87	(37.2)	1298	(36.3)	1.0	(0.8–1.4)
college	105	(44.9)	1624	(45.5)	1.0	(reference)
Annual Household Income:						
<$10,000	52	(22.2)	686	(19.2)	1.2	(reference)
$10,000–$19,999	48	(20.5)	699	(19.6)	1.1	(0.8–1.6)
$20,000–$29,999	50	(21.4)	737	(20.6)	1.1	(0.8–1.6)
$30,000+	84	(35.9)	1373	(38.4)	1.0	(reference)
Maternal Occupation:						
not working	71	(30.3)	1142	(32.0)	1.1	(0.7–1.6)
clerical/sales	85	(36.3)	1119	(31.3)	1.3	(0.9–1.9)
service	26	(11.1)	444	(12.4)	1.0	(0.6–1.6)
factory	9	(3.8)	137	(3.8)	1.1	(0.5–2.3)
professional	43	(18.4)	730	(20.4)	1.0	(reference)
Paternal Occupation:						
not working	18	(7.7)	246	(6.9)	1.3	(0.7–2.2)
clerical/sales	35	(15.0)	618	(17.3)	1.0	(0.6–1.5)
service	40	(17.1)	480	(13.4)	**1.5**	**(1.0–2.2)**
factory	87	(37.2)	1279	(35.8)	1.2	(0.8–1.7)
professional	54	(23.1)	947	(26.5)	1.0	(reference)
Month of Pregnancy Confirmation:						
1st month	26	(11.2)	511	(14.3)	1.0	(reference)
2nd month	130	(55.8)	1955	(54.7)	1.3	(0.8–2.0)
3rd month	43	(18.5)	663	(18.6)	1.3	(0.8–2.1)
4th month or later	34	(14.5)	416	(11.6)	1.6	(0.9–2.7)

Data are missing for income (77 controls), paternal occupation (2 controls), and pregnancy confirmation month (1 case, 27 controls).

Table 9B.7

Pulmonic Valve Stenosis:

Congenital Defects in First-Degree Relatives

The Baltimore-Washington Infant Study (1981–1989)

	PROBAND			FAMILY	
ID.	Severity	Gender	Noncardiac Diagnosis	Relative	Diagnosis
A. Congenital Heart Defects:					
Concordant Defects:					
1959	severe	M	Noonan	Mother	Pulmonic stenosis
2199	severe	F	—	Brother	Pulmonary atresia
11756	moderate	M	—	Father	Pulmonic stenosis
2586	moderate	M	—	Brother	Pulmonic stenosis
2717	mild	M	—	Sister	Pulmonic stenosis
Partially Concordant Defects:					
1803	severe	F	—	Father	Tetralogy
2346	moderate	F	—	Mother	Ventricular septal defect
11572	moderate	F	—	Father	Ventricular septal defect
2174	moderate	F	—	Half-Brother (mat.)	Ventricular septal defect
3047	mild	F	—	Brother	Tetralogy
Discordant Defects:					
1224	severe	F	—	Father	Aortic stenosis
2128	moderate	M	—	Mother	Mitral valve prolapse
2334	moderate	M	—	Father	Mitral valve prolapse
Unknown Defects:					
10405	severe	F	—	Father	Suspect CHD
				Half-Sister (mat.)	Intestinal atresia
11510	moderate	M	Noonan	Father	Suspect CHD, Noonan
3436	mild	M	—	Half-Brother (mat.)	Suspect CHD
B. Congenital Defects of Other Organs:					
2316	severe	M	Noonan	Brother	Noonan
1534	severe	M	—	Mother	Polydactyly
10034	severe	F	—	Father	Deafness
1537	severe	F	Absent Kidney dysmorphic facies	Sister	Pyloric stenosis
3754	severe	M	—	Half-Sister (mat.)	Sickle cell disease
3251	severe	M	CHARGE	Father	Pectus excavatum
2663	moderate	F	—	Father	Spina bifida occulta
2953	moderate	M	—	Sister	Biliary atresia
3653	mild	F	—	Father	Hypospadias
3661	mild	M	—	Half-Sister (mat.)	Pyloric stenosis

CHD = congenital heart defects.

Genetic and Environmental Factors

Univariate Analysis

Results of the screening analyses of all potential genetic, medical, and environmental, as well as sociodemographic risk factors revealed differences in the risk profiles of the three severity subgroups. As expected the severe pulmonic stenosis group (56 interviewed families) showed a high odds ratio for familial cardiac and noncardiac anomalies; the moderate group showed candidate risk factors among the maternal reproductive and medical factors and a high odds ratio for exposure to degreasing solvents (Table 9B.8). Among the very mild cases (interviews in a sample of 65 infants) only family history of congenital heart disease and nonwhite race appeared as suspect factors.

Multivariate Analysis

Based on the large number of potential risk factors identified, cases of severe and moderate pulmonic stenosis were analyzed separately in the multivariate analysis. In the severe group, family history of congenital heart disease was the strongest independent (not interacting) risk factor with a case-control odds ratio of 6.4, which was significant at the 99% confidence level (Part A of Table 9B.9). The risk associated with a family history of noncardiac anomalies was dependent on race: a significant increase in risk was found only for nonwhite infants. Among home and occupational exposures, hair dyes and house painting were significantly associated with severe pulmonic stenoses cases, but only hair dyes remained significant at the 99% confidence level.

In the group of infants with moderate defects (Part B of Table 9B.9), familial congenital heart disease was a strong risk factor. Maternal thyroid disease and maternal fever during the critical period were associated with this subgroup. Maternal use of cigarettes was also associated with moderate defects, but only for heavy smoking among older mothers, for which the odds ratio reached 12.5. Cases were significantly more likely than controls to be nonwhite. All of the above associations were significant at the 99% confidence level. In addition, degreasing solvents showed an increased risk significant at the 95% confidence level.

Discussion

This epidemiologic study of infants with pulmonic stenosis diagnosed during the decade of the 1980s in a regional birth population has provided important information on etiologic, clinical, and health service aspects of this diagnosis in modern pediatric cardiology practice.

Severe pulmonic valve stenosis is still a rare abnormality with an at-livebirth prevalence of 7 per 100,000, thus, on the average, each of the six pediatric cardiology centers would have seen only about 1 case per year. The prevalence of pulmonic valve stenosis severe enough to require cardiac catheterization or surgery in infancy did not vary over the 9-year time period, while there was an enormous increase in the diagnosis of very mild and moderately severe cases with the introduction of two-dimensional and Doppler echocardiography. This finding supports the initial observation of Taussig in 1947,[406] of the rarity of this malformation, and her later comment that, *after the introduction of cardiac catheterization and the Brock operation a much greater number of cases was observed.*[407] Indeed all of those cases must have represented extensive ascertainment over wide territories of referral and a backlog of cases in children, adolescents, and even adults.

The strong association of severe pulmonic stenosis with Noonan syndrome and the presence of pulmonic stenosis and Noonan syndrome in relatives suggests

Table 9B.8
Pulmonic Valve Stenosis (PS) Univariate Analysis of Potential Risk Factors by Severity of Defect
Baltimore-Washington Infant Study (1981–1989)

Variable	Controls (N=3572)		Severe PS (N=56)			Moderate PS (N=112)			Very Mild PS (N=65)		
	no.	(%)	no.	(%)	O.R.	95% CI	no.	(%)	O.R.	95% CI	no.	(%)	O.R.	95% CI

Variable	no.	(%)	no.	(%)	O.R.	95% CI	no.	(%)	O.R.	95% CI	no.	(%)	O.R.	95% CI
Family History (1° relatives)														
congenital heart disease	43	(1.2)	5	(8.9)	**8.0**	**3.1–21.2**	6	(5.4)	**4.6**	**1.9–4.2**	3	(4.6)	**4.0**	**1.2–13.1**
noncardiac anomalies	165	(4.6)	7	(12.5)	**3.0**	**1.3–6.6**	4	(3.4)	0.8	0.3–2.1	4	(6.2)	1.4	0.5–3.8
Maternal Reproductive History														
previous preterm birth(s)	180	(5.0)	2	(3.6)	0.7	0.2–2.9	11	(9.8)	**2.1**	**1.1–3.9**	5	(7.7)	1.6	0.6–3.9
Illnesses and Medications														
thyroid disease	55	(1.5)	0	—	—	—	5	(4.5)	**3.0**	**1.2–2.7**	0	—	—	—
thyroid hormone	40	(1.1)	0	—	—	—	4	(3.4)	**3.3**	**1.2–9.3**	0	—	—	—
fever	166	(4.7)	1	(1.8)	0.4	0.1–2.7	12	(10.7)	**2.5**	**1.3–4.6**	2	(3.1)	0.7	0.2–2.7
Lifestyle Exposures														
alcohol (any amount)	2101	(58.8)	41	(73.2)	**1.9**	**1.1–3.5**	68	(68.8)	1.1	0.7–1.6	36	(55.4)	0.9	0.5–1.4
Home/Occupational Exposures														
hair dyes	238	(6.7)	12	(21.4)	**3.8**	**2.0–7.3**	12	(10.7)	1.7	0.9–3.1	5	(7.7)	1.1	0.5–2.9
degreasing solvents	19	(0.5)	0	—	—	—	3	(2.7)	**5.1**	**1.5–17.7**	1	(1.5)	2.9	0.4–22.1
painting:														
mother	367	(10.3)	7	(12.5)	1.2	0.6–2.8	7	(6.3)	0.6	0.3–1.3	3	(4.6)	0.4	0.1–1.4
father	757	(21.2)	15	(26.8)	1.4	0.7–2.5	24	(21.4)	1.0	0.6–1.6	9	(13.9)	0.6	0.3–1.2
both parents	158	(4.4)	6	(10.7)	**2.6**	**1.1–6.1**	3	(2.7)	0.6	0.2–1.9	2	(3.1)	0.7	0.2–2.8
Race of Infant														
white	2362	(66.2)	38	(67.8)	1.0	(reference)	59	(52.7)	1.0	(reference)	32	(49.2)	1.0	(reference)
nonwhite	1207	(33.8)	18	(32.2)	0.9	0.8–1.4	53	(47.3)	**1.8**	**1.2–2.6**	33	(50.8)	**2.0**	**(1.2–3.3)**

Risk factors are *maternal* unless labelled as paternal; PS = pulmonary valve stenosis, O.R.=odds ratio.

Table 9B.9
Multivariate Analysis of Pulmonic Stenosis by Severity of Defect
Baltimore-Washington Infant Study (1981–1989)

Variable	Cases	Controls	Odds Ratio	95% CI	99% CI
A. Severe Pulmonic Stenosis (N=56)					
family history of					
congenital heart disease	5	43	**6.4**	**2.3–18.0**	**1.6–24.8**
family history of					
noncardiac anomalies:					
no history, white infant	35	2234	1.0	(reference)	
positive history, white infant	3	131	1.3	0.4–4.3	0.3–6.3
no history, nonwhite infant	14	1173	0.9	0.5–1.7	0.4–2.1
positive history,					
nonwhite infant	4	34	**8.7**	**2.5–29.6**	**1.7–43.4**
hair dyes	12	238	**3.7**	**1.9–7.3**	**1.5–9.0**
house painting					
(by both parents)	6	158	**2.9**	**1.2–6.9**	0.9–9.1
B. Moderate Pulmonic Stenosis (N=112)					
family history of congenital					
heart disease	6	43	**5.2**	**2.1–12.8**	**1.6–16.8**
nonwhite race	53	1207	**2.1**	**1.4–3.1**	**1.3–3.5**
thyroid disease	5	55	**3.4**	**1.3–8.8**	**1.0–11.7**
fever	12	166	**2.9**	**1.5–5.4**	**1.3–6.6**
degreasing solvents	3	19	**5.0**	**1.3–18.7**	0.9–28.1
cigarette smoking					
(number per day):					
0–20 cigarettes,					
mother <35 years old	74	2516	1.0	(reference)	
0–20 cigarettes,					
mother 35+ years old	29	880	1.2	0.6–2.4	0.5–3.0
>20 cigarettes,					
mother <35 years old	6	167	1.3	0.6–3.1	0.4–4.1
>20 cigarettes,					
mother 35+ years old	3	9	**12.5**	**3.2–49.4**	**2.1–75.2**

that this heritable disorder is a major etiologic mechanism, and that it may produce a dome-shaped pulmonary valve as described by Peacock,[338] Abbott,[7] Edwards,[121] and Brock.[62]

In contrast, the mildly obstructed, "dystrophic" pulmonic valve, discovered only by echocardiogram and rarely associated with Noonan syndrome, seems to represent a diagnosis made in high-risk infants characterized by urban residence, small size, nonwhite race and female gender who are evaluated for the possible presence of a heart defect.

Infants characterized as moderate pulmonic stenosis diagnosed only by echocardiography represent a middle group; probably some have significant valvar abnormality but many are due to increased ascertainment. The findings in this subgroup may indicate an additional set of etiologic factors including maternal illnesses and degreasing solvent exposures.

The findings in this case-control study of a "pure" anatomic defect of the pulmonic valve suggest that the malformation group is heterogenous. It may be hypothesized that a large proportion of cases is accounted for by hereditary abnormalities such as the Noonan (and Ehlers-Danlos) disorders, while those without this genetic abnormality may be vulnerable to various environmental toxic exposures. Very mild abnormalities may constitute normal anatomic variants.

Section C

Pulmonary Valve Atresia With Intact Ventricular Septum

Pulmonary atresia with intact ventricular septum is a rare malformation of specific epidemiologic interest because it is suspected that the anomaly may be acquired during intrauterine life.[173] The presence of a well-formed pulmonary artery, the absence of broncho-pulmonary collaterals, and the rarity of associated noncardiac anomalies would support the hypothesis that these severe obstructions arise *after* the early embryonic period. Many cardiologists can recall the time of a sudden repeated occurrence of this severe disorder, which previously and subsequently was very rarely encountered. However, in such an occurrence in the author's (CF) experience, in which seven infants with pulmonary atresia were referred over a period of several summer months, the affected families came from different locations in the region and there was no indication of any common factor in the maternal histories. Recently, serial fetal echocardiograpy has demonstrated the development of pulmonary valve atresia in late embryonic and early fetal life[13] and has again raised the question of intrauterine inflammation, suggested long ago by Peacock,[338] Abbott,[7] and others.

Some authors have considered pulmonary atresia with intact ventricular septum the extreme form of pulmonary valve stenosis with fused cusps and a diaphragm-like membrane.[132] The atresia may be valvar, infundibular, or involve the stem of the main pulmonary artery. The malformation is compatible with fetal life as long as the foramen ovale is open and the arterial duct supplies the pulmonary circulation, but after birth closure of the ductus may lead to circulatory collapse. Lózsádi[261] called this abnormality an *unsolved disease of the newborn,* and noted that a right aortic arch was never present in contrast to pulmonary atresia with ventricular septal defect, in which a right aortic arch is quite common.

Atresia of the pulmonic valve is accompanied by abnormalities of the right ventricle and of the tricuspid valve.[60,163] Clinicians find it useful to distinguish cases with a small defective right ventricle from those with a normal right ventricular chamber, the latter having a more favorable prognosis. However, there is an entire spectrum of abnormalities affecting the right ventricle.[163] Freedom et al[161] consider this anomaly a *global disorder of the right ventricle,* with anomalies of the coronary arteries, and coronary artery-myocardial sinusoidal communications that allow egress of blood from the blind right ventricle.

Study Population

Cardiac Abnormalities

Among the 4390 cases in the Baltimore-Washington Infant Study, pulmonary atresia with intact ventricular septum as a principal diagnosis was previously reported in 73 infants.[342] However, a detailed review prior to the diagnostic categorizations for these risk factor analyses reduced this number to 53 cases who had pulmonary atresia with intact ventricular septum as an isolated cardiac anomaly, except for the presence of a patent arterial duct in some of these very young patients. None of the patients had a right-sided aortic arch. The ISC coding system provides identification of pulmonary atresia with small or normal right ventricle, and almost 80% of these infants had a normal right ventricle—the remainder were either coded as small or the size was not specified.

Noncardiac Anomalies

Only four infants with pulmonary atresia and intact ventricular septum had noncardiac anomalies (Table 9C.1). Two infants had trisomy 18 and two had hydronephrosis, one of the latter with an associated hepatic cyst. No case occurred in a Mendelian or non-Mendelian syndromic association.

Descriptive Analyses

Prevalence by Time, Season, and Area of Residence

The 53 cases out of 906,626 area livebirths represent a prevalence of 5.8 per 100,000 over the 9-year study period. Evaluated by temporal occurrence, there was a reduction from 0.76 to 0.54 and to 0.48 for successive 3-year periods (Table 9C.2). This may be due to fetal diagnosis and pregnancy termination. A seasonal difference was also noted with the highest prevalence in the winter birth quarter and the lowest prevalence in the fall. Prevalence by area of residence showed a possible rural excess.

Plotted by date of conception, there were two possible "time clusters" of cases against a widely dispersed background of rare quarterly occurrences (Figure 9C.1). Four cases were conceived during the fourth quarter of 1980, and 3 in the first quarter of 1981; 5 cases were conceived in the third quarter of 1987. The 1980 cases did not occur in close geographic proximity, (2 in Fairfax, Virginia; 1 in Baltimore; 1 in Anne Arundel County). Of the five cases in the 1987 cluster, four were born to residents of Baltimore, and the fifth was born in rural Queen Anne

Table 9C.1
Pulmonary Valve Atresia with Intact Ventricular Septum (N=53)
Noncardiac Anomalies
Baltimore-Washington Infant Study (1981–1989)

Chromosomal Abnormalities:	2	(3.8%)
Trisomy 18	2	[#2569, 10766]
Syndromes	0	
Multiple Anomalies, Nonclassified:	1	(1.9%)
Hydronephrosis, hepatic cyst	1	[# 11090]
Organ Defects:	1	(1.9%)
Hydronephrosis	1	[# 1353]
No Associated Anomalies	49	(92.5%)

Table 9C.2

Pulmonary Valve Atresia with Intact Ventricular Septum: Prevalence

Baltimore-Washington Infant Study (1981–1989)

	Cases	Area Live Births	Prevalence per 10,000
Total Subjects	53	906,626	0.58
Year of Birth:			
1981–1983	21	274,558	0.76
1984–1986	16	295,845	0.54
1987–1989	16	336,223	0.48
Birth Quarter:			
1st (Jan.–Mar.)	19	219,145	0.87
2nd (Apr.–Jun.)	15	231,777	0.65
3rd (Jul.–Sep.)	13	233,626	0.56
4th (Oct.–Dec.)	6	222,078	0.27
Area of Residence:			
urban	11	208,568	0.53
suburban	33	584,022	0.56
rural	9	112,318	0.80

county. Other than geographic area, there were no common risk factors among the four cases; one mother, 31 years old, had epilepsy and reported taking dilantin to control seizures in the second trimester.

These possible clusters were evaluated as follows: when we adjusted for year we found that variation among quarters was consistent with a Poisson process. When we adjusted for quarter, we found evidence of extra-variation among years relative to a Poisson process—presumably due to the possible clusters mentioned above. For a Poisson process with mean 40/36 (as in this data), the probability of seeing seven or more occurrences in two adjacent quarters is 0.007, and the probability of seeing five or more occurrences in one quarter is 0.008. Although we did not use a formal time-cluster test, the findings suggest that these possible clusters are indeed clusters. Glaz[185] discusses "computational problems" in the scan test for time-clusters.

Diagnosis and Course

The great majority of these infants was diagnosed in the first week of life and nearly 87% were diagnosed within 1 month of birth (Table 9C.3). Most patients underwent cardiac catheterization; one severely cyanotic patient died on the first day of life during transport to a tertiary care center and was diagnosed at autopsy. Surgery was performed in 44 of these infants. The all-cause mortality during the follow-up period was 46%.

Gender, Race, and Twinning

This small group of cases did not differ significantly from the controls by neither gender nor by racial distribution (Table 9C.4). There was only one twin birth among the 53 cases (1.9%), a rate similar to that among controls (1.5% twin births).

PULMONARY ATRESIA WITH INTACT VENTRICULAR SEPTUM (N=45)
DATE OF CONCEPTION
BALTIMORE - WASHINGTON INFANT STUDY (1981-1989)

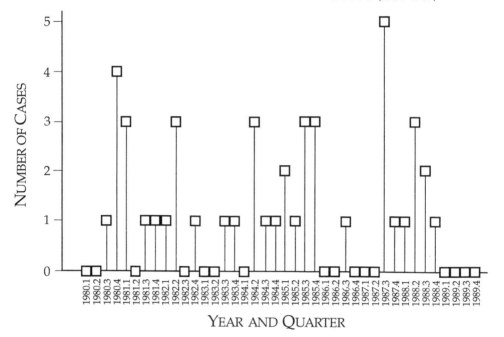

Figure 9C.1: Occurrences of pulmonary atresia with intact ventricular septum are plotted for each year by quarter of conception from 1980 through 1989. Suggestive evidence of clustering for 1980.4 and 1987.3 was provided by statistical analysis.

Information obtained by family interviews available in 45 of the 53 cases (85%) is presented in the sections that follow.

Birthweight and Gestational Age

Among case infants low birthweight and an early gestational age (Table 9C.5) occurred in excess in the control frequencies, and the mean birthweight of the cases was below that of controls. Seven of the case infants were small size for gestational age (15.6% compared with 5.9% for controls).

Sociodemographic Characteristics

There were no significant case-control differences in the distributions of sociodemographic variables (Table 9C.6). The distributions of parental educational levels, occupational categories, and household incomes were similar in the cases and control families.

Table 9C.3
Pulmonary Valve Atresia with Intact Ventricular Septum: Clinical Data
Baltimore-Washington Infant Study (1981–1989)

	Cases N=53 no.	(%)
Age at Diagnosis:		
<1 week	41	(77.4)
1–4 weeks	5	(9.4)
5–24 weeks	4	(7.5)
25–22 weeks	3	(5.7)
Method of Diagnosis:		
echocardiography only	3	(5.7)
echocardiography and cardiac catheterization	5	(9.4)
echocardiography and surgery	1	(1.9)
echocardiography, cardiac catherization, and surgery	31*	(58.6)
autopsy following other methods	12	(22.6)
autopsy only	1	(1.9)
Surgery During First Year of Life	44	(83.0)
Follow-Up:		
not followed	1	(1.9)
followed:	52	—
died in the first year of life	24	(46.1)
alive	28	(53.9)
median age at last visit	12 months	

*Includes 2 cases with cardiac catheterization and surgery without echocardiography.

Table 9C.4
Pulmonary Valve Atresia with Intact Ventricular Septum:
Baltimore-Washington Infant Study (1981–1989)

Characteristic	Cases N=53 no.	(%)	Controls N=3572 no.	(%)
Gender:				
male	29	(54.7)	1817	(50.9)
female	24	(45.3)	1755	(49.1)
Race*:				
white	36	(69.2)	2362	(66.1)
black	13	(25.0)	1109	(31.0)
other	3	(5.8)	98	(2.7)
Twin births	1	(1.9)	53	(1.5)

*Missing race in 1 case and 3 controls.

Table 9C.5
Pulmonary Valve Atresia with Intact Ventricular Septum:
Fetal Growth Characteristics
Baltimore-Washington Infant Study (1981–1989)

Characteristic	Cases no.	(%)	Controls no.	(%)
Number of Families Interviewed	45		3572	
Birthweight (grams)				
<2500	6	(13.3)	252	(7.1)
2500–3500	23	(51.1)	1853	(51.9)
>3500	16	(35.6)	1467	(41.0)
mean±standard error	3177±92		3351±10	
range	1673–4111		340–5273	
Gestational Age (weeks)				
<38	4	(15.9)	339	(9.5)
38+	41	(83.8)	3233	(90.5)
mean±standard error	39.6±0.3		39.6±0.1	
range	33–44		20–47	
Size for Gestational Age				
small (SGA)	7	(15.6)	211	(5.9)
normal	32	(71.1)	2712	(75.9)
large	6	(13.1)	649	(18.2)
Odds Ratio for SGA (95% CI)	2.9	(1.3–6.6)	1.0	(reference)

Potential Risk Factors

Familial Cardiac and Noncardiac Anomalies

The complete absence of familial birth defects in this case group is striking— no family member of any of these infants was reported to have a cardiac or noncardiac congenital anomaly. However, a cardiac defect that may or may not have been congenital was that of a father who had mitral valve prolapse.

Genetic and Environmental Factors

Univariate analysis (Table 9C.7) was performed on 45 cases in comparison to 3572 controls, and included all of the variables on maternal medical and reproductive history and on the exposure histories of both parents. There was a 7.2-fold risk associated with overt diabetes, based on only two affected case mothers, neither of whom reported taking insulin during pregnancy. One of these mothers had epilepsy and took dilantin throughout the pregnancy and was hospitalized for seizures in the second trimester. Another mother with epilepsy had a pre-pregnancy history of seizures, but did not report any seizures or antiepileptic medications during the study pregnancy. The odds ratio for epilepsy, based on these two case mothers was 11.8. A history of influenza showed an almost three-fold risk with seven exposed mothers, none of whom reported fever during the influenza infec-

Table 9C.6
Pulmonary Valve Atresia with Intact Ventricular Septum:
Analysis of Sociodemiographic Characteristics
Baltimore-Washington Infant Study (1981–1989)

Interviewed Families	Cases N=45 no.	(%)	Controls N=3572 no.	(%)	Odds Ratio (95% CI)	
Maternal Marital Status:						
not married	11	(24.4)	990	(27.7)	0.8	(0.4–1.7)
married	34	(75.6)	2582	(72.3)	1.0	(reference)
Maternal Education:						
<high school	6	(13.3)	659	(18.4)	0.9	(0.3–2.2)
high school	22	(48.9)	1265	(35.4)	1.7	(0.9–3.2)
college	17	(37.8)	1648	(46.1)	1.0	(reference)
Paternal Education:						
<high school	8	(17.8)	650	(18.2)	1.2	(0.5–2.9)
high school	21	(46.7)	1298	(36.3)	1.6	(0.9–3.2)
college	16	(35.5)	1624	(45.5)	1.0	(reference)
Annual Household Income:						
<$10,000	6	(13.6)	686	(19.2)	0.6	(0.2–1.5)
$10,000–$19,999	11	(25.0)	699	(19.6)	1.1	(0.5–2.3)
$20,000–$29,999	7	(15.9)	737	(20.6)	0.7	(0.3–1.5)
$30,000+	20	(45.5)	1373	(38.4)	1.0	(reference)
Maternal Occupation:						
not working	12	(26.7)	1142	(32.0)	0.6	(0.3–1.4)
clerical/sales	14	(31.1)	1119	(31.3)	0.8	(0.4–1.7)
service	4	(8.9)	444	(12.4)	0.5	(0.2–1.7)
factory	3	(6.7)	137	(3.8)	1.3	(0.4–4.8)
professional	12	(26.7)	730	(20.4)	1.0	(reference)
Paternal Occupation:						
not working	5	(11.1)	246	(6.9)	2.1	(0.7–6.4)
clerical/sales	6	(13.3)	618	(17.3)	1.0	(0.4–2.9)
service	6	(13.3)	480	(13.4)	1.3	(0.5–3.7)
factory	19	(42.2)	1279	(35.8)	1.6	(0.7–3.5)
professional	9	(20.0)	947	(26.5)	1.0	(reference)
Month of Pregnancy Confirmation:						
1st month	8	(17.8)	511	(14.3)	1.0	(reference)
2nd month	23	(51.1)	1955	(54.7)	0.8	(0.3–1.7)
3rd month	7	(15.5)	663	(18.6)	0.7	(0.2–1.9)
4th month or later	7	(15.5)	416	(11.6)	1.1	(reference)

Data are missing for income (1 case, 77 controls), paternal occupation (2 controls) and pregnancy confirmation month (27 controls).

Table 9C.7
Pulmonary Valve Atresia with Intact Ventricular Septum:
Univariate Analysis of Risk Factors
Baltimore-Washington Infant Study (1981–1989)

Variable	Cases N=45 no.	Cases N=45 (%)	Controls N=3572 no.	Controls N=3572 (%)	Odds Ratio (95% CI)	
Family History (first-degree relatives)						
congenital heart disease	0		43	(1.2)	—	—
noncardiac anomalies	1	(2.2)	165	(4.6)	0.5	(0.1–3.4)
Parental Age						
maternal age:						
<20	8	(17.8)	507	(14.2)	1.6	(0.7–3.6)
20–29	20	(44.4)	2009	(56.2)	1.0	(reference)
30+	17	(37.8)	1056	(29.6)	1.6	(0.8–3.1)
paternal age:						
<20	1	(2.2)	226	(6.3)	0.3	(0.1–2.5)
20–29	23	(51.1)	1724	(48.3)	1.0	(reference)
30+	21	(46.7)	1622	(45.4)	1.0	(0.5–1.8)
Maternal Reproductive History						
number of previous pregnancies:						
none	13	(28.9)	1159	(32.4)	1.0	(reference)
one	13	(28.9)	1097	(30.7)	1.1	(0.5–1.8)
two	9	(20.0)	709	(19.9)	1.1	(0.6–1.6)
three or more	10	(22.2)	607	(17.0)	1.5	(0.7–3.3)
previous miscarriage(s)	8	(17.8)	681	(19.1)	0.9	(0.4–2.0)
previous induced abortion(s)	16	(35.6)	691	(19.3)	**2.3**	**(1.2–4.3)**
Illnesses and Medications						
diabetes:						
overt	2	(4.4)	23	(0.6)	**7.2**	**(1.6–31.4)**
gestational	4	(8.9)	115	(3.2)	**2.9**	**(1.0–8.3)**
influenza	7	(15.6)	278	(7.8)	**2.7**	**(1.2–6.0)**
epilepsy	2	(4.4)	14	(0.4)	**11.8**	**(2.6–53.6)**
Lifestyle Exposures						
alcohol (any amount)	27	(60.0)	2101	(58.8)	1.1	(0.8–1.8)
smoking (cigarettes/day):						
none	29	(64.4)	2302	(64.5)	1.0	(reference)
1–10	4	(8.9)	565	(15.8)	0.6	(0.2–1.7)
11–20	10	(22.2)	529	(14.8)	1.5	(0.7–3.1)
>20	2	(4.4)	176	(4.9)	0.9	(0.2–3.8)
Home and Occupational Exposures						
ionizing radiation (occupational)	0		48	(1.3)	—	—
soldering with lead (father)	7	(15.6)	231	(6.5)	**2.7**	**(1.2–6.0)**
Race of Infant						
white	32	(71.1)	2362	(66.2)	1.3	(0.7–2.4)
nonwhite	13	(28.9)	1207	(33.8)	1.0	(reference)

Risk factors are *maternal* unless labelled as paternal.

Table 9C.8
Pulmonary Valve Atresia with Intact Ventricular Septum:
Maternal Reports of Fever During Pregnancy
Baltimore-Washington Infant Study (1981–1989)

Time Period	Cases N=45 no.	(%)	Controls N=3572 no.	(%)	Odds Ratio (95% CI)	
1st trimester and preceding 3 months*	4	(8.9)	166	(4.6)	2.0	(0.7–5.7)
2nd trimester	3	(6.7)	150	(4.2)	1.6	(0.5–5.3)
3rd trimester	4	(8.9)	115	(3.2)	**2.9**	**(1.1–8.3)**

*Critical period.

tion; there was a two-fold risk of a history of previous induced abortions. It is of note that no maternal chemical exposures appeared as risk factors in this group, although paternal exposure to lead was significantly associated with pulmonary atresia (odds ratio 2.7).

Exposures In the Second and Third Trimesters

Because of a possible mid-gestational or late-gestational origin, we examined odds ratios for maternal illnesses, medications, lifestyle exposures, and home and occupational exposures reported by the mothers as having occurred during the second and third trimesters of pregnancy. The only significant case-control difference was found for maternal reports of fevers in the third trimester (Table 9C.8). Four case mothers (8.9%) and 115 controls (3.2%) reported 1 or more days of fever during the third trimester, representing a nearly three-fold increase in risk, an association statistically significant at the 5% level but not at the 1% level. Of the four case mothers reporting third trimester fevers, one had a fever for 1 day but gave no other details. Two other case mothers, reported 1-day fevers during upper respiratory infections and treatment with oral antibiotics, and the fourth mother had 2 days of fever with a strep throat infection and received oral antibiotics.

Discussion

The evaluation of potential risk factors for pulmonary atresia with intact ventricular septum was limited to univariate analyses because of the small size of the case group on whom interview data were obtained. Nevertheless the findings, even with their limitations, do provide support for a possible late embryonic or fetal origin in association with maternal illnesses. In addition to the importance of maternal illness, it is also important to note that we found no genetic risks nor any other contributing factor (such as sociodemographic disadvantages) that might be surrogates for unmeasured risk factors.

Although the number of affected mothers was small, the elevated odds ratios for maternal illnesses (specifically influenza and third trimester fevers) suggest the need for well-targeted collaborative studies of case occurrences diagnosed in liveborn infants and fetuses by prenatal studies.

Section D

Tricuspid Atresia With Normally Related Great Arteries

Tricuspid atresia, a failure of communication from right atrium to right ventricle, is a rare malformation unique in its morphogenetic heterogeneity and the rise and fall of interest in the embryonic events that lead to this anomaly. Edwards[122] describes an anatomic classification of congenital tricuspid atresia that was proposed by Marie Kühne in 1906; this separates the cases first by the relation of the great vessels to each other, and then by the level and degree of associated pulmonic obstruction. The extensive bibliography quoted began in 1854, with five articles in the 19th century and 18 articles in the 20th century prior to Maude Abbott's Atlas in 1936.[122] In the *Atlas of Congenital Heart Disease*, of 1000 cases Abbott[7] listed 16 cases of tricuspid atresia, ranging in age from 6 weeks to 56 years, with a "hereditary predisposition" in five and an association with "anomalous vessels" in seven. In her earlier monograph, Abbott noted the *favorable compensatory mechanism for the circulation to the lungs* if tricuspid atresia was associated with transposition of the arterial trunks and quoted the remarkable case described by Hedinger in 1915, of a 56-year-old woman *who could dance and climb mountains with ease* and who died *some years later from failing compensation.* In our study, in which transposition of the great arteries is considered a leading anomaly, cases with associated tricuspid atresia were described with the transposition cases in Chapter 5. Only the cases of tricuspid atresia with normally related great arteries are considered here.

It is interesting that some early writers already considered differences in etiology and embryonic timing. Tricuspid stenosis of an inflammatory origin was described by Abbott[7] as a case of fetal endocarditis; such an occurrence would be similar to other obstructive defects of the right side of the heart, which now are believed to occur in mid or late pregnancy.

Modern morphogenetic studies emphasize the need for a systematic assessment of topographical characteristics, including the nature of the atresia (muscular or membranous), visceral and atrial situs, and atrioventricular and ventriculoarterial relationships.[22] It has been suggested that tricuspid atresia is a unilateral atrioventricular connection defect arising from a looping abnormality[18,19] associated with various forms of malalignment of the atrial and ventricular septa.[23]

A developmental distinction of at least two anatomic categories of tricuspid atresia is of importance: 1) a rare type in which there is an imperforate valve, and the parietal myocardium of the right atrium is continuous with that of the right ventricle, 2) a more common type is defined by absence of any connection between the right atrium and right ventricle.[21] This type is illustrated in Figure 2.6B. Both anatomic types result in the same physiologic consequences and are thus not separated into clinical classifications such as that described by Edwards[122] and expanded by Rosenthal.[371]

In a recent study Orie, Anderson, et al[332] conducted a detailed review of morphologic (39 autopsy cases) and two-dimensional echocardiograms (24 cases) of tricuspid atresia at the Children's Hospital of Pittsburgh. The morphologic study revealed that 37 of the 39 hearts (94.9%) had an absent atrioventricular connection and 25 (67.1%) had concordant ventriculoarterial connections comparable to the Baltimore-Washington Infant Study cases. Among the 24 patients whose echocardiograms were studied 20 (83%) had concordant ventriculoarterial connections. The realities of tricuspid atresia with the complexity of associated anatomic alterations highlighted the serious limitations of an attempted categorization of those very complex defects.

    In the Baltimore-Washington Infant Study the clinical reports of tricuspid atresia did not distinguish between the two anatomic types of the atrioventricular separation, but the codes indicated the presence of associated cardiovascular anomalies. The number of cases of tricuspid atresia was too small to be subdivided in any way and could be evaluated only in univariate descriptive analyses.

Study Population

    Thirty-three cases with tricuspid valve atresia with normally related great arteries were enrolled in the Baltimore-Washington Infant Study between 1981 and 1989, representing a regional prevalence of 3.6 per 100,000 livebirths. Thirty families (91%) completed the interview.

Cardiac Abnormalities

    There was considerable anatomic variation: 22 cases had a ventricular septal defect, 8 had pulmonary atresia; and 12 had hypoplasia of the pulmonary artery; only in 3 cases was the tricuspid obstruction reported as the sole cardiac malformation.

Noncardiac Anomalies

    Only three infants with tricuspid atresia had associated noncardiac anomalies: one had frontometaphyseal dysplasia syndrome, another had hydrocephaly, and the third had pyloric stenosis.

Descriptive Analyses

Prevalence by Time, Season, and Area of Residence

    This was a rare defect with a prevalence of 3.6 cases per 100,000 livebirths in the study area (Table 9D.1). Thus over the 9 study years each of the six regional pediatric cardiology centers identified on the average less than one resident case per year. The number of cases enrolled in each of the 3-year periods was fairly constant. Seasonal prevalence varied and was lowest among the births occurring in October through December. Rural residence was extremely rare, represented by only a single case.

Table 9D.1
Tricuspid Valve Atresia: Prevalence
Baltimore-Washington Infant Study (1981–1989)

	Cases	Area Births	Prevalence per 100,000
Total Subjects	33	906,626	3.63
Year of Birth:			
1981–1983	10	274,558	3.64
1984–1986	13	295,845	4.39
1987–1989	10	336,223	2.97
Birth Quarter:			
1st (Jan.–Mar.)	7	219,145	3.19
2nd (Apr.–Jun.)	11	231,777	4.74
3rd (Jul.–Sep.)	11	233,626	4.70
4th (Oct.–Dec.)	4	222,078	1.80
Area of Residence:			
urban	12	208,568	5.75
suburban	20	584,022	3.42
rural	1	112,318	0.89

Diagnosis and Course

Diagnosis was accomplished early with 67% of cases presenting in the first week of life and only 15% of the cases being diagnosed after the first month of life (Table 9D.2). Cardiac catheterization confirmed the diagnosis in the majority, with only 12% diagnosed solely by echocardiography. Twenty-three cases (70%) underwent surgery in the first year of life. At the time of diagnostic update, one third of the cases had died.

Gender, Race, and Twinning

The tricuspid atresia case group was characterized by a male excess (66.7% male) relative to controls (50.9% male). There was no significant difference in the racial distributions (Table 9D.3) and there were no twin births among the cases.

Birthweight and Gestational Age

Information on fetal growth characteristics was provided in the interview for 30 cases (91%) (Table 9D.4). Tricuspid atresia, like other major cardiovascular defects, was associated with high rates of prematurity and low birthweight, and with a nearly five-fold risk of being small for gestational age relative to control infants.

Sociodemographic Characteristics

There were numerous case-control differences in sociodemographic factors (Table9D.5). Significantly more case mothers than control mothers reported being

Table 9D.2
Tricuspid Valve Atresia: Clinical Data
Baltimore-Washington Infant Study (1981–1989)

	Cases N=33 no.	%
Age at Diagnosis		
<1 week	22	(66.7)
1–4 weeks	6	(18.2)
5–24 weeks	5	(15.1)
25–52 weeks	0	
Method of Diagnosis:		
echocardiography only	4	(12.1)
echocardiography and cardiac catheterization	14	(42.4)
echocardiography and surgery	0	
echocardiography, cardiac catherization, and surgery	13*	(39.4)
autopsy following other methods	2	(6.0)
autopsy only	0	
Surgery During First Year of Life	23	(69.7)
Follow-Up:		
not followed	1	(3.0)
followed:	32	
died in the first year of life	11	(34.4)
alive	21	(65.6)
median age at last visit	11 months	

*Includes 1 cardiac catherization and surgery (without echocardiography).

Table 9D.3
Tricuspid Valve Atresia: Infant Characteristics
Baltimore-Washington Infant Study (1981–1989)

Characteristic	Cases N=33 no.	(%)	Controls N=3572 no.	(%)
Gender				
male	22	(66.7)	1817	(50.9)
female	11	(34.4)	1755	(49.1)
Race:				
white	18	(54.5)	2362	(66.1)
black	13	(39.4)	1109	(31.0)
other	2	(6.1)	98	(2.7)
Twin births	0		53	(1.5)

Table 9D.4
Tricuspid Valve Atresia: Fetal Growth Characteristics
Baltimore-Washington Infant Study (1981–1989)

Characteristic	Cases no.	(%)	Controls no.	(%)
Number of Families Interviewed	30		3572	
Birthweight (grams)				
<2500	9	(30.0)	252	(7.1)
2500–3500	16	(53.3)	1853	(51.9)
>3500	5	(16.7)	1467	(41.0)
mean±standard error	2942±131		3351±10	
range	1418–4082		340–5273	
Gestational Age (weeks)				
<38	8	(26.7)	339	(9.5)
38+	22	(73.3)	3233	(90.5)
mean±standard error	38.7±0.4		39.6±0.1	
range	34–42		20–47	
Size for Gestational Age				
small (SGA)	7	(23.3)	211	(5.9)
normal	21	(70.0)	2712	(75.9)
large	2	(6.7)	649	(18.2)
Odds Ratio for SGA (95% CI)	**4.8**	**(1.7–10.3)**	1.0	(reference)

unmarried, having a low household income, low maternal educational attainment, and low paternal educational level. Analysis of SES scores revealed a case excess of families classified with medium (odds ratio 3.0) and low socioeconomic status (odds ratio 6.4), relative to families with higher scores.

Potential Risk Factors

Familial Cardiac and Noncardiac Anomalies

Only a single infant (#1284) had a family history of birth defects. This was a notable family because the mother had frontometaphyseal dysplasia and her three children had congenital heart disease; severe in the two boys and mild in the girl (all had right-sided obstructive defects). The proband and his brother also had frontometaphyseal dysplasia, which was diagnosed radiographically in infancy.

To our knowledge an association of congenital heart disease, and specifically of right-sided obstructions with familial frontometaphyseal dysplasia, has not previously been described.

Genetic and Environmental Factors

There were not enough subjects for multivariate analysis of multiple risk factors. Only univariate case-control comparisons (Table 9D.6) identified possible risk factor associations.

Table 9D.5

Tricuspid Atresia: Analysis of Sociodemiographic Characteristics
Baltimore-Washington Infant Study (1981–1989)

Interviewed Families	Cases N=30 no.	(%)	Controls N=3572 no.	(%)	Odds Ratio (95% CI)	
Maternal Married Status:						
not married	17	(56.7)	990	(27.7)	**3.4**	**(1.7–7.0)**
married	13	(43.3)	2582	(72.3)	1.0	(reference)
Maternal Education:						
<high school	10	(33.3)	659	(18.4)	**4.2**	**(1.5–11.5)**
high school	14	(46.7)	1265	(35.4)	**3.0**	**(1.2–7.9)**
college	6	(20.0)	1648	(46.1)	1.0	(reference)
Paternal Education:						
<high school	11	(36.7)	650	(18.2)	**2.7**	**(1.2–6.5)**
high school	9	(30.0)	1298	(36.3)	1.1	(0.5–2.8)
college	10	(33.3)	1624	(45.5)	1.0	(reference)
Annual Household Income:						
<$10,000	12	(41.4)	686	(19.2)	**3.4**	**(1.3–8.8)**
$10,000–$19,999	6	(20.7)	699	(19.6)	1.7	(0.6–5.0)
$20,000–$29,000	4	(13.8)	737	(20.6)	1.1	(0.3–3.6)
$30,000+	7	(24.1)	1373	(38.4)	1.0	(reference)
Maternal Occupation:						
not working	7	(23.3)	1142	(32.0)	0.9	(0.3–2.8)
clerical/sales	11	(36.7)	1119	(31.3)	1.4	(0.5–4.1)
service	5	(16.7)	444	(12.4)	1.6	(0.5–5.7)
factory	2	(6.7)	137	(3.8)	2.1	(0.4–11.1)
professional	5	(16.7)	730	(20.4)	1.0	(reference)
Paternal Occupation:						
not working	4	(13.3)	246	(6.9)	2.2	(0.6–7.6)
clerical/sales	4	(13.3)	618	(17.3)	0.9	(0.3–3.0)
service	5	(16.7)	480	(13.4)	1.4	(0.4–4.5)
factory	10	(33.3)	1279	(35.8)	1.1	(0.4–2.8)
professional	7	(23.3)	947	(26.5)	1.0	(reference)
Month of Pregnancy Confirmation:						
1st month	2	(6.9)	511	(14.3)	1.0	(reference)
2nd month	18	(62.1)	1955	(54.7)	2.4	(0.5–10.2)
3rd month	5	(17.2)	663	(18.6)	1.9	(0.4–10.0)
4th month or later	4	(13.8)	416	(11.6)	2.5	(0.4–13.5)

Data are missing for income (1 case, 77 controls), paternal occupation (2 controls), and pregnancy confirmation month (1 case, 27 controls).

The case-control odds ratio for family history was not significantly different from unity.

Mothers of cases were significantly younger than mothers of controls, with a nearly three-fold odds ratio associated with teenage mothers. There was no corresponding increase in the odds ratio for teenage fathers.

Two case mothers (6.7%) reported having a previous stillbirth, (1% among controls), yielding a seven-fold odds ratio. Significant case excesses were ob-

Table 9D.6
Tricuspid Atresia Univariate Analysis of Risk Factors
Baltimore-Washington Infant Study (1981–1989)

Variable	Cases N=30 no.	Cases N=30 (%)	Controls N=3572 no.	Controls N=3572 (%)	Odds Ratio (95% CI)	
Family History (first-degree relatives)						
congenital heart disease	1	(3.3)	43	(1.2)	2.8	(0.4–21.2)
noncardiac anomalies	1	(3.3)	165	(4.6)	0.7	(0.1–5.3)
Parental Age						
maternal age:						
<20	10	(33.3)	507	(14.2)	**2.8**	**(1.3–6.4)**
20–39	14	(46.7)	2009	(56.2)	1.0	(reference)
30+	6	(20.0)	1056	(29.6)	0.8	(0.3–2.1)
paternal age:						
<20	2	(6.7)	226	(6.3)	0.8	(0.2–3.5)
20–29	19	(63.3)	1724	(48.3)	1.0	(reference)
30+	9	(30.0)	1622	(45.4)	0.5	(0.2–1.1)
Maternal Reproductive History						
number of previous pregnancies:						
none	9	(30.0)	1159	(32.4)	1.0	(reference)
one	7	(23.3)	1097	(30.7)	0.8	(0.3–2.2)
two	9	(30.0)	709	(19.9)	1.6	(0.6–4.1)
three or more	5	(16.7)	607	(17.0)	1.1	(0.4–3.2)
previous miscarriage(s)	2	(6.7)	681	(19.1)	1.3	(0.3–5.7)
previous stillbirth	2	(6.7)	35	(1.0)	**7.2**	**(1.7–31.5)**
Illnesses and Medications						
diabetes:						
overt	1	(3.3)	23	(0.6)	5.3	(0.7–40.7)
gestational	1	(3.3)	115	(3.2)	1.0	(0.1–7.7)
influenza	8	(26.7)	278	(7.8)	**4.3**	**(1.9–9.8)**
fever	6	(20.0)	166	(4.7)	**5.1**	**(2.1–12.7)**
narcotic medications	3	(10.0)	93	(2.6)	**4.2**	**(1.2–13.9)**
Lifestyle Exposures						
alcohol (any amount)	17	(56.7)	2101	(58.8)	0.9	(0.4–1.9)
smoking (cigarettes/day):						
none	23	(76.7)	2302	(64.5)	1.0	(reference)
1–10	0		565	(15.8)	—	—
11–20	4	(13.3)	529	(14.8)	0.8	(0.3–2.2)
>20	3	(10.0)	176	(4.9)	1.7	(0.5–5.7)
cocaine:						
mother	1	(3.3)	42	(1.2)	2.9	(0.4–21.8)
father	4	(13.3)	111	(3.1)	**4.8**	**(1.6–14.0)**
marijuana:						
mother	1	(3.3)	263	(7.4)	0.4	(0.1–3.2)
father	10	(33.3)	555	(15.5)	**2.7**	**(1.3–5.8)**
Home and Occupational Exposures						
ionizing radiation (occupational)	0		48	(1.3)	—	—
Race of Infant						
white	15	(50.0)	2362	(66.2)	1.0	(reference
nonwhite	15	(50.0)	1207	(33.8)	2.0	(0.9–4.0)

Risk factors are *maternal* unless labelled as paternal.

served in maternal reports of influenza, fever, and use of narcotic medications during the critical period. Of the three mothers who reported taking narcotic medications, two used a cough syrup containing codeine for relief of influenza symptoms, and the third mother used propoxyphene (Darvon) to relieve arthritic pain. Among lifestyle exposures, cocaine and marijuana use among fathers (odds ratio 4.8), but not among mothers, was significantly associated with tricuspid atresia. No case-control differences in home and occupational exposures were detected.

Discussion

Descriptive information on the small group of cases with tricuspid atresia and normally related great arteries revealed some infant and maternal characteristics that were similar to those observed in the major structural defects of looping abnormalities and outflow tract defects with normally related great arteries. The infants were small with a mean birthweight of less than 3000 grams, one third of the babies were of low birthweight and also of low gestational age, and the infants had an almost five-fold risk of being small for gestational age compared to controls. Univariate analyses showed an almost three-fold risk of the anomaly in teenage mothers, significantly elevated odds ratios for febrile illness, use of narcotic medications, and paternal use of cocaine and marijuana. There was a six-fold increase over controls of a low socioeconomic score. These findings are similar to those in looping abnormalities (Chapter 4) and suggest modifiable risk factors.

The recent dichotomization of this malformation into looping abnormalities and obstructed tricuspid valves could not be evaluated for lack of anatomic information. The diagnostic information was explored to the extent possible by comparing this group of tricuspid atresia with normally related great arteries to the small group of tricuspid atresia with transposed great arteries. In both groups there was a male predominance (67%) but infants with transposition did not have fetal growth retardation, their mothers had no history of previous abnormal reproductive outcomes, no influenza or fever, no lifestyle or environmental exposures, and none fell into the low socioeconomic score group. There was, however, a statistically significant elevated risk for overt diabetes, which was based on two of the 18 case mothers.

Although these comparisons are based on very small numbers of cases they do suggest a distinction of the case groups with and without transposition. Only a multi-institutional collaborative effort could assemble the number of cases needed to evaluate these differences and to identify possible preventive interventions.

Chapter 10

Atrial Septal Defect

From: Ferencz C, Loffredo CA, Correa-Villaseñor A, Wilson PD, eds: *Genetic & Environmental Risk Factors of Major Cardiovascular Malformations: The Baltimore-Washington Infant Study 1981–1989.* Armonk, NY: Futura Publishing Co., Inc; ©1997.

Atrial septal defect is a malformation that is compatible with a long and active life. This defect has been amenable to surgical closure since 1950 and most recently by catheter stent techniques.[357] Patients of all ages with atrial septal defect have been studied in cardiology centers, and as a total group are believed to have a favorable outcome following cardiac surgery.

In the individual patient with an atrial septal defect, the evolution of clinical findings is closely tied to the progressive pathophysiologic changes during growth. A healthy infant without any manifestations of congenital heart disease may develop a slight pulmonic systolic murmur, as the pulmonary vascular resistance diminishes after birth sufficiently to permit left to right shunting across the defect. The sizes of the defect and of the shunt may both increase with growth. When this occurs, the classic physical findings are those of excessive blood flow through the right side of the heart: a palpable parasternal "right ventricular lift," increasing loudness of the systolic murmur, and a delay of the pulmonic component of the second heart sound. If the flow is excessive, a mid-diastolic murmur develops over the tricuspid valve due to the rapid inflow of blood from right atrium to right ventricle. Together with x-ray evidence of an enlarging heart and pulmonary vascular markings, and electrocardiographic evidence of moderate right ventricular hypertrophy, these clinical findings can establish the diagnosis and indicate the need for surgery in childhood before cardiac failure ensues. Until recently the diagnosis was usually made in children at school age or even in asymptomatic adolescents or young adults. Since pediatric cardiology started to focus on infants and since the introduction of the use of noninvasive diagnostic echocardiography, atrial septal defect has been recognized at earlier and earlier ages. Recognition of an atrial septal defect in infancy may be based on cardiovascular symptoms requiring surgery for closure of a defect. However the diagnosis is now increasingly based on echocardiographic evaluation in the course of routine physical examinations—evaluation of other illnesses or malformations. Thus secundum atrial septum defects diagnosed in infancy may be a selectively ascertained condition with a wide range of associated abnormalities.

Defects of the atrial septum are of various anatomic types. The most common defects are those of the ostium secundum in the position of the foramen oval. In fetal life flap patency of the foramen ovale shunts oxygenated blood from the inferior vena cava (ie, the placenta) into the left side of the heart and then into the systemic circulation. The defect may represent true absence of the septal tissues, but may also represent a "stretching" of the foramen due to an increase in pressure that can occur with severe associated right-sided malformations such as Ebstein's malformation, tricuspid atresia, or severe pulmonary stenosis. In association with these severe defects secondary patency of the foreman ovale is not counted as a separate malformation in this study. A different segment of atrial septal defects are ostium primum defects, ie, communications in the lower part of the septum often involving mitral valve abnormalities. This endocardial cushion type of atrial septal defects has been considered a partial form of atrioventricular septal defect (Chapter 6). A third type of atrial septal defect occurs posteriorly between the right atrium and the sinus venosus so that some of the right pulmonary veins may carry arterialized blood directly into the right atrium. Finally, in some patients the entire atrial septum may be absent (single atrium).

In an epidemiologic study of atrial septal defect, the different etiologic categories must be considered separately. In this study we defined an atrial septal defect as an isolated cardiovascular abnormality of the septum secundum; the association of patent arterial duct was not considered a separate anomaly in this infant population. We selected for risk factor evaluation only the cases with an isolated cardiac defect ie, those without associated noncardiac abnormalities that may have brought the patient to medical attention. Our definition of atrial septal defects in infancy as *isolated* ostium secundum defects is narrow and tries to focus on the cardiac defect itself. The results of this study therefore represent a specific population-based evaluation of "pure" ostium secundum defect diagnosed in infancy and studied in relation to genetic and environmental risk factors. However, the descriptive tables (Table 10.1-Table 10.9) compare and contrast these cases with those with noncardiac anomalies.

Study Population

Cardiac Abnormalities

In the Baltimore-Washington Infant Study during the years 1981 to 1989, 407 cases with defects of the atrial septum were enrolled (Figure 10.1), of which 291 represented ostium secundum atrial septal defects. Septum primum defects were analyzed in Chapter 6, which focuses on atrioventricular septal defects. Atrial defects that closed spontaneously, sinus venosus defects, and common atrium were not included in these analyses.

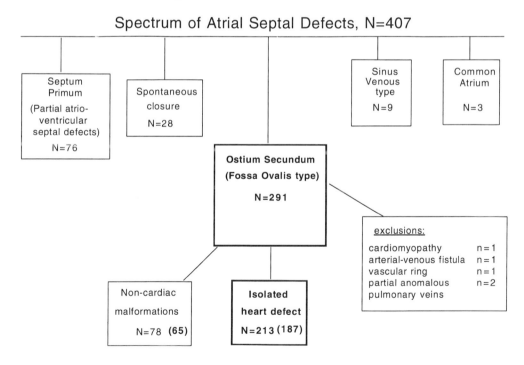

Spectrum of Atrial Septal Defects, N=407

Numbers in parentheses indicate interviewed families.

Figure 10.1: Most infants with atrial septal defects had ostium secundum defects, among whom 213 had isolated heart defects and were analyzed for risk factors. Septum primum defects were analyzed in Chapter 6. Sinus venosus defects, common atrium, and defects that spontaneously closed were excluded from analysis.

Noncardiac Anomalies

Among the 291 cases 213 were isolated heart defects and 78 (26.8%) had associated noncardiac malformations (Table 10.1) Thiry-six were found in infants with Down syndrome and an additional six had other chromosomal anomalies, including three with full or partial trisomy 13. Mendelian syndromes occurred in

Table 10.1
Atrial Septal Defect (N=291)
Noncardiac Anomalies
Baltimore-Washington Infant Study (1981–1989)

Chromosomal Abnormalities:	N=42	(14.4%)
Down syndrome	36*	
Trisomy 13	2	[# 1651, 1896]
Partial trisomy 13	1	[#1330]
Trisomy 11q	1	[#2231]
Deletion 7	1	[#2797]
Inversion 1	1	[#1870]
Syndromes	N=16	(5.5%)
a. Mendelian Syndromes:	7	
Goldenhar	1	[#1026]
Holt-Oram	1	[#1554]
Marden-Walker	1	[#1650]
Noonan	1	[#1747]
Prader-Willi	1	[#3064]
Carnitine deficiency	1	[#10394]
Acro-renal syndrome	1	[#10809]
b. Non-Mendelian Associations:	9	
VACTERL	4	[#1240, 1450, 3023, 10644]
Hydrocephaly-VACTERL	1	[#3656]
Fetal alcohol syndrome	2	[#2754, 10053]
Cytomegalovirus	1	[#1698]
Rubella	1	[#1907]
Multiple, Nonclassified Anomalies:	N=2	(0.7%)
Gastroschisis, cystic thymus	1	[#11271]
Spina bifida, diaphragmatic hernia, Agenesis of 1 kidney	1	[#11565]
Organ Defects:	N=18	(6.2%)
Hydrocephaly	3	[#3478, 3774, 11581]
Spina bifida	1	[#11200]
Craniosynostosis	1	[#11451]
Choanal atresia	1	[#3159]
Tracheoesophageal fistula	1	[#10796]
Omphalocele	1	[#1885]
Imperforate anus	2	[#1615, 2718]
Pyloric stenosis	1	[#10213]
Bilateral renal agenesis	1	[#11946]
Hydronephrosis	3	[#1440, 1639, 1685]
Pectus excavatum	1	[#1934]
Cystic lung	1	[#10101]
Hemivertebrae	1	[#1213]
No Associated Anomalies	N=213	(73.2%)

*See Appendix C for list of subject identification numbers.

seven infants but only a single case had the Holt-Oram syndrome. Among five cases with non-Mendelian associations, four were associated with pregnancy exposures (maternal alcohol, cytomegalovirus, and rubella infection). The VACTERL, hydrocephaly, two other multiple but nonclassified anomalies, as well as most of the 18 single-organ defects were of a severe nature—some potentially lethal. This confirms the expectation that in a large segment of infant cases the cardiac defect was of secondary importance.

Descriptive Analyses

Prevalence by Time, Season, and Area of Residence

The 291 cases of atrial septal defect represent a prevalence of 3.21 per 10,000 livebirths; the prevalence of isolated defects was 2.35/10,000 (Table 10.2). There was a moderate increase in prevalence observed during the study years, probably due to the greater use of two-dimensional echocardiography. There was a small variation in prevalence by birth quarter. The diagnosis of atrial septal defect was made more frequently in infants with urban residence than in those from suburban or rural areas.

Diagnosis and Course

Enrollment and outcome characteristics are shown in Table 10.3. The effect of noncardiac anomalies on the recognition of the cardiac defect is best shown in the

Table 10.2
Atrial Septal Defect: Prevalence
Baltimore-Washington Infant Study (1981–1989)

	All Cases	Isolated Cases	Area Live Births	prevalence per 10,000 All Cases	Isolated Cases
Total Subjects	291	213	906,626	3.21	2.35
Year of Birth:					
1981–1983	70	44	274,558	2.55	1.60
1984–1986	101	78	295,845	3.41	2.64
1987–1989	117	90	336,223	3.48	2.68
Birth Quarter:					
1st (Jan.–March)	62	46	219,145	2.83	2.10
2nd (April–June)	85	59	231,777	3.67	2.55
3rd (July–Sept.)	79	56	233,626	3.38	2.40
4th (Oct.–Dec.)	65	52	222,078	2.93	2.34
Area of Residence:					
urban	90	73	208,568	4.32	3.50
suburban	167	116	584,022	2.86	1.99
rural	34	24	112,318	3.03	2.14

Table 10.3
Atrial Septal Defect (ASD): Clinical Data
Baltimore-Washington Infant Study (1981–1989)

	ASD, total N=291		Isolated ASD N=213		ASD with Noncardiac Anomalies N=78	
	no.	(%)	no.	(%)	no.	(%)
Age at Diagnosis:						
<1 week	22	(7.6)	9	(4.2)	14	(17.9)
1–4 weeks	66	(22.7)	35	(16.4)	31	(39.7)
5–24 weeks	132	(45.4)	108	(50.7)	23	(29.5)
25–52 weeks	71	(24.4)	61	(28.6)	10	(12.8)
Method of Diagnosis:						
echocardiography only	253	(86.9)	188	(88.3)	65	(83.3)
echocardiography and cardiac catheterization	19	(6.5)	13	(6.1)	6	(7.7)
echocardiography and surgery	5	(1.7)	3	(1.4)	2	(2.6)
echocardiography, cardiac catheterization, and surgery	5	(1.7)	5	(2.3)	0	
autopsy following other methods	1	(0.3)	1	(0.5)	0	
autopsy only	8	(2.7)	3	(1.4)	5	(6.4)
Surgery During First Year of Life for Atrial Septal Defect	10	(3.4)	8	(3.8)	2	(2.6)
Follow-Up:						
not followed	36	(12.4)	32	(15.0)	4	(5.1)
followed:	255		181		74	
died in the first year of life	18	(7.1)	8	(4.4)	10	(13.5)
alive	237	(92.9)	173	(95.6)	64	(86.5)
median age at last visit	**10 months**		10 months		10 months	

distribution by age of diagnosis. Over half (57.6%) of the infants with associated anomalies were recognized to have an atrial septal defect within the first month of life (in contrast to only about 20% of isolated cases). In most of the infants the atrial septal defect was diagnosed by echocardiography only, with similar frequencies in those with and without noncardiac anomalies. However, the diagnosis was somewhat more frequently made by autopsy only in the multiple malformation group. All cause mortality during the follow-up period was three times higher in the group of infants with associated noncardiac anomalies (13.5%) than in the isolated subgroup (4.4%). Fifteen percent of the infants registered with isolated defects had no follow-up, compared to only 5% of those with associated noncardiac anomalies.

Among all the infants with atrial septal defect, 18 died during the first year of life. Four of these cases were ascertained in the community search of pathology records in the regional hospitals, and two were ascertained in the medical examiner's office (Table 10.4). In seven of these infants the atrial septal defect occurred as an isolated malformation; they died of cardiac failure, aspiration, or sepsis. In

Table 10.4

Atrial Septal Defect: Mortality in First Year of Life (N=18)

Baltimore-Washington Infant Study (1981–1989)

ID	Clinical Cardiac Diagnosis	Ascertainment	Age at Death	Autopsy +/-	Final Diagnosis	Cause of Death
Isolated Cases:						
3829	—	Medical Examiner	1 day	+	Large ASD	visceral congestion
3830	—	Medical Examiner	3 days	+	Large ASD	visceral congestion
2270	—	Community Search	6 months	+	Large ASD	aspiration
10519	ASD	Center	2 months	−	ASD	operative death
11400	ASD	Center	7 months	+	ASD	Group B Strep. sepsis
11088	ASD, PDA	Center	8 days	+	ASD, premature	renal failure
1361	ASD, PDA	Center	7 months	+	ASD, premature (widely patent foramen ovale)	perforated colon, intracerebral hemorrhage
Cases with Noncardiac Defects:						
1327	—	Community Search	4½ months	+	ASD, Down	sudden death, aspiration
11372	ASD, PDA	Center	10½ months	+	ASD, Down	pneumococcal sepsis
1896	ASD, Trisomy 13	Center	1½ months	−	ASD, Trisomy 13	apnea
1701	ASD, Airway cyst	Center	?	?	ASD, airway cyst	airway obstruction
1650	ASD, Syndrome	Center	7 days	−	ASD, premature, syndrome	Mardan-Walker Syndrome (respiratory)
1450	ASD, VACTER	Center	1 day	+	ASD, multiple anomalies, dysplastic kidneys	renal failure
11200	—	Center	5 months	−	ASD, meningomyelocele, spondylo-costal dysplasia	neurosurgery
11271	—	Community Search	3 months	+	ASD, gastroschisis, bowel perforation	gastroinestinal surgery
11565	—	Center	1 day	+	ASD, hydrocephalus, Arnold-Chiari, diaphragmatic hernia, lung hypoplasia	multiple malformations
11946	—	Community Search	1 day	+	ASD, Bilateral renal agenesis	non-cardiac malformation
1330	ASD, PDA	Center	6 months	−	ASD, partial trisomy 13	holoprosencephaly

ASD = atrial septal defect; PDA = patent ductus arteriosus.

all of the other deceased infants, death was due to the associated major malformation. These findings emphasize the potential lethality of the cardiac defect itself, but ascribe most of the fatalities to associated conditions.

Gender, Race, and Twinning

Gender and race distributions are shown in Table 10.5 and reveal a female excess that was more pronounced among the isolated cases than among the cases with associated malformations. The racial distribution approximated that of the control population with more black infants among the isolated cases. There were more twins among the cases than among controls especially among the isolated cases, indicating another source of preferential ascertainment.

Birthweight and Gestational Age

About one quarter of the cases were of low birthweight, and about one quarter were of low gestational age (Table 10.6), but there was a considerable difference between the two subgroups, with greater levels of prematurity, low birthweight, and especially of small-for-gestational-age in those with associated noncardiac anomalies (odds ratio 5.6 for the latter, relative to controls). However, even among those with isolated atrial septal defect there was an increase in the proportion who were small for gestational age (odds ratio 2.7).

Sociodemographic Characteristics

An analysis of sociodemographic characteristics of the families revealed few significant case-control differences (Table 10.7). Only a low level of maternal edu-

Table 10.5
Atrial Septal Defect: Infant Characteristics
Baltimore-Washington Infant Study (1981–1989)

Characteristic	All Cases N=291		Isolated Cases N=213		Cases with Noncardiac Anomalies N=78		Controls N=3572	
	no.	(%)	no.	(%)	no.	(%)	no.	(%)
Gender:								
male	101	(34.7)	67	(31.5)	33	(42.3)	1,817	(50.9)
female	190	(65.3)	146	(68.5)	45	(57.7)	1,755	(49.1)
Race*:								
white	170	(59.2)	122	(57.3)	48	(64.9)	2,362	(66.1)
black	89	(34.1)	80	(37.6)	18	(24.3)	1,109	(31.0)
other	19	(6.6)	11	(5.1)	8	(10.8)	98	(2.7)
Twin births	10	(3.4)	8	(3.8)	2	(2.6)	53	(1.5)

*Data missing are in 4 cases and 3 controls.

Table 10.6
Atrial Septal Defect (ASD): Fetal Growth Characteristics
Baltimore-Washington Infant Study (1981–1989)

Characteristic	ASD Total no. (%)	Isolated ASD no. (%)	ASD with Noncardiac Anomalies no. (%)	Controls no. (%)
Number of Families Interviewed	252	187	65	3572
Birthweight (grams)				
<2500	59 (23.5)	39 (20.9)	20 (31.3)	252 (7.1)
2500–3500	136 (54.2)	104 (55.6)	32 (50.0)	1853 (51.9)
>3500	56 (22.3)	44 (23.5)	12 (18.7)	1467 (41.0)
mean±standard error	2910±52	2939±60	2826±102	3351±10
range	496–4608	496–4608	528–4394	340–5273
Gestational Age (weeks)				
<38	68 (27.0)	46 (24.6)	22 (33.9)	339 (9.5)
38+	184 (73.0)	141 (75.4)	43 (66.1)	3233 (90.5)
mean±standard error	38.1±0.2	38.1±0.3	38.0±0.4	39.6±0.1
range	19–44	19–44	24–44	20–47
Size for Gestational Age				
small (SGA)	44 (17.5)	27 (14.4)	17 (26.2)	211 (5.9)
normal	172 (68.3)	134 (71.7)	38 (58.5)	2712 (75.9)
large	36 (14.3)	26 (13.9)	10 (15.3)	649 (18.2)
Odds Ratio for SGA (95% CI)	3.4 (2.4–4.8)	2.7 (1.7–4.1)	5.6 (3.2–10.0)	1.0 (reference)

cation was associated with atrial septal defect, but neither marital status, income, nor occupation were associated with case-control status.

Potential Risk Factors

Familial Cardiac and Noncardiac Anomalies

Atrial septal defect was a recurrent cardiac defect in the families of 10 case infants, four with a concordant diagnosis of atrial septal defect in at least one first-degree relative (Table 10.8). In one of these infants who also had choanal atresia, the father, brother, sister, and two paternal uncles each had an atrial septal defect, a dominant transmission from the paternal grandmother (Figure 10.2). The pedigree of this family has been previously reported.[143] It was the only family with atrial septal defect in multiple affected members. It is notable that the older individuals, including the oldest sibling, had third-degree heart block, while the younger sibling had only a prolonged atrioventricular conduction time. Other forms of congenital heart disease in other families included patent arterial duct, pulmonary stenosis, and a ventricular septal defect in three of the fathers. Among the siblings, two half-brothers are of interest, one with Ebstein's malformation and

Table 10.7

Atrial Septal Defect: Analysis of Sociodemiographic Characteristics

Baltimore-Washington Infant Study (1981–1989)

Interviewed Families	Cases N=252 no.	(%)	Controls N=3572 no.	(%)	Odds Ratio (95% CI)	
Maternal Marital Status:						
not married	81	(32.1)	990	(27.7)	1.2	(0.9–1.6)
married	171	(67.9)	2582	(72.3)	1.0	(reference)
Maternal Education:						
<high school	57	(22.6)	659	(18.4)	**1.4**	**(1.0–2.0)**
high school	95	(37.7)	1265	(35.4)	1.2	(0.9–1.7)
college	100	(39.7)	1648	(46.1)	1.0	(reference)
Paternal education:						
<high school	48	(19.0)	650	(18.2)	1.1	(0.8–1.6)
high school	96	(38.1)	1298	(36.3)	1.1	(0.8–1.5)
college	108	(42.9)	1624	(45.5)	1.0	(reference)
Annual Household Income:						
<$10,000	62	(25.3)	686	(19.2)	1.3	(0.9–1.8)
$10,000–$19,999	38	(15.5)	699	(19.6)	0.8	(0.5–1.2)
$20,000–$29,999	50	(20.4)	737	(20.6)	1.0	(0.7–1.4)
$30,000+	95	(38.8)	1373	(38.4)	1.0	(reference)
Maternal Occupation:						
not working	92	(36.5)	1142	(32.0)	1.2	(0.8–1.7)
clerical/sales	73	(29.0)	1119	(31.3)	1.0	(0.7–1.4)
service	28	(11.1)	444	(12.4)	0.9	(0.6–1.5)
factory	10	(4.0)	137	(3.8)	1.1	(0.5–2.2)
professional	49	(19.4)	730	(20.4)	1.0	(reference)
Paternal Occupation:						
not working	27	(10.7)	246	(6.9)	1.5	(0.9–2.4)
clerical/sales	34	(13.5)	618	(17.3)	0.7	(0.5–1.1)
service	30	(11.9)	480	(13.4)	0.8	(0.5–1.3)
factory	91	(36.1)	1279	(35.8)	1.0	(0.7–1.3)
professional	70	(27.8)	947	(26.5)	1.0	(reference)
Month of Pregnancy Confirmation:						
1st month	39	(15.9)	511	(14.3)	1.0	(reference)
2nd month	131	(53.3)	1955	(54.7)	0.9	(0.6–1.3)
3rd month	53	(21.5)	663	(18.6)	1.0	(0.7–1.6)
4th month or later	23	(9.3)	416	(11.6)	1.0	(reference)

Data are missing for income (7 cases, 77 controls), paternal occupation (2 controls), and pregnancy confirmation month (6 cases, 27 controls).

one with total anomalous pulmonary venous return. Among the 178 full-siblings and 96 maternal half-siblings, four and two respectively (2.2% and 2.1%) had congenital heart disease (control frequencies 0.5% and 0.6% respectively).

The importance of a genetic factor is further suggested by familial noncardiac defects. In one family both mother and infant had Holt-Oram syndrome, present also in the maternal grandmother who had an atrial septal defect as well; in another family the mother had cleft lip, and in the third family the mother had von

Table 10.8
Atrial Septal Defect
Congenital Defects and Hereditary Disorders in First-Degree Relatives
The Baltimore-Washington Infant Study (1981–1989)

Proband	Sex	Diagnosis	Relative	Diagnosis
		*A. Congenital Heart Defects:*		
2576	F		Mother	Atrial septal defect
11061	F		Mother	Atrial septal defect
3159	F	choanal atresia	Father	Atrial septal defect/ heart block
			Sister	Atrial septal defect/ heart block
			Brother	Atrial septal defect/ heart block
			Pat. Uncle	Atrial septal defect, 1° heart block
			Pat. Uncle	Atrial septal defect
			Pat. Gmother	Atrial septal defect
2245	F		Sister	Atrial septal defect/ duodenal atresia
3364	F		Father	Patent arterial duct
10644	M	VACTERL	Father	Pulmonic stenosis
10791	F		Father	Ventricular septal defect
11371	M		Maternal Half-Brother	Ebstein's anomaly
2027	F		Maternal Half-Brother	Total anomalous pulmonary venous return
1670	M		Brother	Aortic valve anomaly, cutis laxa
2742	F		Mother	Mitral valve prolapse
10565	M		Mother	Mitral valve prolapse
10994	F		Mother	Mitral valve prolapse
10102	M		Father	Mitral valve prolapse
10394	M	carnitine deficiency	Father	Mitral valve prolapse
		B. Congenital Defects of Other Organs:		
1554	M	Holt-Oram	Mother	Holt-Oram
			Maternal Grandmother	Atrial septal defect and Holt-Oram
3769	F		Mother	von Willebrand
11001	F		Mother	Cleft lip
3152	M	Down syndrome	Father	Polydactly

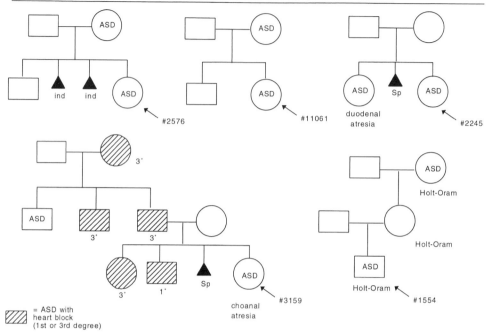

Figure 10.2: Five families had an occurrence of atrial septal defects among relatives of the affected proband. In one of these families, the pedigree (#3159) shows members of three generations affected with atrial septal defect and varying degrees of heart block. Reprinted with permission of the publisher. In another family (#1554), Holt Oram syndrome with or without atrial septal defect occurred in three generations.

Willebrand disease. Interestingly, no noncardiac defects were reported in any sibling and only one father was affected, reportedly with polydactyly.

Genetic and Environmental Factors

Univariate Analysis

In order to minimize the possibility of analyzing a heterogenous anatomic group, the risk factor analysis focused on the subgroup of infants with isolated heart defects (N = 187 interview participants). Results of the univariate case-control comparison are shown in Table 10.9. A history of congenital heart disease among first-degree relatives carried significant risk, as did maternal age of 30+ years, three or more previous pregnancies, previous preterm births, and induced abortions. Compared to controls, there was a significant excess of case mothers who reported gestational diabetes, though not of overt diabetes or insulin use during pregnancy. Urinary tract infections, bleeding during pregnancy, and use of corticosteroid agents were also significantly associated with the heart defect. Four of the six case mothers who reported taking corticosteroids used oral prednisone or steroid inhalers for the relief of asthma symp-

Table 10.9
Atrial Septal Defect, Isolated:
Univariate Analysis of Risk Factors
Baltimore-Washington Infant Study (1981–1989)

Variable	Cases N=187 no.	(%)	Controls N=3572 no.	(%)	Odds Ratio (95% CI)	
Family History (first-degree relatives)						
congenital heart disease	8	(4.3)	43	(1.2)	**4.2**	**(2.0–8.6)**
noncardiac anomalies	6	(3.2)	165	(4.6)	0.7	(0.3–1.6)
Parental Age						
maternal age:						
<20	29	(15.5)	507	(14.2)	1.2	(0.8–1.9)
20–29	93	(49.7)	2009	(56.2)	1.0	(reference)
30+	65	(34.8)	1056	(29.6)	**1.3**	**(1.0–1.8)**
paternal age:						
<20	13	(7.0)	226	(6.3)	1.1	(0.6–2.1)
20–29	88	(47.1)	1724	(48.3)	1.0	(reference)
30+	86	(45.9)	1622	(45.4)	1.0	(0.8–1.4)
Maternal Reproductive History						
number of previous pregnancies:						
none	53	(28.3)	1159	(32.4)	1.0	(reference)
one	49	(26.2)	1097	(30.7)	1.0	(0.7–1.5)
two	39	(20.9)	709	(19.9)	1.2	(0.8–1.8)
three or more	46	(24.6)	607	(17.0)	**1.7**	**(1.1–2.5)**
previous miscarriage(s)	37	(19.8)	681	(19.1)	1.0	(0.7–1.5)
previous preterm birth(s)	21	(11.2)	180	(5.0)	**2.4**	**(1.5–3.8)**
previous induced abortion(s)	56	(30.0)	691	(19.3)	**1.8**	**(1.3–2.5)**
Illnesses and Medications						
diabetes:						
overt	3	(1.6)	23	(0.6)	2.5	(0.7–8.5)
gestational	14	(7.5)	115	(3.2)	**2.4**	**(1.4–4.3)**
influenza	12	(6.4)	278	(7.8)	0.8	(0.4–1.5)
urinary tract infection	38	(20.3)	467	(13.1)	**1.7**	**(1.2–2.5)**
bleeding during pregnancy	39	(20.9)	534	(15.0)	**1.5**	**(1.0–2.2)**
corticosteroids	6	(3.2)	23	(0.6)	**5.1**	**(2.1–12.7)**
Lifestyle Exposures						
alcohol (any amount)	117	(62.6)	2101	(58.8)	1.2	(0.9–1.6)
smoking (cigarettes/day):						
none	109	(58.3)	2302	(64.5)	1.0	(reference)
1–10	38	(20.3)	565	(15.8)	1.4	(0.9–2.1)
11–20	27	(14.4)	529	(14.8)	1.1	(0.7–1.7)
>20	13	(7.0)	176	(4.9)	1.6	(0.9–2.8)
cocaine (father)	13	(7.0)	111	(3.1)	**2.3**	**(1.3–4.2)**
smoking (father), cigarettes/day:						
none	97	(52.7)	2058	(58.0)	1.0	(reference)
1–10	21	(11.4)	510	(14.4)	0.9	(0.5–1.4)
11–20	38	(20.7)	638	(18.0)	1.3	(0.9–1.9)
>20	28	(15.2)	342	(9.6)	**1.7**	**(1.1–2.7)**
Home and Occupational Exposures						
ionizing radiation (occupational)	3	(1.6)	48	(1.3)	1.2	(0.4–3.9)

Table 10.9—Continued
Atrial Septal Defect, Isolated:
Univariate Analysis of Risk Factors
Baltimore-Washington Infant Study (1981–1989)

Variable	Cases N=187 no.	(%)	Controls N=3572 no.	(%)	Odds Ratio (95% CI)	
paternal exposures:						
paint stripping	20	(10.7)	250	(7.0)	**1.6**	**(1.0–2.6)**
miscellaneous solvents	23	(12.3)	267	(7.5)	**1.7**	**(1.1–2.7)**
extreme cold temperature (occupational)	4	(2.1)	9	(0.3)	**8.7**	**(2.6–28.4)**
laboratory work with viruses	3	(1.6)	15	(0.4)	**3.9**	**(1.1–13.5)**
Race of Infant						
white	106	(56.7)	2362	(66.2)	1.0	(reference)
nonwhite	81	(43.3)	1207	(33.8)	**1.5**	**(1.1–2.0)**

Risk factors are *maternal* unless labelled as paternal.

toms, as did most of the 23 controls; the remaining subjects used topical medications. While no maternal home or occupational exposures showed significant case-control differences, several paternal exposures during the 6-month prepregnancy period were reported in excess among the cases including paint stripping, miscellaneous solvents (eg, paint thinners, photographic solutions, printing solvents), occupational exposures to extremely cold temperatures ($<0°$ F), and exposures to laboratory viruses by molecular biologists and other researchers. The latter two exposures were reported by four and three case fathers respectively, and were extremely rare among the controls ($<1\%$). Occupations reported to involve extremely cold temperatures included stock handlers working in freezers, ice manufacturers, and roofers working outdoors during the winter months; most exposures occurred daily. In addition, there was a significant association of paternal use of cocaine, and of paternal cigarette smoking (>20 cigarettes per day).

Multivariate Analysis

The multivariate analysis of potential risk factors for isolated atrial septal defect (Table10.10) revealed a strong effect of family history. Mothers with one or more previous preterm births were also at increased risk. A modest increase in risk (odds ratio 1.6) among women 35 years of age and older, but neither maternal age nor bleeding in pregnancy maintained significance at the 99% confidence level. Other maternal factors in the model that maintained significance at the 1% level included gestational diabetes and use of corticosteroids. Paternal factors included occupational exposures to extremely cold temperatures as well as cocaine use. Of the sociodemographic associations only race remained as a significant factor, with a modest increase in risk among nonwhites. No effect modifiers or confounding variables were found.

Table 10.10
Atrial Septal Defect, Isolated (N=187)
Multivariate Model
Baltimore-Washington Infant Study (1981–1989)

Variable	Adjusted Odds Ratio	95% CI	99% CI
family history of congenial heart disease	3.9	1.8–8.4	1.4–10.7
previous preterm birth	2.1	1.2–3.4	1.1–3.9
maternal age 35+	1.6	1.0–2.5	0.9–2.9
bleeding during pregnancy	1.5	1.0–2.1	0.9–2.4
gestational diabetes	2.4	1.3–4.4	1.1–5.2
urinary tract infection	1.6	1.1–2.3	0.9–2.6
corticosteriods	4.8	1.8–12.7	1.4–17.1
paternal occupational exposures:			
extreme cold temperatures	8.0	2.3–28.1	1.6–41.5
laboratory work with viruses	3.9	1.0–14.5	0.7–21.9
cocaine (paternal)	2.3	1.2–4.2	1.0–5.0
race (nonwhite)	1.5	1.1–2.1	1.0–2.3

Discussion

A notable finding among the infants with atrial septal defects is the large proportion diagnosed with associated severe noncardiac anomalies that accounted for their early death. However, among infants with isolated atrial septal defects early death also occurred due to congestive heart failure. This finding was also noted by Benson and Freedom[38] who could find no explanation for this occurrence.

The risk factor analysis revealed a prominent role of family history of congenital heart disease, maternal use of corticosteroids, and paternal exposure to extremely cold temperatures. Also of interest are the findings of a sex ratio favoring females, the association of the defect with intrauterine growth retardation, and temporal trends in prevalence, some of which suggest the possibility of detection bias.

Diagnosis and Prevalence

Atrial septal defects are associated with low mortality, especially among cases free of other major birth defects. Historically, the diagnosis was rarely accomplished in infancy, but in our study the increasing prevalence of infants diagnosed with the defect from 1981to 1989 reflects the increasing diagnostic use of echocardiography. Prevalence did not vary substantially by season of birth, but was high in urban areas compared to both suburban and rural areas of the region, suggesting a possible detection bias favoring the diagnosis among infants born in or near tertiary care centers located in major cities.

The possibility that the variation in prevalence with calendar time and region may reflect variation in the use of echocardiography is further supported by the low rates of invasive diagnostic procedures (ie, cardiac catheterization) and of surgical interventions, indicating the predominance of mild clinical malformations.

Infant Characteristics

Females predominated in this case group, especially among infants with isolated heart defects. This observation is consistent with previous reports from
other studies but its significance is unclear: is it a consequence of increased gestational survival of affected females relative to affected males? Could it be a bias
resulting from increased likelihood of detection among newborn females? The latter possibility was raised in the previous chapter on pulmonic stenosis, in which
we speculated that the increased rates of low birthweight and small-for-gestational-age in females (even among controls) may lead to increased medical attention relative to male newborns. In the group of infants with atrial septal defects,
low birthweight and prematurity were associated with both isolated cases and
multiply malformed cases. Compared to controls, isolated cases had a 412 gram
deficit in mean birthweight, and multiply malformed cases were 525 grams lighter
on average. Odds ratios for growth retardation (ie, small-for-gestational-age) were
also striking: a 2.7-fold odds ratio for isolated atrial septal defect and a 5.6-fold
odds ratio for atrial septal defect with noncardiac anomalies. Tikkanen and
Heinonen[418] report similar odds ratios among their cases with atrial septal defects.

Genetic Factors

Over one quarter (27%) of the infants with atrial septal defect in our study had
noncardiac anomalies; half of these had Down syndrome. Among the non-Down
syndrome patients, a wide array of Mendelian syndromes, multiple anomalies, and
organ defects were described. Some of these may lead to early ascertainment of
the cardiac defect and some may represent etiologic clues to the origins of some
cases of atrial septal defect. The association of the defect with a family history of
congenital heart disease in siblings and parents is well known. For isolated atrial
septal defect our analysis revealed a nearly four-fold adjusted odds ratio for family history, and we described five families in which the proband and affected relatives were concordant, including a three-generation pedigree of individuals who
had combinations of atrial septal defect and heart block.

Similar observations have been reported in the literature. Bosi and Pease,[50,339]
among others, published three- and four-generation pedigrees similar to our case,
in which multiple family members had atrial septal defects with atrioventricular
conduction defects. LiVolti[255] reports on Sicilian families with similar lesions in
which an autosomal dominant form of atrial septal defect affected four generations
in one family. Recurrence risks have been estimated in two recent studies,[368,448]
which prospectively followed probands over many decades to determine rates of
cardiac defects in their offspring. The two studies were remarkably consistent in
observing 10.7% and 11.3% rates of affected offspring. Concordance of the defects
in the parent-offspring pairs was also high, as it was in our study, with atrial septal defects diagnosed in 17 of 35 pairs (combining both prospective studies). In a
retrospective case-control study similar in design to our own, Tikkanen and
Heinonen[418] report a 10-fold odds ratio for the association with congenital heart
defects in the father, as well as elevated odds ratios for associations with other
types of birth defects in affected relatives.

Environmental Factors

Very few published epidemiologic studies have evaluated possible environmental risk factors for atrial septal defect. Olshan[329] examined birth certificates for paternal occupations in a case-control study arising from a malformation registry in British Columbia. Firemen in that study had an increased risk of having offspring with atrial septal defects, quantified by an odds ratio of 5.6 (95% confidence interval 1.6 to 21.8) for all atrial septal defects and 6.8 (1.4 to 33.2) for isolated cases. We did not observe this association in our data, but the number of fathers employed as firemen in our study was quite small (3 cases and 26 controls: odds ratio 1.6; 95% confidence interval 0.5 to 5.5). Tikkanen and Heinonen[418] studied 50 cases of atrial septal defects born in Finland in 1982 and 1983. In their study, maternal alcohol consumption (any amount) was associated with a two-fold odds ratio (95% confidence interval 1.1 to 3.4), as was maternal occupational exposures to "chemicals" during the first trimester (adjusted odds ratio 1.9; 95% confidence interval 1.1 to 3.4). These results differed from those in our study: we did not detect a significant alcohol association and we did not assess an aggregated "chemical" category of exposure. Results of our analysis at a restrictive level of significance did show association with maternal medical conditions and therapies, especially previous preterm birth, gestational diabetes and corticosteroid medications, as well as with paternal exposures such as cocaine use and occupational exposures to extreme cold. The latter association have not been reported before and warrant further study.

Chapter 11

Patent Arterial Duct

From: Ferencz C, Loffredo CA, Correa-Villaseñor A, Wilson PD, eds: *Genetic & Environmental Risk Factors of Major Cardiovascular Malformations: The Baltimore-Washington Infant Study 1981–1989.* Armonk, NY: Futura Publishing Co., Inc; ©1997.

Introduction

The arterial duct between the pulmonary artery and the aorta is an essential component of the fetal circulation. It is the duct through which blood ejected from the right ventricle is shunted into the aorta and thus bypasses the nonaerated lungs. With the infant's first breath, the pulmonary resistance falls abruptly; the arterial duct closes within a few hours after birth when the lungs have expanded.

In a remarkably comprehensive historical review, Olley[328] has brought to life numerous scientific contributions over the centuries that elucidated the anatomy, physiology, embryology, and gross and microscopic characteristics of the normal ductus, as well as the alterations associated with its persistent postnatal patency.

This history begins with Galen (?130–200 A.D.) who related the pulse to the heart and is therefore a founder of cardiology. He also described the fetal circulation and noted the dramatic readjustment of the circulation at birth. Harvey (1578–1657) is part of this history, as he recognized that in fetal life blood flows from the pulmonary artery to the aorta. The postnatal blood flow through a persistently patent duct is from the high-pressure aorta into the low-resistance pulmonary vascular bed; it is often a tumultuous flow which gives rise to bounding, "collapsing" pulses and a "machinery-like" continuous murmur in the pulmonary area, a process "exquisitely" described by Gibson 100 years ago.[328] Prior to the surgical era this anomaly had serious consequences in childhood and adulthood with congestive heart failure, an increased risk of obstructive pulmonary vascular disease, and bacterial endocarditis.

The new era in pediatric cardiology opened on August 26, 1938, when Robert Gross ligated a patent arterial duct in a 7½ year-old girl at the Boston Children's Hospital.[190] This child was seen by Hubbard at the age of 3 years and remained a notable patient in Boston until the age of 35 years when Paul Dudley White[444] noted that she was *fully active with no symptoms, had married and had two sons.*

Patency of the arterial duct also has a unique history in the early recognition of its etiologic heterogeneity. In 1960 an increased frequency of this anomaly was noted at high altitudes[14,328] and this was soon followed by the recognition of two other etiologic factors: prematurity[126] and maternal rubella infection.[189]

In preterm births, persistent patency of the arterial duct can be considered a physiologic phenomenon. Gittenberger[182,183] demonstrated histological immaturity of the ductal tissue in pathologic sections taken from preterm infants. As more and more premature infants survived in neonatal intensive care nurseries, ductal patency was recognized as a major contributor to postnatal congestive cardiac failure.[311] A national collaborative randomized clinical trial tested the efficacy of indomethacin treatment in comparison to surgical ligation of the ductus but reported no significant differences in survival.[179] Now patency of the arterial duct has become the concern of neonatology,[170] and many cases do not come to the attention of pediatric cardiologists.

Maternal rubella infection as a cause of ductal patency was recognized by an "alert" ophthalmologist in Australia.[189] Twenty years later, Rose et al[369] were surprised that the Toronto Heart Registry failed to note the 1964 epidemic of rubella by an increased ascertainment of patent arterial duct cases. However, in a retrospective evaluation, a small "epidemic curve" was identified, which constituted only about a tenth of all patent ductus cases. The routine immunization of children against rubella in the past 20 years has been a major preventive intervention, and ductal patency due to maternal rubella is now rarely observed. However ductal patency in full-term infants still occurs and raises further etiologic questions.

A genetic origin of ductal patency is evident in reports of families with multiple-affected members across several generations, but the results of systematic studies remain uncertain. Sanchez-Cascos and Garcia Sagrade[383] studied 119 families hoping to settle the question of relative importance of genetic factors versus teratogens. They demonstrated a female excess among cases but an excess in abnormal dermatoglyphs in males, consistent with the multifactorial model. Nora[325] performed a meta-analysis of second-generation studies (children of parents with congenital heart disease) and concluded that a *multilevel genetic-environmental interaction* was present. None of these studies made attempts to correlate the findings with various confounding variables such as repeated premature births in the families. However it seems clear that the appropriate measurements of genetic influences have not yet been identified.

In this epidemiologic study we tried to eliminate prematurity as an etiologic factor, hoping that the evaluation of families of full-term infants may offer a unique opportunity to explore other possible risk factors for this anomaly.

Study Population

The descriptive and risk factor analyses reported in this chapter are limited to those infants with patent arterial duct who were born at full term (38+ weeks of gestation) and in whom persistent patency of the arterial duct constituted the only cardiovascular malformation.

There were 80 infants with "pure" patent ductus, and 62 families were interviewed. Data for the 80 enrolled cases are shown in Table 11.1 through Table 11.5, while Tables 11.6 and beyond give data obtained in the 62 interviews.

In the hierarchical allocation of cases (see Chapter 2) a patent duct was a primary diagnosis only in the absence of other structural anomalies, septal defects, or obstructive anomalies, however, it occurred frequently in association with those cardiac defects.

The flow chart (Figure 11.1) illustrates the frequency of this abnormality. To create a "pure" diagnostic group, we excluded cases who had patent duct with delayed closure and those with associated cardiac anomalies. We did not study patent arterial duct as a secondary diagnosis.

Noncardiac Anomalies

Noncardiac malformations were present in 26% of the 80 cases with "pure" arterial duct patency (Table 11.1). Chromosome anomalies were found in nine patients (11.3%), including Down syndrome in seven of the infants. Spondylocostal dysplasia, a Mendelian syndrome, and the VACTERL and Goldenhar associations were each reported in a single case. A multiple but nonclassified sequence of anomalies was present in a male infant (#2317) reported to have Peter's anomaly with congenital deafness. In eight patients, a variety of nonsyndromic organ defects were present.

Patent Arterial Duct, N=512

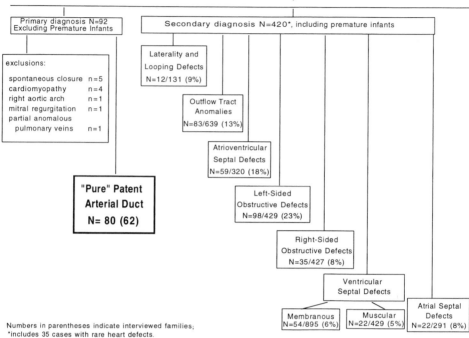

Numbers in parentheses indicate interviewed families;
*includes 35 cases with rare heart defects.

Figure 11.1: Patent arterial duct was diagnosed as a primary cardiac defect in only 92 full-term infants. In another 420 infants, some of whom were born prematurely, patent arterial duct was secondary to other cardiac defects.

Descriptive Analyses

Prevalence by Time, Season, and Area of Residence

The 9-year prevalence of patent arterial duct was 0.88 cases per 10,000 live-births. Prevalence of the defect increased from 0.69 in the first 3 years of the study to 0.92 in 1987 through 1989 (Table 11.2). Prevalence of the defect was lowest (0.68) among first quarter births (January to March) and greatest in the last quarter (1.05). Prevalence in urban areas was nearly twice as high as in suburban and rural areas.

Diagnosis and Course

Diagnosis of patent arterial duct as a single cardiac defect was made through-out the first year of life, with roughly equal proportions diagnosed in the first 4 weeks, the 5th through 24th weeks, and beyond the 24th week of life (Table 11.3). Nearly 40% of cases were diagnosed solely by echocardiography. Of the remain-der, all but two had cardiac catheterization and/or surgery, and two cases were discovered at autopsy. Surgery was performed in 44 cases (55%) during the first year of life. Ninety-five percent of all patients survived the first year of life. In the

Table 11.1
Patent Arterial Duct (N=80)
Noncardiac Anomalies
Baltimore-Washington Infant Study (1981–1989)

Chromosomal Abnormalities:	N=9 (11.3%)
Down syndrome	7 [#1008, 1210, 2749, 3105, 3749, 10566, 10764]
Trisomy 13	1 [# 1329]
Unspecified	1 [# 2955]
Syndromes	N=3 (3.8%)
a. Mendelian Syndromes:	1
Spondylocostal dysplasia	1 [#10099]
b. Non-Mendelian Associations:	2
VACTERL	1 [#1048]
Goldenhar	1 [#3689]
Multiple, Nonclassified Anomalies:	1 (1.3%)
Peter's anomaly, deafness	1 [#2317]
Organ Defects:	N=8 (10.0%)
Pyloric stenosis	2 [#1526, 1792]
Aniridia	1 [#2127]
Sequestration of lung	1 [#10653]
Cleft lip and palate	1 [#11071]
Pectus excavatum	1 [#1057]
Dysplastic kidney	1 [#3614]
Hypospadias	1 [#1739]
No Associated Anomalies	N=59 (73.8%)

Table 11.2
Patent Arterial Duct: Prevalence
Baltimore-Washington Infant Study (1981–1989)

	Cases	Area Births	Prevalence per 10,000
Total Subjects	80	906,626	0.88
Year of Birth:			
1981–1983	19	274,558	0.69
1984–1986	24	295,845	0.81
1987–1989	30	336,223	0.92
Birth Quarter:			
1st (Jan.–March)	15	219,145	0.68
2nd (April–June)	20	231,777	0.91
3rd (July–Sept.)	22	233,626	0.94
4th (Oct.–Dec.)	23	222,078	1.04
Area of Residence:			
urban	29	208,568	1.39
suburban	43	584,022	0.75
rural	8	112,318	0.71

Table 11.3
Patent Arterial Duct: Clinical Data
Baltimore-Washington Infant Study (1981–1989)

	Cases N=80 no.	(%)
Age at diagnosis:		
<1 week	6	(7.5)
1–4 weeks	24	(30.0)
5–24 weeks	27	(33.8)
25–52 weeks	23	(28.8)
Method of Diagnosis:		
echocardiography only	31	(38.8)
echocardiography and cardiac catheterization	3	(3.8)
echocardiography and surgery	12	(15.0)
echocardiography, cardiac catheterization, and surgery	32*	(41.6)
autopsy following other methods	0	
autopsy only	2	(2.5)
Surgery During First Year of Life	44	(55.0)
Follow-Up:		
not followed	7	(8.8)
followed:	73	
died in the first year of life	4	(5.5)
alive	69	(94.5)
median age at last visit	10 months	

*Includes 1 cardiac catheterization only, 1 surgery only, 5 cardiac catheterization and surgery.

four infants who died in the first year of life, (Table 11.4) the causes of death were associated disorders.

Gender, Race, and Twinning

Compared to controls there was a slight excess of females in this case group [56% versus 49% (Table 11.5)], and a slight excess of black infants. Only one case of patent arterial duct occurred in a twin birth, a proportion similar to controls; the co-twin was unaffected.

Birthweight and Gestational Age

Fetal growth characteristics of cases and controls are displayed in Table 11.6. By case definition, all infants were full-term births and are therefore compared to full-term controls. Despite the exclusion of premature infants, those with patent ductus were more likely than controls to have low birthweight, a reduced mean birthweight, and to be small for gestational age (SGA), with a nearly three-fold elevation in risk of being SGA.

Table 11.4
Patent Arterial Duct: Mortality
Baltimore-Washington Infant Study (1981–1989)

ID	Gender	Race	Gestational Age (wks)	Birth Weight (gm)	Age at Diagnosis	Diagnosis Method	Surgery	Noncardiac defects	Cause and Age at Death
1329	M	W	40	2991	0 wks	autopsy	no	Trisomy 13	respiratory arrest - Day 1
11511	F	B	40	4082	0 wks	autopsy	no	no	pneumonia - Day 3
11683	M	B	40	2892	1 wk	echo only	no	no	Group B streptococcal meningitis - Day 28
10099	M	B	38	2580	5 wks	echo + surgery	yes	spondylocostal dysplasia	respiratory distress - Day 7

Table 11.5

Patent Arterial Duct: Infant Characteristics

Baltimore-Washington Infant Study (1981–1989)

Characteristic	Cases N=80 no.	(%)	Controls N=3572 no.	(%)
Gender:				
male	35	(43.8)	1,817	(50.9)
female	45	(56.2)	1,755	(49.1)
Race:*				
white	45	(56.3)	2,362	(66.1)
black	30	(37.5)	1,109	(31.0)
other	2	(2.5)	98	(2.7)
Twin births	1	(1.3)	53	(1.5)

*Missing race in 3 cases and 3 controls.

Table 11.6

Patent Arterial Duct: Fetal Growth Characteristics

Baltimore-Washington Infant Study (1981–1989)

Characteristic	Cases no.	(%)	Controls* no.	(%)
Number of families interviewed	62		3231	
Birthweight (grams)				
<2500	6	(9.7)	81	(2.5)
2500–3500	39	(62.9)	1702	(52.6)
>3500	17	(27.4)	1448	(44.8)
mean±standard error	3220±76		3450±9	
range	2070–5103		1376–5273	
Gestational Age (weeks)				
<38	0		0	
38+	62	(100)	3231	(100)
mean±standard error	39.8±0.2		39.9±0.1	
range	38–44		38–47	
Size for Gestational Age				
small (SGA)	9	(14.5)	173	(5.4)
normal	47	(75.8)	2456	(76.0)
large	6	(9.7)	602	(18.6)
Odds Ratio for SGA (95% CI)	2.9	(1.4–6.1)	1.0	(reference)

*Includes only full-term control (38+ weeks of gestation).

Potential Risk Factors

Familial Cardiac and Noncardiac Anomalies

Congenital heart defects in first-degree relatives of the proband were reported in only two families (Table 11.7). In one family, a sister of the proband had tetralogy of Fallot, and in another family a female proband with Down syndrome had a maternal half-brother with bicuspid aortic valve. Noncardiac anomalies were reported in five other first-degree relatives. Among three patients with an isolated patent arterial duct, one had a father with thalassemia (also affecting his three siblings), another had a father with horseshoe kidneys, and the third a brother with cystic fibrosis—all genetic abnormalities. Two case infants with noncardiac defects had siblings affected with nonconcordant malformations. There were no affected mothers in this case group.

Sociodemographic Characteristics

Indications of socioeconomic disadvantage were noted in that a significantly elevated odds ratio (2.1) was detected among mothers who did not complete high school; a similar result was observed among fathers (Table 11.8). Mothers who did not work during pregnancy (including unemployed workers, students, and housewives) were at increased risk relative to professional working mothers, but no differences were observed in paternal occupational categories.

Table 11.7
Patent Arterial Duct:
Congenital Defects and Hereditary Disorders in First-Degree Relatives
The Baltimore-Washington Infant Study (1981–1989)

Proband	Sex	Diagnosis	Relative	Diagnosis
A. Congenital Heart Defects:				
2820	M	—	Sister	Tetralogy
2749	F	Down	Maternal Half-Brother	Bicuspid aortic valve
B. Congenital Defects of Other Organs:				
1997	M	—	Father (2 Pat. Uncles, 1 Pat. Aunt)	Thalassemia
10085	F	—	Father	Horseshoe kidney
1856	M	—	Brother	Cystic fibrosis
3614	M	dysplastic kidney	Brother	Pyloric stenosis
2955	M	unspecified chromosome anomaly	Maternal Half-Sister	Spina bifida

Table 11.8

Patent Arterial Duct: Analysis of Sociodemiographic Characteristics

Baltimore-Washington Infant Study (1981–1989)

Interviewed Families	Cases N=62 no.	(%)	Controls N=3572 no.	(%)	Odds Ratio (95% CI)	
Maternal Marital Status:						
not married	17	(27.4)	990	(27.7)	1.0	(0.6–1.7)
married	45	(72.6)	2582	(72.3)	1.0	(reference)
Maternal Education:						
<high school	18	(29.0)	659	(18.4)	**2.1**	**(1.1–4.0)**
high school	24	(38.7)	1265	(35.4)	1.5	(0.8–2.7)
college	20	(32.3)	1648	(46.1)	1.0	(reference)
Paternal Education:						
<high school	18	(29.0)	650	(18.2)	**2.0**	**(1.0–3.6)**
high school	22	(35.5)	1298	(36.3)	1.0	(0.7–2.2)
college	22	(35.5)	1624	(45.5)	1.0	(reference)
Annual Household Income:						
<$10,000	12	(19.4)	686	(19.2)	1.0	(0.5–2.1)
$10,000–$19,999	17	(27.4)	699	(19.6)	1.5	(0.8–2.7)
$20,000–$29,999	11	(17.7)	737	(20.6)	0.9	(0.4–1.8)
$30,000+	22	(35.5)	1373	(38.4)	1.0	(reference)
Maternal Occupation:						
not working	28	(45.2)	1142	(32.0)	**2.6**	**(1.1–5.9)**
clerical/sales	19	(30.6)	1119	(31.3)	1.9	(0.8–2.7)
service	8	(12.9)	444	(12.4)	1.9	(0.7–5.2)
factory	0		137	(3.8)	—	
professional	7	(11.3)	730	(20.4)	1.0	(reference)
Paternal Occupation:						
not working	3	(4.8)	246	(6.9)	0.6	(0.2–2.2)
clerical/sales	3	(4.8)	618	(17.3)	**0.3**	**(0.1–0.9)**
service	10	(16.1)	480	(13.4)	1.1	(0.5–2.4)
factory	29	(46.8)	1279	(35.8)	1.2	(0.7–2.2)
professional	17	(27.4)	947	(26.5)	1.0	(reference)
Month of Pregnancy Confirmation:						
1st month	7	(11.3)	511	(14.3)	1.0	(reference)
2nd month	33	(53.2)	1955	(54.7)	1.3	(0.6–2.9)
3rd month	14	(22.6)	663	(18.6)	1.5	(0.6–3.8)
4th month or later	8	(12.9)	416	(11.6)	1.4	(0.5–3.9)

Data are missing for income (77 controls), paternal occupation (2 controls), and pregnancy confirmation month (27 controls).

Genetic and Environmental Factors

Univariate Analysis

Compared to controls, a greater proportion of case mothers reported having gestational diabetes (odds ratio 2.6, just barely significant), but no case mothers reported overt diabetes and none took insulin. (Table 11.9). There were no other univariate case-control differences in familial and parental characteristics.

Table 11.9
Patent Arterial Duct:
Univariate Analysis of Risk Factors
Baltimore-Washington Infant Study (1981–1989)

Variable	Cases N=62 no.	(%)	Controls N=3572 no.	(%)	Odds Ratio (95% CI)
Family History **(first-degree relatives)**					
congenital heart disease	2	(3.2)	43	(1.2)	2.7 (0.6–11.5)
noncardiac anomalies	5	(8.1)	165	(4.6)	1.8 (0.7–4.5)
Parental Age					
maternal age:					
<20	7	(11.2)	507	(14.2)	0.8 (0.3–1.7)
20–29	37	(59.7)	2009	(56.2)	1.0 (reference)
30+	18	(29.0)	1056	(29.6)	1.0 (0.6–1.7)
paternal age:					
<20	4	(6.4)	226	(6.3)	1.1 (0.4–3.1)
20–29	28	(45.2)	1724	(48.3)	1.0 (reference)
30+	30	(48.4)	1622	(45.4)	1.2 (0.7–2.0)
Maternal Reproductive **History**					
number of previous pregnancies:					
none	20	(32.2)	1159	(32.4)	1.0 (reference)
one	12	(19.4)	1097	(30.7)	0.6 (0.3–1.3)
two	12	(19.4)	709	(19.9)	1.0 (0.5–2.0)
three or more	18	(29.0)	607	(17.0)	1.7 (0.9–3.3)
previous miscarriage(s)	15	(24.2)	681	(19.1)	1.4 (0.8–2.6)
Maternal Illnesses and **Medications**					
diabetes:					
overt	0		23	(0.6)	— —
gestational	5	(8.1)	115	(3.2)	**2.6 (1.0–6.6)**
influenza	4	(6.4)	278	(7.8)	0.8 (0.3–2.2)
Lifestyle Exposures					
alcohol (any amount)	33	(53.2)	2101	(58.8)	0.8 (0.5–1.3)
smoking (cigarettes/day):					
none	42	(67.7)	2302	(64.5)	1.0 (reference)
1–10	9	(14.5)	565	(15.8)	0.9 (0.4–1.8)
11–20	6	(9.7)	529	(14.8)	0.6 (0.3–1.5)
>20	5	(8.1)	176	(4.9)	1.6 (0.6–4.0)
Home and Occupational **Exposures**					
ionizing radiation (occupational)	0		48	(1.3)	— —
Race of Infant					
white	39	(61.9)	2362	(66.2)	0.8 (0.5–1.4)
nonwhite	24	(38.1)	1207	(33.8)	1.0 (reference)

Multivariate Analysis

In the multivariate analysis (Table 11.10), mothers with gestational diabetes who were 30 years or older were nearly five times more likely than younger mothers without diabetes to have an affected infant. Neither younger mothers with gestational diabetes nor older mothers without diabetes showed increased risk relative to younger mothers without diabetes.

Salient findings in the proband and maternal histories are shown in Table 11.11 for the five case mothers who reported having gestational diabetes. Four of these mothers were over 30 years of age; older maternal age was associated with multiparity, repeated miscarriages, maternal obesity, and some other indications of a complicated pregnancy. One mother had subfertility and the pregnancy occurred following the administration of clomiphene. All five mothers appeared to have had multiple factors that could have constituted potential risks for adverse pregnancy outcomes.

**Maternal Exposures During the Second
and Third Trimesters**

In order to examine the possibility that mid- or late-gestational exposures might be associated with patent arterial duct, we examined the data for associations of maternal illnesses, medications, lifestyle exposures, and home and occupational exposures reported in the second and third trimesters. Only corticosteroid medications during the second trimester were significantly associated. Two case mothers (3.2%) and 13 controls (0.4%) reported using corticosteroids (odds ratio 9.1; 95% confidence interval 2.0 to 41.3) during this time period. Of the two case mothers, one (#10653) was a gestational diabetic (see Table 11.11) who was severely asthmatic and took prednisone (4 pills per day) and used steroid inhalers (3 times per day) throughout the pregnancy; her female child (41 weeks gestation, 3430 grams) was born with lung sequestration in addition to the patent arterial duct. The other case mother (#11071) was treated

Table 11.10

Patent Arterial Duct (N=62)

Multivariate Analysis of Potential Risk Factors

Baltimore-Washington Infant Study (1981–1989)

Variable	Cases	Controls	Adjusted Odds Ratio	95% CI	99% CI
Maternal Age and					
Gestational Diabetes					
nondiabetic, <30 years	43	1883	1.0	(reference)	
nondiabetic, 30+ years	14	1574	0.8	0.4–1.5	0.3–1.8
gestational diabetes, <30 years	1	67	0.9	0.2–6.3	0.1–11.7
gestational diabetes, 30+ years	4	48	**4.7**	**1.6–13.7**	**1.2–19.1**

for mononucleosis during the second trimester with multiple medications including prednisone (4 pills per day) to control swelling; the infant (39 weeks gestation, 3056 grams) was born with a missing ear lobe and cleft lip and palate in addition to her heart defect.

Discussion

During the years 1981 to 1989, the Baltimore-Washington Infant Study enrolled 80 full-term infants with a diagnosis of "pure" patent arterial duct, constituting only 1.8% of the total population of infants with congenital heart disease and a prevalence of 8/100,000 livebirths. The exclusion of premature infants was based on gestational age calculated from the infant's birth date and the expected date of delivery as reported by the mother. In spite of this constraint in admitting cases for study, the excess proportion with low birthweight and small size for gestational age suggests that cases in which prematurity was the etiologic factor may still have been included.

Prevalence and Relative Frequency

The rarity of patent arterial duct in this case population is in contrast to earlier experiences in pediatric cardiology where it constituted one of the most common anomalies. Among the 1000 cases of congenital heart disease in Abbott's *Atlas of Cardiovascular Malformations*[6] 92 patients (9.2% of all cases) had simple patency of the ductus ranging in age from 2 weeks to 66 years. Included in the atlas is a "heart sign record" by HN Segall, illustrating the continuous murmur with its systolic accentuation and transmission throughout the chest. Patent arterial duct remained a common malformation in the Toronto Heart Registry (1948–1949) with a population occurrence of 1 in 3850 livebirths[175] in comparison to the frequency in the Baltimore-Washington Infant Study (1981–1989) of 1 in 12,500 births. Comparison of prevalence data from other studies indicates marked variations in case definitions and even differences in the definition of prematurity by various authors.

Potential Risk Factors

As reported in pediatric cardiology textbooks and other studies, we found an excess of females in this diagnostic group, but the difference from controls was not statistically significant. Evaluation of risk factors in this small case population provided surprising results: there was no evidence of familial aggregation, nor of any association with the many variables tested with the exception of gestational diabetes in older mothers. Gestational diabetes (ie, diabetes only during the pregnancy) seemed to be an indicator of a complex set of factors including socioeconomic disadvantage affecting this and perhaps also previous pregnancies. Thus of all diagnostic groups studied, this group has provided no clear directions toward future studies.

Table 11.11
Patent Arterial Duct: Gestational Diabetes
Baltimore-Washington

					Case Infant				
ID	Sex	Race	Gestational Age (wks)	Birth Weight (g)	Non-cardiac Anomalies	Method of Diagnosis	Age at Diag-nosis	Surgery	Alive
1526	Male	White	38	2495	Pyloric stenosis	E,C,S	5 weeks	yes	yes
11204	Male	White	38	4196	none	E only	15 weeks	no	yes
11410	Male	White	40	3175	none	E only	5 weeks	no	yes
3749	Female	White	40	3118	Down syndrome	E,S	1 day	yes	yes
10653	Female	White	41	3430	Lung seques-tration	E,C,S	9 days	yes	yes

E = echocardiography; C = catheterization; S = surgery.

Salient Infant and Reported Maternal Characteristics
Infant Study (1981–1989)

				Mother	
Age	Pregnancy Confir- mation	Years of Edu- cation	Occupation	Previous Pregnancies	Health Data
29	8 weeks	11	sewing machine operator father - auto mechanic	4 total 1st = 7 lbs 12 oz 2nd = 6 lbs 9 oz 3rd = miscarriage 4th = miscarriage	210 lbs, 4'11" 1800 calorie diet hypertension (3rd trimester) hospitalized for last 2½ weeks had 9–10 ultrasounds
34	6 weeks	16	housewife father - stock broker	3 total 1st-2nd = miscarriages 3rd = 8lbs 5 oz	131 lbs, 5'6" bronchitis (1st trimester) took amoxicillin
35	1 week	11	office clerk father-lab tech	0	142 lbs, 5'3" 3rd trimester - insulin (hospitalized) took Clomid into early pregnancy had 10 sonograms Chlamydia and antibiotics yeast infection, flu, endometriosis
37	5 weeks	12	housewife father - fireman	6 total 1st = 7 lbs 4 oz 2nd-4th = miscarriages 5th = 8 lbs 4 oz 6th = 6 lbs 14 oz	152 lbs, 5'7" dietary control of diabetes lost weight in pregnancy
43	23 weeks	14	housewife father - English professor	7 total 1st-2nd = miscarriages 3rd = 7 lbs 3 oz 4th = 7 lbs 11 oz 5th = 6 lbs 3 oz 6th = 6 lbs 2 oz 7th = 6 lbs 2 oz	170 lbs, 5'4" chronic asthma-many medicines had D&C 4–6 months prior to pregnancy

Chapter 12

Updates and Comments on Previous Publications

Section A:
Total Anomalous Pulmonary Venous Return

Section B:
Cardiomyopathy

Section C:
Ebstein's Malformation of the Tricuspid Valve

From: Ferencz C, Loffredo CA, Correa-Villaseñor A, Wilson PD, eds: *Genetic & Environmental Risk Factors of Major Cardiovascular Malformations: The Baltimore-Washington Infant Study 1981–1989.* Armonk, NY: Futura Publishing Co., Inc; ©1997.

The systematic evaluations of three specific phenotypes of cardiovascular malformations were initiated some time before the entire case population had been assembled. These three phenotypes are total anomalous pulmonary venous return,[92] cardiomyopathies,[148] and Ebstein's malformation of the tricuspid valve,[93] and they include the full set of enrolled cases.

With the broadened appreciation of characteristics of infants, their families, and their environments that we gained from the diagnosis-specific evaluations represented in this book, it is reasonable to revisit those three earlier studies with updates on additional cases, where appropriate, and to present the material in the format of the foregoing chapters. Additionally in these three malformation groups, which had low hierarchical allocations, we will also describe their occurrence as secondary diagnoses.

Section A

Total Anomalous Pulmonary Venous Return

Total anomalous pulmonary venous return was the first diagnostic phenotype to undergo comprehensive evaluation of familial and environmental factors,[92] with the analysis of 41 cases and 2801 controls registered in the first 6 years of the study. In that analysis, we gained valuable experience regarding our analytic methods and the many methodological issues that were discussed in detail in the article. Because of the importance of some of the findings on familial and environmental risk factors, we reevaluated this diagnostic group with the addition of the 19 cases subsequently enrolled in the final 3 years of the study.

Anomalous return of pulmonary veins to the right, instead of the left, atrium constitutes a severe form of congenital heart disease that is not compatible with prolonged survival. Prior to operative correction of the abnormality, an inexorable downhill course characterized the short life of affected infants and children with progressive right heart failure. This is a unique anomaly described by Clark[82] as "failure of targeted growth," in which the primitive pulmonary venous sinus fails to become part of the left atrium. This failure of communication is common in the complex hearts of laterality and looping abnormalities: there is total anomalous pulmonary venous return in the asplenia syndrome and partial anomalous pulmonary venous return in the so-called polysplenia syndrome (see Chapter 4). The cases considered in this chapter are those in which total anomalous pulmonary venous return was the clinical diagnosis in the absence of clinically evident laterality defects.

The morphogenesis of the normal embryonic sequence of the targeted union of the common pulmonary vein and the left atrium was described by Neill[315] 40 years ago, and since then little progress has been made regarding the nature of targeted growth.[85] There is, however, general agreement that the variations and the types of anomalous connections are secondary to the primary error of failed pulmonary vein-left atrial communication.[31] A genetic component in the etiology of this defect is evident in the associations of the anomaly with heritable disorders and in the recurrence of different anatomic forms in siblings, cousins, and even across successive generations.[92]

This reevaluation of epidemiologic findings in total anomalous pulmonary venous return at the end of the series of diagnosis-specific studies provides the opportunity to address questions not previously considered, such as the possible similarity of risk factors to those found for laterality and looping defects and for other major diagnoses. It also provides an opportunity to provide the updated description of associated genetic abnormalities in the cases and in their families, and to reassess possible associations with lead, pesticide, and solvent exposures.

Study Population

The diagnosis of total anomalous pulmonary venous return to the right atrium and its tributaries in patients with normally positioned heart and viscera (situs solitus) was made in 60 infants (1.4% of all cases) with an overall prevalence of 6.6 per 100,000 livebirths during the time from 1981 to 1989. These cases, additional cases occurring as secondary diagnoses in conjunction with major cardiac defects, and cases of partial anomalous pulmonary venous return are shown on Figure 12A.1. For total anomalous pulmonary venous return, three quarters of all cases were primary diagnoses.

Cardiac Abnormalities

Table 12A.1 shows the site of anomalous drainage of the pulmonary veins among the 60 cases. The left superior vena cava, the coronary sinus, and vessels below the diaphragm were the most common drainage sites.

Noncardiac Anomalies

Fourteen infants (23%) with a primary diagnosis of total anomalous pulmonary venous return had associated noncardiac anomalies (Table 12A.2). Two of

Total and Partial Anomalous Pulmonary Venous Return
Baltimore-Washington Infant Study (1981-1989)

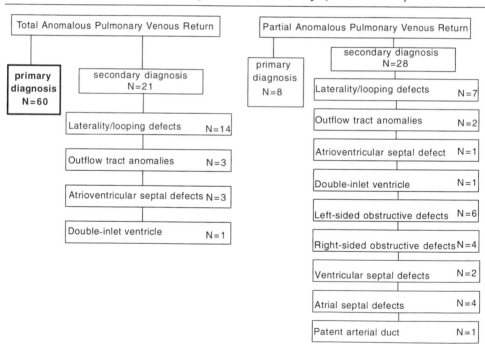

Figure 12A.1: Total anomalous pulmonary venous return occurred as a primary diagnosis in 60 infants, the subjects included in the descriptive and analytic report of this chapter. An additional 21 infants has this anomaly as a secondary diagnosis, 14 of whom had defects of laterality and cardiac looping. Partial anomalous pulmonary venous return was diagnosed in 36 infants, 8 with the anomaly as a primary diagnosis and 28 with a secondary diagnosis.

Table 12A.1

Total Anomalous Pulmonary Venous Return: Site of Anomalous Drainage
Baltimore-Washington Infant Study (1981–1989)

Site of Anomalous Drainage of the Pulmonary Veins	Previously Reported (1981–86)	Added Cases (1987–89)	Total (1981–89)
Type unknown	0	1	1
Atresia of common vein	1	0	1
Right atrium	2	1	3
Coronary sinus (CS)	5	6	11
Right superior vena cava	8	0	8
Left superior vena cava (LSVC)	12	5	17
Mixed: CS and LSVC	1	3	4
Below diaphragm	12	3	15
Total	41	19	60

Table 12A.2

Total Anomalous Pulmonary Venous Return (N=60)
Noncardiac Anomalies
Baltimore-Washington Infant Study (1981–1989)

Chromosomal Abnormalities:	N=2	(3.3%)
Trisomy 22	1	[# 3759]
Bisatellited marker chromosome (unknown origin)	1	[# 3835]
Syndromes	N=5	(8.3%)
a. Mendelian Syndromes:	4	
Holt-Oram	2	[#2046, 10436]
Alagille	1	[# 2745]
Townes-Brocks	1	[# 10209]
b. Non-Mendelian Associations:	1	
Prune belly	1	[# 10424]
Multiple, Nonclassified Anomalies:	N=1	(1.7%)
Occipital encephalocele, anophthalmia, absent optic nerves, agenesis of corpus callosum, brain teratoma, cleft palate, Bicornuate uterus	1	[# 1933]
Organ Defects	N=6	(10%)
Agenesis of right lung	1	[# 3825]
Bilateral tri-lobed lungs	1	[# 3342]
Bilateral pulmonary emphysema (day 1 of life)	1	[# 10840]
Anal atresia	1	[# 2833]
Hemivertebrae	1	[# 1522]
Bicornuate (uterus)	1	[# 10350]
No Associated Anomalies	N=46	(76.7%)

these infants had chromosomal anomalies and five had syndromes, including two with the Holt-Oram syndrome. One case (#1933) had multiple severe anomalies of the central nervous system and other organs, which could not be classified as any known syndrome. Among six cases with nonsyndromic anomalies of organs there were three with anomalies of the lungs; in one (case #3342) a trilobed lung found at autopsy may indicate a "hidden" laterality defect.

Descriptive Analyses

Prevalence by Time, Season, and Area of Residence

The prevalence of cases with total anomalous pulmonary venous return varied slightly during the study period (Table 12A.3). Prevalence declined in successive birth quarters and was lowest in the October to December birth quarter. There was a lower prevalence in suburban areas compared to urban and rural areas.

Diagnosis and Course

Early diagnosis was accomplished in the majority of infants in this group, with nearly half diagnosed in the first week of life (Table 12A.4). The most common diagnostic method was the combination of echocardiography, catheterization, and surgery, but nearly one third of the cases were examined at autopsy. The great majority (92%) had surgery during the first year of life. All cases were accounted for in the follow-up period; the mortality was high (40%) with an overall survival rate of 60%.

Table 12A.3
Total Anomalous Pulmonary Venous Return: Prevalence
Baltimore-Washington Infant Study (1981–1989)

	Cases	Area Births	Prevalence per 10,000
Total Subjects	60	906,626	0.66
Year of Birth:			
1981–1983	20	274,558	0.73
1984–1986	17	295,845	0.57
1987–1989	22	336,223	0.65
Birth Quarter:			
1st (Jan–Mar)	18	219,145	0.83
2nd (Apr–Jun)	17	231,777	0.73
3rd (Jul–Sep)	14	233,626	0.60
4th (Oct–Dec)	11	222,078	0.50
Area of Residence:			
urban	22	208,568	1.05
suburban	27	584,022	0.46
rural	11	112,318	0.98

Table 12A.4
Total Anomalous Pulmonary Venous Return: Clinical Data
Baltimore-Washington Infant Study (1981–1989)

	Cases N=60 no.	(%)
Age at diagnosis:		
<1 week	29	(48.3)
1–4 weeks	18	(30.0)
5–24 weeks	13	(21.7)
25–52 weeks	0	
Method of Diagnosis:		
echocardiography only	0	
echocardiography and surgery	9	(15.0)
echocardiography, cardiac	35	(58.3)
catheterization, and surgery		
cardiac catheterization and surgery	3	(5.0)
autopsy following other	12	(20.0)
methods		
autopsy only	1	(1.7)
Surgery During First Year of Life	55	(91.7)
Follow-Up:		
not followed	0	
followed:	60	
died in the first year of life	24	(40.0)
alive	36	(60.0)
median age at last visit	9 months	

Gender, Race, and Twinning

Compared to controls, there was no significant difference in the gender distribution of infants with total anomalous pulmonary venous return (Table 12A.5). The proportion of black infants among the cases (16.7%) was only half that of controls (31%). Two cases occurred in twins (3.3%), a rate twice that of the controls. Neither co-twin had a heart defect or other developmental anomalies.

Birthweight and Gestational Age

Ten of the 56 cases (18%) of total anomalous pulmonary venous return had birthweights under 2500 grams, a much greater proportion than among the controls (7%). The mean birthweight of cases was 3022 grams compared to 3351 in the controls (Table 12A.6). The cases were also more likely to be premature (16% compared to 9.5%) and the odds ratio for being small for gestational age was significantly elevated.

Table 12A.5
Total Anomalous Pulmonary Venous Return: Infant Characteristics
Baltimore-Washington Infant Study (1981–1989)

Characteristic	Cases N=60 no.	(%)	Controls N=3572 no.	(%)
Gender:				
male	29	(48.3)	1,817	(50.9)
female	31	(51.7)	1,755	(49.1)
Race*:				
white	45	(75.0)	2,362	(66.1)
black	10	(16.7)	1,109	(31.0)
other	5	(8.3)	98	(2.7)
Twin births	2	(3.3)	53	(1.5)

*Missing race in 3 controls.

Table 12A.6
Total Anomalous Pulmonary Venous Return: Fetal Growth Characteristics
Baltimore-Washington Infant Study (1981–1989)

Characteristic	Cases no. (%)	Controls no. (%)
Number of Families Interviewed	56	3572
Birthweight (grams)		
<2500	10 (17.9)	252 (7.1)
2500-3500	33 (58.9)	1853 (51.9)
>3500	13 (23.2)	1467 (41.0)
mean±standard error	3022±78	3351±10
range	1584–4096	340–5273
Gestational Age (weeks)		
<38	9 (16.1)	339 (9.5)
38+	47 (83.9)	3233 (90.5)
mean±standard error	39.0±0.3	39.6±0.1
range	31–42	20–47
Size for Gestational Age		
small (SGA)	10 (17.9)	211 (5.9)
normal	43 (76.8)	2712 (75.9)
large	3 (5.4)	649 (18.2)
Odds Ratio for SGA (95% CI)	3.5 (1.7–6.9)	1.0 (reference)

Potential Risk Factors

Familial Cardiac and Noncardiac Anomalies

Congenital heart defects were reported in three families of cases with total anomalous pulmonary venous return, and congenital defects in other organ systems were reported in six additional families (Table 12A.7). Of interest are two families in whom the Holt-Oram syndrome was present in both the probands and relatives: in one family (#2046) the mother and sister each had an atrial septal defect with Holt-Oram, and in another family (#10436) the father had Holt-Oram without congenital heart disease.

Genetic and Environmental Factors

In our previous publication[92] we reported several genetic and environmental risk factors in association with total anomalous pulmonary venous return. These factors included maternal exposures to paints and paint strippers, lead solder, and pesticides (the effects of the latter exposure were modified by family history of cardiac and noncardiac anomalies). The univariate case-control comparisons are shown in Table 12A.8 and results of the multivariate analysis for the entire study period (1981–1989) with respect to these same risk factors are shown in Table 12A.9. Painting and soldering were significant risk factors at the 5% but not 1% significance levels: the magnitudes of these odds ratios are smaller than those reported in the earlier study (which reported odds ratio of 3.90 and 15.5, respectively; both odds ratios significant at the 1% level). Evaluation of the interaction of family history and pesticide exposure reveals that only the combination of familial cardiac anomalies with pesticide exposure reached statistical significance at

Table 12A.7
Total Anomalous Pulmonary Venous Return:
Congenital Defects and Hereditary Disorders in First-Degree Relatives
Baltimore-Washington Infant Study (1981–1989)

Proband	Sex	Diagnosis	Relative	Diagnosis
A. Congenital Heart Defects:				
2046	F	Holt-Oram	Mother	Atrial septal defect, Holt-Oram
			Sister	Atrial septal defect, Holt-Oram
3636	F	—	Father	Aortic valve stenosis
3683	M	—	Father	Mitral valve prolapse
B. Congenital Defects of Other Organs:				
2006	F	—	Mother	Deafness
3690	M	—	Mother	Pyloric stenosis
10350	F	bicornuate uterus	Mother	Cleft palate
10436	F	Holt-Oram	Father	Holt-Oram
3342	F	bilateral tri-lobed lungs	Father	Polydactyly
1933	F	multiple anomalies*	Sister	Bicornuate uterus
			Sister	Polydactyly

*See Table 12A.2.

Table 12A.8
Total Anomalous Pulmonary Venous Return
Univariate Analysis of Risk Factors
Baltimore-Washington Infant Study (1981–1989)

Variable	Cases N=56 no.	(%)	Controls N=3572 no.	(%)	Odds Ratio (95% CI)	
Family History (first-degree relatives)						
congenital heart disease	2	(3.6)	43	(1.2)	3.0	(0.7–12.9)
noncardiac anomalies	7	(12.5)	165	(4.6)	**3.0**	**(1.3–6.6)**
Parental Age						
maternal age:						
<20	9	(16.1)	507	(14.2)	1.3	(0.6–2.8)
20–29	27	(48.2)	2009	(56.2)	1.0	(reference)
30+	20	(35.7)	1056	(29.6)	1.4	(0.8–2.5)
paternal age:						
<20	4	(7.1)	226	(6.3)	1.4	(0.5–4.0)
20–29	22	(39.3)	1724	(48.3)	1.0	(reference)
30+	30	(53.6)	1622	(45.4)	1.4	(0.8–2.5)
Maternal Reproductive History						
number of previous pregnancies:						
none	15	(26.8)	1159	(32.4)	1.0	(reference)
one	18	(32.1)	1097	(30.7)	1.3	(0.6–2.5)
two	11	(19.6)	709	(19.9)	1.2	(0.5–2.6)
three or more	12	(21.4)	607	(17.0)	1.5	(0.7–3.3)
previous miscarriage(s)	12	(21.4)	681	(19.1)	1.2	(0.6–2.2)
Maternal Illnesses and Medications						
diabetes:						
overt	1	(1.8)	23	(0.6)	2.8	(0.4–21.1)
gestational	0		115	(3.2)	—	
influenza	5	(8.9)	278	(7.8)	1.2	(0.5–2.9)
phenothiazines	3	(5.4)	29	(0.8)	**6.9**	**(2.0–23.4)**
corticosteroids	2	(3.6)	23	(0.6)	**5.7**	**(1.3–24.8)**
Lifestyle Exposures						
alcohol (any amount)	29	(51.8)	2101	(58.8)	1.4	(0.7–2.5)
smoking (cigarettes/day):						
none	39	(69.6)	2302	(64.5)	1.0	(reference)
1–10	5	(8.9)	565	(15.8)	0.5	(0.2–1.3)
11–20	8	(14.3)	529	(14.8)	0.9	(0.4–1.9)
>20	4	(7.1)	176	(4.9)	1.3	(0.5–3.8)
smoking (paternal):						
none	30	(53.6)	2058	(58.0)	1.0	(reference)
1–10	8	(14.3)	510	(14.4)	1.1	(0.5–2.3)
11–20	7	(12.5)	638	(18.0)	0.8	(0.3–1.7)
>20	11	(19.6)	342	(9.6)	**2.2**	**(1.1–4.4)**
Home and Occupational Exposures						
ionizing radiation (occupational)	1	(1.8)	48	(1.3)	1.3	(0.2–9.8)
pesticides	22	(39.3)	926	(25.9)	**1.8**	**(1.1–3.2)**
painting	11	(19.6)	367	(10.3)	**2.1**	**(1.1–4.2)**

Table 12A.8—Continued
Total Anomalous Pulmonary Venous Return
Univariate Analysis of Risk Factors
Baltimore-Washington Infant Study (1981–1989)

Variable	Cases N=56 no.	(%)	Controls N=3572 no.	(%)	Odds Ratio (95% CI)	
paint stripping	4	(7.1)	85	(2.4)	3.2	(1.1–8.9)
soldering with lead	2	(3.6)	17	(0.5)	7.7	(1.7–34.4)
miscellaneous solvents (paternal)	9	(16.1)	267	(7.5)	2.4	(1.1–4.9)
arts and crafts painting (paternal)	3	(5.4)	46	(1.3)	4.3	(1.3–14.4)
Race of Infant						
white	41	(73.2)	2362	(66.1)	1.4	(0.8–2.5)
nonwhite	15	(26.8)	1207	(33.9)	1.0	(reference)

Table 12A.9
Total Anomalous Pulmonary Venous Return: Multivariate Model
Baltimore-Washington Infant Study (1981–1989)

Risk Factor	Exposed Subjects Cases (N=56)	Controls (N=3572)	Odds Ratio	95% CI	99% CI
painting/paint stripping	13	411	2.0	1.1–3.8	0.9–4.7
soldering with lead	2	17	6.8	1.5–31.5	0.9–50.8
interaction of family history with pesticide exposure:					
not exposed to pesticides:					
no family history	30	2494	1.0	(reference)	
noncardiac anomalies in family	4	126	2.4	0.8–7.0	0.6–9.7
cardiac anomalies in family	0	26	*		
exposed to pesticides:					
no family history	18	874	1.7	0.9–3.0	0.8–3.6
noncardiac anomalies in family	2	35	4.6	1.0–20.1	0.7–31.8
cardiac anomalies in family	2	17	9.0	2.0–41.1	1.2–66.0

*Term dropped from model (no cases).

the 1% level; however the odds ratio of 9.0 and wide confidence intervals (1.2 to 66.0) were based upon only two cases in this exposure category.

Discussion

Sixty infants with a primary diagnosis of total anomalous pulmonary venous return were enrolled in the Baltimore-Washington Infant Study from 1981 to 1989. A wide range of noncardiac anomalies affecting one quarter of the cases was present; half of these were of known genetic origin. Prevalence of these anomalies declined somewhat during the study period, and was lower among fourth quarter births (October to December) and among suburban regions of the area. Most cases were identified early in infancy and required surgery. Compared to controls, case infants were more likely to be white, to come from twin births, and to be small for gestational age. Family history of cardiac and noncardiac anomalies and maternal environmental exposures to toxicants (solvents, lead, and pesticides) were identified as possible risk factors. The findings in this small size of this case group will require confirmation in larger, multicenter studies of this major malformation.

Section B

Cardiomyopathy

Cardiomyopathies in the absence of malformations are not usually considered congenital abnormalities of the heart, as they represent a variety of known etiologies ranging from an inflammatory response to infections to the effects of maternal diabetes and to specific heritable metabolic disturbances of the myocardium. Because of the increasing clinical interest in diagnosing specific forms of cardiomyopathies in infancy and the puzzle posed by Kawasaki's disease, the Baltimore-Washington Infant Study included the enrollment of infants with cardiomyopathy to be categorized as the lowest position in the hierarchical order.

In a previous publication,[148] we reviewed and categorized the cases diagnosed during the first 6 years of the study and also highlighted the therapeutic and research horizons that made evident the need for a better understanding of the etiologies of the various types of cardiomyopathies. Certain findings in the earlier study suggest a reevaluation of this case group, which has been expanded by 30 additional cases enrolled between April 1, 1987 and December 31, 1989.

Findings and implications of interest included:

- the association of heritable syndromic disorders, including metabolic errors, in one fifth of the cases, a greater proportion than has previously been reported, which could be further evaluated;
- the association of cardiomyopathies with low birthweight, prematurity, and fetal growth retardation, which might be further examined according to the clinical pathologic types of cardiomyopathies; and
- the prominent role of maternal diabetes, both overt and gestational, a well-known association with infantile cardiomyopathy, which could be further examined with respect to maternal characteristics and previous pregnancy outcomes.

Accordingly, we present in this review a comprehensive account of infants with isolated myocardial disease, within the perspectives of this work, including all cases with primary diagnoses of cardiomyopathy as well as cases in which cardiomyopathy was considered to be secondary to a structural cardiovascular malformation. We also explore the case-control frequency distributions of parental exposures to toxic agents that might be expected to affect the myocardium.

Study Population

Eighty-six infants were diagnosed with a cardiomyopathy as a primary cardiovascular malformation in the Baltimore-Washington Infant Study during 1981 to 1989 (Figure 12B.1). The largest subgroups were the dilated, congestive phenotype (with 24 affected infants), hypertrophic cardiomyopathies (13 with obstructive and 22 with nonobstructive types), and tumors (10 infants, 6 of whom had tuber-

Cardiomyopathy, N=114

```
                    primary diagnosis N=86 (77 interviews)          secondary diagnosis N=28

  Dilated, congestive          Tumors            outflow tract
        N=24                     N=10            anomalies N=2

                                                 lefted-sided
                 Hypertrophic                     obstructive
                                                    defects
                                                     N=10
    obstructive        Nonobstructive
       N=13                N=22                    right-sided
                                                  obstructive
                                                    defects
                                                     N=4
                     Other Types:

                  endocardial                    ventricular
                    fibroelastosis    N=6       septal defect
                  myocarditis         N=5
                  glycogen storage    N=2     membranous    muscular
                  infarction          N=2        N=4          N=1
                  mucocutaneous       N=2
                    lymph node syndrome                    atrial septal
                                                           defect N=1

                                                             patent
                                                             arterial
                                                            duct N=3

  *Includes: interrupted aortic arch (1), peripheral pulmonary artery stenosis (1),      rare heart
  partial anomalous pulmonary venous return (1).                                          defects*
                                                                                            N=3
```

Figure 12B.1: One hundred fourteen infants were diagnosed with cardiomyopathies: 86 with primary lesions and 28 with cardiomyopathy secondary to other heart defects, half of whom had left-sided or right-sided obstructive lesions. All known types of cardiomyopathies were represented in this small population.

ous sclerosis). Rarer types included endocardial fibroelastosis ($N = 6$), myocarditis ($N = 5$), glycogen storage disease ($N = 2$), myocardial infarction ($N = 2$), and mucocutaneous lymph node syndrome ($N = 2$), which is also known as Kawasaki's disease.

An additional 28 cases had a cardiomyopathy secondary to a structural cardiac anomaly, 10 of whom had left-sided obstructive defects such as aortic valve stenosis.

Noncardiac Anomalies

There were 31 cases (36%) with associated noncardiac anomalies in the group of infants with primary cardiomyopathy. A wide range of malformations were described, including metabolic and other genetic syndromes in 21 cases. The noncardiac anomalies are displayed in Table 12B.1, arranged by the type of cardiomyopathy.

Among the 24 cases with dilated, congestive cardiomyopathy, 7 had associated noncardiac anomalies, including 2 with blood disorders, 3 with metabolic syndromes, 1 with Coxsackie virus embryopathy, and 1 case with multiple (non-classified) anomalies partially concordant with rubella embryopathy.

Table 12B.1

Cardiomyopathies: Noncardiac Anomalies

Baltimore-Washington Infant Study (1981–1989)

Type Of Cardiomyopathy	N	Isolated Heart Defect no. (%)	Cases with Associated Noncardiac Anomalies no. (%)	type	Case ID Number
All Cases:	**86**	**55 (64.0)**	**31 (36.0)**	—	—
Dilated, congestive	24	17 (70.8)	7 (29.2)	thalassemia	[#1235]
				methemoglobinemia	[#1800]
				tryosinemia	[#2286]
				carnitine deficiency	[#3645]
				lactic acidosis	[#10150]
				Coxsackie embryopathy	[#2260]
				multiple, nonclassified*	[#10133]
Hypertropic:					
obstructive	13	8 (61.5)	5 (38.5)	partial trisomy	[#2370]
				Noonan	[#2756]
				Costello	[#1498]
				Beckwith-Wiedemann	[#3764, 11469]
nonobstructive	22	14 (63.6)	8 (36.4)	Down	[#11026]
				Noonan	[#10545]
				Waardenburg	[#1635]
				carnitine deficiency	[#1828, 2532]
				Smith-Lemli-Opitz	[#2701]
				hydronephrosis	[#2864]
				diaphragmatic hernia	[#2139]
Tumors	10	3 (30.0)	7 (70.0)	tuberous sclerosis	[#1214, 2677, 3597, 3714 10484, 10918]
				hypospadias	[#10533]
Endocardial fibroelastosis	6	5 (83.3)	1 (16.7)	Coxsackie embryopathy	[#1431]
Glycogen storage	2	0	2 (100)	glycogen storage	[#2527, 11690]
Mucutaneous lymph node syndrome	2	2 (100)	0	—	[#1487, 1921]
Mycardial Infarction	2	1 (50.0)	1 (50.0)	cerebral atrophy	[#11562]
Myocarditis	5	5 (100)	0	—	[#2596, 10004, 10699, 10815, 11341]

*Cataracts, deafness, abnormal ureters, hydronephrosis, (rubella embryopathy).

Among the 10 cases with myocardial tumors, six had the autosomal dominant disease tuberous sclerosis. Noncardiac anomalies among the remaining rarer forms of cardiomyopathy included a case with endocardial fibroelastosis who had Coxsackie viral embryopathy, and a case with a myocardial infarction who also had cerebral atrophy.

Descriptive Analyses

Prevalence by Time, Season, and Area of Residence

Overall, the prevalence of cardiomyopathies was 0.95 cases per 10,000 live-births. There was no clear trend in prevalence by year of birth (Table 12B.2). During the second and third birth quarters (April through September) the prevalence of the defect was lower than in the first and fourth quarters. Prevalence in the urban areas of the study region was over twice that of suburban areas and 60% higher than in rural areas.

Diagnosis and Course

Less than half of the cases with cardiomyopathies were diagnosed in the first 4 weeks of life (Table 12B.3), and a quarter of the infants were diagnosed after 6 months of age. The most frequent method of diagnostic confirmation was echocardiography alone (55%). Twelve cases were diagnosed at autopsy, including two solely by autopsy. Surgery in the first year of life was rare, and 60 of the cases (74%) were alive at diagnostic update.

Table 12B.2
Cardiomyopathy: Prevalence
Baltimore-Washington Infant Study (1981–1989)

	Cases	Area Births	Prevalence per 10,000
Total subjects	86	906,626	0.95
Year of Birth:			
1981–1983	24	274,558	0.87
1984–1986	36	295,845	1.22
1987–1989	24	336,223	0.71
Birth Quarter:			
1st (Jan–Mar)	23	219,145	1.05
2nd (Apr–Jun)	20	231,777	0.86
3rd (Jul–Sep)	18	233,626	0.77
4th (Oct–Dec)	25	222,078	1.13
Area of Residence:			
urban	33	208,568	1.58
suburban	42	584,022	0.72
rural	11	112,318	0.98

Table 12B.3
Cardiomyopathy: Clinical Data
Baltimore-Washington Infant Study (1981–1989)

	Cases N=86 no.	(%)
Age at Diagnosis:		
<1 week	27	(31.4)
1–4 weeks	16	(18.6)
5–24 weeks	21	(24.4)
25–52 weeks	22	(25.6)
Method of Diagnosis:		
echocardiography only	47	(54.7)
echocardiography and catheterization	24	(27.9)
echocardiography, catheterization, and surgery	3	(3.5)
autopsy following other methods	10	(11.6)
autopsy only	2	(2.3)
Surgery During First Year of Life	6	(7.0)
Follow-Up:		
not followed	5	(5.8)
followed:	81	
died in the first year of life	21	(25.9)
alive	60	(74.1)
median age at last visit	11 months	

Gender, Race, and Twinning

Compared to controls there were no significant differences in the distributions of cases by gender and race (Table 12B.4). There were three cases born within twin sets (3.5%), a rate more than double that of control infants (1.5%). No co-twin was affected with any type of congenital anomaly.

Birthweight and Gestational Age

Twice as many cases as controls had birthweights under 2500 grams, and the mean birthweight of the cases was 3287 grams (Table 12B.5). Cases were three times more likely than controls to be premature (28.6% versus 9.5%), but the proportion of cases classified as small for gestational age was not significantly different from that of controls when evaluated as an odds ratio with 95% confidence limits.

Sociodemographic Characteristics

There were several sociodemographic characteristics that distinguished the families of infants with primary cardiomyopathies from controls (Table 12B.6). Both mothers and fathers of cases were significantly more likely than control parents to have low educational attainment; one quarter of the case parents did not

Table 12B.4
Cardiomyopathy: Infant Characteristics
Baltimore-Washington Infant Study (1981–1989)

Characteristic	Cases N=86 no.	(%)	Controls N=3572 no.	(%)
Gender				
male	49	(57.0)	1,817	(50.9)
female	37	(43.0)	1,755	(49.1)
Race*:				
white	52	(61.2)	2,362	(66.1)
black	31	(36.5)	1,109	(31.0)
other	2	(2.4)	98	(2.7)
Twin births	3	(3.5)	53	(1.5)

*Missing race in 1 case and 3 controls.

Table 12B.5
Cardiomyopathy: Fetal Growth Characteristics
Baltimore-Washington Infant Study (1981–1989)

Characteristic	Cases no.	(%)	Controls no.	(%)
Number of Families Interviewed	77		3572	
Birthweight (grams)				
<2500	11	(14.5)	252	(7.1)
2500–3500	36	(47.4)	1853	(51.9)
>3500	16	(38.2)	1467	(41.0)
mean±standard error	3287±106		3351±10	
range	652–5075		340–5273	
Gestational Age (weeks)				
<38	22	(28.6)	339	(9.5)
38+	55	(71.4)	3233	(90.5)
mean±standard error	38.3±0.4		39.6±0.3	
range	26–43		20–47	
Size for Gestational Age				
small (SGA)	8	(10.4)	211	(5.9)
normal	48	(62.3)	2712	*75.9)
large	21	(27.3)	649	(18.2)
Odds Ratio for SGA (95% CI)	1.9	(0.9–3.9)	1.0	(reference)

Table 12B.6
Cardiomyopathy: Analysis of Sociodemographic Characteristics
Baltimore-Washington Infant Study (1981–1989)

Interviewed Families	Cases N=77 no. (%)		Controls N=3572 no. (%)		Odds Ratio (95% CI)
Maternal Marital Status:					
not married	26	(33.8)	990	(27.7)	1.3 (0.8–2.1)
married	51	(66.2)	2582	(72.3)	1.0 (reference)
Maternal Education:					
<high school	20	(26.0)	659	(18.4)	**2.4 (1.3–4.4)**
high school	36	(46.8)	1265	(35.4)	**2.2 (1.3–3.8)**
college	21	(27.2)	1648	(46.1)	1.0 (reference)
Paternal Education:					
<high school	19	(24.7)	650	(18.2)	**1.9 (1.0–3.5)**
high school	33	(42.9)	1298	(36.3)	**1.7 (1.0–2.8)**
college	25	(32.5)	1624	(45.5)	1.0 (reference)
Annual Household Income:					
<$10,000	15	(19.4)	686	(19.2)	1.3 (0.6–2.5)
$10,000–$19,999	19	(24.7)	699	(19.6)	1.6 (0.8–3.0)
$20,000–$29,999	19	(24.7)	737	(20.6)	1.5 (0.8–2.8)
$30,000+	23	(29.9)	1373	(38.4)	1.0 (reference)
Maternal Occupation:					
not working	31	(40.3)	1142	(32.0)	**2.2 (1.0–4.7)**
clerical/sales	22	(28.6)	1119	(31.3)	1.6 (0.7–3.5)
service	14	(18.2)	444	(12.4)	**2.6 (1.1–6.0)**
factory	1	(1.3)	137	(3.8)	0.7 (0.1–4.7)
professional	9	(11.7)	730	(20.4)	1.0 (reference)
Paternal Occupation:					
not working	10	(13.0)	246	(6.9)	**3.9 (1.6–9.4)**
clerical/sales	10	(13.0)	618	(17.3)	1.5 (0.6–3.7)
service	12	(15.6)	480	(13.4)	**2.4 (1.0–5.5)**
factory	35	(45.5)	1279	(35.8)	**2.6 (1.3–5.3)**
professional	10	(13.0)	947	(26.5)	1.0 (reference)
Month of Pregnancy Confirmation:					
1st month	12	(14.6)	511	(14.3)	1.0 (reference)
2nd month	39	(50.4)	1955	(54.7)	0.8 (0.4–1.6)
3rd month	12	(22.6)	663	(18.6)	0.8 (0.3–1.7)
4th month or later	14	(12.4)	416	(11.6)	1.4 (0.7–3.1)

Data are missing for income (1 case, 77 controls), paternal occupation (2 controls) and pregnancy confirmation month (27 controls).

complete high school. There were no significant case-control differences in household income, but several occupational categories had elevated odds ratios; for both mothers and fathers, nonworkers, and those in service occupations were at increased risk relative to professionals. Additionally, fathers in factory-related occupations were at increased risk.

Potential Risk Factors

Familial Cardiac and Noncardiac Anomalies

Cardiac and/or noncardiac anomalies in first-degree relatives are shown in Table 12B.7. Other than the one father reported to have a patent arterial duct (case #1424), no first-degree relatives had cardiovascular malformations. In four families, the mothers of the probands reported having noncardiac anomalies, including two mothers who had agenesis of one kidney; one father had hydronephrosis.

Three additional families are noteworthy for the congenital anomalies reported among second-and third-degree relatives (Figure 12B.2). In the family of case #1649, the maternal grandmother and two of her three daughters and both of her sons each had idiopathic hypertrophic subaortic stenosis, as did the proband, but the mother of the proband was apparently unaffected. In another family (case #1431), the proband was diagnosed with endocardial fibroelastosis and the Coxsackie virus; a paternal aunt had tetralogy of Fallot, a maternal aunt had Down syndrome without a heart defect, and a maternal uncle had a ventricular septal defect. In the third family (case #3344) there were two maternal uncles affected with a familial form of cardiomyopathy (unspecified type) with mental retardation. This combination of defects has been highlighted as a possible indication of hyperphenylalanemia.[248,249]

Genetic and Environmental Factors

Univariate Analysis

There were no significant case-control differences in the proportion of first-degree relatives with congenital anomalies (Table 12B.8). There were increased risks associated with young parental age for both mothers and fathers, with odds ratios of 1.8 and 2.4, respectively for parental age under 20 years. There were no

Table 12B.7

Infants with Cardiomyopathies:
Congenital Defects and Hereditary Disorders in First-Degree Relatives
Baltimore-Washington Infant Study (1981–1989)

Proband	Sex	Diagnosis	Relative	Diagnosis
		A. Congenital Heart Defects:		
1424	M	endocardial fibroelastosis	Father	Patent arterial duct
		B. Congenital Defects of Other Organs:		
2313	F	dilated, congestive	Mother	Deafness
2636	M	hypertrophic, obstructive	Mother	Kidney agenesis
2139	M	hypertrophic, nonobstructive	Mother	Kidney agenesis, bicornuate uterus
10918	M	tuberous sclerosis	Mother	Polydactyly
			Paternal Cousin	deafness
1635	F	hypertropic, nonobstructive, Waardenburg	Father	Hydronephrosis
			Maternal Uncle	Pyloric stenosis

Infants with Cardiomyopathy:
Second-Degree Relatives with Cardiac and Non-Cardiac Anomalies
Baltimore-Washington Infant (1981-1989)

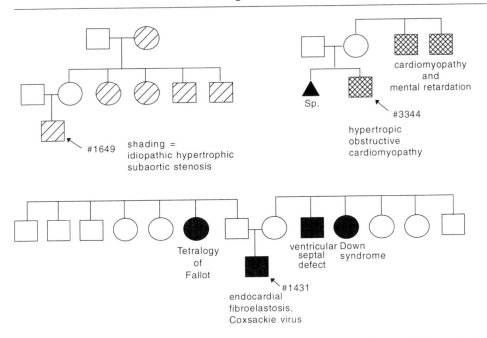

Figure 12B.2: Unusual familial patterns of anomalies including cardiomyopathies and structural cardiac defects.

significant case-control differences in the number of previous pregnancies and miscarriages, but there was an elevated odds ratio (2.5) for mothers who reported one or more previous preterm births.

Maternal overt diabetes was a very strong risk factor for cardiomyopathy (odds ratio 13.0), with six affected case mothers, five of whom reported taking insulin during pregnancy. There were no other significant case-control differences in illnesses and medications reported by mothers.

Among lifestyle exposures, neither alcohol, cigarette smoking, nor recreational drugs were significantly associated with case status. Alcohol consumption showed no association, whether analyzed as a continuous variable (maximum number of drinks per day), a binary variable (drinkers versus nondrinkers), or by frequency of consumption (daily, once or twice a week, once a month). Two paternal home and occupational exposures possibly involving lead and other metals were significantly associated: 12 case fathers were involved in auto body repair work and 10 were involved in welding.

Multivariate Analysis

Three variables were significantly associated with cardiomyopathy in the multivariate analysis (Table 12B.9): maternal overt diabetes, a maternal history of previous premature births, and paternal auto body repair work. Of these vari-

Table 12B.8
Cardiomyopathy
Univariate Analysis of Risk Factors
Baltimore-Washington Infant Study (1981–1989)

Variable	Cases N=77 no.	(%)	Controls N=3572 no.	(%)	Odds Ratio (95% CI)
Family History (first-degree relatives)					
congenital heart disease	1	(1.3)	43	(1.2)	1.1 (0.1–7.9)
noncardiac anomalies	5	(6.5)	165	(4.6)	1.4 (0.6–3.6)
Parental Age					
maternal age:					
<20	17	(22.1)	507	(14.2)	**1.8 (1.0–3.2)**
20–29	38	(49.4)	2009	(56.2)	1.0 (reference)
30+	22	(28.5)	1056	(29.6)	1.1 (0.6–1.9)
paternal age:					
<20	9	(11.7)	226	(6.3)	**2.4 (1.1–5.1)**
20–29	29	(37.7)	1724	(48.3)	1.0 (reference)
30+	39	(50.6)	1622	(45.4)	1.4 (0.9–2.3)
Maternal Reproductive History					
number of previous pregnancies:					
none	26	(33.8)	1159	(32.4)	1.0 (reference)
one	26	(33.8)	1097	(30.7)	1.1 (0.6–1.8)
two	7	(9.1)	709	(19.9)	0.4 (0.2–1.0)
three or more	18	(23.3)	607	(17.0)	1.3 (0.7–2.4)
previous miscarriage(s)	14	(18.2)	681	(19.1)	0.9 (0.5–1.7)
previous preterm birth(s)	9	(11.7)	180	(5.0)	**2.5 (1.2–5.1)**
Maternal Illnesses and Medications					
diabetes:					
overt	6	(7.8)	23	(0.6)	**13.0 (5.2–33.0)**
gestational	3	(3.9)	115	(3.2)	1.2 (0.4–3.9)
influenza	4	(5.2)	278	(7.8)	0.6 (0.2–1.8)
Lifestyle Exposures					
alcohol (any amount)	39	(50.6)	2101	(58.8)	0.7 (0.5–1.1)
smoking (cigarettes/day):					
none	47	(61.0)	2302	(64.5)	1.0 (reference)
1–10	13	(16.9)	565	(15.8)	1.1 (0.6–2.1)
11–20	13	(16.9)	529	(14.8)	1.2 (0.6–2.2)
>20	4	(5.2)	176	(4.9)	1.1 (0.4–3.1)
Home and Occupational Exposures					
ionizing radiation (occupational)	0		48	(1.3)	—
auto body repair work (paternal)	12	(15.6)	236	(6.6)	**2.6 (1.4–4.9)**
welding (paternal)	10	(13.0)	228	(6.4)	**2.2 (1.1–4.3)**
Race of Infant					
white	46	(59.7)	2362	(66.2)	0.8 (0.5–1.2)
nonwhite	31	(40.3)	1207	(33.8)	1.0 (reference)

Table 12B.9
Multivariate Analysis of Cardiomyopathy (N=77)
Baltimore-Washington Infant Study (1981–1989)

Risk Factor	Odds Ratio	95% CI	99% CI
overt diabetes	11.5	4.4–29.8	3.3–40.0
previous preterm birth	2.2	1.0–4.5	0.8–5.7
paternal auto body repair work	2.5	1.3–4.8	1.1–5.9

ables, diabetes was the strongest risk factor (odds ratio 11.5) and remained significant at the 1% level, as did paternal auto body repair work. Previous preterm birth was significant only at the 5% level.

In addition to these risk factors, maternal and paternal age were also weakly associated with cardiomyopathy. In the absence of paternal age in the model, maternal age under 20 years had an elevated odds ratio of 1.7 ($P = .05$), but this association declined to 1.5 in the presence of paternal age under 20 years in the model ($P = .26$). Similarly, paternal age under 20 years had an odds ratio of 2.0 ($P = .05$) when maternal age was not in the model, but taking maternal age into account reduced the paternal odds ratio to 1.5 ($P = .35$). There was no significant interaction between the parental age variables.

Discussion

The reevaluation of this heterogeneous case group of infants has yielded information of interest that goes beyond the observations made in the previous publication.[148] Findings strikingly illustrate the major role of genetic factors in the origin of cardiomyopathies—among the 86 cases 31 (36%) had associated noncardiac conditions, 22 of which were hereditary/metabolic anomalies or were associated with chromosomal or Mendelian syndromes. In 10 infants inflammatory origins were due to Coxsackie or rubella virus or the undefined virus of Kawasaki disease and myocarditis. In addition, among the 77 infants whose families were interviewed, 9 were infants of mothers with diabetes (6 with overt and 3 with gestational diabetes). Cardiomyopathy is a well-recognized manifestation in the infants of mothers with diabetes, and it carries a favorable prognosis. However, in this group the all-cause mortality was high (26%) even during the restricted 1-year follow-up period. A notable finding is the evidence of sociodemographic disadvantage with predominantly urban residence, poor education, and working status of the parents. The predominance of teen-aged parents and past history of preterm births are factors that may indicate deficiencies in the health care of the mother before and during pregnancy.

Of special note is the absence of the initially expected associations of cardiomyopathy with lifestyle variables such as the use of alcohol or recreational drugs. The only toxic exposure that remains significant is the paternal exposure to lead in auto body repair work or welding.

An evaluation of cardiomyopathy confined to the first year of life should not be considered equivalent to the usual clinical evaluations of this cardiac diagnostic group. Cases seen at older ages may well show associations of infectious and toxic agents. The metabolic abnormalities predominant in this infant group may in general be associated with a poor prognosis for long-term survival. The findings of this evaluation should have a two-pronged impact on future research, one directed to the early recognition of the disease and the amelioration of adverse pregnancy outcomes and the other directed to long-term follow-up and the evaluation of postnatal therapies of the affected infants.

Section C

Ebstein's Malformation of the Tricuspid Valve

This unusual congenital anomaly of the tricuspid valve was first described by Wilhelm Ebstein[119,120] and was named after him by Arnstein.[25]

A previous report on the genetic and environmental risk factors of Ebstein's malformation[93] includes all of the cases of the Baltimore-Washington Infant Study and describes the results of case-control analyses, comparing this diagnostic group to all other cases with cardiovascular malformations and to controls. The findings indicate a strong role for genetic and multiple environmental factors in the genesis of this anomaly, which must be further evaluated in collaborative studies.

This chapter presents the information previously described in a format consistent with that used for other diagnostic subgroups in this book, and highlights the important genetic and environmental risk factors that characterize this small case group. Additional considerations take into account some findings of the diagnosis-specific analyses described in previous chapters.

Study Population

There were 47 infants with Ebstein's anomaly ascertained in the study region during 1981 to 1989.

Two infants in our case population had Ebstein's anomaly as a secondary malformation. One of these (#10925) had corrected (L-loop) transposition of the great arteries, the other (#11265) had a supracristal ventricular septal defect.

Cardiac Abnormalities

Eighteen (39%) of the infants had associated cardiac malformations, the most common of which were pulmonary atresia (N = 8) and pulmonic valve stenosis (N = 4). Four infants had an associated left-sided obstructive lesion (coarctation of the aorta, N = 2; aortic stenosis, N = 1; bicuspid aortic valve, N = 1), and two had ventricular septal defects.

Noncardiac Anomalies

Nine infants (19%) had associated noncardiac anomalies (Table 12C.1), including two with chromosomal anomalies. Malformation syndromes diagnosed in four infants, included Mendelian syndromes (cobalophilin, Apert, and Noonan) and the CHARGE association. Three other infants had nonsyndromic malformations including biliary atresia, hydronephrosis and thrombocytopenia.

Table 12C.1
Ebstein's Anomaly (N=47) Noncardiac Anomalies
Baltimore-Washington Infant Study (1981–1989)

Chromosomal Abnormalities:	N=2	(4.2%)
Trisomy 18	1	[# 11686]
Chromosome 2p+	1	[# 10197]
Syndromes	**N=4**	**(8.5%)**
a. Mendelian Syndromes:	3	
Cobalophilin	1	[# 1636]
Apert	1	[# 2082]
Noonan	1	[# 10171]
b. Non-Mendelian Associations:	1	
CHARGE	1	[# 1816]
Multiple, Nonclassified Anomalies:	**N=1**	**(2.1%)**
Biliary atresia, cleft lip and palate	1	[# 10372]
Organ Defects:	**N=2**	**(4.2%)**
Hydronephrosis	1	[# 1864]
Thrombocytopenia	1	[# 10422]
No Associated Anomalies	**N=38**	**(80.9%)**

Descriptive Analyses

Prevalence by Time, Season, and Area of Residence

The overall prevalence of Ebstein's anomaly was 0.52 cases per 10,000 live-births (Table 12C.2). During the first 3 years (1981 to 1983) and the last 3 years of the study (1987 to 1989), the prevalence was 0.62 per 10,000, but the anomaly was only half as prevalent (0.30) during 1984 to 1986. The analysis of prevalence by birth quarters revealed a lower prevalence rate during the first and fourth quarters relative to the second and third quarters. The anomaly was less prevalent in rural areas than in urban and suburban areas of the region.

Diagnosis and Course

The vast majority of these infants (80.9%) were diagnosed within the first week of life, but six cases were diagnosed between 5 and 52 weeks (Table 12C.3). Echocardiography was the sole method of diagnosis in 22 cases (46.8%), and 12 cases were diagnosed at autopsy (4 of whom were diagnosed only at autopsy). Eleven infants had cardiac surgery for associated conditions, and 30 cases were alive among the 44 who had a diagnostic update (68.2%).

Gender, Race, and Twinning

Nearly half (49%) of the infants were male, a similar proportion to that of the controls (Table 12C.4). However, the cases were much more likely than controls to be white. Four of the cases were from multiple births, a much greater propor-

Table 12C.2
Ebstein's Anomaly: Prevalence
Baltimore-Washington Infant Study (1981–1989)

	Cases	Area Births	Prevalence per 10,000
Total Subjects	47	906,626	0.52
Year of Birth:			
1981–1983	17	274,558	0.62
1984–1986	9	295,845	0.30
1987–1989	21	336,223	0.62
Birth Quarter:			
1st (Jan–Mar)	7	219,145	0.32
2nd (Apr–Jun)	14	231,777	0.60
3rd (Jul–Sep)	16	233,626	0.68
4th (Oct–Dec)	10	222,078	0.45
Area of Residence:			
urban	13	208,568	0.62
suburban	30	584,022	0.51
rural	4	112,318	0.36

Table 12C.3
Ebstein's Anomaly: Clinical Data
Baltimore-Washington Infant Study (1981–1989)

	Cases N=47 no.	(%)
Age at Diagnosis:		
<1 week	38	(80.9)
1–4 weeks	3	(6.4)
5–24 weeks	5	(10.6)
25–52 weeks	1	(2.1)
Method of Diagnosis:		
echocardiography only	22	(46.8)
echocardiography and cardiac catheterization	6	(12.8)
echocardiography, cardiac catheterization, and surgery	7	(14.9)
autopsy, following other methods	8	(17.0)
autopsy only	4	(8.5)
Surgery During First Year of Life	11	(23.4)
Follow-Up:		
not followed	3	(6.4)
followed:	44	
died in the first year of life	14	(31.8)
alive	30	(68.2)
median age at last visit	11months	

Table 12C.4
Ebstein's Anomaly: Infant Characteristics
Baltimore-Washington Infant Study (1981–1989)

Characteristic	Cases N=47 no.	(%)	Controls N=3572 no.	(%)
Gender:				
male	23	(48.9)	1,817	(50.9)
female	24	(51.1)	1,755	(49.1)
Race*:				
white	41	(87.2)	2,362	(66.1)
black	6	(12.8)	1,109	(31.0)
other	0		98	(2.7)
Twin births	4	(8.5)	53	(1.5)

*Missing race in 3 controls; odds ratio for white race = 3.5 (95% CI 1.5–8.2); odds ratio for twin birth = 6.2 (95% CI 2.1–17.8).

tion than among controls; three of these cases were twins and one case was from a set of triplets. One of the twin cases, a male, had a stillborn dizygotic twin brother who also had Ebstein's anomaly and a hypoplastic left ventricle. All other co-twins were developmentally normal. The odds ratio for multiple births was 8.2 (95% confidence interval 2.6 to 25.3).

Birthweight and Gestational Age

One quarter of the cases had birthweights below 2500 g (Table 12C.5), more than three times the proportion among controls. The proportion of cases born prematurely (23%) was also in excess relative to controls (9.5%). Deficits in fetal growth are evident in the four-fold odds ratio for being small for gestational age.

Sociodemographic Characteristics

There were some case-control differences in the socioeconomic characteristics of the families. Case mothers were older (56.8% over 30 years versus 34.2% in controls). They were also more likely than controls to be married (86% versus 72%) and to have started prenatal care during the first 8 weeks of pregnancy (89% versus 69%). Fifty-five percent of the case families had income of $30,000 or more, compared to 38% of controls. Parents of cases and controls did not differ in educational levels or in paternal occupation, though case mothers were more likely than control mothers to be working during pregnancy (91% versus 80%), but there were no significant differences in occupational subcategories.

Table 12C.5
Ebstein's Anomaly: Fetal Growth Characteristics
Baltimore-Washington Infant Study (1981–1989)

Characteristic	Cases no.	(%)	Controls no.	(%)
Number of Families Interviewed	44		3572	
Birthweight (grams)				
<2500	11	(25.0)	252	(7.1)
2501–3500	17	(38.6)	1853	(51.9)
>3500	16	(36.4)	1467	(41.0)
mean±standard error	3075±148		3351±10	
range	640–5216		340–5273	
Gestational Age (weeks)				
<38	10	(22.7)	339	(9.5)
38+	34	(77.3)	3233	(90.5)
mean±standard error	38.2±0.6		39.6±0.1	
range	26–43		20–47	
Size for Gestational Age				
small (SGA)	9	(15.6)	211	(5.9)
normal	27	(61.4)	2712	(75.9)
large	8	(18.2)	649	(18.2)
Odds Ratio for SGA (95% CI)	**4.1**	**(2.0–8.6)**	1.0	(reference)

Potential Risk Factors

Familial Cardiac and Noncardiac Anomalies

Cardiac anomalies in first-degree relatives were reported in four families (9%, compared to 1.2% among controls). One infant's mother had an atrial septal defect (#3168); in the other three families, siblings of the index case were affected. A male index case (#10489) had two brothers with Ebstein's anomaly, one of which was his stillborn dizygotic twin brother (Family A in Figure 12C.1). A tricuspid valve anomaly, possibly Ebstein's, was reported in a stillborn sibling who preceded the index case (#2920) (Family B in 12C.1), and hypoplastic left heart syndrome was the autopsy-confirmed diagnosis in two previous siblings of a third case (#3107) (Family C in 12C.1). These occurrences may indicate either a genetic or a common environmental effect during pregnancy, with damage to the right or left atrioventricular valve.

Univariate and Multivariate Analysis

The unadjusted case-control comparisons for genetic and environmental risk factors are shown in Table 12C.6. A modified form of the multiple logistic regression results that appeared in our previous publication on this anomaly are shown

Ebstein's Malformation of the Tricuspid Valve
Baltimore-Washington Infant Study (1981-1989)

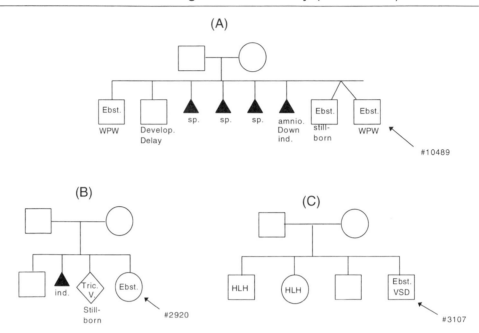

Figure 12C.1: Pedigrees are shown for three families in which probands with Ebstein's anomaly of the tricuspid valve had relatives with congenital heart defects.
Abbreviations: Ebst. = Ebstein's anomaly; WPW = Wolff-Parkinson-White syndrome; Tric.V. = tricuspid valve atresia; HLH = hypoplastic left heart; VSD = ventricular septal defect. Reprinted with permission from the publisher.

in Table 12C.7, which combines the information from Tables .5 and .6 of that report.[93] The analysis first examined genetic and environmental factors (Model A), then examined these factors while stratifying on previous pregnancies (Model B), and finally examined the subset of isolated/simplex cases (Model C).

The analysis of genetic and environmental factors (Part A of Table 12C.7) revealed that the risk of Ebstein's anomaly among liveborn infants increased five-fold with a family history of congenital heart disease, 10% per year of maternal age. It increased five-fold with exposure to benzodiazepines, nearly four-fold with exposure to varnishing, and three-fold with exposure to marijuana. Each of the corresponding 99% confidence intervals included 1.0.

In order to assess possible confounding by previous adverse reproductive outcomes, a separate analyses was stratified by the number of previous pregnancies (Part B of Table 12C.7). The results revealed an odds ratio of 6.4 for family history, 2.0 for one previous miscarriage and 3.9 for mothers who had two or more miscarriages, 5.4 for benzodiazepine exposure, and 3.4 for varnishing. Maternal age and use of marijuana did not improve the fit of the model. White race was not a confounder for the other factors in the model, but had an independent effect on the risk of Ebstein's anomaly (odds ratio = 2.9). Socioeconomic status exhibited neither an independent association nor evidence of any confounding of the other variables.

Table 12C.6
Ebstein's Anomaly Univariate Analysis of Risk Factors
Baltimore-Washington Infant Study (1981–1989)

Variable	Cases N=44 no. (%)		Controls N=3572 no. (%)		Odds Ratio (95% CI)
Family History (first-degree relatives)					
congenital heart disease	3	(6.8)	43	(1.2)	**6.0 (1.8–20.1)**
noncardiac anomalies	1	(2.3)	165	(4.6)	0.5 (0.1–3.7)
Parental Age					
maternal age:					
<20	2	(4.5)	507	(14.2)	0.5 (0.1–2.0)
20–29	18	(40.9)	2009	(56.2)	1.0 (reference)
30+	24	(54.6)	1056	(29.6)	**2.6 (1.4–4.8)**
paternal age:					
<20	1	(2.3)	226	(6.3)	0.5 (0.1–3.9)
20–29	15	(34.1)	1724	(48.3)	1.0 (reference)
30+	26	(59.0)	1622	(45.4)	**1.8 (1.0–3.5)**
Maternal Reproductive History					
number of previous pregnancies:					
none	7	(15.9)	1159	(32.4)	1.0 (reference)
one	12	(27.3)	1097	(30.7)	1.8 (0.7–4.6)
two	16	(36.4)	709	(19.9)	**3.7 (1.5–9.1)**
three or more	9	(20.5)	607	(17.0)	2.4 (0.9–6.6)
previous miscarriage(s)	18	(40.9)	681	(19.1)	**3.2 (1.7–5.9)**
Maternal Illnesses and Medications					
diabetes:					
overt	0		23	(0.6)	—
gestational	2	(4.5)	115	(3.2)	1.4 (0.7–2.7)
influenza	3	(6.8)	278	(7.8)	0.9 (0.3–3.0)
benzodiazepines	3	(6.8)	35	(1.0)	**7.4 (2.2–25.0)**
gastrointestinal medications	4	(9.1)	115	(3.2)	**3.0 (1.1–8.5)**
Lifestyle Exposures					
alcohol (any amount)	28	(63.6)	2101	(58.8)	1.2 (0.7–2.2)
smoking (cigarettes/day):					
none	30	(68.2)	2302	(64.5)	1.0 (reference)
1–10	6	(13.6)	565	(15.8)	0.7 (0.3–1.8)
11–20	3	(6.8)	529	(14.8)	0.5 (0.1–1.5)
>20	5	(11.4)	176	(4.9)	2.2 (0.8–6.7)
marijuana	7	(15.9)	263	(7.4)	**2.4 (1.0–5.4)**
Home and Occupational Exposures					
ionizing radiation (occupational)	0		48	(1.3)	—
varnishing	5	(11.4)	106	(3.0)	**4.2 (1.6–10.8)**
miscellaneous solvents	3	(6.8)	64	(1.8)	**4.0 (1.2–13.3)**
Race of Infant					
white	36	(81.8)	2362	(66.1)	**2.3 (1.1–50.0)**
nonwhite	8	(18.2)	1207	(33.9)	1.0 (reference)

Table 12C.7

Ebstein's Anomaly: Multivariate Models
Baltimore-Washington Infant Study (1981–1989)

Models and Variables	Cases no. (%)		Controls no. (%)		Adjusted Odds Ratio	95% CI	99% CI
Model A. Genetic and Environmental Factors Only (N=44 cases and 3,572 controls)							
family history of CVM	3	(6.8)	43	(1.2)	5.3	1.5–18.1	1.0–26.6
maternal age (per year)	—		—		1.1	1.0–1.1	1.0–1.2
benzodiazepines	3	(6.8)	35	(1.0)	5.3	1.5–18.5	1.0–27.4
varnishing	5	(11.4)	106	(3.0)	3.6	1.4–9.3	1.0–12.6
marijuana	7	(15.9)	263	(7.4)	2.8	1.2–6.5	0.9–8.4
Model B: Stratified By Previous Pregnancies (N=44 cases and 3,572 controls)							
family history of CVM	3	(6.8)	43	(1.2)	6.4	1.8–22.2	1.2–32.9
previous miscarriages:							
0	24	(54.5)	2891	(80.9)	1.0	(reference)	—
1	16	(36.4)	550	(15.4)	2.0	1.2–3.3	1.0–3.9
2+	4	(9.1)	131	(3.7)	3.9	1.4–11.0	1.0–15.1
benzodiazepines	3	(6.8)	35	(1.0)	5.4	1.5–19.1	1.0–28.4
varnishing	5	(11.4)	106	(3.0)	3.4	1.3–9.1	1.0–12.4
white race	39	(88.6)	2362	(66.2)	2.9	1.2–7.0	0.9–9.2
Model C: Isolated/Simplex Subset (N=34 cases and 3,317 controls)							
previous miscarriages:							
0	21	(61.8)	2691	(81.1)	1.0	(reference)	—
1	10	(29.4)	506	(15.3)	2.1	1.1–3.9	0.9–4.7
2+	3	(8.8)	120	(3.6)	4.3	1.3–15.0	0.9–22.0
marijuana	7	(20.6)	244	(7.3)	3.6	1.6–8.5	1.2–11.1

CVM = cardiovascular malformations.

Finally, the model that best fitted the data for the isolated/simplex subset of the cases (Part C of Table 12C.7) included only two factors: number of previous miscarriages and maternal use of marijuana. Only the latter variable was significant at the 1% level.

For varnishing there was a sufficient number of exposed subjects to permit analysis of exposure frequency (ie, never exposed, exposed once a month or less, exposed at least once per week). With the "never exposed" category as the reference, the odds ratio was nearly three-fold greater among those exposed infrequently (odds ratio 2.9) and nearly seven-fold greater among those exposed at least once per week (odds ratio 6.9). The statistical test for a trend in these odds ratios has a P value of 0.0002.

Discussion

Ebstein's malformation of the tricuspid valve is a unique anomaly characterized by a downward displacement of the attachment of the tricuspid valve into the inflow portion of the right ventricle. The abnormal leaflets and attachments are such that the tricuspid valve hangs like a curtain into the right ventricle, incorporating a greater or lesser portion of it to the right atrium and diminishing the trabecular outlet portion of the right ventricle, resulting in obstruction of blood flow into the pulmonary artery. The abnormal valve leaflets are frequently perforated, allowing blood to pass from the atrial side of the valve, which includes a portion of the right ventricle, directly into the pulmonary artery. This large right atrium often becomes a chamber ineffective in propelling the blood forward, so that the foramen ovale leading into the left atrium remains patent and shunts venous blood from the right atrium to the left, resulting in cyanosis and signs of right-sided heart failure.

The malformation varies greatly in severity, with an extensive spectrum of manifestations ranging from severe disturbances in fetal and neonatal life to virtually symptomless survival throughout a long and active adult life. Among the 6300 hearts studied at autopsy by Bharati and Lev,[41] 67 had Ebstein's malformation, 3 were stillborn, 40 were infants under 1 year of age, 11 were children and adolescents, and the oldest patient was 84 years old. An 85-year old man with Ebstein's anomaly was reported by Seward[392] to have had a relatively illness-free life until the age of 77 when he was operated for carcinoma of the prostate. Two years later he began to have signs of congestive failure, and he died 6 years later just before his 85th birthday.

In reported case series tabulated in our previous paper, we demonstrated the gradual increase with time in the proportion of infant cases. The classic study of Yater and Shapiro[456] did not enroll infants and in the International Cooperative Study 35 of the 505 patients (7%) were infants.[439] Since then the severe forms of the anomaly in fetal life and in the neonatal period have become recognized.[76,77,199,241] Our case population represents this high-risk group in whom, with one exception, the diagnosis was established under 6 months of age.

The morphogenesis and developmental relationships of Ebstein's malformation to other heart defects remain to be elucidated. It is notable that the unique pathologic picture was recognized as a congenital anomaly distinct from the many forms of valvar abnormalities seen 130 years ago, indicating the astute observations of pathologic changes by the noted clinician, Wilhelm Ebstein[119] who described that malformation. Clark[82] suggested abnormal cell death as a possible mechanism of this anomaly, a hypothesis supported by the evidence of myocardial fibrosis.[77] Recently Bharati and Lev[41] raised the hypothesis that Ebstein's anomaly may be a connective tissue disorder related to Marfan's syndrome. These authors observed that mitral anomalies are also associated in some of the cases. This association of mitral abnormalities with the distinct abnormality of the tricuspid valve is of interest in view of the occurrence of hypoplastic left heart syndrome in the siblings of two of our cases.

Further considerations may be added taking into account the results of the multivariate analyses and the findings in other malformation groups, most notably left-sided obstructive defects.

Family history of cardiovascular malformations, history of previous adverse pregnancy outcomes, and maternal exposure to solvents all occurred as risk factors in Ebstein's malformation *and* in left-sided defects. It is tempting to consider Ebstein's malformation with associated left-sided cardiac defects as comparable manifestations warranting the exploration of the same genetic and environmental influences on pathogenesis.

Furthermore, one may try to bring into the same etiologic horizon the apparent effect of mood-altering substances (benzodiazapines and marijuana) with previous reports of a relationship between Ebstein's and therapeutic lithium intake.[436] Given the rarity of Ebstein's malformation, the coexistent lithium therapy in several instances could hardly have been due to chance. No lithium intake occurred in any of the 44 case mothers evaluated here, allowing us to state only that lithium is not a major etiologic factor for Ebstein's malformation. Neither this retrospective study, nor studies of small series of pregnant women could establish a lack of an association between Lithium and Ebstein's malformation. For a prospective (cohort) study of mothers exposed and not exposed to lithium, a sample size of 7000 mothers in each group would be necessary to detect the relative risk of 20 (power $= 80\%, \alpha = 0.05$, one-sided) that has been claimed to exist. Withholding this potentially harmful agent during pregnancy has been a wise public health measure, but has made it impossible to pursue the relationship by epidemiologic approaches.[145]

Ebstein's malformation of the tricuspid valve is a severe adverse pregnancy outcome, which must be further evaluated in histopathologic studies as well as in future collaborative molecular-epidemiologic studies.

PART III

SYNTHESIS

Chapter 13

The Infant With Congenital Heart Disease

From: Ferencz C, Loffredo CA, Correa-Villaseñor A, Wilson PD, eds: *Genetic & Environmental Risk Factors of Major Cardiovascular Malformations: The Baltimore-Washington Infant Study 1981–1989.* Armonk, NY: Futura Publishing Co., Inc; ©1997.

Introduction

The Baltimore-Washington Infant Study enrolled 4390 infants with congenital heart disease. These were liveborn infants of residents of a defined geographic area in whom a structural abnormality of the cardiovascular system was identified by multiple sources of ascertainment. By dividing the cases into categories and subcategories of cardiac diagnoses and evaluating associated noncardiac abnormalities and size at birth, we have defined the total "portrait" of congenital heart disease and gained extensive new information, which has been presented in the previous chapters.

The Baltimore-Washington Infant Study case information does not include the anatomic details of the cardiac abnormalities or the hemodynamic consequences by which clinicians judge their patients, as they concentrate principally or even uniquely on the circulatory system. The epidemiologic image presents the specific features under evaluation as retrieved one facet at a time from coded information that is consistent for cases and controls (Figure 13.1).

In this search for causes the primary focus is also on the cardiovascular system described in terms of the cardiac anatomy. However, in addition, each cardiac diagnosis is viewed within a triad of two other abnormal developmental outcomes: noncardiac anomalies and low birthweight (Figure 13.2). This strategy captures the teratogenic setting of abnormal hearts. Over the years this approach gained paramount importance with the beginnings of an understanding of the molecular determination of developmental events. Thus the foregoing diagnosis-specific chapters presented information that may lead to new perspectives and possibly

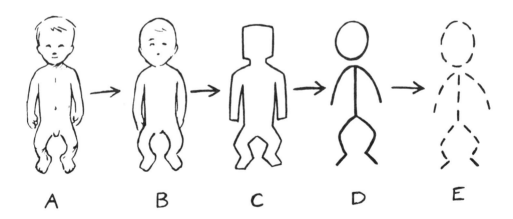

A   B   C   D   E

Diagram illustrating the distortion of the information in a reporting system, from the infant to the coded data. (A) shows the actual infant, (B) is the doctor's picture of it and what is written down in the medical records, (C) is the content of the report form to the surveillance registry, (D) is the interpretation of that form in the registry, and (E) is the coded data which are stored in a computer.

Figure 13.1: The epidemiologic view of the infant (Form E) conveys information on specific features under investigation and lacks the clinician's detailed view of the "actual infant" (Form A).
Reproduced from Källén B, (1988) *Epidemiology of Human* Reproduction, Boca Raton, Florida: CRC Press, Inc.; with permission from the author and the publisher.

Figure 13.2: The teratogenic setting of cardiovascular defects includes impaired fetal growth and noncardiac anomalies.

new directions for future molecular research. Additional ideas for productive future investigations are derived also from these summation chapters.

Prevalence

Measures of disease occurrence represent a time-honored pillar of epidemiology in the search for causes by defining the characteristics of populations and persons at greatest risk of a disease.

Birth defects constitute a special disease group because the occurrence of new cases (incidents) cannot be counted due to prenatal events, which select out very serious defects by death of the embryo and also some mild anomalies by "healing" (Figure 13.3). For this reason the best measure available is a count of the survivors as a proportion of livebirths.

The Baltimore-Washington Infant Study took place in a critical period of change for pediatric cardiology with the introduction and expanding use of two-dimensional diagnostic echocardiography in liveborn infants as well as in fetuses. This resulted in a notable change over time in the prevalence of certain malformations (Table 13.1). The cardiac diagnostic groups shown in Table 13.1 are *not the same* as reported in our earlier book[152] because we now limit this discussion to "pure" diagnoses selected for risk factor analyses. There is, however, no great change in the findings.

The application of two-dimensional echocardiographic diagnosis affected the frequency of diagnostic subgroups in two different ways:

1. Increases in the prevalence at livebirth occurred in certain types of congenital heart disease that were diagnosed in the first year of life instead of in subsequent periods (ie, infants with tetralogy of Fallot, aortic stenosis, and atrial septal defects). Indeed for these three diagnoses, as expected, the prevalence increased over the three successive 3-year enrollment periods. More dramatic was the increase of

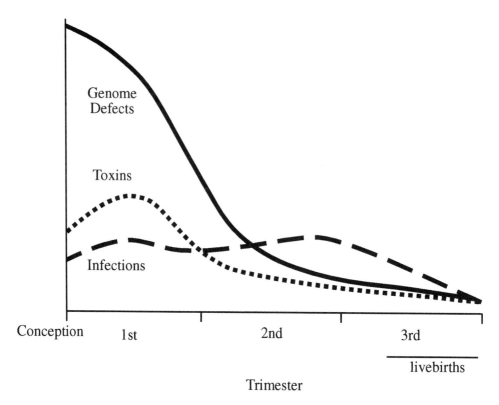

Figure 13.3: Anomalies arising during intrauterine life may heal or be embryo lethal, preventing the estimation of the true occurrence (incidence) of human malformations.
Reproduced from Ferencz C and Neill CA, (1992) Cardiovascular malformations: Prevalence at livebirth. Chapter 2 in Freedom RM, Benson LN, Smallhorn JF (eds:) *Neonatal Heart Disease,* London: Springer-Verlag; with permission from the editors and the publisher.

clinically mild defects such as ventricular septal defects and mild pulmonic stenosis, conditions that are asymptomatic in infants. The dramatic increases for all of these defects have been described in their respective chapters, with note that the frequency of occurrence of the severe forms of the defects changed little over time.

2. Decreases in the prevalence at livebirth were found for conditions that can be recognized in routine prenatal evaluations such as maternal amniocentesis after 35 years of age, and fetal echocardiography of high risk pregnancies. Indeed, we observed decreases over time in the prevalence at livebirth for atrioventricular septal defect with Down syndrome, for hypoplastic left heart syndrome, and also for pulmonary atresia—probably due to interruption of the pregnancy after fetal diagnosis. Variations due to health service practices might be expected to differ in urban areas compared to suburban or rural areas. This was true to some extent with the highest prevalence of mild defects in urban cases, particularly ventricular septal defects of the membranous type, pulmonic valve stenosis, and atrial septal defect (Table 13.2). For malformations detectable by fetal echocardiography, the urban prevalence was lowest, probably due to greater access; ex-

Table 13.1
Prevalence of Congenital Heart Disease
by Diagnostic Group and Time Period
Baltimore-Washington Infant Study (1981–1989)

"Pure" Diagnostic Group	N	Prevalence per 10,000 Livebirths			
		Total	1981–83	1984–86	1987–89
Laterality/Looping Defects	131	1.44	1.27	1.96	1.10
Outflow Tract Anomalies:					
Transposition Group	239	2.64	2.70	2.60	2.56
TGA-IVS	115	1.27	1.23	1.39	1.16
Normal Great Arteries Group	400	4.41	3.82	4.53	4.55
Tetralogy of Fallot	236	2.60	2.15	2.77	2.61
Atrioventricular Septal Defects:					
Down Syndrome	210	2.32	2.29	2.67	1.87
Nonchromosomal	88	0.97	0.76	1.25	0.83
Ventricular Septal Defects:					
Membranous	895	9.87	8.09	9.73	11.21
Muscular	429	4.73	0.84	3.54	8.95
Left-Sided Obstructive Defects:					
Hypoplastic Left Heart	162	1.78	2.37	1.79	1.31
Coarctation of the Aorta	126	1.39	1.78	1.22	1.14
Aortic Valve Stenosis	74	0.81	0.44	1.11	0.83
Bicuspid Aortic Valve	67	0.74	0.66	0.98	0.59
Right-Sided Obstructive Defects:					
Pulmonary Valve Stenosis	341	3.78	2.22	4.43	4.31
Pulmonary Atresia	53	0.58	0.76	0.54	0.48
Tricuspid Valve Atresia	33	0.36	0.36	0.44	0.30
Atrial Septal Defect	291	3.21	2.55	3.41	3.48
Patent Arterial Duct	80	0.88	0.69	0.81	0.92
Cardiomyopathy	86	0.95	0.87	1.22	0.71
TAPVR	60	0.66	0.73	0.57	0.68
Ebstein's Anomaly	43	0.47	0.62	0.27	0.53
Area Births (1981–1989)	906,626	—	274,558	295,845	336,223

amples are atrioventricular septal defect with Down syndrome and pulmonary atresia. The low urban prevalence of coarctation of the aorta and aortic valve stenosis may be due to population factors (see below).

In evaluating prevalence by gender and race, we noted the increased prevalence of males for some cardiac malformations and of females for others; this is a well-known phenomenon, but so far it is without cogent explanations. The findings in the Baltimore-Washington Infant Study are shown as the crude odds ratio for male infants in the various diagnostic groups in comparison to that for controls (Figure 13.4). As expected, we found a male excess in transposition of the great arteries and in left-sided obstructive defects. However, as noted in Chapter 5 (on outflow tract defects), the male excess was statistically significant only in transposition with intact ventricular septum (TGA-IVS). Among left-sided defects the male

Table 13.2

Prevalence by Urban, Suburban, and Rural Areas of Residence

Baltimore-Washington Infant Study (1981–1989)

"Pure" Diagnostic Group	N	Prevalence per 10,000 Live Births		
		Urban	Suburban	Rural
Laterality/Looping Defects	131	2.21	1.20	1.33
Outflow Tract Anomalies:				
Transposition Group	239	2.46	2.74	2.72
TGA-IVS	115	1.01	1.34	1.42
Normal Great Arteries Group	400	5.30	4.10	4.72
Tetralogy of Fallot	236	3.83	3.10	3.20
Atrioventricular Septal Defects:				
Down syndrome	210	1.96	2.09	2.40
Nonchromosomal	88	1.00	1.01	0.71
Ventricular Septal Defects:				
Membranous	895	9.35	6.40	6.32
Muscular	429	1.92	1.74	1.87
Left-Sided Obstructive Defects:				
Hypoplastic Left Heart	162	2.30	1.51	2.31
Coarctation of the Aorta	126	0.77	1.51	1.96
Aortic Valve Stenosis	74	0.38	0.96	0.89
Bicuspid Aortic Valve	67	0.86	0.74	0.53
Right-Sided Obstructive Defects:				
Pulmonary Valve Stenosis	341	6.71	2.86	3.03
Pulmonary Atresia	53	0.53	0.56	0.80
Tricuspid Valve Atresia	33	0.57	0.34	0.09
Atrial Septal Defect	291	4.32	2.86	3.03
Patent Atrial Duct	80	1.39	0.75	0.71
Cardiomyopathy	86	1.58	0.72	0.98
TAPVR	60	1.05	0.46	0.98
Ebstein's Anomaly	47	0.62	0.51	0.36
Area Births (1981–1989)	906,626	208,568	584,022	112,318

excess was significant only in aortic stenosis (AoS). The female excess in atrioventricular septal defect with Down syndrome was significant, as was that in muscular ventricular septal defect and atrial septal defect. Gender differences are known to occur in multifactorial disorders.[67] In earlier reports we noted that the coexistence of transposition and hemophilia in some families (also reported elsewhere in the literature) raises the possibility of an X-linked abnormality in this malformation.[138,143]

A great surprise was found in prevalence by race. In Table 13.3 we note in bold type those case groups in which the prevalence of whites and nonwhites differed by more than 50%. Most striking was the predominance of whites with aortic valve stenosis ("pure" diagnosis) with a nine-fold difference, but the case number is small. For coarctation of the aorta, the prevalence of white infants was also in excess, but it was not in hypoplastic left heart syndrome or bicuspid aortic valve, conditions that are related to aortic stenosis and coarctation. An excess of white infants with Ebstein's anomaly is also well known, but perhaps the magnitude of

COMPARISON OF GENDER ODDS RATIOS BY DIAGNOSTIC GROUP
BALTIMORE - WASHINGTON INFANT STUDY (1981-1989)

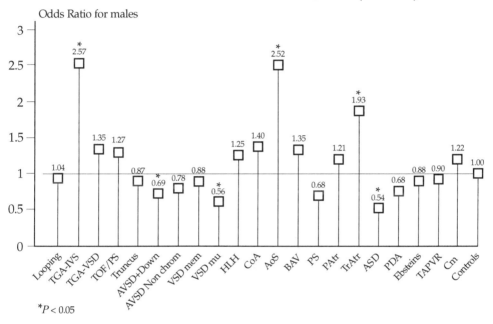

*P < 0.05

Figure 13.4: Gender inequalities in specific diagnostic subgroups are shown as unadjusted case-control odds ratios for male gender. A statistically significant male excess was found in transposition with intact ventricular septum, aortic stenosis, and tricuspid atresia with normally related great arteries. A female excess was found in atrioventricular septal defect with Down syndrome, muscular ventricular septal defect, and atrial septal defect.

Abbreviations: TGA-IVS = transposition with intact ventricular septum; TGA-VSD = transposition with ventricular septal defect; TOF/PS = tetralogy of Fallot; AVSD = atrioventricular septal defect; Nonchrom = nonchromosomal; VSDmem = ventricular septal defect, membranous; VSDmu = ventricular septal defect, muscular; HLH = hypoplastic left heart; CoA = coarctation of the aorta; AoS = aortic stenosis; BAV = bicuspid aortic valve; PS = pulmonic stenosis; Patr = pulmonary atresia with intact ventricular septum; TrAtr = tricuspid atresia; ASD = atrial septal defect; PDA = patent arterial duct; TAPVR = total anomalous pulmonary venous return; Cm = cardiomyopathy.

Table 13.3

Prevalence by Diagnostic Group and Race

Baltimore-Washington Infant Study (1981–1989)

"Pure" Diagnostic Group	N	Prevalence per 10,000 Live Births		% Difference	Race Group in Excess (<50%)
		Whites	Nonwhites		
Laterality/Looping Defects	131	1.28	1.64	27	
Outflow Tract Anomalies:					
Transposition Group	239	2.80	2.14	34	
TGA-IVS	115	1.42	0.98	45	
Normal Great Arteries Group	400	4.18	4.69	12	
Tetralogy of Fallot	236	3.09	3.42	11	
Atrioventricular Septal Defects:					
Down syndrome	210	2.42	1.98	22	
Nonchromosomal	88	0.98	0.98	0	
Ventricular Septal Defects:					
Membranous	895	8.47	10.82	28	
Muscular	429	4.83	3.74	29	
Left-Sided Obstructive Defects:					
Hypoplastic Left Heart	162	1.74	1.76	1	
Coarctation of the Aorta	**126**	**1.67**	**0.82**	**106**	whites
Aortic Valve Stenosis	**74**	**1.16**	**0.13**	**866**	whites
Bicuspid Aortic Valve	67	0.77	0.66	17	
Right-Sided Obstructive Defects:					
Pulmonary Valve Stenosis	**341**	**2.92**	**5.07**	**73**	nonwhite
Pulmonary Atresia	53	0.61	0.50	22	
Tricuspid Valve Atresia	**33**	**0.31**	**0.47**	**52**	nonwhites
Atrial Septal Defect	291	2.90	3.40	17	
Patent Arterial Duct	80	0.77	1.00	31	
Cardiomyopathy	**86**	**0.31**	**1.67**	**439**	nonwhites
TAPVR	**60**	**0.77**	**0.47**	**64**	whites
Ebstein's Anomaly	**47**	**0.70**	**0.19**	**268**	whites
Area Births (1981–1989)	906,626	585,493	317,811	—	

this difference has not been recognized, nor has the great predominance of cardiomyopathy among nonwhite infants received attention.

The Malformation Triad

Cardiovascular Malformations

In the majority of infants (75%), the cardiovascular defect formed the only malformation, but the proportion of isolated congenital heart disease varied considerably among the various diagnostic groups (Table 13.4).

Isolated cardiac defects made up more than 90% of the cases in transposition of the great arteries, pulmonic stenosis, and muscular ventricular septal defect, suggesting that the teratogenic insult in these cases was directed to the heart and maybe the heart alone, with few exceptions.

Teratogens usually lead to multiple congenital anomalies and not to isolated cardiac defects, but virtually all syndromes have their forme fruste, and sometimes an isolated cardiac defect may be the only phenotypic manifestation of a syndrome.

As the proportion of isolated cases decreased in the various diagnostic categories, a broader systemic teratogenic effect on development was evident. There was one single cardiovascular diagnostic group, atrioventricular septal defects, in which cases with isolated cardiac anomalies were clearly in a minority (23%). Thus, this specific cardiac malformation is intimately related to abnormal development of other organ systems (Figure 13.5).

Noncardiac Anomalies

Among all cases of cardiovascular malformations approximately one fourth were found in infants with multiple congenital anomalies. Most of these cases were of genetic origin including those in infants with chromosomal syndromes and those in infants with Mendelian syndromes. A listing of all diagnostic groups with the number and proportion of cases with noncardiac defects is also shown in Table 13.4. Noncardiac defects are divided into four major categories: chromosome anomalies, Mendelian and non-Mendelian syndromes, multiple nonclassified anomalies, and single-organ defects.

Distribution of Isolated Heart Defects by Major Diagnostic Groups
Baltimore-Washington Infant Study (1981-1989)

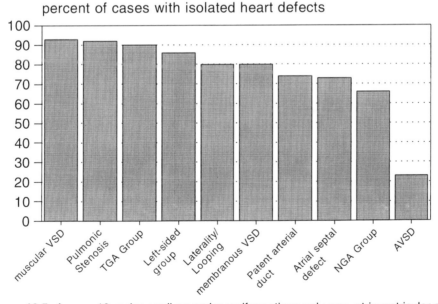

percent of cases with isolated heart defects

Figure 13.5: Among 10 major cardiovascular malformations only one, atrioventricular septal defect, is rare as an isolated cardiac defect.
Abbreviations: AVSD = atrioventricular septal defect; NGA = normal great arteries; TGA = transposed great arteries; VSD = ventricular septal defect.

Table 13.4
"Pure" Diagnoses and Noncardiac Anomalies—Summary Table
Baltimore-Washington Infant Study (1981–1989)

Pure Diagnostic Group	Total Cases	Isolated Cardiac Anomalies n (%)	Chromosome Anomalies n (%)	Noncardiac Anomalies				Multiple, Nonclassified, Anomalies n (%)	Single Organ Defects n (%)
				Syndromes n (%)		Mendelian	Non-Mendelian		
Laterality/Looping	131	105 (80.1)	3 (2.3)	12 (9.2)		5	7	2 (1.5)	9 (6.9)
Outflow Tract Anomalies:									
Transposition Group	239	215 (90.0)	2 (0.8)	8 (3.3)		3	5	3 (1.3)	11 (4.6)
IVS	115	105 (91.0)	0	2 (1.7)		1	1	1 (0.9)	7 (6.1)
VSD	68	64 (94.1)	1 (1.5)	1 (1.5)		0	1	1 (1.5)	1 (1.5)
DORV	31	24 (77.4)	1 (3.2)	3 (9.7)		1	2	1 (3.2)	2 (6.5)
Complex	25	22 (88.0)	0	2 (8.0)		1	2	0	1 (4.0)
Normal Great Arteries Group	400	265 (66.3)	43 (10.8)	44 (11.0)		13	31	9 (2.3)	39 (9.8)
Tetralogy-PS	236	160 (67.8)	28 (11.9)	17 (7.2)		6	11	5 (2.1)	26 (11.4)
Tetralogy-PAtr	60	44 (73.3)	5 (8.3)	7 (11.7)		1	6	0	4 (6.7)
DORV	30	15 (50.0)	6 (20.0)	3 (10.0)		2	1	2 (6.7)	4 (13.3)
Truncus	44	26 (59.1)	1 (2.3)	13 (29.5)		3	10	0	4 (9.1)
Supracristal VSD	22	13 (59.1)	2 (9.1)	4 (18.2)		1	3	2 (9.1)	1 (4.5)
AP Window	8	7 (87.5)	1 (12.5)	0		0	0	0	0
Atrioventricular Septal Defects:									
Total Group	320	73 (22.8)	225 (70.3)	17 (5.3)		7	10	0	5 (1.6)
with Down Syndrome	210	0	210 (100)	0		0	0	0	0
without Down Syndrome	110	71 (64.5)	16 (14.5)	15 (13.6)		7	8	2 (1.8)	6 (5.4)

Ventricular Septal Defects:							
membranous	895	718 (80.2)	87 (9.7)	31 (3.5)	14	17	4 (0.4)	55 (6.1)
muscular	429	398 (92.8)	11 (2.6)	7 (1.6)	4	3	0 (0)	13 (3.0)
Left-Sided Obstructive Defects:							
Hypoplastic Left Heart	162	138 (85.2)	9 (5.6)	8 (4.9)	4	4	1 (0.6)	6 (3.7)
Coarctation of Aorta	126	112 (88.9)	10 (7.9)	4 (3.2)	2	2	0	0
Aortic Stenosis	74	61 (82.4)	7 (9.5)	0	0	0	0	6 (8.1)
Bicuspid Aortic Valve	67	56 (83.6)	5 (7.4)	2 (3.0)	0	2	0	4 (6.0)
Right-Sided Obstructive Defects:							
Pulmonic Stenosis (total)	341	313 (91.8)	4 (1.2)	14 (4.1)	13	1	2 (0.6)	8 (2.3)
severe	64	55 (84.4)	0	8 (12.5)	7	1	1 (1.6)	1 (1.6)
moderate	125	116 (92.8)	3 (2.4)	3 (2.4)	3	0	0	3 (2.4)
mild	152	143 (94.1)	1 (0.7)	3 (2.0)	3	0	1 (0.7)	4 (2.6)
Pulmonary Atresia	53	49 (92.5)	2 (3.8)	0	0	0	1 (1.9)	1 (1.9)
Tricuspid Atresia	33	30 (90.9)	0	1 (3.0)	1	0	0	2 (6.1)
Other Lesions:							
Atrial Septal Defect	291	213 (73.2)	42 (14.4)	16 (5.5)	7	9	2 (0.7)	18 (6.2)
Patent Arterial Duct	80	59 (73.8)	9 (11.3)	3 (3.8)	1	2	1 (1.3)	8 (10.0)
TAPVR	60	46 (76.7)	2 (3.3)	5 (8.3)	4	1	1 (1.7)	6 (10.0)
Ebstein's Anomaly	47	38 (80.9)	2 (4.2)	4 (8.5)	3	1	1 (2.1)	2 (4.2)
Cardiomyopathy	86	55 (64.0)	2 (2.3)	23 (26.7)	21	3	1 (1.2)	5 (5.8)
Total CVM	3834	2944 (76.8)	465 (12.1)	199 (5.2)	102	97	28 (0.7)	198 (5.2)
Controls	3572	3511 (98.3)	3 (0.1)	22 (0.6)	3	19	0	36 (1.0)

AP = aortic-pulmonary; DORV = double outlet right ventricle; IVS = intact ventricular septum; Patr = pulmonary atresia; PS = pulmonic stenosis; VSD = ventricular septal defect; CVM = cardiovascular malformations; TAPVR = total anomalous pulmonary venous return.
*Free of all birth defects.

Chromosomal Syndromes

During the years 1981 to 1989, when this study was performed, chromosomal defects were defined as abnormalities microscopically visible by standard cytogenetic examination. The molecular cytogenetic techniques that were introduced during and after this period changed this traditional definition.

The most common type of chromosomal pathology in the Baltimore-Washington Infant Study was trisomy 21 (Down syndrome), which was diagnosed in 354 infants; 210 of these patients had an atrioventricular septal defect (Table 13.5). This was the only example in which the number of cases of a cardiac defect in association with any specific nosology was larger than the number of cases with an isolated defect. Although it seems likely that chromosome 21 has to contain the gene(s) which, when present in duplicate in the infant, lead to atrioventricular septal defect, no such genes have been found so far. The genes coding the polypeptide chain of collagen VI that lie on 21q may be good candidates,[157] but the known familial cases of atrioventricular septal defect were not linked to chromosome 21.[98] Korenberg's studies,[231] which identified a potential candidate region for cardiac defects on chromosome 21 also indicate the important role of environment and chance in the development of specific outcomes.

Chromosome defects were also common in outflow tract defects with normal great arteries, membranous ventricular septal defect, and patent arterial duct, as well as the left-sided obstructive heart defects, which were associated with the Turner syndrome (monosomy X). These later patients predominantly had coarctation of the aorta and aortic stenosis.

Some correlations, although not so striking, between the type of cardiac defects and the type of chromosomal defects were also found in trisomy 13 and trisonomy 18 (Table 13.5). Among the 20 cases of trisomy 13, half had outflow tract defects with normally related great arteries, one fifth had left heart obstructions, and only a single case of membranous ventricular septal defect was observed among the trisomy 13 cases over the 9-year research period. In contrast, in trisomy 18, membranous ventricular septal defect made up almost one third of the 38 cases, atrioventricular

Table 13.5
Distribution of Trisomies Among Major Cardiac Diagnostic Groups
Baltimore-Washington Infant Study (1981–1989)

Diagnostic Group	Trisomy 13		Trisomy 18		Trisomy 21	
	n	(%)	n	(%)	n	(%)
Outflow tract anomalies with						
Normally-related great arteries	10	(50.0)	7	(18.4)	22	(6.2)
Left-sided obstructive defects	4	(20.0)	4	(10.5)	4	(1.1)
Atrial septal defect	3	(15.0)	0		36	(10.2)
Atrioventricular septal defects	0		9	(23.7)	210	(59.3)
Membranous VSD	1	(5.0)	12	(31.6)	64	(18.1)
Pulmonic stenosis	0		0		2	(0.6)
Other diagnostic groups	2	(10.0)	6	(15.8)	16	(4.5)
Total	20	(100)	38	(100)	354	(100)

VSD = ventricular septal defect.

septal defect made up about one quarter, and outflow tract defects with normally related great arteries made up only about one fifth. These comparative diagnostic distributions show that genes on chromosome 13 have an etiologic role in the genesis of ventricular outflow tract defects while chromosome triplicates in trisomies 18 and 21 are related to inflow defects at the atrioventricular junction.

Deletions/translocations were reported with increasing frequency over the study years, in a total of 34 cases. Certainly, deletions of some genes that are important for a specific stage of cardiogenesis can lead to very strong associations between the type of cardiac defect and the form of the chromosomal pathology. Atrioventricular septal defect in del(8)(p23-pter) and hypoplastic left heart in del (11)(q23-qter)[113,201] may be examples of this kind.

Syndromic malformations constituted about 8% of the total group of cardiovascular malformations.[56] Among the cases with pure diagnoses 5.2% had associated syndromes (see Table 13.4), a proportion 12-fold greater than the occurrence of such defects among our controls. The proportions of cardiac diagnoses are shown for the five most frequent syndromic malformations in Table 13.6.

The outflow tract anomalies group with normal great arteries had the greatest proportion of syndromic malformations; these were described in detail in a previous report.[265] None of the 18 infants with DiGeorge syndrome was reported to have a chromosome 22q11 deletion. Pulmonic stenosis occurred in the majority of the Noonan syndrome cases; atrial septal defect occurred in only one of the 10 cases with Holt-Oram syndrome. The presence with Holt-Oram syndrome of severe defects, such as outflow tract and atrioventricular septal defects and anomalous pulmonary venous return, suggests that the designation of atrial septal defect as the classic association was derived from observations on adults. Emphasis on the occurrence of complex cardiovascular malformations in this syndrome is important for appropriate family counseling.[397]

Table 13.6

**Distribution of Cardiac Diagnoses Within the 5 Most Frequent Syndromes
Baltimore-Washington Infant Study (1981–1989)**

Diagnostic Group	VACTERL	DiGeorge	Noonan	Goldenhar	Holt-Oram
Laterality and looping anomalies	0	0	0	2 (15%)	0
Outflow tract anomalies:					
Transposed great arteries	3 (12%)	1 (6%)	0	0	0
Normal great arteries	7 (28%)	16 (88%)	1 (6%)	2 (15%)	2 (20%)
Atrioventricular septal defects	1 (16%)	0	0	2 (15%)	1 (10%)
Membranous VSD	4 (16%)	1 (6%)	2 (12%)	3 (23%)	2 (20%)
Muscular VSD	0	0	0	1 (8%)	2 (20%)
Left-sided obstructive defects	4 (16%)	0	0	1 (8%)	0
Pulmonic stenosis	0	0	10 (58%)	0	0
Atrial septal defect	5 (20%)	0	1 (6%)	1 (8%)	1 (10%)
Patent arterial duct	1 (4%)	0	0	1 (8%)	0
Cardiomyopathy	0	0	2 (12%)	0	0
Ebstein's anomaly	0	0	1 (6%)	0	0
TAPVR	0	0	0	0	2 (20%)
Total	25 (100%)	18 (100%)	17 (100%)	13 (100%)	10 (100%)

TAPVR = total anomalous pulmonary venous return; VSD = ventricular septal defect.

The ratio of nonchromosomal syndromic and multiple defects to single noncardiac defects is shown in Table 13.7. For the total case group, these two types of anomalies occurred with approximately equal frequency. There were, however, three case groups: outflow tract defects with normally related great arteries, nonchromosomal atrioventricular septal defects, and valvar pulmonic stenosis, in which multiple associated noncardiac anomalies were in excess. For ventricular septal defects of both membranous and muscular type, multiple anomalies represented a minority of the associated defects. These findings again emphasize the systemic embryonic origin of the major "inflow" and "outflow" defects.

The numerous Mendelian syndromes that include a cardiovascular defect may be subdivided into two groups: a) syndromes with a relatively stable association with a specific cardiovascular defect (ie, Williams syndrome, Noonan syndrome), and b) syndromes with a very wide and uncharacteristic spectrum (ie, Meckel syndrome, 3C syndrome). Involvement of the specific genes that are important for cardiogenesis is presumed for the first group. Defects of the elastin gene mapped to 7q11.23 explain supravalvular aortic stenosis, which is typical for the Williams syndrome.[232,334] The pathogenesis of the broad spectrum of cardiovascular malformations in the other group of syndromes may be better explained by nonspecific disturbances of very basic processes of cellular homeostasis such as changes in the speed of cell divisions or levels of cellular adhesivity.[264]

New Mendelian syndromes involving cardiovascular defects are continuously described.[11,48,125,200,251,451] At least one new syndrome (HOMAGE syndrome) was delineated in this study,[265] however, the subset of all new syndromes in the total pool of cardiac patients is very small.

For the large group of non-Mendelian syndromes and associations with cardiovascular defects such as VACTERL association, Ivemark's syndrome, Golden-

Table 13.7
Ratio of Multiple Associated Malformations
to Single Associated Malformations
Baltimore-Washington Infant Study (1981–1989)

Pure Diagnostic Group	Total Subjects	Cases with Syndromes and Multiple Defects	Cases with Single Associated Defects	Ratio (Multiple/Single)
Laterality and looping anomalies	131	14	9	1.56
Outflow tract anomalies:				
Transposed great arteries	239	11	11	1.00
Normal great arteries	400	53	39	1.36
Atrioventricular septal defect[1]	88	17	5	3.40
Membranous VSD	895	35	55	0.64
Muscular VSD	429	7	13	0.54
Left-sided obstructive defects	429	15	16	0.94
Pulmonic stenosis	341	16	8	2.00
Atrial septal defect	291	18	18	1.00
Total	3243	186	174	1.07
Controls	3572	11	47	0.23

[1]Nonchromosomal cases only; VSD = ventricular septal defect.

har syndrome, and Cornelia De Lange syndrome the etiology remains unknown. Environmental and nontraditional genetic (uniparental disomy, mitochondrial mutations, duplication of the gene, etc.) factors may play a role in the origin of the cardiovascular defect in these entities.

Small Size at Birth: Premature and Growth Retarded

Small size at birth due to either prematurity and intrauterine growth retardation, represents a major cause of infant mortality and morbidity in the United States. Comprehensive population studies have identified many maternal and environmental factors that can contribute to low birthweight.[57,228]

For infants with congenital heart disease, small size at birth constitutes a major handicap. The dramatic effect of low birthweight on the all-cause mortality of infants with congenital heart disease was previously demonstrated in the Baltimore-Washington Infant Study data[342] and is shown again in Figure 13.6. There is a remarkably consistent inverse relationship of birthweight to all-cause mortality for the total case group—irrespective of the type of cardiovascular malformation,

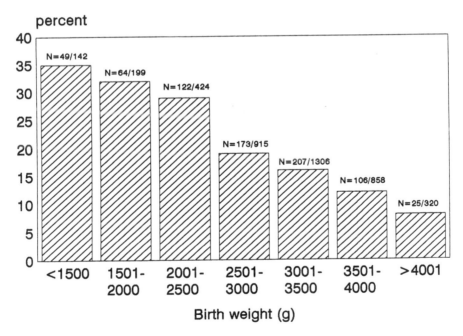

Baltimore-Washington Infant Study (1981-1989)
One Year Deaths from All Causes by Birth weight

Figure 13.6: All-cause mortality of infants with cardiovascular malformations (all cases combined) is inversely related to birthweight.
Reproduced from Ferencz C, Rubin JD, Loffredo CA, Magee CA (eds:) (1993) *Epidemiology of Congenital Heart Disease: The Baltimore-Washington Infant Study 1981–1989*. Mt. Kisco, New York: Futura Publishing Company, Inc.; with permission from the publisher.

associated noncardiac anomalies, or other factors. Of the infants who weighed less than 1500 g, about one third died, and there was a continuing decrease in mortality as birthweight increased. Mortality was lowest among infants whose birthweight exceeded 4000 g (8%).

Major congenital anomalies including congenital heart disease have been shown to be associated with a birthweight deficit. Both the malformation and the low birthweight may be independent results of the same primary defect. Rosenthal[372] used the Baltimore-Washington Infant Study data with added anthropometric measurements to suggest that in intrauterine life hemodynamic alterations due to the cardiac malformation may affect size and growth patterns.

With regard to diagnoses-specific differences, we previously reported that among infants with outflow tract defects,[144] there was a striking difference between cases with and without transposition of the great arteries. The former case group approached control values for birthweight while in the latter group there was diminished birthweight with evidence of both prematurity and intrauterine growth retardation, and with a striking effect of associated noncardiac anomalies. Notable was the contrast between the two small subgroups of double-outlet right ventricle, in which those with transposition had normal growth and those with normally related great arteries had the greatest growth deficit.

In the data on the "pure" diagnostic subgroups of cardiovascular malformations presented in this volume, we described growth characteristics for each malformation and we were struck by the importance of prematurity in some diagnoses but not in others. Table 13.8 shows the diagnostic groups ordered by decreasing mean birthweight, the proportion of infants weighing less than 2500 g, the mean gestational age, and the proportion under 38 weeks of gestation. Table 13.9 shows case-control odds ratios for being small for gestational age in each specific analytic group.

In the control population about 6% of the infants were small for gestational age according to the population standard of Brenner[61] and 90% of the control infants were born at term. For the infants with congenital heart disease, the outflow tract defect subgroups with normally related great arteries had the highest odds ratios for intrauterine growth retardation, closely followed by the atrioventricular septal defect groups and looping abnormalities as well as hypoplastic left heart syndrome. In these groups, with the exception of the atrioventricular septal defect-Down syndrome group, growth retardation affected both full-term and premature infants.

An evaluation of the isolated congenital heart disease versus the multiple malformation groups—with regard to birthweight, gestational age, the effects of infants' sex and race, and maternal smoking during pregnancy—is illustrated in graphic form in Figure 13.7. For each diagnostic group, two columns, one representing infants with isolated congenital heart disease and one representing those with noncardiac anomalies, show the crude mean birthweight followed by the mean birthweight after adjustment for gestational age, and the final value adjusted also for gender, race, and maternal smoking. There was a striking effect of associated noncardiac anomalies in every diagnostic group, including the small number of cases with transposition of the great arteries. Adjustment for gestational age in every diagnostic subgroup raised the value of the adjusted mean birthweight, indicating a major effect of prematurity. This phenomenon was most striking in the subgroups with noncardiac anomalies. The additional effects of the other adjust-

Table 13.8

Birth Weight and Gestational Age: Summary Table†

Baltimore-Washington Infant Study (1981–1989)

Pure Diagnostic Group	Total Subjects*	Birth Weight Mean (g)	Birth Weight % <2500g	Gestational Age Mean (wks)	Gestational Age % <38 wks
Controls	3572	3351	7.1	39.6	9.5
Transposition, Intact V. Septum	106	3428	3.8	39.8	8.5
Transposition, V. Septal Defect	60	3368	6.7	39.8	8.3
Muscular VSD	163	3328	13.5	39.4	12.3
Bicuspid Aortic Valve	60	3312	8.5	39.2	15.0
Aortic Valve Stenosis	65	3296	12.3	38.4	15.4
Transposition, DORV	27	3292	7.4	39.6	11.1
Cardiomyopathy	77	3287	14.5	38.3	28.6
Patent Arterial Duct	62	3220	9.7	(includes only full-term infants)	
Pulmonary Atresia, Intact V. Septum	45	3177	13.3	39.6	8.9
Coarctation of the Aorta	120	3152	14.2	39.3	15.1
Membranous VSD	640	3083	18.2	39.0	16.0
Ebstein's Anomaly	44	3075	25.0	38.2	22.7
Hypoplastic Left Heart	138	3070	14.5	39.1	17.4
Truncus Arteriosus	38	3054	21.1	38.7	26.3
TAPVR	56	3021	17.9	39.0	16.1
Laterality/Looping Defects	112	2999	20.5	38.8	25.0
AVSD with Down Syndrome	190	2998	13.7	38.7	19.0
Pulmonary Valve Stenosis	233	2991	22.3	38.4	24.0
Nonsyndromic AVSD	76	2977	23.7	38.9	14.5
Tricuspid Atresia	30	2941	30.0	38.7	26.7
Atrial Septal Defect	252	2910	23.5	38.1	27.0
Tetralogy of Fallot	204	2904	25.5	38.3	21.1
Normal Great Arteries, DORV	27	2706	44.4	38.3	18.5

*Interviewed families; + ordered by decreasing mean birthweight; AVSD = atrioventricular septal defect; DORV = double-outlet right ventricle; TAPVR = total anomalous pulmonary venous return; V = ventricular; VSD = ventricular septal defect.

ment factors (gender, race, and maternal smoking) varied but was most evident for the group of isolated outflow tract defects with normal great arteries, isolated membranous ventricular septal defects, isolated atrial septal defects, and in general in both left-and right-sided defects. In fact, among cases with associated noncardiac anomalies the role of these additional factors was relatively minor compared to the impact of prematurity, except for atrioventricular septal defects.

These graphs indicate that small size at birth for infants with cardiac malformations was associated with the presence of noncardiac anomalies and that these infants were likely to be born prematurely. It is notable also that only one group, the isolated atrial septal defect group, had adjusted mean birthweight standard error bars that overlapped those of the control population, indicating no case-control difference. Other diagnoses that came close to the control values were isolated membranous ventricular septal defect and isolated left-sided and right-sided obstructive lesions.

Table 13.9

Case-Control Comparison of Infants Classified as Small-for-Gestational-Age and Large-for-Gestational-Age Baltimore-Washington Infant Study (1981–1989)

Diagnostic Group	Total Interviews	Small-for-Gestational-Age			Large-for-Gestational Age		
		N	OR	(95% CI)	N	OR	(95% CI)
Laterality and Looping Defects	112	20	**3.5**	**(2.1–5.7)**	12	0.6	(0.3–1.2)
Outflow Tract Anomalies:							
Transposition of the Great Arteries:							
Intact ventricular septum	106	5	0.8	(0.3–2.0)	18	0.9	(0.5–1.5)
Ventricular septal defect	60	6	1.8	(0.8–4.2)	8	0.7	(0.3–1.6)
Double-outlet right ventricle	27	2	1.3	(0.3–5.4)	5	1.0	(0.4–2.8)
Normally-Related Great Arteries:							
Tetralogy of Fallot	204	49	**5.0**	**(3.5–7.1)**	27	0.9	(0.6–1.3)
Truncus arteriosus	38	7	**3.6**	**(1.6–8.3)**	4	0.6	(0.2–1.8)
Double outlet right ventricle	27	15	**19.9**	**(9.2–43.1)**	2	0.8	(0.2–3.8)
Atrioventricular Septal Defects:							
Down syndrome	190	32	**3.2**	**(2.2–4.8)**	12	**0.3**	**(0.2–0.6)**
Nonsyndromic	76	15	**3.9**	**(2.2–7.0)**	7	0.5	(0.2–1.2)
Left-Sided Obstructive Defects:							
Hypoplastic left heart	138	30	**4.4**	**(2.9–6.8)**	17	0.8	(0.5–1.3)
Coarctation of the aorta	120	14	**2.1**	**(1.2–3.7)**	18	0.9	(0.5–1.4)
Aortic valve stenosis	65	5	1.3	(0.5–3.3)	14	1.3	(0.6–2.3)
Biscuspid aortic valve	60	4	1.2	(0.8–2.1)	14	1.4	(0.7–3.9)
Right-Sided Obstructive Defects:							
Pulmonary valve stenosis	233	31	**2.4**	**(1.6–3.6)**	24	**0.6**	**(0.4–0.9)**
Pulmonary atresia, intact septum	45	7	**2.9**	**(1.3–6.6)**	6	0.8	(0.3–1.9)
Tricuspid valve atresia	30	8	**4.8**	**(1.7–10.3)**	2	0.4	(0.1–1.7)
Septal Defects:							
Ventricular septal defect, membranous	640	112	**3.4**	**(2.7–4.3)**	84	0.8	(0.6–1.0)
Ventricular septal defect, muscular	163	17	**1.9**	**(1.1–3.1)**	34	**0.6**	**(0.4–0.8)**
Atrial septal defect	252	44	**3.4**	**(2.4–4.8)**	**36**	0.9	(0.6–1.3)
Controls	3572	211	1.0	(reference)	649	1.0	(reference)

Bold type indicates statistically significant case-control odds ratio; OR = odds ratio; CI = confidence interval.

Implications of the Malformation Triad

The Baltimore-Washington Infant Study data illustrate the close interrelationship of cardiac and noncardiac anomalies and growth abnormalities. The healthiest infants in terms of growth are those with isolated transposition, isolated membranous ventricular septal defect, isolated atrial septal defect, and the isolated forms of left-sided and right-sided obstructive defects. The most severely compromised infants are those with multiple noncardiac malformations, including those with outflow tract defects with normally related great arteries, and those with atrioventricular septal defects alone and with noncardiac anomalies. These

CASE-CONTROL COMPARISONS OF MEAN BIRTH WEIGHTS

EFFECTS OF CARDIAC DIAGNOSIS, ASSOCIATED NON-CARDIAC ANOMALIES,
GESTATIONAL AGE, GENDER, RACE AND MATERNAL SMOKING

Baltimore-Washington Infant Study (1981-1989)

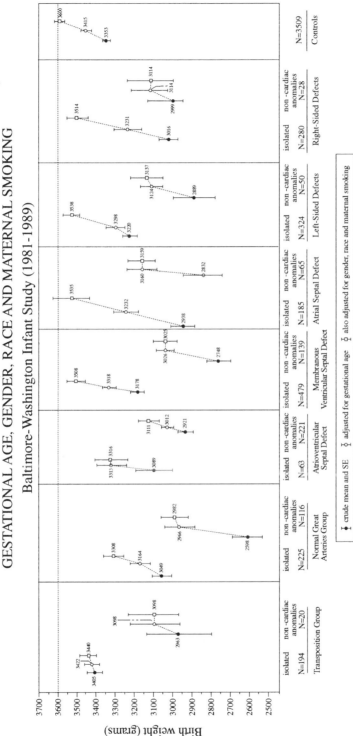

Figure 13.7: A comparison was made of the mean birthweights of infants in the major diagnostic subgroups for isolated cases and for those with noncardiac anomalies. Multiple linear regression adjusted the birthweight for gestational age, and then for infant gender and race and for maternal smoking during pregnancy, as well as for interactions among these variables. The reference line, indicating the adjusted mean birthweight of controls, is closely approached by four of the isolated subgroups but not by subgroups with noncardiac anomalies, nor by defects of primary cardiogenesis.

are the infants with severe intracardiac defects indicative of errors in primary morphogenesis.

Outstanding in all these comparisons are the infants with atrioventricular septal defects, a cardiac malformation that occurs as an isolated anomaly in only 23% of the cases. A strong association with trisomies 21 and 18 distinguishes these infants, as does a striking rarity of this anomaly in the five syndromes that commonly occur in association with other cardiac defects (Table 13.6). Furthermore, patients with atrioventricular septal defects also have the highest ratio of multiple to single noncardiac anomalies and also have marked growth retardation due in part to prematurity.

Atrioventricular abnormalities in our study include those with partial (ostium primum and atrioventricular type ventricular septal defects) and those with complete "atrioventricular canal" defects. This group was subdivided into those with Down syndrome and those without chromosome abnormalities. These divisions were guided by the number of cases available for analysis, a critical issue in advancing any interpretation of the findings. The group of isolated atrioventricular septal defects was small in this 9-year regional case population and invites collaborative case collections to determine clinical, anatomic, morphogenetic, and molecular characteristics. If these isolated cases, less than one quarter of the total, represent forme fruste of the larger groups then we should know what pathologic features distinguish either a "chromosome" or a "syndromic" origin. Some attempt at this distinction has already been made in regard to associated cardiac defects,[72,111,277] but we have no clear understanding of this important cardiac anomaly at this time.

In contrast to the tremendous attention given to the origin of outflow tract anomalies for almost 100 years stimulated by the intriguing "spiral" arrangement of the great arteries, the atrioventricular canal was simply an endocardial cushion defect and part of the septation process. Work over the past 2 decades has cast light on the special characteristics of the extracellular matrix and the transformation of endothelial cells into mesenchymal cells in the appropriate formation of the mitral and tricuspid valves.[278-281] These research directions, guided by increasing knowledge of the cascade of regulatory genes, have set an example for other targeted studies of cardiac development. It is our hope that the information derived from this epidemiologic study will further contribute to this understanding.

The evaluation of birthweight, gestational age, and size for gestational age as indicators of the fetal life course was the first measure that led us to distinguish outflow tract defects in those with and without transposition.[144] This striking distinction in fetal growth was also noted in the small groups of cases of double-outlet right ventricle and of tricuspid atresia each with and without transposition. A possible explanation for the former is found in the morphogenetic studies of Männer,[274] which assigned double-outlet right ventricle with transposition to outflow tract defects and those with normally related great arteries to a range of inflow defects, ie, defects of the atrioventricular junction. The small case group of tricuspid atresia with transposition (N = 18) (not included in Table 13.8) revealed a normal mean birthweight (3335 g) and only a two-fold but nonsignificant elevation of the odds ratio for being small for gestational age. This is in contrast to the babies with tricuspid atresia with normally related great arteries (mean birthweight 2941 g) who had an almost five-fold risk of intrauterine growth retardation. Other concepts presented by Männer et al[275] may provide further explanations, as the experimental evidence indicates a spectrum of anomalies of the atrioventricular junction at the site of the early reflexion of the cardiac tube to form first a C-shape and then the final S-shape of the looped heart. Looping defects therefore are not confined to laterality defects and "corrected transpositions" may be but a part of complex anomalies such as double-outlet right ventricle and tricuspid atresia. Clarification of our understanding of the critical influences that result in cardiovascular and noncardiac malformations will arise from the conjunction of new genetic, morphogenetic and epidemiologic information.

Chapter 14

Risk Factor Analysis: A Synthesis

From: Ferencz C, Loffredo CA, Correa-Villaseñor A, Wilson PD, eds: *Genetic & Environmental Risk Factors of Major Cardiovascular Malformations: The Baltimore-Washington Infant Study 1981–1989.* Armonk, NY: Futura Publishing Co., Inc; ©1997.

Introduction

Etiologic research must take many directions and encompass both concrete evidence and justifiable hypotheses that have already have been found to link disease with presumably causative factors. The Baltimore-Washington Infant Study aimed to develop hypotheses regarding a modest range of human malformations as a focus for future multidisciplinary etiologic studies. The systematic evaluation of congenital heart disease by anatomically defined "pure" diagnoses emphasized a "clean slate" on which to record the genetic fetal, maternal, and environmental factors that may constitute the "teratogenic dose" by which development has been disturbed.[152] This composite effect of many factors is captured in the epidemiologic approach, which evaluates characteristics of cases and controls, each within the context of the infant's family and the total "ecology" in which the family lives and works.[235] (Figure 14.1)

The initial case-control analyses of the Baltimore-Washington Infant Study data were performed on all cardiovascular malformations as a group and were summarized in a monograph.[152] Using the group of all cardiovascular malformations, we identified associations between cardiac defects and family history of cardiac defects, maternal diabetes, and a number of conditions and exposures that had not been previously reported. Examples of these conditions and exposures include previous reproductive problems, maternal use of certain medications, and exposures to teratogenic agents in personal lifestyle and in home and occupational activities. Some of these associations were restricted to cases with

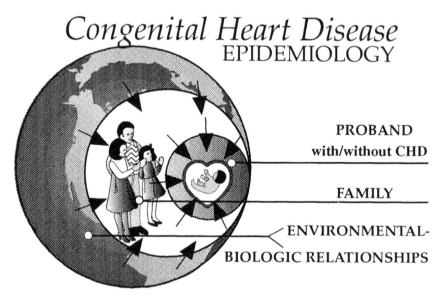

Congenital Heart Disease
EPIDEMIOLOGY

PROBAND
with/without CHD

FAMILY

ENVIRONMENTAL-
BIOLOGIC RELATIONSHIPS

Figure 14.1: A comprehensive epidemiologic evaluation of cases and controls views the proband within the context of the family and of the environment, which may result in biologic-environmental interactions in parents and/or probands.
Reproduced from Ferencz C and Correa-Villaseñor A. (1995) Overview: The epidemiologic approach to the study of congenital cardiovascular malformations. In Clark EB, Markwald RR, and Takao A. (eds:) *Developmental Mechanisms of Heart Disease,* with permission from Futura Publishing Company, Inc.

a family history of birth defects or associated anomalies in the proband. Other associations, however, were limited to subsets of cases defined as "free of genetic factors" by the absence of a family history of birth defects and other associated anomalies in the proband.

Those earlier observations provided support for the role of genetic and environmental factors in the origin of cardiac defects. However, a basic question regarding etiology remained: would the risk factors observed for all cardiac defects as a group be evident for the individual diagnostic groups? An affirmative answer to this question would suggest that any anatomic variation among the diagnostic groups could be a function of the effect of exposure to any of these risks during embryogenesis rather than a function of the *type* of teratogenic exposure or morphogenetic mechanism. A negative answer on the other hand, would suggest etiologic or morphogenetic heterogeneity. Such heterogeneity might be expected since fetal sensitivity to toxic agents could vary by hereditary variations in metabolism, relative affinity of an agent for cellular receptors, and critical stages of morphogenetic processes where an agent might exert adverse effects.

In those analyses with all diagnostic groups combined, it was impossible to detect any underlying heterogeneity of risk factors among the diagnostic groups. Furthermore, because of the lumping of all groups into one, if such heterogeneity was present, some important risk factors that are unique to individual groups could have gone undetected. Here we sought to determine if this happened.

Accordingly, we have conducted analyses within cardiac diagnostic groups that are homogenous on the basis of anatomy. This approach allowed us to 1) define and compare the risk profiles for individual diagnostic groups, 2) determine the extent to which various types of environmental factors were associated with various cardiac malformations, and 3) identify the cardiac diagnostic groups that account for the associations we had previously noted for all cardiovascular malformations as a group.

Table 14.1 is a summary of the risk factors by the cardiac diagnostic groups examined in this volume. The risk factor categories include: family history of cardiac defects, history of maternal reproductive disorders, maternal illnesses, use of medications, lifestyle factors, and occupational and environmental factors.

Family History of Cardiovascular Malformations

Family history of cardiovascular malformations was statistically associated with the cardiac malformation in the offspring for 11 cardiac diagnostic groups (Table 14.2). In a case-control study of birth defects and family history, there is always the concern that a statistical association may be found due to differences in surveillance between affected and unaffected families. If an association with family history is due primarily to detection bias, then one might expect to see more and stronger associations with family history for the more severe malformations. However, we did not find this to be the case; we observed associations of family history of cardiac defects with mild defects, as well as with severe defects in the offspring, and conversely we observed no association with family history even for some of the more severe defects. Furthermore, among the diagnostic groups associated with a family history of cardiac defects there was also evidence of a complex etiology, since family history was only one of several risk factors identified.

Table 14.1

Summary of Potential Risk Factors ($P<.01$) for Specific Cardiovascular Malformations from Logistic Regression Analyses The Baltimore-Washington Infant Study, (1981–1989)

Cardiovascular Malformation	Family History	Maternal Illnesses	Age/Reproductive History	Drugs	Lifestyle Factors	Home & Occupation	Paternal
Laterality and Looping Defects (N=104)	CVM, NCA	diabetes	—	antitussives	—	—	smoking
Total Anomalous Pulmonary Venous Return (N=56)	CVM	—	—	—	—	pesticides	—
Transposition of Great Arteries, intact ventricular septum (N=106)	—	influenza	—	benzodiazepines ibuprofen	—	solvents	—
Transposition of Great Arteries, with ventricular septal defect (N=60)	—	—	—	—	smoking	—	—
Tetralogy of Fallot (N=204)	—	—	miscarriages	clomiphene	—	paints	anesthesia paints
Atrioventricular Septal Defect with Down Syndrome (N=190)	—	—	age>34 years	ibuprofen	—	—	—
Atrioventricular Septal Defect, nonchromosomal (N=76)	CVM	diabetes	stillbirths	antitussives diuretics	—	varnishes	radiation varnishing
Membranous Ventricular Septal Defect (N=640)	CVM	diabetes	—	anesthesia ibuprofen metronidazole	cocaine	auto body repair	cocaine marijuana

Hypoplastic Left Heart (N=138)	CVM	—	—	—	—	solvents	—
Aortic Stenosis (N=55)	—	gestational diabetes	—	—	—	—	—
Coarctation of the Aorta (N=120)	CVM	—	—	clomiphene macrodantin	—	solvents	—
Bicuspid Aortic Valve (N=60)	CVM	—	age>34 years	ibuprofen	—	—	—
Pulmonic Valve Stenosis, severe (N=56) and moderate (N=112)	CVM, NCA	thyroid disease fever	—	—	smoking	hair dyes	—
Muscular Ventricular Septal Defect (N=153)	—	—	—	gastrointestinal medications acetaminophen	alcohol	—	—
Atrial Sepal Defect, isolated (N=187)	CVM	gestational diabetes	preterm births	corticosteroids	—	—	cocaine cold temperature
Patent Arterial Duct (N=62)	—	gestational diabetes	age>30 years	—	—	—	—
Ebstein's Anomaly (N=44)	CVM	—	miscarriages	benzodiazepines	marijuana	varnishes	—
Cardiomyopathy (N=77)	—	diabetes	—	—	—	—	auto body repair

CVM = cardiovascular malformations; NCA = noncardiac anomalies.

Table 14.2
Family History of Birth Defects: Summary of Significant Odds Ratios
Baltimore-Washington Infant Study (1981–1989)

Group	N	Affected Families	Adjusted Odds Ratio	95% CI	99% CI
Congenital Heart Defects Among Family Members					
Laterality and looping defects	104	5	4.5	1.7–11.8	1.5–13.5
Atrioventricular septal defect, nonchromosomal	76	5	7.0	2.6–18.9	1.9–25.8
Membranous ventricular septal defect	640	22	3.1	1.9–5.3	1.6–6.2
Hypoplastic left heart syndrome	138	7*	4.8	2.1–10.8	1.6–14.0
Coarctation of the aorta	120	8	4.6	2.0–10.8	1.5–13.9
Aortic valve stenosis	65	3	3.7	1.1–12.4	0.8–14.1
Bicuspid aortic valve	60	5	7.2	2.7–19.1	2.0–25.8
Pulmonic valve stenosis (severe)	56	5	6.4	2.3–18.0	1.6–24.8
Pulmonic valve stenosis (moderate)	112	6	5.2	2.1–12.8	1.6–16.8
Atrial septal defect (isolated)	187	8	3.9	1.8–8.4	1.4–10.7
Ebstein's anomaly	44	3	6.4	1.8–22.2	1.2–32.9
Controls	3572	43	1.0	reference	
Noncardiac Anomalies Among Family Members					
Laterality and looping defects	104	11	2.5	1.3–4.8	1.0–5.9
Pulmonic valve stenosis (severe):	56				
nonwhites		4	8.7	2.5–29.6	1.7–43.4
whites		3	1.3	0.4–4.3	0.3–6.3
Controls	3572	165	1.0	reference	

*Six of these seven infants were born during 1981–1985.

Familial Aggregation of Cardiac and Noncardiac Abnormalities

Associated cardiac and noncardiac disorders in first-degree relatives are de-scribed and discussed for each specific cardiac diagnostic group in Chapters 4–12. We have previously published details of our finding for the individual diagnostic case groups,[56] the most important pedigrees,[143] and a summation of information from this and from other studies.[146] As the family history information was obtained by questionnaire, under-reporting was probable, especially for second-degree rel-atives. However, reporting the presence of a malformation in the nuclear family of the proband was probably quite accurate, although some doubt may remain about the nature of the anomaly unless confirmed by medical records. We had an un-usual opportunity to estimate diagnostic concordance between siblings by the in-dependent ascertainment of a second infant with congenital heart disease into the study. This occurred in 20 instances over the 9 study years. Comparisons of the di-agnoses of the two siblings are shown on Table 14.3, and reveal a high degree of concordance of cardiac and noncardiac defects.

Genes that are known to be associated with cardiovascular abnormalities ap-peared among the subjects and families of this study population. There was a sin-gle family that showed the transmission of an atrial septal defect across three gen-erations associated in the older members with heart block. This case (#3159) has

Table 14.3

Diagnostic Concordance in Doubly-Ascertained Sibships Baltimore-Washington Infant Study (1981–1989)

First Ascertained Sibling				Second Ascertained Sibling			
Year of Birth	Noncardiac Anomalies	BWIS ID	Cardiac Diagnosis	Cardiac Diagnosis	BWIS ID	Noncardiac Anomalies	Year of Birth
Concordant Heart Defects:							
1984	ambiguous genitalia	2230	tetralogy of Fallot	tetralogy of Fallot, dextrocardia	2780	ambiguous genitalia	1985**
1981	—	1047	atrioventricular septal defect	atrioventricular septal defect	3274	Down	1986
1984	Ellis Van Creveld	2548	atrioventricular septal defect	atrioventricular septal defect	11553	Ellis Van Creveld	1989
1988	oro-facial-digital	10535	common atrium	common atrium	11762	oro-facial-digital	1989+
1983	—	1893	membranous VSD	membranous VSD	2957	—	1985
1981	—	1033	hypoplastic left heart	hypoplastic left heart	1889	—	1981
1982	—	1715	pulmonic valve stenosis	pulmonic valve stenosis, VSD	2350	cataracts	1984
1981	—	1291	muscular VSD	membranous VSD	2159	—	1983
Partially Concordant Heart Defects							
1987	—	3674	TGA with VSD	membranous VSD	10989	—	1988
1985	—	2686	tetralogy of Fallot	membranous VSD	3261	—	1986
1986	—	3198	common arterial trunk	tetralogy of Fallot	10816	—	1988
1987	—	3673	hypoplastic left heart	coarctation of aorta	11085	—	1989
1984	—	2458	hypoplastic left heart	aortic valve stenosis	10145	—	1987
1981	hydronephrosis	1353	pulmonary atresia with intact ventricular septum	pulmonic valve stenosis	2199	—	1984
1982	Ritcher-Schinzel	1777	VSD and atrial septal defect	atrioventricular septal defect	2860	Down	1985
1986	—	3199	muscular VSD	atrioventricular septal defect	10672	—	1988
Discordant Heart Defects:							
1983	—	1829	tetralogy of Fallot	patent arterial duct	2820	—	1984
1984	—	2197	tetralogy of Fallot	hypoplastic left heart	10912	—	1988
1983	—	1906	common atrial trunk	muscular VSD	2853	—	1985
1987	Williams	10031	supravalvar pulmonic stenosis	TGA, intact ventricular septum	10736	—	1988

**Homage Syndrome, +oro-facial-digital syndrome, type I (X-linked dominant); TGA=transposed great arteries; VSD = ventricular septal defect.

previously been published.[143] This association may represent a special type of atrial septal defect. Similar examples are found also in the literature.[269,319,458]

A new syndrome (HOMAGE) was described in Chapter 5 for two sisters with tetralogy of Fallot with laterality/looping defect in the index case.[265] There was a single family in which we can suspect the presence of a chromosome 22q11 deletion, although this diagnosis was *never* made in any of our patients. Case #10207 had DiGeorge syndrome and pulmonary atresia and her sister had tetralogy of Fallot. In families with the Noonan syndrome there were two (#11510 and #2316) in which both the patient and a family member were so diagnosed. In two other families (#1959 and #1996) the patient had Noonan syndrome; the first with pulmonic stenosis and the second with a membranous ventricular septal defect. In both instances, the mother had pulmonary stenosis and was not previously suspected of having the Noonan syndrome (see Chapter 9). These families with known "cardiogenic" genes (or deletions) constitute only a small proportion of the total interviewed families.

Genetic Heterogeneity and Genetic Liability

Beginning with the classic studies of Polani and Campbell[350] and Lamy,[239] many cardiovascular defects have been described in families with a vertical (parent-child) transmission and in families with affected sibs of nonaffected parents.* The occurrence of a cardiac defect in more than one child of healthy parents does not necessarily indicate the presence of recessive inheritance; multifactorial inheritance or gonadal mosaicism in one of the parents, or even common environmental exposures may also explain such an occurrence.

Morphologically, the same malformation may result from mutations in different genes. Among cardiovascular malformations, the gene for total anomalous pulmonary venous return was mapped to the proximal part of chromosome 4q[46] although there are some familial cases for which no linkage to this region had been shown, indicating that at least one more gene has to be involved in the origin of this defect. This shows that morphogenesis of any structure depends on many different genes. In other words, genetic heterogeneity is not an exception, but a common rule. Moreover, variable forms of an anomaly should not be regarded as separate nosologic entities, as they may be different manifestations of the effect of a teratogen or gene. For instance, hypoplastic left heart, coarctation of aorta, aortic valve stenosis, and bicuspid aortic valve may be manifestations of one teratologic series. In the Baltimore-Washington Infant Study most affected relatives of the probands with hypoplastic left heart had one or another of these abnormalities (see Chapter 8). Such families were also reported in the literature.[178,448]

Genetic liability has both specific and nonspecific components. It is apparent that the first-degree relatives of index cases with cleft lip and palate will have different forms of facial clefts more frequently than will first-degree relatives of healthy persons. But the frequency of most other types of multifactorial defects in these relatives is also increased,[205] suggesting a nonspecific component of genetic liability. Therefore, one should expect an increased frequency of different multifactorial defects in the relatives of patients with cardiovascular defects.

In the Baltimore-Washington Infant Study this phenomenon explains the oc-

---

References 35, 47, 86, 113, 114, 124, 131, 252, 354, 388, 399, 400, 432, 448.

currence of different multifactorial defects in such families as: case #1447 (L-transposition in the proband, coloboma iris in the patient's sister, cleft lip and palate in maternal uncle); case #10184 (heterotaxy in the proband, cleft lip and palate in the mother, and pyloric stenosis in a maternal uncle) (Chapter 4); and case #3767 (tetralogy of Fallot in the index case, cleft palate in the mother, and polydactyly in the maternal aunt) (Chapter 5).

Among the cases of trisomy 21 with atrioventricular septal defect, three unusual families have previously been described.[143,152] In these families, atrioventricular septal defect without Down syndrome was found in two mothers (families #1458 and #1994) and in one half-sister (#3274) (see Chapter 6). The origin of trisomy 21 may not depend on the maternal cardiac defect, but the atrioventricular septal defect is likely to occur in a liable person.

An unexpectedly large number of relatives of the index cases had genetically determined blood disorders. In most of these cases the probands were either heterozygous carriers of the mutant gene (if one of the parents is affected) or heterozygous (a high probability if sib is affected). It seems possible that microanomalies of the erythrocytes or thrombocytes might alter blood viscosity and flow and might be involved in the origin of some defects.[138,154]

Environmental Risk Factors

In our previous monograph[152], many risk factors were detected in the analysis of all cardiac defects as a group, the stronger of these associations (defined as

Table 14.4
Summary of Significant Odds Ratios (.01 significance level) From Our Previously Published
Multivariate Analyses of All Cardiac Malformations Combined*

	All cases (3377)		Cases without Genetic Risk Factors (2203)	Cases with known Genetic Risk Factors (1171)
	Model without Maternal Reproductive Variables	Model with Maternal Reproductive Variables		
Family history of cardiac malformations	2.20	2.18	**	**
Diabetes	2.97	3.11	2.83	3.06
Maternal Age 30+ years	1.28	—	—	1.83
3+ previous pregnancies	—	1.29	—	1.65
Previous induced abortions	—	—	1.18	—
Diazepam	2.14	2.31	—	—
Metronidazole	—	—	—	3.89
Ibuprofen	—	—	—	1.73
Paternal cocaine	1.62	1.62	—	1.65
Hair dyes	—	—	1.34	—
Solvents (misc)	—	—	1.64	—
Auto body repair	—	—	—	3.40

*Ferencz C, Rubin JD, Loffredo CA, Magee CA eds: *Epidemiology of Congenital Heart Disease: The Baltimore-Washington Infant Study.* New York: Futura Publishing Co., Inc.; 1993.
**Not analyzed.

statistically significant at the 1% level) are summarized in Table 14.4. In the following sections of this chapter we will compare the results in Table 14.4 with those described in this book for specific cardiac diagnostic groups.

Maternal Diabetes

In the analyses of all cardiovascular malformations as a group, overt diabetes mellitus of the mother appeared to be the strongest association, with an approximately three-fold risk in each of the multivariate models that took into account the presence and absence of genetic factors. In that analysis, no other reported illness showed such an association.

Among the diagnostic subgroups evaluated in this report, diabetes (Table 14.5) was strongly associated with the following defects: laterality/looping defects (odds ratio 8.3), the transposition group (odds ratio 3.8), the group of outflow tract defects with normal great arteries (odds ratio 5.4), nonchromosomal atrioventricular septal defects (odds ratio 10.6), membranous ventricular septal defect (odds ratio 2.9), hypoplastic left heart syndrome (odds ratio 3.9), and cardiomyopathy (odds ratio 11.5). Within the outflow tract group the only phenotype with a sufficient sample size for adjusted analysis was tetralogy of Fallot, which showed only a weak association (odds ratio 1.8). In unadjusted analyses of the smaller diagnostic subgroups, complex transposition, common arterial trunk and double-outlet right ventricle had the highest odds ratios (odds ratio 16.2, 13.2, and 9.9 respectively). With the exception of the often benign cardiopathy of infants of mothers with diabetes, the findings indicate an adverse effect of diabetes on primary cardiogenesis. The above associations between cardiac defects and overt diabetes were not evident for gestational diabetes.

In the previous monograph,[152] we conducted analyses to examine the possibility of modification of the associations between diabetes and cardiac anomalies by a marker of genetic burden, maternal age, race, and socioeconomic status, and we found evidence of such effect modification only by our marker for genetic burden. Similar results were found in new analyses. Table 14.6. For outflow tract anomalies with normally related great arteries (see Chapter 5), the association with maternal diabetes was stronger in the presence of noncardiac defects in the proband/and or anomalies in first degree relatives (multiple anomalies/multiplex subset), and was further strengthened once subjects with chromosomal anomalies were excluded from analysis. For membranous ventricular septal defect (see Chapter 7), we observed a similar modification of the association with diabetes in subjects in the multiple anomalies/multiplex subset (Table 14.6). These findings are concordant with the known multiple teratogenic effects of maternal diabetes and indicate that the risk of such effects may be increased in the presence of familial birth defects.

Diabetes constitutes a serious hazard to the health of the mother and her offspring. Insulin therapy made possible the successful completion of a pregnancy, but only the recent team approach to high-risk pregnancies has brought about a safe course for mother and infant. The infant of a mother with diabetes can suffer from a variety of metabolic disturbances as well as from organic disease, including cardiomyopathy with ventricular hypertrophy, and is at an increased risk for congenital malformations of the cardiovascular, central nervous, and urinary systems.[348] The cardiac and noncardiac anomalies observed in infants with cardiovascular malformations born of mothers with diabetes in the Baltimore-Washington Infant Study's first 6 years of enrollment were detailed in a previous publication.[153]

Table 14.5
Summary Table: Maternal Diabetes
Baltimore-Washington Infant Study (1981–1989)

Group	N	Number of Diabetics[1]	Crude Odds Ratio (95%CI)	Adjusted Odds Ratio (95%CI)
Laterality/Looping Defects	104	5	**7.8 (2.9–20.9)**	**8.3 (3.0–23.0)**
Outflow Tract Anomalies:				
(A) Transposition group	214	5	**3.7 (1.4–9.8)**	**3.8** (1.4–10.2)
intact ventricular septum	106	1	1.5 (0.2–11.0)	—[2]
ventricular septal defect	60	0	—	—
double-outlet right ventricle	27	2	**12.3 (2.8–55.2)**	—[2]
complex types (tricuspid/ pulm. atresia)	21	2	**16.2 (3.6–73.8)**	—[2]
(B) Normal great arteries group	341	12	**5.6 (2.8–11.4)**	**5.4 (2.5–10.8)**
tetralogy of Fallot (TOF)	204	4	**3.1 (1.1–9.0)**	**1.8 (1.0–3.2)**
TOF with pulmonary atresia	50	3	**9.9 (2.9–33.9)**	—[2]
double outlet right ventricle	27	1	5.9 (0.8–45.6)	—[2]
common arterial trunk	38	3	**13.2 (3.8–46.1)**	—[2]
supracristal VSD	17	0	—	—
Atrioventricular Septal Defects:				
with Down syndrome	190	1	0.8 (0.1–6.1)	—[2]
nonchromosomal	76	5	**10.9 (4.0–29.4)**	**10.6 (3.7–30.6)**
Ventricular Septal Defects:				
membranous	640	11	**2.7 (1.3–5.6)**	**2.9 (1.4–6.1)**
muscular	163	2	1.9 (0.4–8.2)	—[2]
Left-sided Obstructive Defects:				
hypoplastic left heart syndrome	138	3	**3.4 (1.0–11.5)**	**3.9 (1.2–13.2)**
coarctation of the aorta	120	1	1.3 (0.2–9.7)	—[2]
aortic valve stenosis	65	1	2.4 (0.3–18.1)	—[2]
bicuspid aortic valve	60	0	—	—
Right-sided Obstructive Defects:				
pulmonic valve stenosis	234	2	1.3 (0.3–5.7)	—[2]
pulmonary atresia	45	2	**7.2 (1.6–31.4)**	—[2]
tricuspid atresia	30	0	—	—
Ebstein's anomaly	42	0	—	—
Other Heart Defects:				
atrial septal defect	187	3	2.5 (0.7–8.5)	—[2]
patent arterial duct	62	0	—	—
TAPVR	56	0	—	—
cardiomyopathy	77	6	**13.0 (5.2–33.0)**	**11.5 (4.4–29.8)**
Controls	3472	23	**1.0 (reference)**	**1.0 (reference)**

[1]Overt diabetes; [2]too few exposed subjects for multivariate analysis; TAPVR = total anomalous pulmonary venous return.

Table 14.6

Diabetes Association in Simplex and Multiplex Families

Baltimore-Washington Infant Study (1981–1989)

			adjusted odds ratio (99% CI) for diabetes		
				Multiple Anomalies/Multiplex Subset	
Diagnostic Group	All Cases	Isolated/ Simplex Subset	All	Excluding Chromosome Anomalies	Cases with Chromosome Anomalies
Outflow Tract Anomalies with Normally Related Great Arteries	N=341	N=209	N=132	N=94	N=38
	5.4	**5.0**	**6.3**	**8.9**	**(0 cases)***
	(2.0–13.6)	(1.5—15.8)	(1.7–23.8)	(2.3–33.8)	
Membranous Ventricular Septal Defect	N=640	N=459	N=181	N=118	N=63
	2.9	**1.9**	**3.9**	**6.6**	**(0 cases)***
	(1.1–7.6)	(0.8–4.9)	(1.1–14.6)	(1.8–25.0)	

*No mothers reported overt diabetes.

The recent report from the Joslin Diabetes Service at the Brigham and Women's Hospital Boston,[188] which systematically and uniformly evaluated 860 fetuses/infants of mothers with diabetes compared to 96,178 deliveries to women without diabetes during the same period in the same institution, revealed that the relative risk for bilateral renal agenesis was 111 (95% confidence interval 32 to 385), 11.5 for neural tube defects (95% confidence interval 6.0 to 22.0), and 4.5 for defects of the heart and vessels (95% confidence interval 2.7 to 7.5). These findings establish the serious impact of diabetic embryopathy on the early development of organ systems. Despite the results of these studies, the mechanism of abnormal morphogenesis in infants of mothers with diabetes remains unknown.

Gestational diabetes occurs due to the various hormone changes in pregnancy, resulting in a relative insufficiency of insulin production. In our evaluation of all cardiovascular malformations as a group, we did not find gestational diabetes to be a risk factor. However, in our diagnosis-specific evaluations, we did find gestational diabetes as a significant risk factor for some defects of secondary cardiogenesis, in which the fundamental architecture of the heart is normal—atrial septal defect, aortic valve stenosis, and patent arterial duct. It would be unwise to draw a rigid distinction between the teratogenic effect of prepregnancy and pregnancy-related diabetes, since the gestational hormonal events have not yet occurred in the early phases of embryogenesis when major cardiac malformations are determined.

Maternal Influenza and Fever

In some of the cardiac subgroups, we found associations with influenza and with fever (Table 14.7). In general, the associations of cardiac defects with such illnesses were based on small numbers of affected case mothers. Influenza alone was associated with transposition of the great arteries with intact ventricular septum

Table 14.7
Associations Involving
Maternal Illnesses (not including diabetes)
Baltimore-Washington Infant Study (1981–1989)

Illness	Diagnostic Group	exposed Cases	N=3572 Controls	Adjusted Odds Ratio	95% CI	99% CI
Influenza	Transposition with Intact Ventricular Septum (N=106)	14	210	**2.2**	1.2–4.1	1.0–4.9
	Tricuspid Valve Atresia (N=30)	8	278	**4.3**(univariate)	1.9–9.8	1.5–12.6
Fever	Pulmonic Stenosis, moderate (N=112)	12	166	**2.9**	1.5–5.4	1.3–6.6
	Tricuspid Valve Atresia (N=30)	6	166	**5.1**(univariate)	2.1–12.7	1.6–16.9

(odds ratio 2.2)(see Chapter 5). In the presence of both influenza and ibuprofen, the risk of transposition of the great arteries with intact ventricular septum, was further increased (odds ratio was 5.5).

Reports of maternal influenza alone were also associated in univariate analyses with the right-sided obstructive lesion, tricuspid atresia. Small sample size did not permit multivariate analysis of this case group. We found that maternal fever was also associated with right-sided obstructive lesions, moderate pulmonic stenosis and tricuspid valve atresia.

These associations of cardiac defects with influenza and fever were based on self reports and were rare. Thus, it was not possible to disentangle the individual effects of the underlying infections and fever. However, the similarity of the cardiac diagnostic groups (ie, right-sided lesions) suggests the possibility of infectious processes with an affinity for the right side of the heart. However evidence of an inflammatory process in the genesis of cardiac malformations has not been established. Associations between birth defects and nonspecific acute respiratory illnesses have been noted in several ecologic studies.*

Maternal Reproductive History

A history of previous abnormal reproductive outcomes ranging from spontaneous abortions to stillbirths and preterm births was only weakly associated with the total group of cardiac defects (see Table 14.4), but was associated with four subgroups of cardiac defects: tetralogy of Fallot, nonchromosomal atrioventricular septal defect, isolated atrial septal defect, and Ebstein's anomaly. Of these, only two associations remained significant at the 0.01 level. (see Table 14.1).

Of interest is the fact that two of the cardiac diagnostic groups associated with previous reproductive problems in our study also exhibited associations

---

References 87, 193, 236, 243, 268, 452, 459.

with maternal overt diabetes mellitus. This concordance between abnormal reproductive outcomes and overt diabetes for the same cardiac diagnostic groups is consistent with the hypothesis that both are outcomes of a common factor such as the metabolic disturbances due to maternal diabetes.

Maternal Use of Prescription and Nonprescription Pharmaceuticals

Great attention was given to the design for recording the maternal drug history, as in recent years suspicion of teratogenicity has been cast upon various therapeutic agents such as bendectin, hormones, and anticonvulsants. Our own study on maternal hormone therapy and congenital heart disease[147] illustrates the difficulties in the lack of specificity of defining etiologic agents and the possible confounding effects of behavioral and sociodemographic aspects related to drug-taking.

In this study the questionnaire was designed to maximize recall of drugs taken, and we coded drug categories as well as the individual drugs within them. Ongoing surveillance over the 9 years of the study identified changes in the formulation of many products and used the study infant's date of birth to assign corrected drug codes for analysis.[380]

In a more detailed study of this question, Rubin et al[379] found that half of the mothers in the Baltimore-Washington Infant Study took at least one medication during the first trimester. The results of multivariate analysis in that report showed associations with multiple factors. In the present report we take these into account by adjusting medication-associated risks for illnesses, socioeconomic status, and the other potential confounders.

In the analysis of all cardiac defects as a group[152] (see Table 14.4), only three medications were identified as risk factors: diazepam, metronidazole, and ibuprofen. The latter two medications were associated only with cases that had genetic risk factors.

In the present report, we are able to identify associations between specific cardiac defects and additional pharmaceuticals (Table 14.8). These associations have been adjusted for possible confounders, including the underlying medical condition.

Antitussives

Use of this category of drugs occurred infrequently among control mothers (0.6%), and was associated with increased risks of laterality and looping defects (see discussion in Chapter 4), and with nonchromosomal atrioventricular septal defect. Our analyses took into account possible confounding by influenza and other febrile illnesses.

Clomiphene

Clomiphene, which increases the release of pituitary gonadotrophins, is used to stimulate ovulation and enhance fertility. Although use of clomiphene during the 3 months prior to conception was reported by few control mothers (0.53%) in the Baltimore-Washington Infant Study, we found an association of maternal use of

Table 14.8

Medication Associations

Baltimore-Washington Infant Study (1981–1989)

Medication	Diagnostic Group	N	Exposures		Adjusted Odds Ratio	95% CI	99% CI
			Cases	Controls			
antitussives	Laterality/Looping	104	3	23	6.3	1.8–21.6	1.3–31.5
	AVSD, nonsyndromic	76	3	23	8.9	2.6–30.6	1.7–45.2
benzodiazepines	Outflow Tract Anomalies:						
	Transposition Group	214	8	35	3.3	1.6–6.6	1.3–8.2
	Normal Great Arteries Group	341	8	35	2.3	1.0–5.0	0.8–6.5
	Transposition Group - isolated/ simplex	189	7	35	4.1	1.9–8.6	1.5–10.8
	Ebstein's Anomaly	44	3	35	5.3	1.5–18.5	1.0–1.2
clomiphene	Tetralogy of Fallot	204	7	20	3.2	1.6–6.3	1.3–7.8
corticosteroids	Atrial Septal Defect	187	6	63	4.8	1.8–12.7	1.4–17.1
*diurectics***	AVSD, nonsyndromic	76	4	23	7.3	2.1–25.0	1.4–37.0
ibuprofen	TGA with intact ventricular septum	106	10	119	2.5	1.2–4.9	1.0–6.1
	AVSD with Down syndrome	190	15	119	2.4	1.3–4.2	1.1–5.0
	Biscupid aortic valve	60	7	119	4.1	1.8–9.3	1.4–11.9
macrodantin	Coarctation of aorta	120	4	13	6.7	2.0–22.0	1.4–31.8
metronidazole	NGA group	341	4	8	6.0	1.8–20.7	1.2–30.4
	Membranous VSD, multiple anomalies/ multiplex	118	3	7	12.2	3.0–50.2	1.9–78.0

**triamterene-hydrochlorothiazide combination.

AVSD = atrioventricular septal defect; NGA = outflow tract anomalies with normally related great arteries; TGA = transposed great arteries; VSD = ventricular septal defect.

this drug with tetralogy of Fallot (odds ratio 3.2; enhanced in the multiple anomalies/multiplex subset, odds ratio 6.0). We found that these associations of clomiphene with this cardiac defect remained after we took into account possible case-control differences in reproductive problems such as previous stillbirths, spontaneous abortions, and previous preterm births (see Chapter 5).

Corticosteroids

Oral and inhaled corticosteroids (eg, prednisone, cortisone) used in the treatment of asthma by about 2% of control mothers were associated with isolated atrial septal defect. It may be relevant that corticosteroids are known to constrict the fetal ductus arteriosus in laboratory animals,[304] we observed no associations (positive or negative) between use of corticosteroids during pregnancy and a patent arterial duct.

Ibuprofen

In our previous analyses, ibuprofen, an anti-inflammatory agent, was associated with all cardiovascular malformations as a group. In the subgroup analyses, ibuprofen was significantly associated with transposition with intact ventricular septum (odds ratio 2.5). In addition, ibuprofen was associated with atrioventricular septal defect with Down syndrome (odds ratio 2.4) and bicuspid aortic valve (odds ratio 4.1). In the latter associations, ibuprofen was taken to relieve headaches and other aches and pains that did not confound or modify the effect of ibuprofen (see Chapters 6 and 8). Recently, ibuprofen has become an over the counter medication and its use might be increasing, providing further opportunities to evaluate the possible association between ibuprofen and cardiac defects.

Diazepam, Benzodiazepines

The third drug found to be associated with all cardiovascular malformations as a group was diazepam. This association was present in all models except the one limited to cases with genetic risk factors (see Table 14.4). In the individual diagnostic group analyses, we used the category of benzodiazepines that included diazepam, and found associations significant at the 0.05 level with outflow tract defects with and without transposition and Ebstein's anomaly. Although some of these associations did not reach statistical significance at the 0.01 level, there was consistency across the outflow tract subgroups.

Metronidazole

One of the prescription drugs found to be associated with all cardiovascular malformations was metronidazole, a *trichomonacide,* (0.2% use in control mothers).

In one subgroup analysis, metronidazole was associated with outflow tract defects with normally related great arteries (odds ratio 6.0) and membranous ventricular septal defect (multiple anomalies/multiplex subset) (odds ratio 12.2). No associations between birth defects and prenatal exposure to metronidazole have been reported before. In particular, no such associations were observed in several studies summarized by Schardein.[387]

Salicylates

At the start of the study we hypothesized a possible coteratogenic effect of salicylates with other xenobiotic exposures, as demonstrated in experimental work.[136,165,224] We did not observe associations of specific cardiac diagnostic groups with salicylates. This result did not appear to be due to small numbers of exposed mothers, as among control mothers the frequency of use of salicylates was 14.6% during the critical period. The available information also allowed us to examine salicylate intake as a continuous variable, but we did not observe any positive findings with that variable.

Lifestyle Factors

A number of lifestyle factors were of concern at the start of the study, including alcohol consumption, cigarette smoking, recreational drugs, caffeine consumption, and use of hair dyes. The results for these factors are shown in Table 14.9.

Table 14.9
Alcohol, Smoking, and Recreational Drugs: Summary Table
Baltimore-Washington Infant Study (1981–1989)

Substance	Diagnostic Group	Users Cases	Users Controls	Adjusted Odds Ratio	95% CI	99% CI
Alcohol	Muscular VSD, small defects	31	859	1.8[1984–86]	1.0–3.3	0.9–3.9
	(N=73)	23	459	2.6[1987–89]	1.4–4.8	1.1–5.8
Cigarettes, >20/day	Pulmonic stenosis, moderate (N=112)	3	9	12.5[age>34 years]	3.2–49.4	2.1–75.2
	Normal Great Arteries Group (N=341) Transposition with Ventricular Septal Defect (N=60):	6	9	5.1[history of preterm births]	1.8–14.6	1.3–20.3
	21–39/day	3	136	2.1	1.2–3.9	1.0–4.6
	40+/day	4	40	4.5	1.4–14.9	1.0–21.5
Marijuana	Ebstein's Anomaly* (N=34)	7	244	3.6	1.6–8.5	1.2–11.1
Cocaine	Membranous VSD (N=459)*	15	39	2.4	1.3–4.4	1.1–5.4
Hair Dyes	Pulmonic stenosis, severe (N=56)	12	238	3.7	1.9–7.3	1.5–9.0

Effect modification is indicated by superscripts on odds ratios; VSD = ventricular septal defect. *isolated/simplex subset.

Alcohol

Even though alcohol consumption is one of the more common exposures during pregnancy[298] we found no strong evidence of association of between alcohol consumption and cardiac defects, consistent with earlier reports. Two thirds of control mothers reported drinking any alcoholic beverages during the critical period. Even with such common use of alcohol, the only association we found was between maternal alcohol consumption and small muscular ventricular septal defect, and this association was strongest during the period of 1987 to 1989, when ascertainment of this case group increased markedly with wider use of more sensitive methods of diagnosis. Analysis by the greatest number of alcoholic beverages consumed at any occasion ("binge drinking") during the critical period did not reveal any associations or evidence of a trend in the risk of a cardiac defect with exposure. Thus, our analyses of alcohol consumption and cardiac defects have not provided compelling evidence of an association.

Cigarette Smoking

Among maternal lifestyle factors during the critical period, cigarette smoking was another common exposure—during the critical period of pregnancy, 35% of control mothers smoked cigarettes. Maternal cigarette smoking of more than one pack per day was associated with two cardiac diagnoses: transposition with ventricular septal defect and pulmonic stenosis. The more prominent associations were in subgroups characterized by the presence of another factor (ie, a cofactor). Our analyses suggest that if smoking is associated with cardiac defects, this may occur only among selected susceptible groups (eg, older mothers or mothers with a history of preterm births). This hypothesis is consistent with recent observations of gene-environment interactions in the development of other structural defects.[203,300]

Recreational Drugs

Compared to alcohol consumption and cigarette smoking, use of illicit or "recreational" drugs by control mothers during the critical period was reported with lower frequency—marijuana in 7%, cocaine in 1%, and phencyclidine (PCP) in 0.2%. Use of other recreational drugs such as heroine, amphetamines, methadone, and lysergic acid diethylamide (LSD) were extremely rare. Evaluations of maternal use of marijuana and cocaine revealed associations between cocaine and membranous septal defect and between marijuana and the isolated/simplex subset of Ebstein's anomaly. Paternal exposures, among which we did find some additional associations, are discussed separately below.

Caffeine

Another lifestyle factor examined in this study was caffeine consumption. Caffeine exposure as a potential coteratogen was examined as a continuous variable. In the Baltimore-Washington Infant Study, we did find that caffeine consumption during the critical period was common among mothers of study subjects, ranging from about 12% for cocoa, 18% for coffee, 30% for tea, and 35% for cola. However, we observed no associations between cardiac defects and caffeine consumption or the caffeine dose. This is consistent with observations from other case-control studies of caffeine intake and birth defects. For instance, a large nationwide case-control study[370] reports no evidence of association between cardiac defects as a group and maternal ingestion of caffeine of more than 8 mg/kg per day of tea, coffee, or cola.

Hair Dyes

Information collected on use of hair dyes in the Baltimore-Washington Infant Study was limited to any use, without specific measures of exposure such as color, product type (ie, permanent versus semipermanent), or frequency of use. Use of hair dyes during the critical period was reported by 7% of control mothers. Analysis by specific cardiac groups showed that this association was limited to severe pulmonic stenosis (see Chapter 9).

Home and Occupational Exposures

Many environmental agents were examined in the Baltimore-Washington Infant Study.

The search for teratogens in the home environment was based on information derived from studies of carcinogenesis. Numerous toxic materials are known to be present in homes, home maintenance and improvement activities, crafts and hobbies, and cleaning activities. For these home-based exposures, the mother is the person at greatest risk. The background and resources used in the development of environmental items in the questionnaire, as well as the descriptive profiles for reported parental jobs and for specific agents in the workplace, home, and avocational settings, are described in our first book.[273]

In the Baltimore-Washington Infant Study, the common household exposures among control mothers were pesticides (24%), paints (9%), paint stripping (2%), and varnishing (2.5%). Household exposures to other chemicals or metals were rare. Maternal occupational exposures were also rare. The results of the individual diagnostic group analyses for associations with these factors are shown in Table 14.10.

Solvents

Analysis of all cardiac defects as a group identified associations only with maternal exposure to miscellaneous solvents and to auto body repair work.[152] In the individual diagnostic group analyses we found associations between maternal exposure to solvents and transposition of the great arteries with intact ventricular septum (odds ratio 3.4), hypoplastic left heart (odds ratio 3.4), and all left-sided obstructive defects as a group (odds ratio 6.0). These associations were with different types of solvents. Furthermore, we found an association between exposure of both parents to varnishes and nonchromosomal atrioventricular septal defects (odds ratio 5.6). A score that aggregated many possible solvent sources weighted by exposure frequency was associated with coarctation of the aorta.

Table 14.10
Home and Occupational Exposures
Baltimore-Washington Infant Study (1981–1989)

Exposure Class	parent	Diagnostic Group	N	Exposures Cases	Exposures Controls	Adjusted Odds Ratio	95% CI	99% CI
Organic Solvents:								
degreasing agents	mom	All left-sided obstructive defects*	89	3	17	6.0	1.7–21.3	1.2–31.6
degreasing or misc. agents	mom	Hypoplastic left heart	138	9	80	3.4	1.6–6.9	1.3–8.6
solvent score (90+)	mom	Coarctation of the aorta	120	5	68	3.2	1.3–7.9	1.0–10.2
degreasing agents	mom	Pulmonic Stenosis, moderate	112	3	17	5.0	1.3–8.7	0.9–29.1
miscellaneous agents	mom	Transposition-intact septum	108	8	64	3.4	1.5–7.5	1.2–9.7
paints	both	Tetralogy of Fallot (isol./simplex)	128	16	188	2.7	1.5–4.8	1.2–5.7
	mom	TAPVR	56	13	411	2.0	1.1–3.8	0.9–4.7
varnishes	both	AVSD, nonchromosomal	76	3	106	5.6	1.7–18.9	1.1–28.2
	mom	Ebstein's anomaly	44	5	106	3.6	1.4–9.3	1.0–12.6
Pesticides								
	mom	Membranous VSD (all)	640	195	926	1.3	1.0–1.5	0.9–1.6
	mom	TAPVR	56	2	17	9.0**	2.0–41.1	1.2–66.0
Lead/Metals								
soldering with lead	mom	TAPVR	56	2	17	6.8	1.5–31.5	0.9–50.8
auto body repair	mom	Outflow tract, NGA	341	4	10	3.5	1.1–12.9	0.7–19.0
	mom	membranous VSD*	181	3	10	4.7	1.2–19.3	0.7–29.9

AVSD = atrioventricular septal defect; NGA = normal great arteries; TAPVR = total anomalous pulmonary venous return; VSD = ventricular septal defect.
*Multiple anomalies/multiplex subset. **Interaction of pesticides and family history of heart defects.

The findings of a possible association between maternal exposure to solvents and cardiac defects are consistent with other recent case-control studies conducted in different populations and using different methods.[89,393,416] Our results indicate that associations with maternal exposures to solvents are likely to be most pronounced for left-sided obstructive cardiac defects.[256,258,260] Of interest is the finding that most of the potential exposures to solvents among mothers of study subjects occurred at home, a relatively understudied site of potential exposures in studies of pregnancy outcome.

Pesticides

Congenital malformations in humans have been previously linked with pesticide exposure. More recently, an association between conotruncal malformations and work in the agricultural industry has been suggested.[10]

A previous analysis in the Baltimore-Washington Infant Study[92] identifies associations between reports of pesticide use during pregnancy and total anomalous pulmonary venous return. For this defect, an interaction was noted with family history, and the association was strengthened by the presence of cardiac family history (odds ratio = 9.0, from analysis in Chapter 12). This observation demonstrates the need to consider genetic-environmental interactions in the search for environmental teratogens.

Lead and Other Metals

In as early as 1947, an observation was made in a case series report that 61 of 645 fathers of children with congenital heart disease had worked in lead trades such as painting or battery plants.[116] Zierler et al,[461] however, using a case-control design and records of routine water analysis of samples from public taps, found no association between lead and coarctation of the aorta, patent ductus arteriosus, conotruncal defects, or ventricular septal defect.

In the Baltimore-Washington Infant Study, various possible occupational and domestic sources of lead were considered singly and as a group.[94] An association was noted in univariate analysis only between reports of lead soldering and total anomalous pulmonary venous return[92] and double-outlet right ventricle. Maternal reports of repairing automobiles, indicative of metal, paint, and solvent exposures, were associated with outflow tract defects with normally related great arteries and with membranous ventricular septal defect.

Paternal Factors

A hypothesis of interest at the start of the Baltimore-Washington Infant Study was the possible relationship between cardiac defects and paternal factors such as age, use of recreational drugs, and household and occupational exposures. Interest in paternal factors stemmed from concerns raised by reports of adverse reproductive effects (eg, decrease in sperm quantity and quality and infertility) related to paternal exposures, the paucity of information on the frequency of paternal exposures in the general population, and the possible significance of such exposures.

In this study, we were able to determine the exposures that were incurred by control fathers during the 6 months before the index pregnancy (Table 14.11). In general, these exposures occurred with greater frequency among control fathers than among control mothers, and the data permitted more elucidating analyses.

Analysis of all cardiovascular malformations as a group identified only one association with paternal factors, ie, paternal cocaine use. However analyses by specific cardiac diagnostic groups revealed that the association with paternal cocaine use was limited to two cardiac diagnostic groups: membranous ventricular septal defects as a group (Chapter 7), and the small group of tricuspid atresia (Chapter 9). The analyses of paternal exposures by individual cardiac diagnostic groups also revealed associations with one additional lifestyle factor and four occupational exposures not evident for all cardiac defects as a group: heavy smoking, exposure to varnishes, auto body repair, ionizing radiation and extreme cold temperature at work. These results for paternal factors underscore the marked variation in risk profiles between cardiac subgroups.

Comments on Environmental Factors

The evaluation of potential effects on the fetus from the maternal use of pharmaceuticals during pregnancy and from parental exposures to household and occupational chemicals has been of longstanding interest, given the relatively common use of some products and drugs during pregnancy and the teratogenicity observed with selected agents. Evaluations of possible associations between environmental agents and birth defects, however, have proven to be a major research challenge.

The design of the Baltimore-Washington Infant Study was based on careful consideration of the epidemiologic methodology and teratologic knowledge, past experiences with birth defects epidemics and population surveillance efforts, as well as of the clearly enunciated "principles of teratogenesis" derived from experimental studies. All of these aspects, described in the first volume of the Baltimore-Washington Infant Study report[152], justified the design and methods of this regional case-control study.[141,151] In this study we hoped to overcome: 1) difficulties in collecting reliable data on the large number of drugs and other exposures of pregnant women during the periconceptional period, 2) difficulties in differentiating the effects of pharmaceutical agents from those of the underlying conditions for which the agent was administered, and 3) possible confounding or effect modification of any association by other factors that might have been encountered at the same time.

The Baltimore-Washington Infant Study, designed to evaluate the possible relationship of cardiovascular malformations to a wide range of factors, has led to the elucidation of many aspects of the epidemiology and morphogenesis of congenital heart disease. First, this study demonstrates that cardiovascular malformations constitute a heterogenous group of phenotypes that vary not only in stage of abnormal development and anatomic features but also in underlying risk factors and genetic susceptibility. The degree of heterogeneity is illustrated by the fact that no two phenotypes were exactly the same in their sets of risk factors. Nonetheless, there were certain risk factors that were common to several phenotypes: notably family history of cardiovascular malformations and maternal overt

Table 14.11
Paternal Exposures
Baltimore-Washington Infant Study (1981–1989)

Substance	Diagnostic Group	Cases	Controls	Adjusted Odds Ratio	95% CI	99% CI
Cigarettes, >20/day	Laterality/looping (N=104)	20	342	**5.6**	2.5–12.9	1.9–16.6
Cocaine	Membranous VSD (N=640)	44	111	**1.9**	1.3–2.9	1.1–3.3
	Tricuspid atresia (N=30)	4	111	**4.8**(univariate)	1.6–14.0	1.2–19.5
Organic Solvents:						
varnishes	All left-sided obstructive defects (N=89)*	16	240	**2.4**	1.4–4.3	1.2–5.1
auto body repair	Cardiomyopathy (N=77)	12	236	**2.5**	1.3–4.8	1.1–5.9
Ionizing Radiation (occupational):						
	AVSD-nonchromosomal (N=76)	3	32	**5.7**	1.7–19.3	1.1–28.5
Temperature Extremes (occupational):						
cold	Atrial septal defect (N=187)	4	9	**8.0**	2.3–28.1	1.6–41.5

VSD = ventricular septal defect; AVSD = atrioventricular septal defect.
*Multiple anomalies/multiplex subset.

diabetes. Although the former had been considered an important risk factor for several phenotypes, the specific associations with maternal diabetes were not previously recognized. The associations of these two factors for so many cardiac phenotypes suggests that they exert their effects at an early stage of development, ie, at the stage of primary cardiogenesis.

In addition, the various associations we observed with maternal and paternal factors, including prescription and nonprescription drugs, lifestyle factors, household exposures, and occupational exposures, indicate that environmental factors also play a role in the occurrence of cardiovascular malformations in the population. For some cardiac phenotypes, the associations with environmental factors were evident only in the presence of another cofactor. However, the findings did not suggest that cardiac malformations as a group or by phenotypes result from specific coteratogenic exposures. Instead, the demonstration of a wide array of environmental risk factors may indicate nonspecific and overlapping teratogenic effects.

Chapter 15

Research Implications

From: Ferencz C, Loffredo CA, Correa-Villaseñor A, Wilson PD, eds: *Genetic & Environmental Risk Factors of Major Cardiovascular Malformations: The Baltimore-Washington Infant Study 1981–1989.* Armonk, NY: Futura Publishing Co., Inc; ©1997.

*A brief description of an immense subject involves inevitable defects. But
the sketch of a landscape should not be expected to contain all the details
of a photograph.*

—Alex Carrel (1873–1944) *Man the Unknown*
Harper and Bros. Publishers, New York, 1935

Introduction

The Baltimore-Washington Infant Study was undertaken for an important reason: there was a gap of knowledge on the fundamental descriptive information on
congenital heart disease, which prevented focused evaluations of presumptive
teratogenic agents.[147] This volume documents that needed descriptive information. With this publication, the Baltimore-Washington Infant Study is concluded.

We are extremely fortunate that the research funding of this epidemiologic
study was uninterrupted, indicating to us that potential benefits were expected to
arise from this work. The reality of a large and long epidemiologic study did not
parallel the exciting new research opportunities experienced by our colleagues in
the biologic sciences who over the same time period witnessed dramatic advances
in embryology, morphogenesis, and developmental genetics. Now a new generation of investigators has stepped in to recognize the applicability of epidemiologic
data to the new horizons of developmental studies.[30,259,372]

It is immensely gratifying that the epidemiologic study achieved—as it was
supposed to do—some clarification of the giant "jigsaw puzzle" of abnormal cardiogenesis (Figure 15.1). We defined etiologic categories by risk factors, which sur-
mount the anatomic phenotypes and distinguish abnormalities of early, *primary
cardiogenesis* from those of *secondary cardiogenesis,* which occur after the basic
architecture of the heart has been established.

When the Baltimore-Washington Infant Study began to enumerate a high proportion of cases with multiple malformations, we were surprised because it was
assumed that most cardiovascular defects are isolated defects and can be studied
separately within the field of teratology. This is still a common belief not only with
respect to the heart, but also with respect to other major anomalies such as neural
tube defects or renal or alimentary tract disorders. In pediatric cardiology, the "af-
fairs of the heart" remain all-consuming (Figure 15.2).

The recent explosive progress in developmental studies, however, has made
the concept of isolated principal defects outdated and abnormalities affecting the
total embryo a matter of course. Molecular, biochemical, and cellular changes represent systemic processes, and newly created strains of laboratory animals simu-
late complex human anomalies.

An etiologic search requires sensitivity to a new universe of information. At
the same time it requires common sense: molecular studies will be beneficial only
if they can be applied to patients and to populations. We must recognize, categorize, and prevent disease through the clinical application of this new knowledge.[290]

Phenotypic variations of congenital heart defects form the basis of this epidemiologic investigation. The genetic approach assumed that these variations are
influenced by single genes or by the interactions among multiple genes of the individual. However, the expression of the genotype and the phenotype must take
into account environmental (epigenetic) modifications:

AN ETIOLOGIC GROUPING OF MAJOR CARDIOVASCULAR MALFORMATIONS
Baltimore-Washington Infant Study (1981-1989)

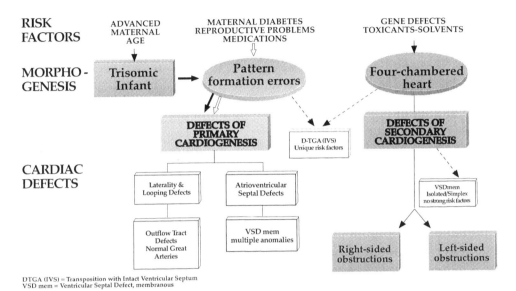

DTGA (IVS) = Transposition with Intact Ventricular Septum
VSD mem = Ventricular Septal Defect, membranous

Figure 15.1: The determination of risk factors defined the major etiologic categories of cardiovascular malformations and distinguished abnormalities of early cardiogenesis (primary defects) from those that occur after the normal architecture of the heart has been established (secondary defects).

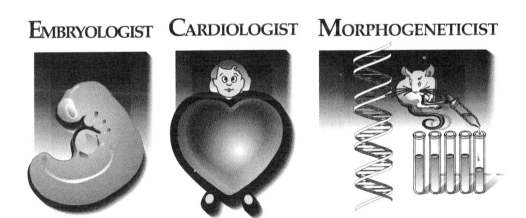

Figure 15.2: Fragmented views of the study of cardiovascular development by embryologists, cardiologists, and morphogeneticists. The interrelated etiologic horizons of altered cardiogenesis demand an integrated disciplinary approach.

The phenotype is, thus, the product of ontogenetic development rather than the mere consequence of the genetic constitution of the zygote. Ontogenesis takes place within a framework of conditions that form the scenario for the processes realizing the phenotype.[454]

This observation has major implications, as many questions regarding the genotype-phenotype correlation can be addressed by an epidemiologic approach. What is the role of heredity in congenital heart defects? What role do extrinsic factors play? What are the interactions of genetic and environmental factors? Possible efforts to refine these questions into testable hypotheses are addressed as follows under the headings of disciplines for which we foresee important implications for future research.

Cardiology and Cardiac Morphogenesis

The cardiologists who helped to assemble and evaluate the Baltimore-Washington Infant Study cases were notable exceptions in their specialty because they realized the importance of a search for causes. One must hope that pediatric cardiologists of the future will join in that search, with comprehensive evaluation of each patient directed not only toward the echocardiographic visualization of the cardiovascular malformations, but also toward the precise description of noncardiac anomalies, the family history, the environmental history, and most importantly, the collection of blood samples for the genetic analyses of proband and family. High-quality clinical records are needed for use in multidisciplinary evaluations.

The results of this epidemiologic study are of more interest to those who study *morphogenesis* (ie, the normal and abnormal development of the heart) than they are to clinical cardiologists. In virtually every diagnostic group described in this volume there were findings that have broader implications for studies of developmental events.

Early defects associated with *laterality and looping* anomalies were of a very great phenotypic variety, indicating that infants in whom these defects were observed at livebirth represent the "tip of an iceberg" of severe cardiac and noncardiac anomalies. Moreover, the effect of abnormal lateralization or looping was also detected in simple clinical entities with presumed situs solitus such as total anomalous pulmonary venous return, which included some cases with autopsy findings of trilobed left lung and splenic anomalies. This indicates that the anomalous connection of pulmonary veins was not simply a "failure of targeted growth,"[82] but part of a greater disorganization of embryonic events. The origin during the looping phase of transposition of the great arteries in experimental animals[347] and of some forms of tricuspid atresia[18,19] further show that the phases of cardiac development cannot be rigidly divided according to sequential hallmarks of chamber formations. For example, a diversity of risk factors was apparent for the group of hypoplastic left heart syndrome cases, suggesting that in some instances the malformation originated very early, possibly during early looping, and that it may be comparable to the various forms of atresia of the tricuspid valve.

In the so called "conotruncal" anomalies, our findings clearly demonstrate a dichotomy of those with and without *transposition of the great arteries.* The transposition component seems to be quite distinct because in all subcategories (ie, those with intact ventricular septum, those with defects of ventricular septal defect, those with associated double-outlet right ventricle, and those with tricuspid atresia) there was evidence of normal fetal growth and a rarity of associated noncardiac anomalies and familial congenital heart disease. The unique risk factors found for transposition with intact ventricular septum (effects of influenza and ibuprofen therapy), as well as the unique risk factor of heavy parental smoking in transposition of the great arteries with ventricular septal defect, suggest that these variables may represent some unknown set of agents not identified in our inquiry. This may very likely be a nutritional effect, as analyses of the food frequency information collected in the final 3-year segment of the Baltimore-Washington Infant Study population confirms this anatomic dichotomy of "cono" and "truncal" malformations due to differential vitamin A effects.[51]

Outflow tract defects with normally related great arteries are errors of primary cardiogenesis and are component parts of a triad of anomalies that also includes noncardiac defects and fetal growth retardation. For infants with this anomaly a positive family history was rare, but it demonstrated the known relationship of the various diagnostic subentities to each other and with membranous ventricular septal defects, that constitutes a spectrum of outflow tract anomalies. Defects with the malformation triad also include nonchromosomal atrioventricular septal defect. Among risk factors, *maternal diabetes* is of greatest importance. This risk factor is also linked with a history of previous pregnancy problems. This association suggests a longstanding alteration in the mother, which antedated the metabolic disturbances observed during the course of the index pregnancy.

It is interesting that infants born to mothers with diabetes and infants with autosomal trisomies have the same spectrum of heart defects. This fact allows us to suppose some common steps in the pathogenesis of heart defects produced by maternal diabetes and by additional chromosomal material.

Comparison of cardiac manifestations and risk factors for trisomic infants of maternal and paternal origin, respectively, should be undertaken in collaborative studies. The importance of clarifications of these relationships cannot be overstated.

Past morphogenetic studies have focused principally on the developmental sequences that achieve a four-chambered heart. Our findings suggest that the subsequent events in development are also of great importance, as about one half of the referred infants with congenital heart disease had defects secondary to a normally formed heart. A notable malformation is transposition of the great arteries with intact ventricular septum in which, to the best available knowledge, the architecture of the heart itself is actually normal.

Another group of great interest is that with obstructions in the aortic tract. These malformations were found to be associated with toxic exposures, mostly solvents, but in a setting of familial aggregations and chromosome defects (monosomy X), which suggests the possibility of genetic susceptibilities.

Dysmorphology and Genetics
by Iosif W. Lurie, M.D., Ph.D.

Recent advances in molecular genetics have considerably changed traditional perceptions of hereditary disorders. The frontier between the mutation of a gene and a chromosomal anomaly as well as that between isolated and multiple congenital abnormalities is not as strong as it was several years ago.

Isolated cardiovascular malformations were previously thought to be the result of (a) environmental (exogenic) events, (b) purely genetic defects (mutations), and (c) complexes of genetic and environmental effects.

(a) Exogenic Events: many teratogens can lead to a cardiovascular malformation, but in most instances the final result will be not isolated cardiac defects but syndromes of multiple congenital anomalies. However, virtually all syndromes can have their forme fruste, and sometimes an isolated cardiac defect may be the only phenotypic manifestation of the syndrome. It is likely that the total contribution of purely exogenic forms in the pool of isolated cases is not high.

(b) Genetic Defects: variability in manifestations also may apply to the presumed purely genetic group of isolated cardiac defects. Some defects are known to be inherited such as the dominant Mendelian transmission of atrial septal defect with conduction abnormalities, and hypertrophic obstructive cardiomyopathies. However such cases also constitute only a small part of all isolated cardiovascular malformations.

(c) Genetic and Environmental Factors: the vast majority of isolated cardiac defects is believed to belong to the so-called "multifactorial" group in which different environmental events are necessary to convert a hereditary predisposition (liability) based on the cumulative action of many genes into a final defect. Thus the exogenic risk factors found in the Baltimore-Washington Infant Study may indicate the realization of a genetic liability.

Mutations and Microdeletions

Mutations affect one gene, and in most cases will stop the production of a normal protein, while with chromosomal abnormalities there is an excess (in complete or partial trisomies) or absence (in deletions) of tens and hundreds of normal genes. This is a major difference. Some syndromes that appeared to be monogenous have been found to be due to microdeletions, which could not previously have been detected by routine cytogenetic investigations. Williams syndrome, due to a small deletion of 7q11.23[232] may be such an example. Deletions may actually involve several genes, but only one or two of these may be important for morphogenesis.

Molecular mechanisms of mutations are variable. In most cases replacement of one nucleotide base can stop the synthesis of a normal protein or result in abnormal splicing. In some disorders (for example, in Duchenne muscular dystrophy) the mutations are large deletions of several exons within the gene. A duplication of a normal gene may produce an abnormal phenotype. Duplication of the PMP22 gene as a cause of Charcot-Marie-Tooth disease is the best known example.[314] All of these examples show that molecular mechanisms of Mendelian mutations and chromosomal abnormalities may actually be the same.

In patients with chromosomal imbalance, at least two basic mechanisms may be involved in the pathogenesis of abnormalities:

(1) loss (or excess) of several tens or hundreds of genes, some of which are important for morphogenesis. For example, the loss of the Sonic

Hedgehog gene on 7q36 leads to holoprosencephaly;[363] loss of the elastin gene on 7q11.23 leads to supravalvular aortic stenosis in persons with Williams syndrome.[334]

(2) almost any cytogenetically visible imbalance results in nonspecific changes of cellular homeostasis such as cellular adhesiveness and speed of cellular division. Abnormalities that destroy the rhythm of developmental processes can change a normal threshold for multifactorial abnormalities, allowing the expression of a primarily "weak" genetic potential of the individual.[264]

If a deleted (or duplicated) segment is relatively small, an isolated (or at least apparently isolated) defect may be the only visible phenotypic manifestation of such an imbalance. In many cases, only detailed investigation of the affected persons can elucidate the problem. For example, Moog[305] describes a man with peripheral pulmonary stenosis. His daughter had pulmonary valve atresia with hypoplasia of the pulmonary arteries and a ventricular septal defect. After another daughter was born with the same cardiac defect (plus an atrial septal defect), a thorough examination of the father showed minimal facial dysmorphism, patchy hypoplasia of iridal stroma, and thoracic butterfly vertebrae, so that a clinical suspicion was formed of the Alagille syndrome, which was then confirmed by cytogenetic study. Certainly, most cases in the Baltimore-Washington Infant Study had detailed cardiological examination, but they did not have detailed ophthalmologic, otorhinolaryngologic, dermatologic, and other specialized examinations, and therefore micromanifestations of syndromes would not have been detected.

Although there has been great interest in recent genetic studies of patients with microdeletions 22q11,[70,118,405] no infant in the Baltimore-Washington Infant Study was reported to have this defect. Some persons with this deletion have the classic DiGeorge syndrome phenotype, others have "partial" DiGeorge syndrome, velocardiofacial syndrome, or association of a cardiac defect with nonspecific craniofacial dysmorphism. Moreover, the same deletions have been found in apparently healthy parents of affected patients and in persons with isolated conotruncal defects or interrupted aortic arch.[250,251,288,330] The size of the subset of apparently isolated cardiovascular defects due to microdeletion of 22q11 will be determined in the nearest future.

Etiologic evaluation of syndromes must define syndromes associated with a specific form of cardiovascular defect and syndromes in which cardiovascular defects are rather common but not a single type is prevalent. The atrioventricular septal defect in patients with trisomy 21 is the best example of a specific association between a chromosome defect and a specific cardiovascular defect; coarctation of the aorta and other left heart lesions are typical for females with Turner syndrome. In such cases one can presume a trisomy or monosomy for any gene involved in a specific stage of cardiogenesis.

The etiology of a large group of syndromes and associations that involve cardiac anomalies remains unknown. If one considers the association of cardiac defects with a single-organ abnormality of the VACTERL type, these forme fruste expressions would greatly add to the count of affected cases.

Nonclassified complexes of developmental anomalies can be considered in three subgroups: (a) random association of two independent malformations in one person, (b) incomplete forms of already known syndromes that do not fit the diagnostic criteria of any nosology (for example, an association of ventricular septal defect and postaxial polydactyly could be components of Meckel, Smith-Lemli-Opitz, pseudotrisomy 13, and many other syndromes), and (c) syndromes that have not yet been delineated.

Recent investigations show that in most cases the same disorder may result from mutations of different genes. The discovery that different mutations within the same gene may lead to different disorders was far more surprising. For example, such disorders as achondroplasia, thanotophoric dysplasia, and Pfeiffer syndrome are due to mutations in the fibroblast growth factor receptor 3 gene.[292] It may be expected that different types of cardiac defects may also result from different mutations of the same gene.

In this context there are important implications for future studies. Analysis of groups of embryologically related cardiovascular defects might be more productive than genetic study of any isolated type of defect. Multiple forms of conotruncal defects in persons with del(22)(q11)[250] and the different types of left-sided obstruction in Turner syndrome may represent the existence of series of cardiovascular defects. Therefore, the whole series, but not a single form, has to be an object of genetic or teratologic study.

A schematic representation of these concepts (Figure 15.3) shows that many forms of cardiovascular malformations may be different stages of the same devel-

Alternate Developmental Pathways

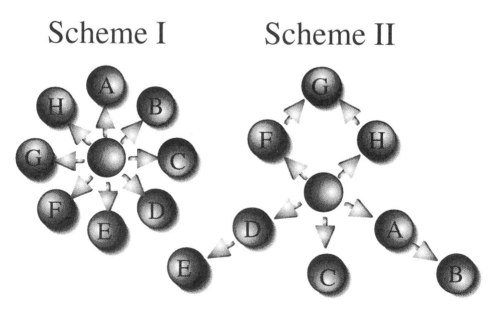

Figure 15.3: Genetic studies indicate alternate developmental pathways.
Scheme I: each final malformation results from a specific alteration of normal morphogenesis.
Scheme II: some abnormalities (as A and B; D and E) are different manifestations of the same morphogenetic defects; other anomalies (such as G) may be outcomes of different morphogenetic defects, (F to G; H to G). See text.

opmental path, as in Scheme 1 of Figure 15.3 (A and B, or D or E). However the problem may be even more complicated. In some cases the same form of cardiac defect, as form G on Figure 15.3 (Scheme II), may be the result of two or more different developmental paths. And it may be impossible to delineate the G forms from the HG and FG paths.

The Baltimore-Washington Infant Study, as did other epidemiologic investigations, showed numerous environmental agents as risk factors for cardiovascular malformations. Historically, even very strong teratogens such as thalidomide do not cause morphological defects in all exposed fetuses. Most likely, genetic factors make some fetuses liable to cardiovascular defects whereas other fetuses remain resistant. When these factors of genetic liability are known, it will be possible to prevent a considerable segment of cardiovascular defects by the exclusion of risk factors in the persons who have the genetic liability to certain defects. The discovery of primary genetic defects leading to cardiovascular anomalies and appropriate genetic counseling of susceptible populations, as well as finding the factors of genetic liability in affected persons and in their parents, will become primary methods for prevention of cardiac and other congenital defects.

Epidemiology

The Baltimore-Washington Infant Study has shown the great importance of evaluating specific diagnostic groups in an etiologic study. Each diagnosis provided some special aspects of the epidemiologic picture. Future studies should not repeat the Baltimore-Washington Infant Study by including all cases of congenital heart disease but should, instead, focus on assembling sufficiently large numbers of cases in specific diagnostic categories to provide new information on risk factors that could not be discovered with the limited size of some of our case groups. Only multicenter clinical-collaborative epidemiologic efforts can achieve this.

Future studies must avoid the investment of time, effort, and funds in studying malformations of no clinical significance; these malformations made up a high proportion of the Baltimore-Washington Infant Study cases and included many instances in which the defect "disappeared" even during the first year of life. Furthermore, benign defects such as isolated muscular ventricular septal defects for which we found no strong risk factors, cannot profitably return rewards for further investments. The Baltimore-Washington Infant Study, which included all cardiovascular defects, has thus performed an important screening function, which should greatly enhance the promise of well-directed future investigations.

Interest should focus on the major malformations, about which we have so much to learn. Among these defects, some may be prevented with a better understanding of the biochemical and nutritional alterations, which if they indeed "caused" the malformation may be limited to subsets of people with specific characteristics that could be identified by specific screening methods. Appropriate attention to genetic-environmental interactions will be useful in targeted high-risk populations and could also be achieved by collaborative epidemiologic studies using molecular markers.

The most important cardiac disorders selected for epidemiologic studies should be the *ventricular inflow and outflow abnormalities*. Defects of the atrioventricular junction (also known as endocardial cushion defects) are clearly of complex developmental origin and are to some degree related to factors on chromosomes 21 and 18. Since it is possible that they are never isolated cardiovascular malformations, except as forme fruste, this possibility should be explored in detail in very careful clinical, genetic, and molecular genetic studies. Similarly, the causative effect of increasing maternal age in nondisjunction must surely have a biologic explanation identifiable by appropriate measurements.

Evaluation of the outflow tract should be performed with detailed clinical and echocardiographic evaluation of morphological variations. The group of *transpositions* should be described in terms of the normal or abnormal architecture of the heart, and cases that are associated with septation defects should be subdivided into those where the ventricular septal defect is small and may even close, in contrast to those where the overall structure of the ventricles is abnormal. In this diagnostic group we should also be alert to the possible association of abnormal clotting factors, a research field in its entirety. Studies of malformations of the outflow tract with *normally related great vessels* must evaluate immune deficiencies and other manifestations of the chromosome 22q11 deletion syndrome to unmask forme fruste of multiple malformations, in which major disorders of the renal and alimentary tract are defects of equal importance to those of the heart.

Further advances in epidemiologic studies will demand that the measures of environmental exposures be improved and enhanced. Biomarkers of heavy metal exposures, for example, in blood, hair, and bone have been validated and are used widely in epidemiologic research. Clearly the issue of biologic susceptibility to xenobiotic agents is of equal importance in birth defects and in cancers, but cancer research has been quick to capitalize on molecular genetic advances in studying metabolic polymorphisms, while birth defects research has lagged seriously behind.

Tremendous advances in molecular genetics have occurred over the years since the start of Baltimore-Washington Infant Study. New knowledge and new molecular tools now permit hypotheses concerning genetic susceptibility to the effects of environmental toxicants to be tested—hypotheses that could not even be imagined just a few years ago. The associations of some specific cardiac phenotypes with specific environmental exposures, in particular, which we have detailed in this book, provide a fertile ground for measuring the role of genetic susceptibility in the etiology of heart defects. Examples of such future studies are described below.

Among the most striking and consistent associations we found are those involving obstructive defects and maternal exposures to organic solvents. This class of home and occupational exposures encompasses a wide range of chemicals with varying toxicity, but most are metabolized by a relatively small number of catalytic enzymes. Some of these enzymes such as those in the cytochrome P-450 family, may bioactivate solvents with relatively low toxicity to produce toxic metabolites. Other enzymes, notably glutathione-S-transferases (GST) and epoxide hydrolase, may confer chemoprotection to cells by catalyzing reactions that yield non-toxic metabolites readily excreted from the body.

Many of these enzymes are genetically polymorphic: the genes that code for these mediators of solvent toxicity are subject to mutations, and the known gene variants can be detected by molecular genetic testing of blood or other tissues. The hypothesis that the infant's genotype at GSTM1 and GSTT1 modifies the risk of congenital heart defects among those whose mothers were exposed to solvents during the vulnerable period of cardiogenesis has been confirmed for certain mal-formations, suggesting further elucidation of differential effects.[258]

Similar hypotheses could be posed concerning birth defects and a range of other maternal exposures. For example, genetic variations in metallothionine, a metal-binding protein, may mediate the effects of exposures to various metals. The risk associated with maternal use of medications such as benzodiazepines may be modified by genetic polymorphisms in drug-metabolizing enzymes such as cy-tochrome P-450 isoenzymes. Expanding knowledge in molecular biology will con-tinue to provide new information on such genetic variations and new tools for de-tecting them in epidemiologic studies. The genetic susceptibility paradigm might also lead to new insights into why some exposures were *not* associated with heart defects in this study. The results presented in this book may guide future research toward new assessments of known risk factors.

Evaluation of a wide horizon of risk factors has revealed a potential impact of our findings on the prevention of cardiovascular malformations. Findings of health care deficiencies in some socioeconomically deprived segments of our population should direct attention to basic concerns that need no new technologies for their resolution.

Need for Multidisciplinary Partnerships

The Baltimore-Washington Infant Study was made possible by the collabora-tive efforts of clinicians and experts in the environmental and biomedical sciences. The findings reported in this volume represent interdisciplinary considerations.

There has been, to date, a surprising delay in understanding congenital heart disease as part of the total developmental process, reflecting the progressive spe-cialization in the fields of biology and medicine. Specialization in these fields has helped to elucidate disease processes. The integration of multidisciplinary efforts was essential in the evaluation of the epidemiologic findings. It was this integration of specialities that greatly enriched the Baltimore-Washington Infant Study.

The future will require a new paradigm for the study and prevention of con-genital heart disease. We hope that the results of this hypothesis-generating case-control study will provide a solid foundation for the work of scientists in the ever-widening spectrum of relevant disciplines.

Appendices

Description of Appendices

A. Glossary of Malformation Syndromes: by I. W. Lurie, M.D., Ph.D.

Multiple malformation syndromes found in the Baltimore-Washington Infant Study 1981–1989

B. Variables:

Categories of variables used in case-control analyses

C. Case Listings:

1. Infants with outflow tract anomalies and associated noncardiac anomalies
2. Infants with Down syndrome cases and atrial or ventricular septal defects

D. Controls Listings:

1. Distribution of potential risk factor variables among 3572 controls
2. Cardiovascular malformations among first-degree relatives of control infants
3. Noncardiac malformations among first-degree relatives of control infants

Appendix A

Multiple Malformations Syndromes Found in the Baltimore-Washington Infant Study (1981–1989)

Name of Syndrome	Manifestations	Cardiovascular Malformation	Etiology, McKusick Number	Birth Defects Encyclopedia (1990)
Aarskog syndrome	Short stature, ptosis, hypertelorism, widow's peak, "shawl" scrotum, hyperextension of proximal interphalangeal joints	Rare	?X-linked recessive #30540	BDE, p. 1–2
Achondroplasia	Rhizomelic micromelia (shortness of proximal segments), large head with depressed nasal bridge and prominent forehead	Rare	Autosomal dominant #10080	BDE, p. 11–12
Acrofacial dysostosis	Antimongoloid palpebral fissures, malar and mandibular hypoplasia, microtia, coloboma of lower lid, cleft palate, underdevelopment of preaxial (in Nager type) or post-axial (in Miller type) structures of upper limbs	Rare	Unknown for Nager type; autosomal recessive for Miller type (#26375)	BDE, p. 44–46
Acrorenal field defect	Underdevelopment of preaxial structures of hands and malformations of kidneys (agenesis, hypoplasia, horseshoe kidney)	Rare	Etiologically heterogeneous field defect	* [Evans et al., 1992]
Alagille syndrome	Hypoplasia of hepatic ducts, butterfly type vertebrae	>50%, mostly peripheral pulmonary artery stenosis	Microdeletion 20p12-p11	BDE, p. 184–185
Apert syndrome	Acrocephaly (shortness of skull in anterior-posterior axis) due to craniosynostosis, almost complete syndactyly of hands and feet	Rare	Autosomal dominant #10120	BDE, p. 37–38

"BBB" syndrome	Hyperterlorism, plagiocephaly, hypospadias	~25%	X-linked recessive #31360	BDE, p. 912–914
Campomelic syndrome	Micromelia with femoral and tibial bending, sex reversal (female genitalia in 46, XY probands), defects of brain and kidneys	~20%	?Autosomal recessive	BDE, p. 252–253
Cantrell pentalogy	Cleft sternum, supraumbilical defect of abdominal wall, defects of pericardium and anterior diaphragm	Almost constant	Etiologically heterogeneous field defect	BDE, p. 1375–1376
Carnitine deficiency	Progressive generalized myopathy due to abnormal beta oxidation of fatty acids	Frequent cardiomyopathy as a result of primary metabolic defect	Autosomal recessive #21214	BDE, p. 1200
CHARGE association	Coloboma of the eye, Heart defects, Atresia of choanae, Retarded growth, Genital and Ear abnormalities	50%–70%, different forms	Etiologically heterogeneous	BDE, p. 308–309
"Cleft sternum - hemangioma" association	Cleft sternum, hemangioma	? Rare	Etiologically heterogeneous field defect	BDE, p. 1594
Cornelia de Lange syndrome	Nanism, mental retardation, microcephaly, synophris, hirsutism, anteverted nostrils, small hands, underdevelopment of upper limbs	20%–30%	Unknown	BDE, p. 486–487
Costello syndrome	Short stature, macrocephaly, redundancy of skin on the neck, palms, soles, papillomata around the mouth, nares, anus	20%–30%, cardiomyopathy	Autosomal dominant	*[Kondo et al., 1993]
Cytomegaly	Prenatal hypoplasia, hepatosplenomegaly, thrombocytopenia, intracranial calcifications	Rather common	Fetal infection	BDE, p. 691–692

Appendix A—Continued

Multiple Malformations Syndromes Found in the Baltimore-Washington Infant Study (1981–1989)

Name of Syndrome	Manifestations	Cardiovascular Malformation	Etiology, McKusick Number	Birth Defects Encyclopedia (1990)
DiGeorge syndrome	A- or hypoplasia of thymus, hypocalcemia, hypoparathyroidism, immunodeficiency	Common, >60%	Etiologically heterogeneous, significant part of cases are due to del(22) (q11.2)	BDE, p. 961–962
Ehlers-Danlos syndrome	Hyperextensibility of joints, hyperlaxity of skin	Rare	There are different types with autosomal dominant, autosomal recessive and X-linked inheritance	BDE, p. 610–611
Ellis-Van Creveld syndrome	Short-limbed dwarfism, postaxial polydactyly, dysplastic nails, genu valgum	>50%, mostly ASD	Autosomal recessive #22550	BDE, p. 322–323
Fanconi anemia	Progressive pancytopenia, hyperpigmentation of skin, a- or hypoplastic thumbs, kidney anomalies	~10%	Autosomal recessive #22765	BDE, p. 1359–1361
Fetal alcohol syndrome	Growth and mental deficiency, microcephaly, short palpebral fissures, midface hypoplasia	Common in offspring of mothers with chronic alcoholism	Result of maternal use of alcohol	BDE, p. 684–685
Frontometaphyseal dysplasia	Hyperostosis, skeletal dysplasia, multiple joint contractures, abnormalities of teeth	15%–20%	X-linked inheritance with severe manifestations in males and mild in females #30562	BDE, p. 749–750
Goldenhar syndrome	Epibulbar dermoid, macrostomia, unilateral defects of external and internal ear, deafness, macrostomia, abnormalities of cervical vertebrae	20%–40%	Etiologically heterogeneous	BDE, p. 1272–1274
Holt-Oram syndrome	Preaxial hand abnormalities, cardiovascular malformations	Common, mostly ASD	Autosomal dominant #14290	BDE, p. 850–851

Syndrome	Features	Frequency	Etiology/Inheritance	Reference
Holzgreve syndrome	Cleft palate, ankyloglosson, a- or hypoplasia of kidneys, caecum mobile	Common	Unknown	*[Thomas et al., 1993]
HOMAGE syndrome	Oligomeganephronia, Ambiguous Genitalia	Only 1 family with tertralogy in sibs	Autosomal recessive	*[Lurie et al, 1995a]
Hydantoin embryopathy	Prenatal hypoplasia, growth deficiency, cleft palate, midface hypoplasia, hypoplastic nails and distal phalanges	Uncommon	Result of maternal use of hydantoin	BDE, p. 714–715
Hydrocephaly-VACTERL syndrome	Complex of VACTERL-association [see below] and hydrocephaly	50%	Heterogeneous, including at least one autosomal recessive and one X-linked recessive form	*[Genuardi et al. 1993]
Ivemark's syndrome	Heterotaxia, abnormality of bowel rotation, asplenia with bilateral trilobed lungs, or polysplenia with bilateral bilobed lungs	Common	Etiologically heterogeneous, including at least one autosomal recessive form	BDE, p. 201–202
Kartagener syndrome	Situs inversus, sinusitis, bronchiectasis, immotile spermatozoa	Frequent dextrocardia	Autosomal recessive #24440	BDE, p. 521–522
Klippel-Feil syndrome	Fusion of cervical vertebrae, sometimes with involvement of thoracic and lumbar vertebrae, scoliosis, Sprengel's deformity, deafness, unilateral renal agenesis, ectopic kidneys	Uncommon	Etiologically heterogeneous	BDE, p. 1015–1016
Larsen syndrome	Multiple joint dislocations, hydrocephaly, prominent forehead, flat face, cleft palate	Common	Etiologically heterogenous, including autosomal recessive and autosomal dominant forms	BDE, p. 1029
Laurence-Moon-Bardet-Biedl syndrome	Oligophrenia, postaxial polydactyly, obesity, retinitis pigmentosa, hypoplastic genitalia	Rare	Autosomal recessive #20990	BDE, p. 215–216

Appendix A—Continued

Multiple Malformations Syndromes Found in the Baltimore-Washington Infant Study (1981–1989)

Name of Syndrome	Manifestations	Cardiovascular Malformation	Etiology, McKusick Number	Birth Defects Encyclopedia (1990)
Lazjuk syndrome	Cleft lip and palate, ectrodactyly	Common	Autosomal dominant	*[Lazjuk et al., 1983]
Marden-Walker syndrome	Multiple joint contractures, psychomotor retardation, blepharophimosis, micrognathia	Uncommon	Autosomal recessive #24780	BDE, p. 1103–1104
Meckel syndrome	Occipital encephalocele, cleft lip/palate, polycystic kidneys, postaxial polydactyly, hepatic fibrosis, ambiguous genitalia	10%–15%	Autosomal recessive #24900	BDE, p. 1113–1114
"Microgastria-limb reduction" complex	Underdevelopment of limbs (mostly upper), microgastria, asplenia or hypoplastic spleen, kidney abnormalities, abnormal lung lobation, intestinal malrotation	40%	Unknown	*[Lurie et al., 1995b]
Noonan syndrome	Short stature, webbed neck, pectus carinatum	~75%, mostly valvular pulmonary stenosis	Autosomal dominant #16395	BDE, p. 1257–1258
Opitz-Frias ("G" syndrome)	Hypospadias, dysphagia, hoarse voice	Uncommon	Autosomal dominant #14541	BDE, p. 755–757
Oro-facio-digital syndrome, type I	Multilobulated tongue, hyperplastic frenulae, cleft lip/palate, alopecia, supernumerary teeth, syn- and brachydactyly, polycystic kidneys	Uncommon	X-linked dominant lethal in males, #31120	BDE, p. 1309–1310
Pena-Shokeir syndrome	Multiple joint ankyloses, camptodactyly, flat face, pulmonary hypoplasia, cystic kidneys, cleft palate, Meckel's diverticulum, hydroureter	Common	Heterogeneous, most cases are inherited as autosomal recessive trait (#20815)	BDE, p. 1374–1375

Peters-plus syndrome	Peters anomaly (mesodermal dysplasia of the anterior eye segment, posterior embryotoxon, corneal opacity, staphyloma, secondary cataracts and glaucoma), rhizomelic micromelia (shortness of proximal segments), narrowing of external auditory canals	Uncommon	Most cases are inherited as autosomal recessive trait	BDE, p. 551–552
Pompe disease	Glycogenosis: absence of α-1,4-glucosidase and secondary accumulation of glycogen in affected tissues	Cardiomyopathy as a result of basic metabolic defect is very common	Autosomal recessive #23230	BDE, p. 796–797
Prader-Willi syndrome	Hypogenitalism, obesity, mental retardation, muscular hypotonia	Uncommon	Microdeletion 15q11 or maternal disomy 15	BDE, p. 1408–1411
Prune belly complex	Underdevelopment of abdominal musculature (with the wrinkled skin of abdominal wall), defects of urinary tract (dysplastic kidneys, hydro- or megaureter)	Uncommon	Etiologically heterogeneous field decect; full picture occurs in males only	BDE, p. 1420–1421
Pseudotrisomy 13	Phenotype similar to trisomy 13 (cleft lip/palate, holoprosencephaly, microcephaly, postaxial polydactyly, abnormal lobation of lungs, anal atresia, renal a- or hypoplasia, ambiguous genitalia) with normal karyotype	Common (50%)	Autosomal recessive	*[Lurie and Wulfsberg, 1993]
Renal-hepatic-pancreatic dysplasia	Polycystic kidneys, fibrosis of pancreas and liver, secondary hypoplasia of the lungs	Uncommon	Autosomal recessive #26320	BDE, p. 1010–1011
Ritscher-Schinzel (3C) syndrome	Dandy-Walker malformation, macro-, brachy- and hydrocephaly, brachydactyly, colobomata of iris, webbed neck, facial dysmorphias	Common (~80%)	Autosomal recessive	*[Marles 1995]

Appendix A—Continued
Multiple Malformations Syndromes Found in the Baltimore-Washington Infant Study (1981–1989)

Name of Syndrome	Manifestations	Cardiovascular Malformation	Etiology, McKusick Number	Birth Defects Encyclopedia (1990)
Roberts syndrome	Tetaphocomelia, microphthalmia, maxillary agenesis, microcephaly, Peters anomaly, cleft lip, polycystic kidneys	Common	Autosomal recessive #26830	BDE, p. 1498–1499
Rubella syndrome	Sensorineural deafness, hepato-splenomegaly, microcephaly, corneal clouding, cataract	Very common, mostly patent ductus arteriosus	Rubella infection of fetus	BDE, p. 723–725
Schinzel-Giedion syndrome	Widely patent sutures and fontanels, hyper-trichosis, shortness of forearms and legs, hydro-nephrosis, cryptorchidism	Common	Autosomal recessive #26915	BDE, p. 1513
Smith-Lemli-Opitz syndrome	Microcephaly, ptosis, epicanthus, cleft palate, postaxial polydactyly, syndactyly 2–3 toes, pyloric stenosis, hypoplasia or cystic dysplasia of kidneys, hydro-nephrosis, hypospadias	Common, ~40%	Autosomal recessive ##26867, 27040	BDE, p. 1570–1572
Spondylocostal dysplasia	Abnormal vertebral segmentation and defects of ribs (without defects of skull or limbs)	Uncommon	Mostly autosomal dominant (#12260)	BDE, p. 1583–1584
Thomas syndrome	Cleft lip and palate, renal agenesis or hypoplasia	Constant	Autosomal recessive	*[Thomas 1993]
Townes-Brocks syndrome	Microtia, "satyr" ear, preaxial poly-dactyly or triphalangeal thumb, anal atresia or stenosis, horse-shoe kidney	~15–20% (mostly in sporadic cases)	Autosomal dominant (#10748)	BDE, p. 155

Syndrome	Features	Frequency	Genetics	Reference
VACTERL association	Vertebral dysgenesis, Anal atresia, Cardiac defects, Tracheo-Esophageal fistula, Renal anomalies, Limb defects	~50%–60%	Unknown	BDE, p. 1743–1744
Van der Woude syndrome	Cleft palate, paramedian pits of the lower lips, club foot	Uncommon	Autosomal dominant #11930	BDE, p. 408–409
Waardenburg syndrome	Dystopia canthorum, deafness, heterochromic irides, premature greying of hair, syndactyly	Uncommon	Various types with autosomal dominant inheritance	BDE, p. 1773–1774
Williams syndrome	"Elfin" face, full lips, anteverted nares, hypercalcemia, mental retardation	Very common, mostly supra-valvular aortic stenosis	Microdeletion in 7q11; autosomal dominant #19405	BDE, p. 1179–1180

REFERENCES:

Birth Defects Encyclopedia. Buyse ML (ed.) (1990): Center for Birth Defects Information Services, Inc., Dover, 1892 p.

Bonnet J, Cordier MP, Ollagnon E, Guillaud MH, Raudrant D, Robert JM, Charvet F. (1983): Hypoplasie rénale, polydactylie, cardiopathie: Un nouveau syndrome? *J Genet Hum* 31:93–105.

Evans JA, Vitez M, Czeizel A. (1992): Patterns of acrorenal malformation associations. *Amer J Med Genet* 44:413–419.

Genuardi M, Chiurazzi P, Capelli A, Neiri G. (1993): X-linked VACTERL with hydrocephalus: The VACTEREL-H syndrome. *Birth Defects* 29(1):235–241.

Kondo I, Tamanaha K, Ashimine K. (1993): The Costello syndrome: Report of a case and review of the literature. *Jpn J Hum Genet* 38:433–436.

Lazjuk GI, Lurie IW, Cherstvoy ED. (1983): "Hereditary Syndromes of Multiple Congenital Abnormalities," Medicine: Moscow, 204 p.

Lurie IW, Kappetein AP, Loffredo CA, Ferencz C. (1995a): Noncardiac malformations in individuals with outflow tract defects of the heart: The Baltimore-Washington Infant Study (1981–1989). *Amer J Med Genet* 59:76–84.

Lurie IW, Magee CA, Sun CCJ, Ferencz C. (1995b): "Microgastria—limb reduction" complex with congenital heart disease and twinning. *Clin Dysmorphology* 4:150–155.

Lurie IW, Wulfsberg EA. (1993): "Holoprosencephaly-polydactyly" (pseudotrisomy 13) syndrome: Expansion of the phenotypic spectrum. *Amer J Med Genet* 47:405–409.

Marles SL, Chodirker BN, Greengerg CR, Chudley AE. (1995): Evidence for Ritscher-Schinzel syndrome in Canadian native Indians. *Amer J Med Genet* 56:343–350.

McKusick VA. (1990) "Mendelian Inheritance in Man", Johns Hopkins Univ. Press, 9th Edition, 2028 p.

Thomas IT, Honore GM, Jewett T, Velvis H, Garber P, Ruiz C. (1993): Holzgreve syndrome: Recurrence in sibs. *Amer J Med Genet* 45:767–769.

Appendix B
Variables Examined in the Case-Control Analysis

Family History cardiac anomalies noncardiac anomalies family history score (0 if none; 1 if cardiac; 2 if noncardiac; 3 if both) *Maternal Reproductive History* hormone pills for menstrual problems fertility medications irregular periods subfertility (trying >12 months, tests, treatments) bleeding during pregnancy number of previous: pregnancies, premature births, miscarriages, induced abortions, stillbirths	*Maternal Medication Categories* aspirin acetaminophen other analgesics narcotics antibiotics sulfonamides other antimicrobials antihistamines phenothiazines barbiturates non-barbiturate tranquilizers antidepressants sympathomimetics parasympatholytics anticonvulsants xanthines diuretics antihypertensives expectorants antitussives gastrointestinal medications progesterones corticosteroids oral contraceptives thyroid hormone insulin xanthine score (mg/day) salicylate score (mg/day)	*Maternal and Paternal Lifestyle Exposures* marijuana, cocaine, smoking, alcohol, caffeinated beverages (maternal only) *Sociodemographic Characteristics* race of infant annual family income maternal marital status maternal education maternal occupation paternal education paternal occupation onset of prenatal care (months since LMP) paternal age maternal age socioeconomic score (SES)[1] area of residence[2]
Home Characteristics type of home (house, row-house, apartment, trailer) gas heat electric heat oil heat kerosene heater frequent use of fireplace for heat gas stove	*Maternal Medical Factors* diabetes: overt, gestational thyroid disease epilepsy hypertension influenza fever urinary tract infection adiposity index (wt/ht^2)	*Maternal and Paternal Home and Occupational Exposures* carpentry art dyes painting textile dyes paint stripping hair dyes varnishing pesticides dry cleaning solvents laboratory chemicals degreasing solvents plastics miscellaneous solvents anesthetic gas autobody repair extreme heat

electric range	*Maternal and Paternal Medical Exposures*	extreme cold
microwave oven	x-rays: (by type),	laboratory viruses
cooking with kerosene,	general anesthesia	ionizing radiation
coal, or wood		lead score
		solvent score
	welding	
	soldering	
	mercury	
	arsenic	
	cadmium	
	jewelry making	
	art paints	
	stained glass	

[1]SES score (0–11) = sum of maternal education (0 for none; 1 for 1–6 years; 2 for 7–9 years; 3 for 10–11 years; 4 for 12 years; 5 for 13–15 years; 6 for 16 or more years) plus annual household income (1 for <$10,000; 2 for $10–$19,000; 3 for $20–$29,000; 4 for $30,000 or more) plus head-of-household occupation (1 for professional; 0 for other). For analysis the following categories were studied: low (0–3), medium (4–7), and high (8–11).

[2]Area of residence: urban (cities of Baltimore and Washington, D.C.), suburban (metropolitan counties of Maryland and Virginia surrounding the 2 cities), rural (remaining counties of Maryland).

Appendix C.1

Identification Numbers of Infants with Outflow Tract Anomalies and Associated NonCardiac Anomalies

Diagnostic Group	ID	Noncardiac Anomaly
Transposed Great Arteries with Intact Ventricular Septum	1991	Schinzel-Giedion
	2294	VACTERL
	11163	pyloric stenosis
	3537	kidney agenesis (left)
	1427	hydronephrosis
	10091	hydronephrosis
	1736	hypospadias
	11051	hypospadias
	3113	polydactyly
	2978	meningomyelocele, cleft palate, clinodactyly
Transposed Great Arteries with Ventricular Septal Defect	10888	trisomy 18
	2116	CHARGE
	2563	polydactyly
	1050	cleft lip and palate, ectopia cordis
Transposed Great Arteries with Double-Outlet Right Ventricle	11783	trisomy 18
	10015	Townes-Brocks
	3691	DiGeorge
	10697	VACTERL
	2302	polydactyly
	10908	accessory spleen
	10711	duplicate ureters (both kidneys), low-set ears, thrombo-cytopenia
Transposed Great Arteries, Complex types	2831	pseudotrisomy 13
	10941	VACTERL
	10077	malrotation

Diagnostic Group	ID	Noncardiac Anomaly
Tetralogy of Fallot with Pulmonic Stenosis	2133	trisomy 13
	3178	trisomy 13
	11000	trisomy 13
	11947	trisomy 13
	10913	trisomy 13
	3112	trisomy 18
	3519	trisomy 18
	10733	trisomy 18
	11303	trisomy 18
	1335	trisomy 21
	1718	trisomy 21
	1813	trisomy 21
	1953	trisomy 21
	2005	trisomy 21
	2011	trisomy 21
	2395	trisomy 21
	2585	trisomy 21
	2587	trisomy 21
	2863	trisomy 21
	3175	trisomy 21
	3204	trisomy 21
	3282	trisomy 21
	3356	trisomy 21
	3540	trisomy 21
	10933	trisomy 21
	11500	trisomy 21
	11760	trisomy 21
	10356	partial trisomy, uncharacterized
	3011	Townes-Brocks
	10423	Townes-Brocks
	1393	BBB
	1551	Noonan
	10660	Smith-Lemli-Opitz
	2230	HOMAGE
	10680	DiGeorge
	2735	DiGeorge
	1940	VACTERL
	2571	VACTERL
	10000	VACTERL

Diagnostic Group	ID	Noncardiac Anomaly
	11381	VACTERL
	11496	VACTERL
	3697	Cantrell
	10170	Cantrell
	1678	Goldenhar
	1944	Klippel-Feil
	1422	microcephaly
	11362	craniosynostosis
	11389	craniosynostosis
	10898	hydrocephaly
	1656	duodenal atresia
	2284	duodenal atresia
	2415	duodenal atresia
	2560	pyloric stenosis
	2441	omphalocele
	3374	omphalocele
	3551	kidney agenesis (left)
	11549	kidney agenesis (left)
	1414	hypospadias
	2805	accessory spleen
	1063	ambiguous genitalia
	1387	polycystic kidney
	2794	cleft palate
	10025	cleft palate
	11297	cleft lip and palate
	2200	choanal atresia
	10109	absence of right ear
	2981	bronchial defect
	1998	thrombocytopenia
	10983	anemia
	10690	tuberous sclerosis
	10117	diaphragmatic hernia
	1697	hermivertebrae, hypospadias
	11202	hydrocephaly, hydronephrosis
	11685	microcephaly, agenesis of corpus callosum
	1966	multiple minor dysmorphic facial features

Diagnostic Group	ID	Noncardiac Anomaly
	3599	multiple minor dysmorphic facial features
Tetralogy of Fallot with Pulmonic Atresia	2377	trisomy 13
	11407	trisomy 13
	3057	trisomy 21
	10792	trisomy 21
	3826	deletion 5p
	3631	Roberts
	1230	DiGeorge
	10207	DiGeorge
	10182	VACTERL
	10757	VACTERL
	11952	Cantrell
	10087	CHARGE
	1463	anal atresia
	11680	kidney agenesis (left)
	1371	hydroureter
	1322	hypospadias
Normally Related Great Arteries, Double-Outlet Right Ventricle	2182	trisomy 13
	10926	trisomy 18
	10667	trisomy 18
	1293	trisomy 21
	11279	trisomy 1q
	2081	tetrasomy 8p
	10129	Smith-Lemli-Opitz
	1357	Schinzel-Giedion
	11548	Klippel-Feil
	10412	iris defect
	2123	Meckel's diverticulum
	1311	hemivertebrae
	2125	accessory spleen
	2814	hydronephrosis, cleft palate
	1795	micrognathia, clinodactyly, low-set ears
Common Arterial Trunk	1319	trisomy 13
	1626	Holt-Oram
	1824	Holt-Oram

Diagnostic Group	ID	Noncardiac Anomaly
	3838	Holzgreve
	3543	Goldenhar
	2181	DiGeorge
	2566	DiGeorge
	2732	DiGeorge
	3106	DiGeorge
	3335	DiGeorge
	10877	DiGeorge
	11269	DiGeorge
	3353	DiGeorge
	11003	DiGeorge
	1248	microcephaly
	1355	Meckel's diverticulum
	2340	butterfly vertebrae
	2638	butterfly vertebrae
Supracristal Ventricular Septal Defect	10391	trisomy 18
	2215	trisomy 21
	3556	Smith-Lemli-Opitz
	1533	DiGeorge
	2570	DiGeorge
	11502	DiGeorge
	10115	hemivertebrae
	2779	tracheoesophageal fistula, cleft palate
	11265	nystagmus, low-set ears
Aortic-Pulmonary Windows	3586	trisomy 13, cleft palate

Appendix C.2

Identification Numbers of Down Syndrome Cases with Atrial or Ventricular Septal Defects
Baltimore-Washington Infant Study (1981–1989)

Diagnostic Group	N	n	(%)	Down Syndrome Identification Numbers							
Membranous Ventricular Septal Defect:											
moderate size defect	567	59	(10.4)	1019	1045	1222	1280	1419	1654	1714	1725
				1759	1846	2091	2157	2160	2244	2276	2315
				2328	2337	2445	2494	2518	2546	2644	2657
				2709	2804	2826	3116	3226	3248	3327	3355
				3441	3457	3640	3686	3775	3801	10056	10166
				10179	10180	10199	10281	10318	10411	10426	10633
				10795	10935	11075	11087	11189	11312	11364	11377
				11530	11535	11763					
small size defect	328	5	(1.5)	3222	3748	10055	10479				

Muscular Ventricular Septal Defect:

moderate size defect	114	5	(4.4)	2194	2776	10625	10777	10828
small size defect	315	3	(1.0)	3277	10127	10830		

Atrial Septal Defect 291 36 (12.4)

1103	1327	1488	1628	1908	1945	2060	2071
2179	2311	2327	2364	2493	2878	2934	3069
3145	3152	3153	3206	3480	3772	10126	10154
10201	10236	10641	10659	10715	10793	10944	11174
11372	11378	11558	11576				

Appendix D.1

Distribution of Potential Risk Factor
Variables Among 3572 Controls
Baltimore-Washington Infant Study (1981–1989)

Variable	Number	Percent
Family History (first-degree relatives)		
congenital heart disease	43	1.2
noncardiac anomalies	165	4.6
Maternal Reproductive History		
hormone pills for menstrual problems	49	1.4
infertility, treated with medications	56	1.6
irregular periods	574	16.1
subfertility	374	10.5
bleeding during pregnancy	534	15.0
number of previous pregnancies:		
none	1159	32.4
one	1097	30.7
two	709	19.9
three or more	607	17.0
previous preterm births	180	5.0
previous miscarriages	681	19.1
previous induced abortions	691	19.3
previous stillbirths	35	1.0
Maternal Illnesses		
overt diabetes	23	0.6
gestational diabetes	115	3.2
thyroid disease	55	1.5
epilepsy	14	0.4
overt hypertension	29	0.8
influenza	278	7.8
fever	166	4.7
urinary tract infection	467	13.1

Variable	Number	Percent
Therapeutic Drugs		
aspirin*	523	14.6
acetaminophen	1485	41.6
other analgesics	199	5.6
narcotics*	93	2.6
antibiotics	350	9.8
sulfonamides	24	0.7
other antimicrobials	28	0.8
antihistamines*	424	11.9
phenothiazines*	29	0.8
barbiturates*	33	0.9
non-barbiturate tranquilizers	48	1.3
antidepressants	6	0.2
sympathomimetics	318	8.9
parasympatholytics*	16	0.5
anticonvulsants*	11	0.3
xanthines*	235	6.6
diuretics	56	1.6
antihypertensives*	22	0.6
expectorants	62	1.7
antitussives	23	0.6
gastrointestinal medications	115	3.2
progesterones	44	1.2
corticosteroids	23	0.6
oral contraceptives*	588	16.5
local contraceptives*	608	17.0
thyroid hormone	40	1.1
insulin	16	0.5
xanthine score (mg/day):*		
0–60	858	24.0
61–120	608	17.0
121–180	591	16.5
181–240	382	10.7
241–300	396	11.1
>300	737	20.6
salicylate score (mg/day):*		
0	3065	85.8
1–500	282	7.9
501–1000	145	4.1
>1000	80	2.2

Variable	Number	Percent
Lifestyle Exposures		
marijuana (mother)	263	7.4
marijuana (father)	555	15.5
cocaine (mother)	42	1.2
cocaine (father)	111	3.1
cigarettes, number per day (mother):		
0	2302	64.5
1–10	565	15.8
11–20	529	14.8
21–39	136	3.8
40+	40	1.1
cigarettes, number per day (father):		
0	2058	58.0
1–10	510	14.4
11–20	638	18.0
21–39	226	6.4
40+	116	3.3
alcohol, frequency (mother):		
never	1471	41.2
<once a week	1229	34.4
once a week	635	17.8
daily	237	6.6
alcohol, frequency (father):*		
never	759	21.4
<once a week	827	23.3
once a week	1063	30.0
daily	897	25.3
alcohol, greatest number of drinks on one occasion (mother):		
0	1473	41.3
1–2	1137	31.8
3–4	653	18.3
5 or more	309	8.6
alcohol, greatest number of drinks on one occasion (father):*		
0	760	21.5
1–2	746	21.0
3–4	880	24.9
5 more more	1155	32.6

Variable	Number	Percent
Housing Characteristics		
type of home:*		
individual house	1479	41.7
rowhouse or townhouse	1009	28.4
apartment	985	27.8
trailer	75	2.1
gas heating*	1748	48.4
electric heating*	1106	31.0
oil heater*	742	20.8
kerosene, coal, or wood heater	202	5.7
frequent use of fireplace for heat	419	11.7
gas stove*	2017	56.5
electric stone*	1536	43.0
microwave oven	1074	30.1
cooking with kerosene, coal, or wood*	8	0.2
Medical Exposures		
mother:		
dental x-rays*	295	8.3
chest x-rays*	80	2.2
skeletal x-rays*	77	2.2
abdominal x-rays*	37	1.0
general anesthesia	105	2.9
father:		
dental x-rays*	693	19.4
chest x-rays*	148	4.1
skeletal x-rays*	197	5.5
abdominal x-rays	49	1.4
general anesthesia	98	2.7
Maternal Home and Occupational Exposures		
carpentry*	140	3.9
house painting, indoors or outdoors	367	10.3
stripping or sanding old paint	85	2.4
varnishing	106	3.0
dry cleaning solvents*	12	0.3
solvents for degreasing guns or motors	19	0.5
solvents for other uses	64	1.8
auto body repair work	10	0.3
welding*	5	0.1
soldering with lead*	17	0.5
mercury*	21	0.6

Variable	Number	Percent
cadmium*	5	0.1
arsenic*	2	0.1
jewelry making*	5	0.1
arts and crafts painting*	166	4.7
stained glass crafts*	10	0.3
art dyes*	9	0.3
textile dyes*	9	0.3
hair dyes	238	6.7
pesticides -	926	25.9
laboratory chemicals*-	57	1.6
plastics manufacturing*	10	0.3
anesthetic gas (occupational)*	26	0.7
drug manufacturing *	39	1.1
extreme heat (occupational)	4	0.1
extreme cold (occupational)*	1	0.03
laboratory viruses*	33	0.9
ionizing radiation (occupational)	48	1.3
lead score:*		
0 (no exposures)	3523	98.6
1–11 (1 or more exposures, each <once a week)	23	0.6
12–89 (1 or more exposures, each 1–2 times a week)	13	0.4
90+ (1 or more exposures, each daily)	13	0.4
solvent score:		
0 (no exposures)	2932	82.1
1–11 (1 or more exposures, each <once a week)	358	10.0
12–89 (1 more more exposures, each 1–2 times a week)	214	6.0
90+ (1 or more exposures, each daily)	68	1.9

Paternal Home and Occupational Exposures

carpentry*	771	21.6
house painting, indoors or outdoors	757	21.2
stripping or sanding old paint	250	7.0
varnishing	268	7.5
dry cleaning solvents*	14	0.4
solvents for degreasing guns or motors	553	15.5
solvents for other uses	267	7.5
auto body repair work*	236	6.6

Variable	Number	Percent
welding	228	6.4
soldering with lead	231	6.5
mercury*	19	0.5
cadmium*	8	0.2
arsenic*	8	0.2
jewelry making	4	0.1
arts and crafts painting*	46	1.3
stained glass crafts*	7	0.2
art dyes*	4	0.1
textile dyes*	4	0.1
hair dyes*	8	0.2
pesticides*	962	26.9
laboratory chemicals*	81	2.3
plastics manufacturing*	21	0.6
anesthetic gas (occupational)*	28	0.8
drug manufacturing*	16	0.5
extreme heat (occupational)*	55	1.5
extreme cold (occupational)	9	0.3
laboratory viruses	15	0.4
ionizing radiation (occupation)	32	0.9
lead score:*		
0 (no exposures)	3055	85.5
1–11 (1 or more exposures, each <once a week)	208	5.8
12–89 (1 or more exposures, each 1–2 times a week)	191	5.3
90+ (1 or more exposures, each daily)	118	3.3
solvent score:*		
0 (no exposures)	2231	62.5
1–11 (1 or more exposures, each <once a week)	627	17.6
12–89 (1 or more exposures, each 1–2 times a week)	455	12.7
90+ (1 or more exposures, each daily)	259	7.2

Sociodemographic Characteristics

race of infant:		
white	2362	66.2
nonwhite	1207	33.8
mother's age:		
<20	507	14.2

Variable	Number	Percent
20–29	2009	56.2
30+	1056	29.6
father's age:		
<20	226	6.3
20–29	1724	48.3
30+	1622	45.4
maternal martial status:		
married	2582	72.3
not married	990	27.7
mother's education:		
<high school	659	18.4
high school	1265	35.4
college	1648	46.1
father's education:		
<high school	650	18.2
high school	1298	36.3
college	1624	45.5
annual household income:		
<$10,000	686	19.2
$10,000–$19,999	699	19.6
$20,000–$29,999	737	20.6
$30,000+	1373	38.4
mother's occupation:		
not working	1142	32.0
clerical/sales	1119	31.3
service	444	12.4
factory	137	3.8
professional	730	20.4
father's occupation:		
not working	246	6.9
clerical/sales	618	17.3
service	480	13.4
factory	1279	35.8
professional	947	26.5
month of pregnancy confirmation:		
1st month	511	14.3
2nd month	1955	54.7
3rd month	663	18.6
4th month or later	416	11.6

*Variables that had no significant case-control difference in any diagnostic group.

Appendix D.2

Cardiovascular Malformations Among First-Degree Relatives of Control Infants*
Baltimore-Washington Infant Study (1981–1989)

ID	Relative	Cardiac Anomaly
Major Defects:		
8351	father	dextrocardia, abdominal situs inversus
4413	brother	total anomalous pulmonary venous return
40157	maternal half-sister	Ebstein's anomaly
5023	mother	bicuspid aortic valve
	father	pyloric stenosis
	paternal uncle	missing fingers and toes, "many gross anomalies" (no details)
Left-Sided Obstructive Defects:		
9498	father	aortic stenosis
60237	father	coarctation of aorta
8398	brother	aortic stenosis
4198	sister	hypoplastic left heart
8207	sister	aortic stenosis
7114	maternal half-sister	coarctation of aorta, Turner syndrome
Right-Sided Obstructive Defects:		
5424	father	pulmonic stenosis
6002	father	pulmonic stenosis
8066	father	pulmonic stenosis
50187	sister	pulmonic stenosis
Ventricular and Atrial Septal Defects:		
9077	father	ventricular septal defect
	maternal aunt	cleft lip and palate

ID	Relative	Cardiac Anomaly
40151	father	ventricular septal defect
9024	maternal half-brother	ventricular septal defect
5231	sister	ventricular septal defect
9426	sister	ventricular septal defect
8058	maternal half-sister	ventricular septal defect
5015	mother	atrial septal defect
6478	mother	atrial septal defect
	paternal aunt	Down syndrome
9171	father	atrial septal defect
7157	sister	atrial septal defect, hydronephrosis

Patent Atrial Duct:

7029	mother	patent arterial duct
40024	mother	patent arterial duct
40392	mother	patent arterial duct
6011	father	patent arterial duct
6012	father	patent arterial duct
4216**	brother	patent arterial duct
9086	brother	patent arterial duct

Miscellaneous Defects:

50346	mother	double aortic arch
40440	father	cardiomyopathy

Operated for Congenital Heart Defects (CHD):

40252	mother	operated for CHD
50305	mother	operated for CHD
40003	paternal half-brother	operated for CHD

Suspected Congenital Heart Defects:

4171	father	possible CHD
8074	father	suspect CHD
60113	father	possible CHD

*if affected, second-degree relatives are also shown.

**Proband has spina bifida (all other probands are free of congenital anomalies).

Appendix D.3

Noncardiac Malformations Among First-Degree Relatives Control Infants Baltimore-Washington Infant Study (1981–1989)

ID	Proband Noncardiac Anomalies	Relative	Family Member Noncardiac Anomalies
Chromosome			
9438	—	brother	Down syndrome
7114	—	maternal half-sister	Turner's syndrome, coarctation of aorta
Syndromes			
5196	—	father	Marfan syndrome
8268	—	father and pat. aunt	Marfan syndrome
6057	—	father	Poland syndrome
5395	—	father	limb-girdle muscular dystrophy
7284	—	mother, mat. grand-father, mat. aunt	Milroy's lymphedema
		paternal aunt	cleft lip and palate
9289	—	sister	Charcot-Marie syndrome
5276	albinism	mother	albinism
7406	—	brother	albinism
40136	—	mat. half-brother	albinism
		mat. cousin	deafness
40196	—	pat. half-sister	Goltz syndrome
Blood disorders			
4048	—	mother	sickle cell disease
5413	—	mother and maternal aunt	sickle cell disease
6208	sickle cell disease	sister, mother and maternal uncle	sickle cell disease
8413	sickle cell disease	father	sickle cell disease
60043	—	brother	sickle cell disease

Proband ID	Noncardiac Anomalies	Family Member Relative	Noncardiac Anomalies
6227	—	mother and maternal aunt	thalassemia
6453	—	mother	thalassemia
8200	—	mother, 3 mat. uncles, both grandparents	thalassemia
9479	—	mother	thalassemia
40035	—	mother and mat. grandmother	thalassemia, mitral valve prolapse
40463	—	father, pat. aunt, uncle and grandfather	thalessemia
50250	—	father and pat. grandfather	thalassemia
40406	—	brother	thalassemia
5113	—	brother and father	anemia
7048	—	father	elliptocytosis
40019	—	mother	thrombocytopenia
		father	rheumatic heart
50131	—	mother	thrombocytopenia
9286	—	brother	thrombocytopenia
40045	anemia	mother	anemia
4463	—	father and paternal aunt	Rh incompatibility
4490	—	father	erythrocytosis (unspecified)
5497	—	father	erythrocytosis (unspecified)
7161	—	mother	Rh incompatibility
60134	—	sister	ABO incompatibility
7003	—	father and 2 paternal aunts, pat. grandfather	angioneurotic edema
9243	—	father	erythrocytosis (unspecified)

Other Heritable Disorders

| 6010 | — | father and paternal grandfather | polycystic kidney |
| 6116 | — | maternal half-sister | cystic fibrosis |

Proband ID	Noncardiac Anomalies	Family Member Relative	Noncardiac Anomalies
7093	—	father and pat. grandfather	neurofibromatosis
4104	—	father and 2 paternal aunts	jaw deformity
60213	—	mother and 2 maternal aunts	familial hyperthryoidism
60246	—	mother and 2 mat. uncles	hypothyroidism
		2 paternal aunts, and grandmother, and 1 pat. cousin	3 kidneys each

Poly/Syndactyly

40158	—	mother	polydactyly and syndactyly
4020	—	mother	syndactyly
6210	polydactyly	mother, maternal half-brother, 2 maternal uncles, maternal aunt, and mat. grandmother	polydactyly
7369	—	mother and 2 maternal aunts	polydactyly
9422	—	mother	polydactyly
60058	—	mother	polydactyly
5350	—	mother, maternal aunt, and maternal grandfather	polydactyly
5068	—	father	polydactyly
4326	—	father	polydactyly
5299	syndactyly	father	syndactyly
6172	—	father	syndactyly
7087	—	father	syndactyly
7296	—	father and pat. grandfather	polydactyly
8497	polydactyly	sister and father	polydactyly
9276	—	father	syndactyly
7444	—	sister	polydactyly
8103	—	sister	polydactyly

Proband ID	Noncardiac Anomalies	Family Member Relative	Noncardiac Anomalies
Central Nervous System			
40107	—	sister	anencephaly
60101	—	sister	anencephaly
5403	—	mother	spina bifida
40293	—	mother	spina bifida
4168	—	father	spina bifida
9472	—	father	spina bifida, mitral valve prolapse
40498	—	father	spina bifida
		pat. grandfather	cleft lip and palate
50192	—	father	spina bifida
9018	—	mother	spina bifida
5131	—	father	hydrocephaly
7076	—	father	hydrocephaly
7085	—	brother	hydrocephaly
		paternal uncle	teratoma
60174	—	brother	hydrocephaly
9231	—	sister	hydrocephaly
Craniofacial			
4067	—	mother	cleft lip and palate
5345	—	mother	cleft palate
7184	—	mother	cleft lip and palate
7498	—	mother	cleft lip and palate
		maternal uncle	cleft lip only
8164	—	mother, maternal uncle, mat. grandmother	cleft palate
8360	—	mother	cleft lip and palate
40073	—	mother	cleft palate
6158	—	father	cleft palate
8122	—	father	cleft lip and palate
8504	—	father	cleft lip
4442	—	brother	cleft palate
6418	—	brother	cleft palate
9054	—	maternal half-sister	cleft palate
40290	—	pat. half-sister	cleft palate
60224	—	brother	cleft lip and palate

Proband ID	Noncardiac Anomalies	Family Member Relative	Noncardiac Anomalies
Sensory			
5163	—	mother	deafness
4337	—	father, paternal aunt	deafness
5245	—	father	deafness
9301	—	brother, father, paternal uncle	deafness
40359	—	brother	deafness
		maternal uncle	Scheueman syndrome
		maternal cousin	hypospadias
5199	—	sister	deafness
Urogenital			
7016	—	mother	hydronephrosis
9200	—	mother	hydronephrosis
60141	—	mother	hydronephrosis
4464	—	mother	hydronephrosis
4410	—	mother	dilated kidney
8411	—	mother	horseshoe kidney
9211	—	mother	agenesis of one kidney
7468	—	mother	duplicated urinary tracts
6218	—	father	hydronephrosis
7072	—	father	hydronephrosis
9204	—	father	duplicated renal veins
5158	—	brother	dysfunctional kidneys
7107	—	brother	hydronephrosis
7146	—	brother	hydronephrosis
4280	—	maternal half-sister	hydronephrosis
6283	—	sister	hydronephrosis
6006	—	maternal half-sister	agenesis of one kidney
7157	—	sister	hydronephrosis, atrial septal defect
7433	—	brother	hypospadias
8087	—	brother	hypospadias

Proband ID	Noncardiac Anomalies	Family Member Relative	Noncardiac Anomalies
40422	—	brother	hypospadias
50278	—	pat. half-brother	hypospadias
		pat. cousin	coarctation of aorta
7288	—	sister	Potter's syndrome

Alimentary Tract

8141	—	mother	intestinal atresia
50294	—	mother	imperforate anus
8109	pectus excava- tum	father	imperforate anus
4183	—	father	pyloric stenosis
5023	—	father	pyloric stenosis
		mother	bicuspid aortic valve
		paternal uncle	missing fingers and toes
5211	—	father	pyloric stenosis
		paternal uncle	deafness
5412	—	father	pyloric stenosis
7014	—	father	pyloric stenosis
40114	—	father	pyloric stenosis
6029	—	brother	pyloric stenosis
6226	—	brother	pyloric stenosis
6360	—	brother	intestinal atresia
7425	—	brother	intestinal atresia
8508	—	mat. half-brother	intestinal atresia
5167	—	brother	intestinal atresia
7360	—	maternal half-sister	biliary atresia

Skeletal

4414	—	mother	pectus excavatum
4284	—	father	pectus excavatum
7063	—	sister	pectus excavatum
7045	—	mother	spondylolisthesis
9168	—	father	spondylolisthesis
50127	—	mother	butterfly vertebrae
5404	—	father	butterfly vertebrae
4220	—	father	missing right lower arm

Proband ID	Noncardiac Anomalies	Family Member Relative	Noncardiac Anomalies
Miscellaneous			
5179	—	mother	omphalocele
6454	—	mother	pituitary adenoma (age 25)
6458	—	mother	polycystic lung
7507	—	mother	polycystic lung
8100	—	mother	vitiligo
40088	—	mother	thyroid hemangioma at birth
		paternal cousin	bone cancer at birth
		paternal uncle	Poland syndrome
4191	—	father	malignant tumor on hip at birth
40438	—	father	deafness, deformed ear, shrunken palate
5280	—	sister	adrenal neuro-blastoma
7229	pyloric stenosis	sister	"Endocrine disorder" (no details)
50058	—	sister	cataract

Excluded "Coagulopathy" (undefined) 40064, 40143, 40160, 40166, 42082, 40284, 50306, 50327, 50337; Leukemia 50091.

REFERENCES

1. Abbott ME. (1907) Congenital Heart Disease. In Osler W (ed:) *Modern Medicine: Its Theory and Practice.* Philadelphia: Lea Brothers & Co.
2. Abbott ME. (1908) Statistics of congenital cardiac disease. *J Med Res* 19:77–81.
3. Abbott ME. (1928) Coarctation of the aorta of the adult type II. Statistical study and historical retrospect of 200 recorded cases with autopsy of stenosis or obliteration of the descending aorta in subjects over the age of two years. *Am Heart J* 3:575–618.
4. Abbott ME. (1930) Notes to Paul Dudley White. Osler Library, McGill University, Montreal, Canada.
5. Abbott ME. (1932) The McGill University exhibit: Development of the heart and the clinical classification of congenital cardiac disease. *Br Med J* 2:1197–1199.
6. Abbott ME. (1936) *Atlas of Congenital Cardiac Disease,* 1st Ed. The American Heart Association, New York, New York.
7. Abbott ME. (1936) Congenital heart disease. *Nelson Looseleaf Medicine* 4:207–321.
8. Abbott ME, Dawson WT. (1924) The clinical classification of congenital cardiac disease. *Internat Clin* 4:156–188.
9. Adam PAJ, Teramo K, Raiha N, Gitlin D, Schwartz R. (1969) Human fetal insulin metabolism early in gestation: Response to acute elevation of the fetal glucose concentration and placental transfer of human insulin-I-131. *Diabetes* 18:409–416.
10. Adams MM, Mulinare J, Dooley K. (1989) Risk factors for conotruncal cardiac defects in Atlanta. *J Am Coll Cardiol* 14(2):432–442.
11. Al-Gazali LI, Aziz SAA, Salem F. (1996) A syndrome of short stature, mental retardation, facial dysmorphism, short webbed neck, skin changes and congenital heart disease. *Clin Dysmorphol* 5:321–327.
12. Allan LD, Cook A, Sullivan I, Sharland GK. (1991) Hypoplastic left heart syndrome: Effects of fetal echocardiography on birth prevalence. *Lancet* 337:959–961.
13. Allan LD, Sharland GK, Milburn A, Lochart SM, Groves AMM, Anderson RH, Cook AC, Fagg NLK. (1994) Prospective diagnosis of 1,006 consecutive cases of congenital heart disease in the fetus. *J Am Coll Cardiol* 23:1452–1458.
14. Alzamora-Castro V, Battilana G, Abugattas R, Sialer S. (1960) Patent ductus arteriosus and high altitude. *Am J Cardiol* 5:761–763.
15. Anderson A, Devine WA, Anderson RH, Debich DE, Zuberbuhler JR. (1990) Abnormalities of the spleen in relation to congenital malformations of the heart: A survey of necropsy findings in children. *Br Heart J* 63:122–128.
16. Anderson RH. (1988) New light on the morphogenesis of atrioventricular septal defects. *Int J Cardiol* 18:79–83.
17. Anderson RH, Baker EJ, Siew YH, Rigby M, Ebels T. (1991) The morphology and diagnosis of atrioventricular septal defects. *Cardiol Young* 1:290–305.
18. Anderson RH, Becker AE, Macartney FJ, Shinebourne EA, Wilkinson JL, Tynan MJ. (1979) Is 'triscuspid atresia' a univentricular heart? *Pediatr Cardiol* 1:51–56.
19. Anderson RH, Becker AE, Tynan M, Macartney FJ, Rigby ML, Wilkinson JL. (1984) The univentricular atrioventricular connexion: Getting to the root of a thorny problem. *Am J Cardiol* 54:822–828.
20. Anderson RH, Devine WA, Uemura H. (1995) Diagnosis of heterotaxy syndrome (letter to the editor). *Circulation* 91:906–908.
21. Anderson RH, McCartney FJ, Shinebourne EA, Tynan M. (1987) Tricuspid atresia. In *Paediatric Cardiology.* Edinburgh: Churchill Livingstone; pp. 675–696.

22. Ando M, Satomi G, Takao A. (1980) Atresia of tricuspid or mitral orifice: Anatomic spectrum and morphogenetic hypothesis. In Van Praagh R, Takao A (eds:) *Etiology and Morphogenesis of Congenital Heart Disease.* Mt. Kisco, NY: Futura Publishing Co., Inc.; pp. 421–487.

23. Ando M, Takao A, Yutani C, Nakano H, Tamura T. (1984) What is cardiac looping? Considerations based on morphologic data. In Nora JJ, Takao A (eds:) *Congenital Heart Disease: Causes and Processes.* Mt. Kisco, NY: Futura Publishing Co., Inc.; pp. 553–577.

24. Ariyuki Y, Utsu M. (1991) A case report: Prenatal diagnosis of Ellis-van Creveld syndrome. *Teratology* 44:26B.

25. Arnstein A. (1927) Eine Seltene Missbildung der Trikuspidalklappe ("Ebsteinsche Krankheit"). *Virchows Arch A* 226:247–254.

26. Bahn RC, Edwards JE, DuShane JW. (1952) Coarctation of the aorta as a cause of death in early infancy. *Pediatrics* 8:192.

27. Bamberger H. (1857) *Lehrbuch der Krankheiten des Herzens.* Wien: Wilhelm Braumuller.

28. Bartelings MM, Gittenberger-de-Groot AC. (1991) Morphogenetic considerations on congenital malformations of the outflow tract. Part 1: Common arterial trunk and tetralogy of Fallot. *Int J Cardiol* 32:213–230.

29. Bartelings MM, Gittenberger-de-Groot AC. (1991) Morphogenetic considerations on congenital malformations of the outflow tract. Part 2: Complete transposition of the great arteries and double outlet right ventricle. *Int J Cardiol* 33:5–26.

30. Beaty TH, Yang P, Munoz A, Khoury MJ. (1988) Effect of maternal and infant covariates on sibship correlation in birth weight. *Genet Epidemiol* 5:241–253.

31. Becker AE, Anderson RH. (1983) Anomalous pulmonary venous return. In *Cardiac Pathology.* New York: Raven Press; pp. 1011–1012.

32. Becker AE, Anderson RH. (1990) Isomerism of the atrial appendages—Goodbye to asplenia and all that. In Clark EB, Takao A (eds:) *Developmental Cardiology: Morphogenesis and Function.* Mt. Kisco, NY: Futura Publishing Co., Inc.; pp. 659–670.

33. Becu L, Sommerville J, Gallo A. (1976) "Isolated" pulmonary valve stenosis as part of more widespread cardiovascular disease. *Br Heart J* 38:472–482.

34. Becu LM, Fontana RS, DuShane JW, Kirklin JW, Burchell HB, Edwards JE. (1956) Anatomic and pathologic studies in ventricular septal defect. *Circulation* 14:349–364.

35. Beekman RH, Robinow M. (1985) Coarctation of the aorta inherited as an autosomal dominant tract. *Am J Cardiol* 56:818–819.

36. Belmont JW, Hawkins E, Hetjmancik JF, Greenberg F. (1987) Two cases of severe lethal Smith-Lemli-Opitz syndrome. *Am J Med Genet* 26:65–67.

37. Benson LN, Freedom RM. (1992) Pulmonary valve stenosis, pulmonary arterial stenosis, and isolated right ventricular hypoplasia. In Freedom RM, Benson LN, Smallhorn JF (eds:) *Neonatal Heart Disease.* London: Springer-Verlag; pp. 645–666.

38. Benson LN, Freedom RM. (1992) Atrial septal defect. In Freedom RM, Benson LN, Smallhorn JF (eds:) *Neonatal Heart Disease.* London: Springer-Verlag; pp. 633–643.

39. Berman W Jr, Yabek SM, Burstein J, Dillon T. (1982) Asplenia syndrome with atypical cardiac anomalies. *Pediatr Cardiol* 3:35–38.

40. Bharati S, Lev M. (1996) The future. In *The Pathology of Congenital Heart Disease: A Personal Experience with More Than 6,300 Congenitally Malformed Hearts.* Armonk, NY: Futura Publishing Co., Inc.; pp. 1541–1554.

41. Bharati S, Lev M. (1996) Ebstein's anomaly. In *The Pathology of Congenital Heart Disease: A Personal Experience with More Than 6,300 Congenitally Malformed Hearts.* Armonk, NY: Futura Publishing Co., Inc.; pp. 815–839.

42. Bialer MG, Penchaszadeh VB, Kahn E, Libes R, Krigsman G, Lesser ML. (1987) Female external genitalia and müllerian duct derivatives in 46, XY infants with Smith-Lemli Opitz syndrome. *Am J Med Genet* 28:723–731.

43. Bing RJ. (1988) The Johns Hopkins: The Blalock-Taussig Era: Perspectives in Biology and Medicine. 32:85–90.

44. Bing RJ. (1992) *Cardiology, the Evolution of the Science and the Art.* Philadelphia: Harwood Academic.

45. Blalock A, Taussig HB. (1945) The surgical treatment of malformations of the heart in which there is pulmonary stenosis or pulmonary atresia. *JAMA* 251(16):2123–2138.

46. Bleyl S, Nelson L, Odelberg SJ, Ruttenberg HD, Otterud B, Leppert M, Ward K. (1995) A

gene for familial total pulmonary venous return maps to chromosome 4p13-q12. *Am J Hum Genet* 56:408–415.

47. Bleyl S, Ruttenberg HD, Carey JC, Ward K. (1994) Familial total anomalous pulmonary venous return: A large Utah-Idaho family. *Am J Med Genet* 52:462–466.

48. Bogdanow A, Goodrich J, Hantman E, Shanske A, Marion R. (1996) A new autosomal dominant trigonoccephaly syndrome with dysmorphic facies and congenital heart defects. *Am J Hum Genet* 59S:A348.

49. Bonnet LM. (1903) Sur la lesion dite stenose congenital de l'aorte dans la region de l'isthme. *Rev Med* 23:108, 225, 335.

50. Bosi G, Sensi A, Calzolari E, Scorrano M. (1992) Familial atrial septal defect with prolonged atrioventricular conduction (letter to editor). *Am J Med Genet* 43(3):641.

51. Botto LD, Loffredo CA, Scanlon KS, Khoury MJ, Ferencz C, Moore CA, Correa-Villaseñor A, Wilson PD. (1996) Cardiac outflow tract defects in the offspring of mothers who took retinol supplements. Proceedings of the DW Smith Workshop on Malformations and Morphogenesis.

52. Boughman JA. (1988) Familial risks of congenital heart defects (letter to editor). *Am J Med Genet* 29:233.

53. Boughman JA, Astemborski JA, Berg KA, Clark EB, Ferencz C and The Baltimore-Washington Infant Study Group. (1988) Variation in expression of congenital cardiovascular malformation within and among families. In Woodhead AD, Bender MA, Leonard RC (eds:) *Phenotypic Variation in Populations: Relevance to Risk Assessment.* New York: Plenum Press; pp. 93–103.

54. Boughman JA, Berg KA, Astemborski JA, Clark EB, McCarter RJ, Rubin JD, Ferencz C. (1987) Familial risks of congenital heart defect assessed in a population-based epidemiologic study. *Am J Med Genet* 26:839–849.

55. Boughman JA, Ferencz C, Neill CA. (1989) Genetic advances in pediatric cardiology. *Curr Opin Cardiol* 4(1):53–59.

56. Boughman JA, Neill CA, Ferencz C, Loffredo C. (1993) The genetics of congenital heart disease. In Ferencz C, Rubin JD, Loffredo CA, Magee CA (eds:) *Epidemiology of Congenital Heart Disease: The Baltimore-Washington Infant Study, 1981–1989.* Mt. Kisco, NY: Futura Publishing Co., Inc.; pp. 123–167.

57. Bracken MB. (1984) *Perinatal Epidemiology.* New York: Oxford University Press.

58. Bracken MB. (1986) Drug use in pregnancy and congenital heart disease in offspring. *N Engl J Med* 314(17):1120.

59. Bracken MB, Holford TR. (1981) Exposure to prescribed drugs in pregnancy and association with congenital malformations. *Obstet Gynecol* 58:336–344.

60. Braunlin EA. (1989) Pulmonary atresia and pulmonary stenosis with hypoplastic right ventricle. In Moller JH, Neal WA (eds:) *Fetal, Neonatal, and Infant Cardiac Disease.* Norwalk, CT: Appleton and Lange; p. 671.

61. Brenner WE, Edelman DA, Hendricks CH. (1976) A standard of fetal growth for the United States of America. *Am J Obstet Gynecol* 126:555–564.

62. Brock RC. (1948) Pulmonary valvulotomy for the relief of congenital pulmonary stenosis: Report of 3 cases. *Brit Med J* 1:1121–1126.

63. Brödel M. (1941) Medical illustration. *JAMA* 117:668–672.

64. Brueckner M, McGrath J, D'Eustachio P, Horwich AL. (1991) Establishment of left-right asymmetry in vertebrates genetically distinct steps are involved. Proceedings of Ciba Foundation Symposium 156, 162:202–218.

65. Brumfitt W, Pursell R. (1973) Trimethoprim-sulfamethoxazole in the treatment of bacteriuria in women. *J Infect Dis* 128(suppl):S657–S663.

66. Burn J. (1984) Monozygotic twinning and congenital heart defects. *Biol Neonate* 45:152.

67. Burn J. (1987) The aetiology of congenital heart disease. In Anderson RH, McCartney FJ, Shinebourne EA, Tynan M (eds:) *Paediatric Cardiology* Volume 1. Edinburgh: Churchill Livingstone; pp. 15–63.

68. Burn J. (1991) Disturbance of morphological laterality in humans. In 1991 Biological Assymetry and Handedness (Ciba Foundation Symposium 162). Wiley, Chichester; pp. 282–299.

69. Burn J. (1995) Clinical genetics of 22q11 2 deletion. In Clark EB, Markwald RR, Takao A (eds:) *Developmental Mechanisms of Heart Disease.* Armonk, NY: Futura Publishing Co., Inc.

70. Burn J, Wilson DI, Cross I, Atif U, Scambler P, Takao A, Goodship J. (1995) The clinical significance of 22q11 deletion. In Clark EB, Markwald RR, Takao A (eds:) *Developmental Mechanisms of Heart Disease.* Armonk, NY: Futura Publishing Co., Inc.; pp. 559–567.
71. Campbell M. (1965) Causes of malformations of the heart. *Br Med J* 2:895–904.
72. Carmi R, Boughman JA, Ferencz C. (1992) Endocardial cushion defect: Further studies of "isolated" versus "syndromic" occurrence. *Am J Med Genet* 43:569–575.
73. Carmi R, Magee C, Neill CA, Karrer FM. (1993) Extrahepatic biliary atresia and associated anomalies: Etiologic heterogenity suggested by distinctive patterns of associations. *Am J Med Genet* 45:683–693.
74. Casey B, Devoto M, Jones KL, Ballabio A. (1993) Mapping a gene for familial situs abnormalities to human chromosome Xq24-q27.1 *Nature Genetics* 5:403–407.
75. Castellanos A, Pereiras R, Garcia A. (1937) "La angio-cardiografia radio-obscura". *Arch Soc Estud Clin Habana* 31:9–10.
76. Celermajer DS, Cullen S, Sullivan ID, Spiegelhalter DJ, Wyse RKH, Deanfield JE. (1992) Outcomes in neonates with Ebstein's anomaly. *J Am Coll Cardiol* 19(5):1041–1046.
77. Celermajer DS, Dodd SM, Greenwald SE, Wyse RKH, Deanfield JE. (1992) Morbid anatomy in neonates with Ebstein's anomaly of the tricuspid valve: Pathophysiologic and clinical implications. *J Am Coll Cardiol* 19(5):1049–1053.
78. Chavez I. (1946) Diego Rivera: Sus frescos en el Instituto Nacional de Cardiologia. Policolor, S. de RL, Mexico, D.F.
79. Cheadle WB. (1900) *Occasional Lectures on the Practice of Medicine, Addressed Chiefly to the Students of St. Mary's Medical School, to Which Are Appended the Harveian Lectures on the Rheumatism of Childhood.* London: Smith, Elder, and Company.
80. Cherstvoy ED, Lazjuk GI, Ostovskaya TI, Shved IA, Kravtzova GI, Lurie IW, Gerasimovich AI. (1984) The Smith-Lemli-Opitz Syndrome. A detailed pathological study as a clue to a etiological heterogeneity. *Virchows Arch A Pathol Anat Histopathol* 404:413–425.
81. Chung CS, Myrianthopoulos NC. (1975) Effect of maternal diabetes on congenital malformations. *Birth Defects* 11(10):23–37.
82. Clark EB. (1987) Mechanisms in the pathogenesis of congenital heart defects. In Pierpont MEM, Moller JH (eds:) *The Genetics of Cardiovascular Disease.* Boston: Martinus-Nijoff; pp. 3–11.
83. Clark EB. (1990) Hemodynamic control of the embryonic circulation. In Clark EB, Takao A, (eds:) *Developmental Cardiology: Morphogenesis and Function.* Mt. Kisco, NY: Futura Publishing Co., Inc.; pp. 291–303.
84. Clark EB. (1990) Growth, morphogenesis, and function: The dynamics of cardiac development. In Moller JH, Neal WA (eds:) *Fetal, Neonatal, and Infant Cardiac Disease.* Norwalk, CT: Appleton & Lange; pp. 3–23.
85. Clark EB, Takao A. (1990) Overview: A focus for research in cardiovascular development. In *Developmental Cardiology: Morphogenesis and Function.* Mount Kisco, New York: Futura Publishing Co., Inc.
86. Clementi M, Notari L, Borghi A, Tenconi R. (1996) Familial congenital bicuspid aortic valve: A disorder of uncertain inheritance. *Am J Med Genet* 62:336–338.
87. Coffey VP, Jessop WJE. (1955) Congenital abnormalities. *Isr J Med Sci* 349:30–46.
88. Cordero JF. (1994) Finding the causes of birth defects. *N Engl J Med* 331(1):48–49.
89. Cordier S, Ha M-C, Ayme S, Goujard J. (1992) Maternal occupational exposure and congenital malformations. *Scand J Work Environ Health* 18:11–17.
90. Corone P, Bonaiti C, Feingold J, Fromont S, Berthet-Bondet D. (1983) Familial congenital heart disease: How are the various types related? *Am J Cardiol* 51(6):942–945.
91. Corone P, Bonaiti C, Feingold J, Fromont S, Berthet-Bondet D. (1983) Familial congenital heart disease: How are the various types related? *Am J Cardiol* 51(6):942–945.
92. Correa-Villaseñor A, Ferencz C, Boughman JA, Neill CA, and The Baltimore-Washington Infant Study (BWIS) Group. (1991) Total anomalous pulmonary venous return: Familial and environmental factors. *Teratology* 44:415–428.
93. Correa-Villaseñor A, Ferencz C, Neill CA, Wilson PD, Boughman JA, and The Baltimore-Washington Infant Study (BWIS) Group. (1994) Ebstein's malformation of the tricuspid valve: Genetic and environmental factors. *Teratology* 50(2):137–147.
94. Correa-Villaseñor A, Loffredo C, Ferencz C, Wilson PD, and The Baltimore-Washington Infant Study (BWIS) Group. (1990) Lead and solvent exposure during pregnancy: Possible risk of cardiovascular malformations (CVM). *Teratology* 41(5):545.

95. Correa-Villaseñor A, McCarter RJ, Downing JW, Ferencz C, and The Baltimore-Washington Infant Study Group. (1991) White-black differences in cardiovascular malformations in infancy and socioeconomic factors. *Am J Epidemiol* 134(4):393–402.

96. Correa-Villaseñor A, McCarter RJ, Downing JW, Ferencz C, and The Baltimore-Washington Infant Study (BWIS) Group. (1992) White-black differences in cardiovascular malformations in infancy and socioeconomic factors (condensation of *Am J Epidemiol* 134:393–402, 1991). In Klaus MH, Fanaroff AA (eds:) *Year Book of Neonatal and Perinatal Medicine 1992.* New York: Mosby Year Book; pp. 246–247.

97. Cournand A, Baldwin JS, Himmelstein A. (1949) *Cardiac Catheterization in Congenital Heart Disease: A Clinical and Physiological Study in Infants and Children.* Commonwealth Fund, New York.

98. Cousineau AJ, Lauer RM, Pierpont ME, Burns TL, Ardinger RH, Patil SR, Sheffield VC. (1994) Linkage analysis of autosomal dominant atrioventricular canal defects: Exclusion of chromosome 21. *Hum Genet* 93:103–108.

99. Crafoord C, Nylin G. (1945) Congenital coarctation of the aorta and its surgical treatment. *J Thoracic Surg* 347–361.

100. Cunniff C, Williamson-Kruse L, Olney AH. (1993) Congenital microgastria and limb reduction defects. *Pediatrics* 91:1192–1194.

101. Curry CJR, Carey JC, Holland JS, Chopra D, Fineman R, Golabi M, Sherman S, Pagon RA, Allanson J, Shulman S, et al. (1987) Smith-Lemli-Opitz syndrome - type II: Multiple congenital anomalies with male pseudohermaphroditism and frequent early lethality. *Am J Med Genet* 26:45–57.

102. Cyran SE, Martinez R, Daniels S, Dignan PSJ, Kaplan S. (1987) Spectrum of congenital heart disease in CHARGE association. *J Pediatr* 110:576–578.

103. Czeizel A. (1987) Familial situs inversus and congenital heart defects. *Am J Med Genet* 28:227–228.

104. Czeizel A, Ludanyi I. (1985) An aetiological study of the VACTERL association. *Eur J Pediatr* 144:331–337.

105. Czeizel A, Mészáros M. (1981) Two family studies of children with ventricular septal defect. *Eur J Pediatr* 136:81–85.

106. Dammann JFJ, Ferencz C. (1956) The significance of the pulmonary vascular bed in congenital heart disease. I. Normal lungs. II. Malformations of the heart in which there is pulmonary stenosis. *Am Heart J* 52(1):7–17.

107. Daumer C, Nerlich A. (1992) Carpenter syndrome with complete situs inversus in a fetus at the 26th week of pregnancy. 4th Meeting of the German Society of Human Genetics, Mainz, 1992. *German Soc Hum Genetics* Abstr 133.

108. Davenport M, Savage M, Mowat P, Howard ER. (1993) Biliary atresia splenic malformation syndrome: An etiologic and prognostic subgroup. *Surgery* 113:662–668.

109. De La Cruz MV, Cayre R, Martinez OA-S, Sadowinski S, Serrano A. (1992) The infundibular interrelationships and the ventriculoarterial connection in double outlet right ventricle. Clinical and surgical implications. *Int J Cardiol* 35:153–164.

110. De La Cruz MV, DaRocha JP. (1956) An ontogenetic theory for the explanation of congenital malformations involving the truncus and conus. *Am Heart J* 51(5):782–805.

111. DeBiase L, DiCiommo V, Ballerini L, Bevilacqua M, Marcelletti C, Marino B. (1986) Prevalence of left-sided obstructive lesions in patients with atrioventricular canal without Down syndrome. *J Thorac Cardiovasc Surg* 91:467–472.

112. Debich DE, Divine WA, Anderson RH. (1990) Polysplenia with normally structured hearts. *Am J Cardiol* 65:1274–1275.

113. Digilio MC, Giannotti A, Marino B, Dallapiccola B. (1993) Atrioventricular canal and 8p-syndrome. *Am J Med Genet* 47:437–438.

114. Digilio MC, Marion B, Cicini MP, Giannotti A, Formigari R, Dallapiccola B. (1993) Risk of congenital heart defects in relatives of patients with atrioventricular canal. *Am J Dis Child* 147:1295–1297.

115. Distefano G, Romeo MG, Grasso S, Mazzone D, Sciacca P, Mollica F. (1987) Dextrocardia with and without situs viscerum inversus in two sibs. *Am J Med Genet* 27:929–934.

116. Dogramaci I, Green H. (1947) Factors in the etiology of congenital heart anomalies. *J Pediatr* 30:295–301.

117. Dost P, Majewski F, Roth H, Reckmann M, Burrig K-F. (1991) Trachealagenesie. Ein Fallbericht *Laryngorhinootologie* 70:158–160.

118. Driscoll DA, Goldmuntz E, Emanuel BS. (1995) Detection of 22q11 deletions in patients with conotruncal cardiac malformations, Di George, velocardiofacial, and conotruncal anomaly face syndromes. In Clark EB, Markwald RR, Takao A (eds:) *Developmental Mechanisms of Heart Disease.* Armonk, NY: Futura Publishing Co., Inc.; pp. 569–575.
119. Ebstein E. (1907) Wilhelm Ebstein's Arbeiten ans den Jahren 1859–1906. *Deutsches Archiv fur Klinische Medizin* 89:367–378.
120. Ebstein W. (1866) Ueber einen sehr seltenen Fall von Insufficienz der Valvula Tricuspidalis, bedingt durch eine angeborene hochgradige Missbildung derselben. *Arch Anat Physiol Wissench Med* 238–254.
121. Edwards JE. (1953) Congenital malformations of the heart and great vessels: Isolated Pulmonary Stenosis. In Gould SE (ed:) *Pathology of the Heart.* Springfield, IL: C.C. Thomas; pp. 393–395.
122. Edwards JE. (1953) Congenital malformations of the heart and great vessels: D. Malformations of the valves. In Gould SE (ed:) *Pathology of the Heart.* Springfield, IL: C.C. Thomas; pp. 369–419.
123. Eichele G. (1993) Retinoids in embryonic development. *Ann N Y Acad Sci* 678:22–36.
124. Eisenberg R, Young D, Jacobsen B, Boito A. (1964) Familial suprevalvular aortic stenosis. *Am J Dis Child* 108:341–347.
125. Ellis IH, Yale C, Thomas R, Garrett C, Winter RM. (1996) Three sibs with microcephaly, congenital heart disease, lung segmentation defects and unilateral absent kidney: a new multiple congenital anomaly (MCA) syndrome? *Clin Dysmorphol* 5:129–134.
126. Ellison RC, Peckham GJ, Lang P, Talner NS, Lerer TJ, Lin L, Dooley KJ, Nadas AS. (1983) Evaluation of the preterm infant for patent ductus arteriosus. *Pediatrics* 71(3):364–372.
127. Elzenga NJ. (1986) *The Ductus Arteriosus and Stenoses of the Adjacent Great Arteries.* Alblasserdam: Grafische Verzorging.
128. Elzenga NJ, Gittenberger-de-Groot AC. (1983) Localised coarctation of the aorta: An age dependent spectrum. *Br Heart J* 49:317–323.
129. Elzenga NJ, Gittenberger-de-Groot AC. (1995) Coarctation and related aortic arch anomalies in hypoplastic left heart syndrome. *Int J Cardiol* 8:379–389.
130. Elzenga NJ, Gittenberger-de-Groot AC, Oppenheimer-Dekker A. (1986) Coarctation and other obstructive aortic arch anomalies: Their relationship to the ductus arteriosus. *Int J Cardiol* 13:289–308.
131. Emanuel R, Somerville J, Inns A, Withers R. (1983) Evidence of congenital heart disease in the offspring of parents with atrioventricular defects. *Br Heart J* 49:144–147.
132. Emmanouilides GC, Baylen BG, Nelson RJ. (1968) Pulmonary atresia with intact ventricular septum. In Adams FH, Emmanouilides GC (eds:) *Moss' Heart Disease in Infants, Children, and Adolescents.* Baltimore: The Williams & Wilkins, Co.; pp. 263–271.
133. Emmanuel R. (1979) Genetics of congenital heart disease. *Br Heart J* 32:281–291.
134. Engle MA. (1977) Dr. Helen B. Taussig, the tetralogy of Fallot, and the growth of pediatric cardiac services in the United States. *Hopkins Med J* 140:147–150.
135. Engle MA, Taussig HB. (1950) Valvular pulmonic stenosis with intact ventricular septum and patent foramen ovale: Report of illustrative cases and analysis of clinical syndrome. *Circulation* 2:481–493.
136. Erickson M, Larsson KS. (1976) Synergistic teratogenicity of maternal salicylate treatment and protein deficiency in mice. *Teratology* 14:371.
137. Eriksson UF. (1984) Congenital malformation in diabetic animal models: A review. *Diabetes Research* 1:57–66.
138. Ferencz C. (1985) The etiology of congenital cardiovascular malformations: Observations on genetic risks with implications for further birth defects research. *J Med* 16:497–508.
139. Ferencz C. (1986) Maude Elizabeth Abbott: Pioneer in congenital heart disease. In Arntzenius AC, Dunning AJ, Snellen HA (eds:) *4th Einthoven Meeting on Past and Present Cardiology.* Assen/Maastricht, The Netherlands: Van Gorcum & Co.
140. Ferencz C. (1989) Origin of congenital heart disease: Reflections on Maude Abbott's work. *Can J Cardiol* 5(1):4–9.
141. Ferencz C. (1993) Congenital heart disease: An epidemiologic and teratologic challenge. In Ferencz C, Rubin JD, Loffredo CA, Magee CA (eds:) *Epidemiology of Congenital Heart Disease: The Baltimore Washington Infant Study 1981–1989.* Mt. Kisco, NY: Futura Publishing Co. Inc.; pp. 1–16.

142. Ferencz C, Boughman JA, Berg KA, and The Baltimore-Washington Infant Study (BWIS) Group. (1987) Genetic alterations of endothelial cells as determinants of cardiovascular maldevelopment. *Teratology* 35(2):33A–34A.

143. Ferencz C, Boughman JA, Neill CA, Brenner JI, Perry LW, and The Baltimore-Washington Infant Study (BWIS) Group. (1989) Congenital cardiovascular malformations: Questions on inheritance. *J Am Coll Cardiol* 14(3):756–763.

144. Ferencz C, Brenner JI, Loffredo C, Kappetein AP, Wilson PD. (1994) Transposition of great arteries: Etiologic distinctions of outflow tract defects in a case-control study of risk factors. In Clark EB, Markwald RR, Takao A (eds:) *Developmental Mechanisms of Heart Disease*. Armonk, NY: Futura Publishing Co., Inc.

145. Ferencz C, Correa-Villaseñor A. (1991) Epidemiology of cardiovascular malformations: The state of the art. *Cardiol Young* 1:264–284.

146. Ferencz C, Correa-Villaseñor A. (1993) Overview and research implications. In Ferencz C, Rubin JD, Loffredo CA, Magee CA (eds:) *Epidemiology of Congenital Heart Disease: The Baltimore-Washington Infant Study 1981–1989*. Mt. Kisco, NY: Futura Publishing Co., Inc.; pp. 249–255.

147. Ferencz C, Matanoski GM, Wilson PD, Rubin JD, Neill CA, Gutberlet R. (1980) Maternal hormone therapy and congenital heart disease. *Teratology* 21(2):225–239.

148. Ferencz C, Neill CA. (1992) Cardiomyopathy in infancy: Observations in an epidemiologic study. *Pediatr Cardiol* 13:65–71.

149. Ferencz C, Neill CA. (1992) Cardiovascular malformations: Prevalence at livebirth. In Freedom RM, Benson LM, Smallhorn JR (eds:) *Neonatal Heart Disease*. New York: Springer-Verlag; pp. 19–29.

150. Ferencz C, Neill CA, and The Baltimore-Washington Infant Study (BWIS) Group. (1990) Cardiomyopathy and cardiac tremors in infancy: An epidemiologic study. *Teratology* 41(5):555.

151. Ferencz C, Rubin JD. (1993) Research design, resources, and methods. In Ferencz C, Rubin JD, Loffredo CA, Magee CA (eds:) *Epidemiology of Congenital Heart Disease: The Baltimore-Washington Infant Study 1981–1989*. Mt. Kisco, NY: Futura Publishing Co., Inc.; pp. 17–32.

152. Ferencz C, Rubin JD, Loffredo C, Magee CA. (1993) *Epidemiology of Congenital Heart Disease: The Baltimore-Washington Infant Study, 1981–1989*. Mt. Kisco, NY: Futura Publishing Co., Inc.

153. Ferencz C, Rubin JD, McCarter RJ, Clark EB. (1990) Maternal diabetes and cardiovascular malformations: Predominance of double outlet right ventricle and truncus arteriosus. *Teratology* 41:319–326.

154. Ferencz C, Rubin JD, McCarter RJ, Wilson PD, Boughman JA, Brenner JI, Neill CA, Perry LW, Hepner SI, Downing JW. (1984) Hematologic disorders and congenital cardiovascular malformations: Converging lines of research. *J Med* 15(5/6):337–354.

155. Flexner S, Flexner JT. (1941) *William Henry Welch and the Heroic Age of American Medicine*. New York: The Viking Press.

156. Fraer L, Marchese S, Juda S, Surti U, Huff D, Sherman F, Martin J, Hill LM. (1992) Prenatal diagnosis of a de novo 8p23.1 distal deletion. *Am J Hum Genet* 51:A408.

157. Francomano CA, Cutting GR, McCormick MK, Chu ML, Timpl R, Hong HK, Antonarakis SE. (1991) The COL6A1 and COL6A2 genes exist in a gene cluster and detect highly informative DNA polymorphisms in the telomeric region of chromosome 21q. *Hum Genet* 87:162–166.

158. Freed M. (1995) Congenital heart defects: Acyanotic and cyanotic. In Gewitz MH (ed:) *Primary Pediatric Cardiology*. Armonk, NY: Futura Publishing Co., Inc.; pp. 208–211.

159. Freedom RM. (1974) Aortic valve and arch anomalies in the congenital asplenia syndrome. *Hopkins Med J* 135:124–135.

160. Freedom RM. (1987) Atresia or hypoplasia of the left atrioventricular and/or ventriculo-arterial junction. In Anderson RM, Macartney FJ, Shinebourne EA, Tynan M (eds:) *Paediatric Cardiology*. Edinburgh, London, Melbourne, and New York: Churchill Livingstone; pp. 737–764.

161. Freedom RM, Burrows PE, Smallhorn JF. (1992) Pulmonary atresia and intact ventricular septum. In Freedom RM, Benson LN, Smallhorn JF (eds:) *Neonatal Heart Disease*. London: Springer-Verlag; pp. 285–307.

162. Freedom RM, Smallhorn J, Trusler GA. (1992) Transposition of the great arteries. In

Freedom RM, Benson LN, Smallhorn JF (eds:) *Neonatal Heart Disease.* London: Springer-Verlag; pp. 179–212.

163. Fricker FJ, Zuberbuhler JR. (1987) Pulmonary atresia with intact ventricular septum. In Anderson RH, Macartney FJ, Shinebourne EA, Tynan M (eds:) *Paediatric Cardiology.* Edinburgh, London, Melbourne, and New York: Churchill Livingstone; pp. 711–720.

164. Friedman WE, Novak V, Johnson AD. (1979) Congenital aortic stenosis in adults. In Roberts WL (ed:) *Congenital Heart Disease in Adults.* Philadelphia: F.A. Davis Co.; pp. 235–251.

165. Fritz H. (1976) The effect of cortisone on the teratogenic action of acetylsalicylic acid and diphenylhydantoin in the mouse. *Experientia* 32:721–722.

166. Frontera-Izquierdo P, Cabezuelo-Huerta G. (1990) Natural and modified history of complete atrioventricular septal defect: A 17 year study. *Arch Dis Child* 65:964–967.

167. Frost SB. (1991) *The Man in the Ivory Tower: F. Cyril James of McGill.* Montreal: McGill-Queen's University Press.

168. Fryns J-P, Moerman P. (1993) Short limbed dwarfism, genital hypoplasia, sparse hair, and vertebral anomalies. A variant of Ellis-van Creveld syndrome? *Am J Med Genet* 30:322–324.

169. Fyler DC. (1992) Pulmonary stenosis. In *Nadas' Pediatric Cardiology.* Philadelphia: Hanley & Belfus/Mosby - Year Book, Inc.; pp. 459–470.

170. Fyler DC. (1992) Patent ductus arteriosus. In *Nadas' Pediatric Cardiology.* Philadelphia: Hanley & Belfus/Mosby - Year Book, Inc.; pp. 525–534.

171. Fyler DC. (1992) Double outlet right ventricle. In *Nadas' Pediatric Cardiology.* Philadelphia: Hanley and Belfus/Mosby -Yearbook, Inc.; pp. 643–648.

172. Fyler DC. (1992) The Nadas years: 1949–1982. In *Nadas' Pediatric Cardiology.* Philadelphia: Hanley & Belfus/Mosby—Year Book, Inc.; pp. 1–4.

173. Fyler DC. (1996) Pulmonary atresia with intact ventricular septum. In *Nadas' Pediatric Cardiology.* Philadelphia: Hanley & Belfus, Inc.; pp. 635–642.

174. Fyler DC, Buckley LP, Hellenbrand WE, Cohn HE, Kirklin JW, Nadas AS, Cartier JM, Breibart MH. (1980) Report of the New England Regional Infant Cardiac Program. *Pediatrics* 65(2)(suppl):375–461.

175. Gardener JH, Keith JD. (1951) Prevalence of heart disease in Toronto children: 1948–1949; Cardiac registry. *Pediatrics* 7:713.

176. Garrod AE. (1899) Cases illustrating the association of congenital heart disease with the "Mongolian" form of idiocy. *Trans Clin Soc Lond* 32:6–10.

177. Gennarelli M, Novelli G, Digilio MC, Giannotti A, Marini B, Dallapiccola B. (1994) Exclusion of linkage with chromosome 21 in families with recurrence of non-Down's atrioventricular canal. *Hum Genet* 94:708–710.

178. Gerboni S, Sabatino G, Mingarelli R, Dallapiccola B. (1993) Coarctation of the aorta, interrupted aortic arch, and hypoplastic left heart syndrome in three generations. *J Med Genet* 30:328–329.

179. Gersony WM, Peckham GJ, Ellison RC, Miettinen OS, Nadas AS. (1983) Effects of indomethazine in premature infants with patent ductus arteriosus: Results of a national collaborative study. *J Pediatr* 102(6):895–906.

180. Gibbons GH, Dzau VJ. (1994) The emerging concept of vascular remodelling. *N Eng J Med* 330(20):1431–1438.

181. Gittenberger-de-Groot AC, Bartelings MM, Poelmann RE. (1995) Overview: Cardiac morphogenesis. In Clark EB, Markwald RR, Takao A (eds:) *Developmental Mechanisms of Heart Disease.* Armonk, NY: Futura Publishing Co., Inc.; pp. 157–168.

182. Gittenberger-de-Groot AC, Moulaert AJM, Hitchcock JF. (1980) Histology of the persistent ductus arteriosus in cases of congenital rubella. *Circulation* 62(1):183–186.

183. Gittenberger-de-Groot AC, Strengers JLM, Mentink M, Poelmann RE, Patterson DF. (1985) Histologic studies on normal and persistent ductus arteriosus in the dog. *J Am Coll Cardiol* 6(2):394–404.

184. Glauser TA, Rorke LB, Weinberg PM, Clancy RR. (1990) Congenital brain anomalies associated with the hypoplastic left heart syndrome. *Pediatrics* 85(6):984–990.

185. Glaz J. (1993) Approximations for the tail probabilities and moments of the scan statistic. *Stat Med* 12:1845–1852.

186. Goldmuntz E, Driscoll D, Budarf ML, Zackai EH, McDonald-McGinn DM, Biegel JA,

Emanuel BS. (1993) Microdeletions of chromosomal region 22q11 in patients with congenital conotruncal cardiac defects. *J Med Genet* 30:807–812.

187. Grant RP. (1962) The morphogenesis of transposition of the great vessels. *Circulation* 26:819–840.

188. Greene MF, Armon KH, Lin A, Benacerraf BR, Holmes LB. (1996) Is there a "diabetic embryopathy"? *Teratology* 53:88.

189. Gregg NMC. (1941) Congenital cataract following German measles in the mother. *Trans Opthalmol Soc Aust* 3:35–46.

190. Gross RE, Hubbard JP. (1939) Surgical ligation of a patent ductus arteriosus. Report of first successful case. *JAMA* 112:729–731.

191. Gross RE, Hufnagel CA. (1945) Coarctation of the aorta experimental studies regarding its surgical correction. *N Eng J Med* 233:287–293.

192. Gutgesell HP, Massaro TA. (1995) Management of hypoplastic left heart syndrome in a consortium of university hospitals. *Am J Cardiol* 76:809–811.

193. Hakosalo J, Saxen L. (1971) Influenza epidemic and congenital defects. *Lancet* 2:1346–1347.

194. Hamilton WF, Abbott ME. (1928) Coarctation of the aorta of the adult type: Complete obliteration of the descending arch at insertion of the ductus. *Am Heart J* 3:381–421.

195. Hammon JW, Lupinetti FM, Maples MD, Merrill WH, Frist WH, Graham TP, Bender HW. (1988) Predictors of operative mortality in critical valvular aortic stenosis presenting in infancy. *Ann Thorac Surg* 45:537–540.

196. Hasegawa T, Yamada K, Yokochi A, Enomoto S. (1981) Two unrelated boys with partial trisomy 1q. *Japan J Hum Genet* 26:183.

197. Herman TE, Siegel MJ, Lee BC, Dowton SB. (1993) Smith-Lemli-Opitz syndrome type II: Report of a case with additional radiographic findings. *Pediatr Radiol* 23:37–40.

198. Hoffman JIE, Rudolph AM. (1965) The natural history of ventricular septal defects in infancy. *Am J Cardiol* 16:634–653.

199. Hornberger LK, Sahn DJ, Kleinman CS, Copel JA, Reed KL. (1991) Tricuspid valve disease with significant tricuspid insufficiency in the fetus: Diagnosis and outcome. *J Am Coll Cardiol* 17(1):167–173.

200. Houlston RS, Ironton R, Temple IK. (1994) Association of atrial-ventricular septal defect, blepharophimosis, anal and radial defects in sibs: A new syndrome? *Genetic Counseling* 5:93–96.

201. Hustinx R, Verloes A, Grattagliano B, Herens C, Jamar M, Soyeur D, Schaaps, J-P, Koulisher L. (1993) Monosomy 11q: Report of two familial cases and review of the literature. *Am J Med Genet* 47:312–317.

202. Hutchins GM, Moore GW, Lipford EH, Haupt HM, Walker MC. (1983) Asplenia and polysplenia malformation complexes explained by abnormal embryonic body curvature. *Path Res Pract* 177:60–76.

203. Hwang S-J, Beaty TH, Panny SR, Street NA, Joseph JM, Gordon S, McIntosh I, Francomano CA. (1995) Association study of transforming growth factor alpha (TGFa) Taql polymorphism and oral clefts: Indication of gene-environmental interaction in a population-based sample of infants with birth defects. *Am J Epidemiol* 141:629–636.

204. Icardo JM, Arrechedera H, Colvee E. (1995) Atrioventricular endocardial cushions in the pathogenesis of common atrioventricular canal: Morphological study in the iv/iv mouse. In Clark EB, Markwald RR, Takao A (eds) *Developmental Mechanisms of Heart Disease.* Armonk, NY: Futura Publishing Co., Inc.; pp. 529–544.

205. Ilyina HG, Lurie IW (1984) Neural tube defects in sibs of children with tracheo-oesophageal dysraphism. *Am J Med Genet* 21:73–74.

206. International Society of Cardiology. (1970) Classification of Heart Disease in Childhood. V.R.B. Offsetdrukkerij, Groningen.

207. Ivemark B. (1955) Implications of agenesis of the spleen on the pathogenesis of conotruncus anomalies in childhood. *Acta Paediatr Scand* 44(104 suppl):1–110.

208. Jackson L, Kline AD, Barr MA, Koch S. (1994) DeLange syndrome: A critical review of 310 individuals. *Am J Med Genet* 49:240–243.

209. Jaffe JH, Martin WR. (1990) Opioid analgesics and antagonists. In Gilman AG, Rall WW, Nies AS, Taylor P (eds:) *Goodman and Gilman's The Pharmacological Basis of Therapeutics.* New York: Pergamon Press; pp. 485–521.

210. Jatene AD, Fontes VG, Paulista PP, Souza LCB, Neger F, Galantier M, Sousa JEMR.

(1976) Anatomic correction of transposition of the great vessels. *J Thorac Cardiovasc Surg* 73:363–370.

211. Johnson AL, Ferencz C, Wiglesworth RW, McRae DL. (1951) Coarctation of the aorta complicated by patency of the ductus arteriosus. *Circulation* 4:242–250.

212. Johnson JA, Aughton DJ, Comstock CH, van Oeyen PT, Higgins JV, Schulz R. (1994) Prenatal diagnosis of Smith-Lemli-Optiz syndrome, type II. *Am J Med Genet* 49:240–243.

213. Källèn B. (1988) Surveillance of adverse reproductive outcomes. In *Epidemiology of Human Reproduction.* Boca Raton: CRC Press; pp. 74–87.

214. Kan JS, White RI, Mitchell SE, Gardner TJ. (1982) Percutaneous balloon valvuloplasty: A new method for treating congenital pulmonary-valve stenosis. *N Eng J Med* 307:540–542.

215. Kaplan S, Daoud GI, Benzing III G, Devine FJ, Glass IH, McGuire J. (1963) Natural history of ventricular septal defect. *Am J Dis Child* 105:581–587.

216. Kappetein AP, Gittenberger-de-Groot AC, Zwinderman AH, Rohmer J, Poelmann RE, Huysmans HA. (1991) The neural crest as a possible pathogenetic factor in coarctation of the aorta and bicuspid aortic valve. *J Thorac Cardiovasc Surg* 102:830–836.

217. Kappetein AP, Zwinderman AH, Bogers AJ, Rohmer J, Huysmans HA. (1994) More than 35 years of coarctation repair: An unexpected high relapse rate. *J Thoracic Cardiol Surg* 107:87–95.

218. Keith JD, Rowe RD, Vlad P. (1958) *Heart Disease in Infancy and Childhood.* Third Ed. New York: MacMillan Publishing Company.

219. Keller BB. (1995) Overview: Functional maturation and coupling of the embryonic cardiovascular system. In Clark EB, Markwald RR, Takao A (eds:) *Developmental Mechanisms of Heart Disease.* Armonk, NY: Futura Publishing Co., Inc.; pp. 367–385.

220. Kessler II, Levine M. (1970) *The Community as an Epidemiologic Research Laboratory: A Casebook of Community Studies.* Baltimore, Maryland: The Johns Hopkins University Press.

221. Khoury MJ, Erickson JD. (1992) Can maternal risk factors influence the presence of major birth defects in infants with Down syndrome? *Am J Med Genet* 43:1016–1022.

222. Kidd BSL, Tyrell MJ, Pickering D. (1971) Transposition 1969. In Kidd BSL, Keith JD (eds:) *Congenital Heart Defects.* Springfield, Illinois: Charles C. Thomas; pp. 127–137.

223. Kidd SA, Lancaster PAL, McCredie RM. (1993) The incidence of congenital heart defects in the first year of life. *J Paediatr Child Health* 29:344–349.

224. Kimmel CA, Butcher RE, Vorhees CV, Schumacher HJ. (1974) Metal-salt potentiation of salicylate-induced teratogenesis and behavioral changes in rats. *Teratology* 10:293–300.

225. Kirby ML. (1983) Neural crest cells contribute to normal aorticopulmonary septation. *Science* 220:1059–1061.

226. Kirby ML, Bockman DE. (1984) Neural crest and normal development: A new perspective. *Anat Rec* 209:1–6.

227. Kitchiner D, Jackson M, Walsh K, Peart I, Arnold R. (1993) The progression of mild congenital aortic valve stenosis from childhood into adult life. *Internat J Cardiol* 42:217–223.

228. Kline J, Stein Z, Susser M. (1989) *Conception to Birth: Epidemiology of Prenatal Development.* New York: Oxford University Press.

229. Koifmann CP, Wajntal A, de Souza DH, Gonzalez CH, Coates MV. (1993) Human situs determination and chromosome constitution 46,XY,ins(7;8)(q22;q12q24). *Am J Hum Genet* 47:568–569.

230. Korenberg JR, Kawashima H, Pulst S. (1990) Molecular definition of region of chromosome 21 that causes features of the Down syndrome phenotype. *Am J Hum Genet* 47:236–246.

231. Korenberg JR, Kurnit DM. (1995) Molecular and stochastic basic of congenital heart disease in Down syndrome. In Clark EB, Markwald RR, Takao A (eds:) *Developmental Mechanisms of Heart Disease.* Armonk, NY: Futura Publishing Co., Inc.; pp. 581–596.

232. Kotzot D, Schmitt S, Bernasconi F, Robinson WP, Lurie IW, Ilyina H, Méhes K, Hamel BCJ, Otten BJ, Hergersberg M, et al. (1996) Uniparental disomy 7 in Silver-Russell syndrome and primordial growth retardation. *Hum Molecular Genet* 4:583–588.

233. Kretzer FL, Hittner HM, Mehta RS. (1981) Ocular manifestations of the Smith-Lemli-Opitz syndrome. *Arch Ophthalmol* 99:2000–2006.
234. Kruetz RV (1981) Untersuchungen uber den Einfluss von trimethoprim auf die intrauterine Entwicklung der Ratte. *Anat Anz Jena* 149:151–159.
235. Kučera J. (1977) Relation between population genetics and population teratology. In Szabo G, Papp Z (eds:) *Medical Genetics.* Amsterdam, Oxford: Excerpta Medica; pp. 459–473.
236. Kurppa K, Holmberg PC, Kuosma E, Aro T, Saxen L. (1991) Anencephaly and maternal common cold. *Teratology* 44:51–55.
237. Lachman MF, Wright Y, Whiteman DAH, Herson V, Greenstein RM. (1991) Brief clinical report: A 46, XY phenotypic female with Smith-Lemli-Opitz syndrome. *Clin Genet* 39:136–141.
238. Laegreid L, Olegard R, Wahlstrom J, Conradi N. (1987) Abnormalities in children exposed to benzodiazepines in utero. *Lancet* 1:108–109.
239. Lamy M, DeGrouchy J, Schweisguth O. (1957) Genetic and non-genetic factors in the etiology of congenital heart disease: A study of 1188 cases. *Am J Hum Genet* 9:17–41.
240. Landing BH. (1984) Five syndromes of pulmonary symmetry, congenital heart disease and multiple spleens. *Pediatr Pathol* 2:125–151.
241. Lang D, Oberhoffer R, Cook A, Sharland G, Allan LD, Fagg N, Anderson RH. (1991) Pathologic spectrum of malformations of the tricuspid valve in prenatal and neonatal life. *J Am Coll Cardiol* 17(5):1161–1167.
242. Lang P, Fyler DC. (1992) Hypoplastic left heart syndrome, mitral atresia, and aortic atresia. In: *Nadas' Pediatric Cardiology.* Philadelphia: Hanley & Belfus; pp. 623–634.
243. Leck I. (1963) Incidence of malformations following influenza epidemics. *Br J Prev Soc Med* 17:70–80.
244. Lenz W. (1988) A short history of thalidomide embryopathy. *Teratology* 38:203–215.
245. Leung MP, McKay R, Smith A, Anderson RH, Arnold R. (1991) Critical aortic stenosis in early infancy. *J Thorac Cardiovasc Surg* 101:526–531.
246. Lev M. (1952) Pathologic anatomy and interrelationship of hypoplasia of the aortic tract complexes. *Lab Invest* 1:61–70.
247. Lev M, Bharati S, Meng CCL, Liberthson RR, Paul MH, Idriss F. (1972) A concept of double-outlet right ventricle. *J Thorac Cardiovasc Surg* 64(2):271–281.
248. Levy HL, Ghavami M. (1996) Maternal phenylketonuria: A metabolic teratogen. *Teratology* 53:176–183.
249. Levy HL, Waisbren SE, Lobbregt D, Allred E, Schuler A, Trefz FK, Schweitzer SM, Sardharwalla IB, Walter JH, Barwell BE, et al. (1994) Maternal mild hyperphenylalaninemia. An international survey of offspring outcome. *Lancet* 344:1589–1594.
250. Levy-Mozziconacci A, Lacombe D, Leheup B, Wernert F, Rouault F, Philip N. (1996) La microdeletion du chromosome 22q11 chez l'enfant: A propos d'une serie de 49 patients. *Arch Pediatr* 3:761–768.
251. Lewin M, Lindsay EA, Jurecic V, Towbin J, Baldini A. (1996) A genetic etiology for interrupted aortic arch type B. *Am J Hum Genet* 59–S:A20.
252. Lewis AJ, Ongley PA, Kincaid OW, Ritter DG. (1969) Supravalvular aortic stenosis. Report of a family with peculiar somatic features and normal intelligence. *Dis Chest* 55:372–379.
253. Lewis DA, Loffredo CA, Correa-Villaseñor A, Wilson PD, Martin GR. (1996) Descriptive epidemiology of membranous and muscular ventricular septal defects in the Baltimore-Washington Infant Study. *Cardiol Young* 6:281–290.
254. Lin AE, Siebert JR, Graham JMJ. (1990) Central nervous system malformations in the CHARGE association. *Am J Med Genet* 37:304–310.
255. LiVolti S, Distefano G, Garozzo R, Romeo MG, Sciacca P, Mollica F. (1991) Autosomal dominant atrial septal defect of ostium secundum type: Report of three families. *Ann Genet* 34(1):14–18.
256. Loffredo C, Ferencz C, Correa-Villaseñor A, and The Baltimore-Washington Infant Study (BWIS) Group. (1991) Organic solvents and cardiovascular malformations in the Baltimore-Washington Infant Study. *Teratology* 43:450.
257. Loffredo C, Rubin JD, Correa-Villaseñor A, Magee CA, Wilson PD. (1993) Preparing the database and creating the variables. In Ferencz C, Rubin JD, Loffredo CA, Magee CA

(eds:) *Epidemiology of Congenital Heart Disease: The Baltimore-Washington Infant Study 1981–1989.* Mt Kisco, NY: Futura Publishing Co., Inc.; pp. 81–90.

258. Loffredo CA. (1996) The interaction of prenatal solvent exposures with genetic polymorphisms in solvent-metabolizing enzymes: Evaluation of risk among infants with congenital heart defects. Doctoral thesis. University of Maryland at Baltimore, School of Medicine.

259. Loffredo CA, Ewing CK. (1996) Use of stored newborn blood spots in research on birth defects: Variation in retrieval rates by type of defect and infant characteristics. *Am J Med Genet* 66:1–8.

260. Loffredo CA, Ferencz C, Rubin JD, Correa-Villaseñor A, Wilson PD, and The Baltimore-Washington Infant Study. (1996) A comparative epidemiologic evaluation of risk factors for hypoplastic left heart syndrome, aortic stenosis, and coarctation of the aorta. *Teratology* 53:115/P48.

261. Lózsádi K. (1983) A veleszületett szivbetegségek kliniko patologiaja (clinical pathology of congenital heart disease). Medicina, Budapest.

262. Luke MJ. (1966) Valvular pulmonic stenosis in infancy. *J Pediatr* 68(1):90–102.

263. Lupinetti FM, Bove EL, Minich LL, Snider AR, Callow LB, Meliones JN, Growley DC, Beekman RH, Serwer G, Dick M II, et al. (1992) Intermediate-term survival and functional results after arterial repair for transposition of the great arteries. *J Thorac Cardiovasc Surg* 103:421–427.

264. Lurie IW, Ferencz C. (1996) "Shifted" threshold may explain diversity of cardiovascular malformations in multiple congenital anomalies syndromes. 3C syndrome as an example. *Am J Med Genet* 66:72–74.

265. Lurie IW, Kappetein P, Loffredo CA, Ferencz C. (1995) Non-cardiac malformations in outflow tract defects of the heart: The Baltimore-Washington Infant Study (1981–1989). *Am J Med Genet* 59:76–84.

266. Lurie IW, Magee CA, Sun C-CJ, Ferencz C. (1995) "Microgastria-limb reduction" complex with congenital heart disease and twinning. *Clin Dysmorphol* 4:150–155.

267. Lynberg MC, Khoury MJ, Lammer EJ, Waller KO, Cordero JF, Erickson JD. (1990) Sensitivity, specificity, and positive predictive valve of multiple malformations in Isotretinoin embryopathy surveillance. *Teratology* 42:513–519.

268. Lynberg MC, Khoury MJ, Lu X, Cocian T. (1994) Maternal flu, fever and the risk of neural tube defects: A population-based case-control study. *Am J Epidemiol* 140: 244–255.

269. Lynch HT, Bachenberg K, Harris RE, Becker W. (1996) Hereditary atrial septal defect: Update of a large kindred. *Am J Dis Child* 132:600–604.

270. Lynxweiler CP, Smith S, Babich J. (1996) Coarctation of the aorta. Report of a case. *Arch Pediatr* 68:203–207.

271. Maestri NE. (1988) The Familial Aggregation of Congenital Cardiovascular Malformations. Doctoral Thesis Johns Hopkins University School of Hygiene and Public Health.

272. Maestri NE, Beaty TH, Clark EB, Connolly M, Boughman JA, Ferencz C. (1987) Sibship aggregation of congenital heart defects in families of cases with blood flow malformations. *Am J Hum Genet Suppl* 41:A258.

273. Magee CA, Loffredo CA, Correa-Villaseñor A, Wilson PD. (1993) Environmental factors in occupations, home, and hobbies. In Ferencz C, Rubin JD, Loffredo CA, Magee CA (eds:) *Epidemiology of Congenital Heart Disease: The Baltimore-Washington Infant Study 1981–1989.* Mt. Kisco, NY: Futura Publishing Co., Inc.; pp. 207–231.

274. Männer J, Seidl E, Steding G. (1995) Embryological observations on the morphogenesis of double-outlet right ventricle with subaortic ventricular septal defect and normal arrangement of the great arteries. *Thorac Cardiovasc Surgeon* 43:307–312.

275. Männer J, Seidl W, Steding G. (1993) Correlation between the embryonic head flexures and cardiac development: An experimental study in chick embryos. *Anat Embryol* 188:269–285.

276. Maraist F, Daley R, Draper AJ, Heimbecker R, Dammann JFJ, Kieffer RJ, King TJ, Ferencz C, Bing RJ. (1951) Physiological studies in congenital heart disease. X. The physiological findings in thirty-four patients with isolated pulmonary valvular stenosis. *Bulletin of the Johns Hopkins Hospital* 88:1–19.

277. Marino B. (1989) Left-sided cardiac obstruction in patients with Down syndrome (Letter). *J Pediatr* 115:834–835.

278. Markwald RR. (1995) Overview: Formation and early morphogenesis of the primary heart tube. In Clark EB, Markwald RR, Takao A (eds:) *Developmental Mechanisms of Heart Disease.* Armonk, NY: Futura Publishing Co., Inc.; pp.149–184.

279. Markwald RR, Fitzharris TP, Bank H, Bernanke DH. (1978) Structural analyses on the matrical organization of glycosaminoglycans in developing endocardial cushions. *Dev Biol* 62:292–316.

280. Markwald RR, Fitzharris TP, Manasek FJ. (1977) Structural development of endocardial cushions. *Am J Anat* 148:85–120.

281. Markwald RR, Mjaatveldt CH, Krug EL. (1990) Induction of endocardial cushion tissue formation by adheron-like molecular complexes derived from myocardial basement membrane. In Clark EB, Takao A (eds:) *Developmental Cardiology: Morphogenesis and Function.* Mt. Kisco, NY: Futura Publishing Co., Inc.; pp.191–204.

282. Maron BJ. (1979) Coarctation of the aorta in the adult. In Roberts WL (ed:) *Congenital Heart Disease in Adults.* Philadelphia: F.A. Davis Co.; pp. 311–319.

283. Martin GR, Perry LW, Ferencz C. (1989) Increased prevalence of ventricular septal defect: Epidemic or improved diagnosis. *Pediatrics* 83(2):200–203.

284. Martin ML, Adams MM, Mortensen ML. (1990) Descriptive epidemiology of selected malformations of the aorta, Atlanta, 1970–1983. *Teratology* 42:272–283.

285. McCarter RJ, Kessler II, Comstock GW. (1987) Is diabetes mellitus a teratogen or a coteratogen? *Am J Epidemiol* 125(2):195–205.

286. McCredie J. (1974) Embryonic neuropathy: A hypothesis of neural crest injury as the pathogenesis of congenital malformations. *Med J Aust* 1:159–163.

287. McDermott HE. (1941) *Maude Abbott: A Memoir.* Toronto: The MacMillan Company of Canada Ltd.

288. McDonald-McGinn DM, Driscoll DA, Emanuel BS, Zackai EH. (1996) The 22q11.2 deletion in African-American patients: An underdiagnosed population. *Am J Hum Genet* 59-S:A20.

289. McNamara DG, Manning JA, Engle MA, Whittemore R, Neill CA, Ferencz C. (1987) Helen Brooke Taussig: 1898–1986. *J Am Coll Cardiol* 10(3):662–671.

290. Méhes K. (1996) Classical clinical genetics in the era of molecular genetics. *Am J Med Genet* 61:394–395.

291. Meinecke P, Padberg B, Laas R. (1990) Agnathia, holoprosencephaly, and situs inversus: A third report. *Am J Med Genet* 37:286–287.

292. Meyers GA, Orlow SJ, Munro IR, Przyleps KA, Jabs EW. (1995) Fibroblast growth factor receptor 3 (FGFR3) transmembrane mutation in Crouzon syndrome with acanthosis nigricans. *Nature Genetics* 11:462–464.

293. Mikkila SP, Janas M, Karikoski R, Tarkkila T, Simola KOJ. (1994) X-linked laterality sequence in a family with carrier manifestations. *Am J Hum Genet* 49:435–438.

294. Miller ME, Smith DW. (1979) Conotruncal malformation complex: Examples of possible monogenic inheritance. *Pediatrics* 63(6):890–893.

295. Mills JL. (1979) Malformations in infants of diabetic mothers occur before the seventh gestational week: Implications for treatment. *Diabetes* 28:292–293.

296. Mills JL. (1982) Malformations in infants of diabetic mothers. *Teratology* 25:385–394.

297. Mills JL, Fishl AR, Kropp RH, Ober CL, Jovanovic LG, Polk BF, and the NICHD-Diabetes in Early Pregnancy Study. (1983) Malformations in infants of diabetic mothers: Problems in study design. *Prev Med* 12:274–286.

298. Mills JL, Graubard BI. (1987) Is moderate drinking during pregnancy associated with an increased risk of malformations? *Pediatrics* 80:309–314.

299. Mills JL, Knopp RH, Simpson JL, Jovanovic-Pederson L, Metzger BE, Holmes LE, Arons JH, Brown Z, Reed GF, Bieber FR, et al. (1988) Lack of relation of increased malformation rates in infants of diabetic mothers to glycemic control during organogenesis. *N Eng J Med* 381:671–676.

300. Mills JL, McPartlin JM, Kirke PN, Lee YJ, Conley MR, Weir DG, Scott JM. (1995) Homocysteine metabolism in pregnancies complicated by neural-tube defect. *Lancet* 345:149–151.

301. Mitchell SC, Berendes HW, Clark WMJ. (1967) The normal closure of the ventricular septum. *Am Heart J* 73(3):334–338.

302. Mitchell SC, Sellmann AH, Westphal MC, Park J. (1971) Etiologic correlates in a study of congenital heart disease in 56,109 births. *Am J Cardiol* 28:653–657.

303. Moller JH. (1994) Fifty years of pediatric cardiology and challenges for the future (Alexander S. Nadas Lecture). *Circulation* 89:2479–2483.
304. Momma K, Nishihara S, Ota Y. (1981) Constriction of the fetal ductus arteriosus by glucocorticoid hormones. *Pediatr Res* 15:19–21.
305. Moog U, Engelen J, Albrechts J, Hoorntge T, Hendrikse F, Schrander-Stumpel C. (1996) Alagille syndrome in a family with duplication 20p11. *Clin Dysmorphol* 5:279–288.
306. Morgagni JBT. (1761) *DeSedibus, et Causis Morborum per Anatomen Indagatis (Translated: The Seats and Causes of Diseases Investigated By Anatomy)* 1st and 2nd Ed. Translation by Benjamin Alexander. typ Remondimiana/New York: Venetiis/Hafner Publishing Co. 1960.
307. Morris CD, Outcalt J, Menashe VD. (1990) Hypoplastic left heart syndrome: Natural history in a geographically defined population. *Pediatrics* 85:977–983.
308. Musewe NN, Reisman J, Benson LN, Wilkes D, Levison H, Freedom RM, Trusler GA, Canny GJ. (1988) Cardiopulmonary adaptation at rest and during exercise 10 years after Mustard atrial repair for transposition of the great arteries. *Circulation* 77(5):1055–1061.
309. Mustard WT. (1964) Successful two-stage correction of transposition of the great vessels. *Surgery* 55:469–472.
310. Mustard WT, Rowe RD, Keith JD, Sirek A. (1955) Coarctation of the aorta with special reference to the first year of life. *Ann Surg* 141:429–436.
311. Nadas AS. (1976) Patent ductus revisited. *N Eng J Med* 295:563–565.
312. Nadas AS. (1987) Report of the joint study on the natural history of congenital heart defects. PS, AS, VSD: Clinical course and indirect assessment. *Circulation* 56:1–87.
313. Natowicz M, Chatten J, Clancy R, Conrad K, Glauser T, Huff D, Lin A, Norwood W, Rorke LB, Uri A, et al. (1988) Genetic disorders and major extracardiac anomalies associated with the hypoplastic left heart syndrome. *Pediatrics* 698–706.
314. Nawrotzki R, Blake DJ, Davies KE. (1996) The genetic basis of neuromuscular disorders. *Trends in Genetics* 12:294–298.
315. Neill CA. (1956) Development of the pulmonary veins: With reference to the embryology of anomalies of pulmonary venous return. *Pediatrics* 18:880–887.
316. Neill CA. (1987) Obituary: Dr. Helen Brooke Taussig, May 24, 1898-May 21, 1986, International Cardiologist. *Int J Cardiol* 14:255–261.
317. Neill CA, Clark EB. (1995) *The Developing Heart. A "History" of Pediatric Cardiology.* Dordrecht: Kluwer Academic Publishers.
318. Neu RL, Gallien JU, Steinberg-Warren N, Wynn RJ, Bannerman RM. (1981) An infant with trisomy 6q21—>6qter. *Ann Genet* 24:167–169.
319. Newbury-Ecob RA, Leanage R, Young I. (1996) Clinical heterogeneity in dominantly inherited atrial septal defect. *J Med Genet* 59–S:A100.
320. Newman NM, Correy JF. (1983) Possible teratogenicity of sulphasalazine. *Med J Aust* 1:528–529.
321. Newman TB. (1985) Etiology of ventricular septal defects: An epidemiologic approach. *Pediatrics* 76(5):741–749.
322. Noonan JA, Nadas AS. (1958) The hypoplastic left heart syndrome: An analysis of 101 cases. *Pediatr Clin North Am* 5:1029–1057.
323. Noonan JA, Nadas AS, Rudolph AM, Harris GBC. (1960) Transposition of the great arteries: A correlation of clinical, physiologic and autopsy data. *N Engl J Med* 263(12):592–596.
324. Nora JJ. (1993) Causes of congenital heart diseases: Old and new modes, mechanisms, and models. *Am Heart J* 125:1409–1419.
325. Nora JJ. (1994) From generational studies to a multilevel genetic-environmental interaction. *J Am Coll Cardiol* 23:1468–1471.
326. Norwood WI, Lang P, Hansen DD. (1983) Physiologic repair of aortic atresia - hypoplastic left heart syndrome. *N Engl J Med* 308:23–26.
327. Olley CA, Baraister M, Grant DB. (1988) A reappraisal of the CHARGE association. *J Med Genet* 25:147–156.
328. Olley PM. (1987) The ductus arteriosus, its persistence and its patency. In Anderson RH, Shinebourne EA, Macartney FJ, Tynan M (eds:) *Paediatric Cardiology.* Edinburgh: Churchill Livingstone; pp. 931–958.

329. Olshan AF, Teschke K, Baird PA. (1990) Birth defects among offspring of firemen. *Am J Epidemiol* 131:312–321.
330. Olson SB, Lawce H, Pillers DM, Duong-Tran J, Rice M, Magenis RE. (1996) Deletion of the Velocardiofacial/DiGeorge syndromes region of chromosome 22q11.2 in fetuses, infants and children ascertained through unexplained cardiac defects. *Am J Hum Genet* 59(suppl):A128.
331. Opitz JM. (1985) The developmental field concept. *Am J Med Genet* 21:1–11.
332. Orie JD, Anderson C, Ettedgui JA, Zuberbuhler JR, Anderson RH. (1995) Echocardiographic-morphologic correlations in tricuspid atresia. *J Am Coll Cardiol* 26:750–758.
333. Ornoy A, Zusman I. (1991) Embryotoxic effects of diabetes on pre-implantation embryos. *Isr J Med Sci* 27:487–492.
334. Osborne LR, Scherer SW, Martindale D, Shi X-M, Heng HHO, Costa T, Pober B, Rommens J, Koop B, Tsui LC. (1996) Identification of genes in a 500 kb region commonly deleted in Williams syndrome. *Am J Hum Genet* 59(suppl):A230.
335. Paris M. (1791) Retrecissement considerable de l'aorte pectorale, observe a l'Hotel-Dieu de Paris. *J de Chirurgie* 2:107–110.
336. Pasquini L, Sanders SP, Parness IA, Cllan SD, VanPraagh S, Mayer JE, VanPraagh R. (1993) Conal anatomy in 119 patients with d-loop transposition of the great arteries and ventricular septal defect: An echocardiographic and pathologic study. *J Am Coll Cardiol* 21(7):1712–1721.
337. Patterson DF, Pexieder T, Schnarr WR, Navratil T, Alaili R. (1993) A single major-gene defect underlying cardiac conotruncal malformations interferes with myocardial growth during embryonic development: Studies in the CTD line of Keeshond dogs. *Am J Hum Genet* 52:388–397.
338. Peacock TB. (1858) *On Malformations of the Human Heart.* London: John Churchill.
339. Pease WE, Nordenberg A, Ladda RL. (1976) Familial atrial septal defect with prolonged atrioventricular conduction. *Circulation* 53(3):759–762.
340. Pederson LM, Tygstrup I, Pederson J. (1964) Congenital malformations in newborn infants of diabetic women: Correlation with maternal diabetic complications. *Lancet* 1:1124–1126.
341. Peoples WM, Moller JH, Edwards JE. (1983) Polysplenia: A review of 146 cases. *Pediatr Cardiol* 4:129–137.
342. Perry LW, Neill CA, Ferencz C, Rubin JD, Loffredo C. (1993) Infants with congenital heart disease: The cases. In Ferencz C, Rubin JD, Loffredo CA, Magee CA (eds:) *Epidemiology of Congenital Heart Disease: The Baltimore-Washington Infant Study 1981–1989.* Mt. Kisco, NY: Futura Publishing Co., Inc.; pp. 33–73.
343. Perry LW, Scott LP, Shapiro SR, Chandra RS, Roberts WC. (1977) Atresia of the aortic valve with ventricular septal defect. A clinicopathologic study of four newborns. *Chest* 72:757.
344. Pexieder T. (1995) Overview: Proper laboratory practice in experimental studies of abnormal cardiovascular development. In Clark EB, Markwald RR, Takao A (eds:) *Developmental Mechanisms of Heart Disease.* Armonk, NY: Futura Publishing Co., Inc.; pp. 169–174.
345. Pexieder T. (1995) Conotruncus and its septation at the advent of the molecular biology era. In Clark EB, Markwald RR, Takao A (eds:) *Developmental Mechanisms of Heart Disease.* Armonk, NY: Futura Publishing Co., Inc.; pp. 227–247.
346. Pexieder T, Bloch D, EUROCAT Working Party on Congenital Heart Disease. (1995) EUROCAT subproject on epidemiology of congenital heart disease: First analysis of the completed study. In Clark EB, Markwald RR, Takao A (eds:) *Developmental Mechanisms of Heart Disease.* Armonk, NY: Futura Publishing Co., Inc.; pp. 655–671.
347. Pexieder T, Rousseil MP, Prados-Frutos JC. (1992) Prenatal pathogenesis of the transposition of the great arteries. In Vogel M, Buhlmeyer K (eds:) *Transposition of the Great Arteries 25 Years after Rashkind Balloon Septostomy.* Darmstadt: Steinkopff Verlag; pp. 11–27.
348. Phibbs RH. (1987) *The Newborn Infant: Pediatrics.* Rudolph AM (ed.) Norwalk, CT: Appleton-Lange.
349. Pinar H, Rogers BB. (1992) Renal dysplasia, situs inversus totalis, and multisystem fibrosis: A new syndrome. *Pediatr Pathol* 12:215–221.

350. Polani PE, Campbell M. (1955) An aetiological study of congenital heart disease. *Ann Hum Genet* 19:209–230.
351. Pradat P. (1992) Epidemiology of major congenital heart defects in Sweden, 1981–1986. *J Epidemiol Community Health* 46:211–215.
352. Pradat P. (1994) Recurrence risk for major congenital heart defects in Sweden: A registry study. *Genetic Epidemiol* 11:131–140.
353. Qureshi F, Jacques SM, Evans MI, Johnson MP, Isada NB, Yang SS. (1993) Skeletal histopathology in fetuses with chondroectodermal dysplasia (Ellis-van Creveld syndrome). *Am J Med Genet* 45:471–476.
354. Raisher BD, Dawton SB, Grant JW. (1991) Father and son with total anomalous pulmonary venous connection. *Am J Med Genet* 40:105–106.
355. Ramsing M, Gillessen KG, Jutting G, Loffing R, Ngo KN, Rehder H. (1991) Fryns syndrome with cystic hygroma and multiple pterygia. 3rd Meeting of German Society of Human Genetics, Ulm, 1991. *German Soc Hum Genetics* Abstr 102.
356. Rashkind WJ. (1979) Pediatric cardiology: A brief historical perspective. *Pediatr Cardiol* 1:63–71.
357. Rashkind WJ. (1983) Transcatheter treatment of congenital heart disease. *Circulation* 67:711–716.
358. Rashkind WJ, Miller WW. (1966) Creation of an atrial septal defect without thoracotomy. *JAMA* 196:991–992.
359. Reifenstein GH, Levine SA, Gross RF. (1946) Coarctation of the aorta. *Am Heart J* 146–168.
360. Rein AJJT, Dollberg S, Gale R. (1990) Genetics of conotruncal malformations: Review of the literature and report of a consanguineous kindred with various conotruncal malformations. *Am J Med Genet* 36:353–355.
361. Rhodes LA, Colan SD, Perry SB, Jonas RA, Sanders SP. (1991) Predictors of survival in neonates with critical aortic stenosis. *Circulation* 84:2325–2335.
362. Roberts WC. (1970) Anatomically isolated aortic valvular diseases: Case against its being of rheumatic etiology. *Am J Med* 49:151.
363. Roessler E, Belloni E, Gaudenz K, Jay P, Berta P, Scherer SW, Tsui L-C, Muenke M. (1996) Mutations in the human Sonic Hedgehog gene cause holoprosencephaly. *Nature Genetics* 14:357–360.
364. Roger H. (1879) Communication congenital du coeur par inocclusion du septum interventriculaire. *Bull-de l Acad de Med* 8:1074.
365. Roguin N, Du Z-D, Barak M, Nasser N, Hershkowitz S, Milgram E. (1995) High prevalence of muscular ventricular septal defect in neonates. *J Am Coll Cardiol* 26:1545–1548.
366. Rokitansky C. (1852) Anomalies and diseases of the heart. In *A Manual of Pathological Anatomy.* London: Translation printed by the Sydenham Society; pp. 141–153.
367. Rokitansky CFV. (1875) *Die Defecte der Scheidewande des Herzens.* Wien: Wilhelm Braumuller.
368. Rose V, Gold RJM, Lindsay G, Allen M. (1985) Congenital heart defect recurrence in offspring. *J Am Coll Cardiol* 6(2):376–382.
369. Rose V, Hewitt D, Milner J. (1972) Seasonal influences on the risk of cardiac malformation: Nature of the problem and some results from a study of 10,077 cases. *Int J Epidemiol* 1(3):235–244.
370. Rosenberg L, Mitchell AA, Shapiro S, Slone D. (1982) Selected birth defects in relation to caffeine-containing beverages. *JAMA* 247(10):1429–1432.
371. Rosenthal A, Dick M III. (1983) Tricuspid atresia. In Adams FH, Emmanouilides GC (eds:) *Moss' Heart Disease in Infants, Children, and Adolescents.* Baltimore: Williams and Wilkins; pp. 271–283.
372. Rosenthal GL. (1996) Patterns of prenatal growth among infants with cardiovascular malformations: Possible fetal hemodynamic effects. *Am J Epidemiol* 143:505–513.
373. Rosenthal GL, Wilson PD, Permutt T, Boughman JA, Ferencz C. (1991) Birth weight and cardiovascular malformations: A population-based study. The Baltimore-Washington Infant Study. *Am J Epidemiol* 133(12):1273–1279.
374. Rowe RD, Freedom RM, Mehrizi A, Bloom KR. (1981) Aortic stenosis. In *The Neonate with Congenital Heart Disease.* Philadelphia, London, Toronto, Sydney: W.B. Saunders Company; pp. 562–576.

375. Rowe RD, Freedom RM, Mehrizi A, Bloom KR. (1981) Aortic atresia. In *The Neonate with Congenital Heart Disease*. Philadelphia, London, Toronto, Sydney: W.B. Saunders Company; pp. 204–220.

376. Rowe RD, Mehrizi A. (1968) *The Neonate with Congenital Heart Disease,* 1st Ed. Philadelphia: W.B. Saunders Company.

377. Rowland TW, Hubbell JPJ, Nadas AS. (1973) Congenital heart disease in infants of diabetic mothers. *J Pediatr* 83(5):815–820.

378. Rubanyi GM. (1992) Endothelium-derived vasoactive factors in health and disease. In Rubanyi GM (ed:) *Cardiovascular Significance of Endothelium-Derived Vasoactive Factors.* Mt. Kisco, NY: Futura Publishing Co., Inc.; pp. 11–19.

379. Rubin JD, Ferencz C, Loffredo C. (1993) The use of prescription and nonprescription drugs in pregnancy. *J Clin Epidemiol* 46:581–589.

380. Rubin JD, Loffredo CA. (1993) Prescription and nonprescription drugs. In Ferencz C, Rubin JD, Loffredo CA, Magee CA (eds:) *Epidemiology of Congenital Heart Disease: The Baltimore-Washington Infant Study 1981–1989.* Mt. Kisco, NY: Futura Publishing Co., Inc.; pp. 181–189.

381. Rutledge JC, Friedman JM, Harrod MJE, Currarion G, Wright CG, Pinckney L, Chen H. (1984) A "new" lethal multiple congenital anomaly syndrome: Joint contractures, cerebellar hypoplasia, renal hypoplasia, urogenital anomalies, tongue cysts, shortness of limbs, eye abormalities, defects of the heart, gallbladder agenesis, and ear malformations. *Am J Med Genet* 19:255–264.

382. Sadler TW. (1990) *Langman's Medical Embryology.* 6th Ed. Baltimore: Williams & Wilkins.

383. Sanchez-Cascos A, Garcia-Sagredo JM. (1975) Genetics of patent ductus arteriosus. *Basic Res Cardiol* 70:456–465.

384. SAS Institute Inc. (1989) *SAS/STAT User's Guides, Version 6,* 4th Ed. Cary, NC: SAS Institute, Inc.

385. Scambler P. (1995) CATCH 22- Can genetics explain the phenotype? In Clark EB, Markwald RR, Takao A (eds:) *Developmental Mechanisms of Heart Disease.* Armonk, NY: Futura Publishing Co., Inc.

386. Schardein JL. (1985) *Chemically Induced Birth Defects.* Shardein JL (ed.) New York: Marcel Dekker.

387. Schardein JL. (1993) *Chemically Induced Birth Defects,* 2nd Ed. New York: Marcel Dekker.

388. Schmidt MA, Ensing GJ, Michels VV, Carter GA, Hagler DJ, Feldt RH. (1989) Autosomal dominant supravalvular aortic stenosis: Large three-generation family. *Am J Med Genet* 32:384–389.

389. Schwarz G. (1978) Estimating the dimension of a model. *Ann Stat* 6:461–464.

390. Segall HN. (1988) *Pioneers of Cardiology in Canada 1820–1970.* Willowdale, Ontario: Hounslow Press.

391. Serraf A, Lacour-Gayet F, Bruniaux J, Touchot A, Losay J, Comas J, Uva MS, Planche C. (1993) Anatomic correction of transposition of the great arteries in neonates. *J Am Coll Cardiol* 22(1):193–200.

392. Seward JB, Tajik AJ, Feist DJ, Smith HC. (1979) Ebstein's anomaly in an 85-year-old man. *Mayo Clin Proc* 54:193–196.

393. Shaw GM, Malcoe LH, Katz E. (1992) Maternal workplace exposures to organic solvents and congenital cardiac anomalies. *J Occup Med Toxicol* 1:371–376.

394. Shaw GM, Malcoe LH, Swan SK, Cummins J, Schulman J, Harris JA. (1990) Risks for congenital cardiac anomalies relative to selected maternal exposures during early pregnancy. *Teratology* 41(5):590.

395. Shen-Schwarz S, Dave H. (1988) Meckel syndrome with polysplenia: Case report and review of literature. *Am J Med Genet* 31:349–355.

396. Shokeir MHK. (1977) The Goldenhar syndrome: A natural history. *Birth Defects* 13:67–83.

397. Sletten LL, Pierpont MEM. (1996) Variation in severity of cardiac disease in Holt-Oram syndrome. *Am J Med Genet* 65:128–132.

398. Smith AT, Sack GHJ, Taylor GJ. (1979) Holt-Oram syndrome. *J Pediatr* 95(4):538–543.

399. Soloff LA, Stauffer HM, Zatuchni J. (1951) Ebstein's disease: Report of the first case diagnosed during life. *Am J Med Sci* 222:554–561.

400. Solymar L, Sabel K, Zetterqvist P. (1987) Total anomalous pulmonary venous connection in siblings: Report on three families. *Acta Paediatr Scand* 76:124–127.
401. Stanger P, Rudoph AM, Edwards JE. (1977) Cardiac malpositions: An overview based on study of sixty five necropsy specimens. *Circulation* 56:161–172.
402. Stoll C, AlembikY, Dott B, Roth MP. (1990) Epidemiology of Down syndrome in 118,265 consecutive births. *Am J Med Genet* (suppl)7:79–83.
403. Stoll C, Alembik Y, Roth MP, Dott B, DeGeeter B. (1989) Risk factors in congenital heart disease. *Eur J Epidemiol* 5(3):382–391.
404. Storch TG, Mannick EE. (1992) Epidemiology of congenital heart disease in Louisana: An association between race and sex and the prevalence of specific cardiac malformations. *Teratology* 46:271–276.
405. Takao A, Momma K, Kondo C, Ando M, Shimizu T, Burn J, Matsuoka R. (1995) Conotruncal anomaly face syndrome. In Clark EB, Markwald RR, Takao A (eds:) *Developmental Mechanisms of Heart Disease.* Armonk, NY: Futura Publishing Co., Inc.; pp. 555–558.
406. Taussig HB. (1947) *Congenital Malformations of the Heart,* 1st Ed. Cambridge, MA: Harvard University Press.
407. Taussig HB. (1960) *Congenital Malformations of the Heart, Volume II: Specific Malformations,* 2nd Ed. The Commonwealth Fund, Cambridge, MA: Harvard University Press.
408. Taussig HB. (1973) Dr. Edwards A. Park: Physician, teacher, investigator, friend. *Hopkins Med J* 132(6):361–376.
409. Taussig HB, Bauersfeld SR. (1953) Follow-up studies in the first 1000 patients operated on for pulmonary stenosis or atresia: Results up to March, 1952. *Ann Intern Med* 38:1–8.
410. Taussig HB, Crocetti A, Eshaghpour E, Keinonen R, Yap KN, Bachman D, Momberger N, Kirk H. (1971) Long-time observations of the Blalock-Taussig operation: I. Results of first operation. *Hopkins Med J* 129(5):243–257.
411. Taussig HB, Kallman CH, Nagel D, Baumgardner R, Momberger N, Kirk H. (1975) Long-time observations on the Blalock-Taussig operation: VIII. 20–28 year follow-up on patients with a tetralogy of Fallot. *Hopkins Med J* 137:13–19.
412. Taussig HB, King JT, Bauersfeld R, Padmavati-lyer S. (1951) Results of operation for pulmonary stenosis and atresia: Report of 1000 cases. *Trans Assoc Am Physicians* 64:67–73.
413. Therkelsen AJ, Hulten M, Jonasson J, Lindsten J, Christensen NC, Iversen T. (1973) Presumptive direct insertion within chromosome 2 in man. *Ann Hum Genet* 36:367–373.
414. Tikkanen J, Heinonen OP. (1990) Risk factors for cardiovascular malformations in Finland. *Eur J Epidemiol* 6:348–356.
415. Tikkanen J, Heinonen OP. (1991) Risk factors for ventricular septal defect in Finland. *Public Health* 105:99–112.
416. Tikkanen J, Heinonen OP. (1992) Congenital heart disease in the offspring and maternal habits and home exposures during pregnancy. *Teratology* 46(5):447–454.
417. Tikkanen J, Heinonen OP. (1992) Risk factors for conal malformations of the heart. *Eur Heart J* 8(1):48–57.
418. Tikkanen J, Heinonen OP. (1992) Risk factors for atrial septal defect. *Eur J Epidemiol* 8:509–515.
419. Tikkanen J, Heinonen OP. (1993) Risk factors for coarctation of the aorta. *Teratology* 47:565–572.
420. Torfs CT, Curry CJ, Harris JA. (1991) The descriptive epidemiology of hypoplastic left heart, coarctation of the aorta, and aortic stenosis. *Teratology* 43(5):448–449.
421. Urioste M, Martinez-Frias ML, Bermejo E, Jimenez N, Romero D, Nieto C, Villa A. (1994) Short rib-polydactyly syndrome and pericentric inversion of chromosome 4. *Am J Med Genet* 49:94–97.
422. Van Mierop LHS, Gessner IH, Schiebler GL. (1972) Asplenia and polysplenia syndrome. *Birth Defects* 8(1):74–82.
423. Van Praagh R. (1972) The segmental approach to diagnosis in congenital heart disease. *Birth Defects* 8(5):4–23.
424. Van Praagh R. (1992) Foreword. In Freedom MR, Benson LN, Smallhorn JF (eds:) *Neonatal Heart Disease.* London: Springer-Verlag; pp. 7–9.
425. Van Praagh R, Geva T, Kreutzer J. (1989) Ventricular septal defects: How shall we describe, name, and classify them? *J Am Coll Cardiol* 14(5):1298–1299.

426. Van Praagh R, Geva T, Van Praagh S. (1990) Segmental situs in congenital heart disease: Recent rare findings. In Clark EB, Takao A (eds:) *Developmental Cardiology: Morphogenesis and Function.* Mt. Kisco, NY: Futura Publishing Co., Inc.; pp. 625–657.
427. Van Praagh R, Layton WM, VanPraagh S. (1980) The morphogenesis of normal and abnormal relationships between the great arteries and the ventricles: Pathologic and experimental data. In Van Praagh R, Takao A (eds:) *Etiology and Morphogenesis of Congenital Heart Disease.* Mt. Kisco, NY: Futura Publishing Co., Inc.; pp. 271–316.
428. Van Praagh R, Leidenfrost RD, Matsuoka R, et al. (1984) Segmental situs in congenital heart disease: Relevance to diagnosis, pathology, embryology, and etiology. In Nora JJ, Takao A (eds:) *Congenital Heart Disease: Causes and Processes.* Mt. Kisco, NY: Futura Publishing Co., Inc.; pp. 173–196.
429. Van Praagh S, Kreutzer J, Alday L, Van Praagh R. (1990) Systemic and pulmonary venous connections in visceral heterotaxy with emphasis on the diagnosis of the atrial situs: A study of 109 postmortem cases. In Clark EB, Takao A (eds:) *Developmental Cardiology: Morphogenesis and Function.* Mt. Kisco, NY: Futura Publishing Co., Inc.; pp. 671–727.
430. Van Praagh S, Santini F, Sanders SP. (1992) Cardiac malpositions with special emphasis on visceral heterotaxy. In Fyler DC (ed:) *Nadas's Pediatric Cardiology.* Philadelphia: Henry & Belfus; pp. 589–608.
431. Van Praagh S, Antoniadis S, Otero-Coto E, Leidenfrost RD, Van Praagh R. (1984) Common atrioventricular canal with and without conotruncal malformations: An anatomic study of 251 postmortem cases. In Nora JJ, Takao A (eds:) *Congenital Heart Diseases: Causes and Processes.* Mt. Kisco, NY: Futura Publishing Co., Inc.; pp. 599–637.
432. Vinh LT, Duc TV, Aicardi J, Thieffry S. (1968) Retour veineux pulmonaire anormal total infra-diaphragmatique familial (in French). *Arch Fr Pediatr* 25:1141–1149.
433. Voiculescu I, Back E, Duncan AMV, Schwaibold H, Schempp W. (1987) Trisomy 22 in a newborn with multiple malformations. *Hum Genet* 76:298–301.
434. Vuillemin M, Reymond C, Krstic R. (1995) Professor Tomas Pexider (6.6.1941–28.10.1995). *Annals of Anatomy* 178:197–199.
435. Warkany J. (1971) Transposition of the great vessels. In *Congenital Malformations: Notes and Comments.* Chicago, IL: Year Book Medical Publishers; pp. 515–525.
436. Warkany J. (1988) Tetratogen update: Lithium. *Teratology* 38:593–596.
437. Warkany J, Roth CB, Wilson JG. (1948) Multiple congenital malformations: A consideration of etiologic factors. *Pediatrics* 1:462–471.
438. Wasz-Hockert O, Simila S, Rosberg G, Vuorenkoski V, Lind J. (1969) El sindrome de Smith-Lemli-Opitz en dos ninas, con special atencion a los patrones de sus gritos de dolor. *Rev Mexicana Pediatr* 38:63–68.
439. Watson H. (1974) Natural history of Ebstein's anomaly of tricuspid valve in childhood and adolescence: An international co-operative study of 505 cases. *Br Heart J* 36:417–427.
440. Waugh, D. (1992) *MAUDIE OF McGILL: Dr. Maude Abbott and the Foundations of Heart Surgery.* Toronto and Oxford: Hannah Institute and Dundurn Press.
441. Weir EK, Joffee HS, Barnard CN, Beck W. (1978) Double outlet right ventricle: Clinical and anatomical spectrum. *Thorax* 33:283–289.
442. Wenink ACG, Zevallos J. (1988) Developmental aspects of atrioventricular septal defects. *Int J Cardiol* 18:65–78.
443. White PD. (1967) Congenital cardiovascular defects. Aortic stenosis undifferentiated as to etiology. In White PD, Donovan A (eds:) *HEARTS: Their Long Follow-up.* Philadelphia, London: W.B. Saunders; pp. 80–82.
444. White PD. (1967) Serious patency of the ductus arteriosus in a girl of 7 1/2 with excellent health 28 years later — The world's first cure. In White PD, Donovan A (eds:) *HEARTS: Their Long Follow-up.* Philadelphia, London: W.B. Saunders Company; pp. 47–49.
445. White PD. (1967) Coarctation of the aorta: "Bright's Disease" cured by the knife. In White PD, Donovan A (eds:) *HEARTS: Their Long Follow-up.* Philadelphia, London: W.B. Saunders Company; pp. 70–75.
446. White PD. (1967) The beginning of heart disease. In White PD, Donovan A (eds:) *HEARTS: Their Long Follow-up.* Philadelphia, London: W.B. Saunders Company; pp. 33–35.

447. White PD, Sprague HB. (1929) The tetralogy of Fallot: Report of a case in a noted musician, who lived to his sixtieth year. *JAMA* 92(10):787–791.
448. Whittemore R, Wells JA, Castellsague X. (1994) A second-generation study of 427 probands with congenital heart defects and their 837 children. *J Am Coll Cardiol* 23:1459–1467.
449. Willett WC, Sampson L, Stampfer MJ, Rosner B, Bain C, Witschi J, Hennekens CH, Speizer FE. (1985) Reproducibility and validity of a semiquantitative food frequency questionnaire. *Am J Epidemiol* 122(1):51–65.
450. Wilson GN, Sout JP, Schneider NR, Zneimer SM, Gilstrap LC. (1991) Balanced translocation 12/13 and situs abnormalities: Homology of early pattern formation in man and lower organisms? *Am J Med Genet* 38:601–607.
451. Wilson LC, Kerr BA, Super M. (1996) A new autosomal dominant syndrome of blepharochalasis, short stature, joint laxity, cardiac and urogenital defects. *Am J Hum Genet* 59–S:A109.
452. Wilson MG, Stein AM. (1969) Teratogenic effects of Asian influenza: An extended study. *JAMA* 210:336–337.
453. Wilson PD, Correa-Villaseñor A, Loffredo C, Ferencz C, and The Baltimore-Washington Infant Study (BWIS) Group. (1993) Temporal trends in prevalence of cardiovascular malformations in Maryland and the District of Columbia, 1981–1988. *Epidemiology* 4(3):259–265.
454. Wolf U. (1995) The genetic contribution to the phenotype. *Hum Genet* 95:127–148.
455. Wood PH. (1950) *Diseases of the Heart and Circulation,* 1st and 3rd Ed. Philadelphia: Lippincott.
456. Yater WM, Shapiro MJ. (1937) Congenital displacement of the tricuspid valve (Ebstein's disease): Review and report of a case with electrocardiographic abnormalities and detailed histologic study of the conduction system. *Ann Intern Med* 11:1043–1062.
457. Zangwill KM, Boal DKB, Ladda RL. (1988) Dandy-Walker malformation in Ellis-van Creveld syndrome. *Am J Med Genet* 31:123–129.
458. Zetterquist P, Turesson I, Johansson BW, Laurell S, Ohlsson NM. (1971) Dominant mode of inheritance in atrial septal defect. *Clin Genet* 2:78–86.
459. Zhang J, Cai WW. (1993) Association of the common cold in the first trimester of pregnancy with birth defects. *Pediatrics* 92:559–563.
460. Zhang KZ, Sun QB, Cheng TO. (1986) Holt-Oram syndrome in China: A collective review of 18 cases. *Am Heart J* 111(3):572–577.
461. Zierler S, Theodore M, Cohen A, Rothman KJ. (1988) Chemical quality of maternal drinking water and congenital heart disease. *Int J Epidemiol* 17(3):589–594.
462. Zlotogora J, Schimmel MS, Glaser Y. (1987) Familial situs inversus and congenital heart defects. *Am J Med Genet* 26:181–184.

INDEX

A

Aarskog syndrome, 396
Abbott, Maude E. Seymour, contributions of, 5–8
Abdominal x-rays, as risk factor, 416
Absence of diaphragm, laterality, looping defects, 45
Absence of ear
  membranous ventricular septal defect, 139
  outflow tract anomaly, 67
Accessory spleen, 408, 409
  outflow tract anomaly, 67
Acetaminophen
  muscular ventricular septal defect, 161
  risk factors, 414
Achondroplasia, 396
  membranous ventricular septal defect, 130
Acro-renal syndrome, atrial septal defect, 270
Acrofacial dysostosis, 396
Acrorenal field defect, 396
Adrenal neuroblastoma, 428
Adreno-genital association, pulmonic valve stenosis, 235
Agenesis of kidney
  atrial septal defect, 270
  membranous ventricular septal defect, 131
  muscular ventricular septal defect, 152
Alagille syndrome, 396
Albinism syndrome, membranous ventricular septal defect, 130
Alcohol use
  aortic stenosis, 209
  atrial septal defect, 279

atrioventricular septal defects, 117
bicuspid aortic valve, 218
cardiomyopathy, 322
coarctation of aorta, 198
Ebstein's anomaly, 331
hypoplastic left heart syndrome, 186
laterality, looping defects, 52
left-sided obstructive defects, 173
muscular ventricular septal defect, 161
pulmonary valve atresia with intact ventricular septum, 256
total anomalous pulmonary venous return, 310
Ambiguous genitalia, 408
  outflow tract anomaly, 67
Anal atresia, 409
  membranous ventricular septal defect, 131
  outflow tract anomaly, 66
Analgesics, risk factors, 414
Anemia, 408, 423
  outflow tract anomaly, 67
Anencephaly, 424
  atrioventricular septal defects, 115
  membranous ventricular septal defect, 139
Anesthesia
  gas, as risk factor, 417, 418
  outflow tract anomaly, 98
Aniridia, patent arterial duct, 289
Antibiotics, risk factors, 414
Anticonvulsants, risk factors, 414
Antidepressants
  left-sided obstructive defects, 173
  risk factors, 414
Antihistamines, risk factors, 414
Antihypertensives, risk factors, 414

Antimicrobials, risk factors, 414
Antitussives
    atrioventricular septal defects, 117
    muscular ventricular septal defect,
        161
    risk factors, 414
Aortic valve stenosis, prevalence of,
    341, 342
Apert syndrome, 396
    membranous ventricular septal de-
        fect, 130, 139
Arsenic, as risk factor, 417, 418
Art dyes, as risk factor, 417, 418
Arts, crafts painting, as risk factor,
    417, 418
Aspirin, risk factors, 414
Atrial septal defect, 267–283
    atrioventricular septal defects, 115
    descriptive analyses, 271–275
        birthweight, gestational age, 274
        diagnosis, course, 271–274
        gender, race, twinning, 274
        prevalence by time, season, area
            of residence, 271
        sociodemographic characteris-
            tics, 274–275
    diagnosis, prevalence, 281
    environmental factors, 283
    genetic factors, 282
    infant characteristics, 282
    multivariate analysis, 280
    potential risk factors, 275–281
        familial cardiac, noncardiac
            anomalies, 275–278
        genetic, environmental factors,
            278–281
        univariate analysis, 278–280
    prevalence of, 341, 342
    study preparation, 269–271
        cardiac abnormalities, 269
        noncardiac anomalies, 270–271
Atrioventricular septal defect, 21,
    103–122, 108
    descriptive analyses, 108–113
        birthweight, gestational age, 111
        diagnosis, course, 109
        gender, race, twinning, 109
        prevalence by time, season, area
            of residence, 108–109

    sociodemographic characteris-
        tics, 111
    lifestyle exposures, 119
    maternal illnesses, medications,
        118–119
    parental home, occupational expo-
        sures, 119–120
        multivariate analysis, 119–120
    potential risk factors, 114–120,
        115
        familial cardiac, noncardiac
            anomalies, 114
        genetic factors, environmental
            factors, 114
        maternal age, reproductive his-
            tory, 114–118
        univariate analyses, 114
    prevalence, 341, 342
    study population, 105–108
        cardiac abnormalities, 105–106
        chromosomal, mendelian disor-
            ders, 107
        non-mendelian associations,
            other noncardiac defects,
            107–108
        noncardiac anomalies, 106–107
Auto body repair work
    cardiac malformation, 367
    cardiomyopathy, 322
    membranous ventricular septal de-
        fect, 145
    outflow tract anomaly, 91, 93, 98
    as risk factor, 416

B
Barbiturates, risk factors, 414
Bartter syndrome, membranous ven-
        tricular septal defect, 130
"BBB" syndrome, 397
Benzodiazepines
    atrioventricular septal defects, 117
    Ebstein's anomaly, 331
    left-sided obstructive defects, 173,
        224
    outflow tract anomaly, 91, 92, 93, 95,
        97, 98
Bicuspid aortic valve, 22, 427
    atrioventricular septal defects, 115
    prevalence of, 341, 342

Bilateral renal agenesis, atrial septal defect, 270
Biliary atresia, 427
Bleeding, during pregnancy
  atrial septal defect, 279
  outflow tract anomaly, 93, 98
Bone cancer at birth, 428
Branchial cyst, laterality, looping defects, 51
Bronchial defect, 408
  outflow tract anomaly, 67
Butterfly vertebrae, 410, 427

C
Cadmium, as risk factor, 417, 418
Campbell, Maurice, contributions of, 11–12
Campomelic syndrome, 397
Cantrell syndrome, 409
  outflow tract anomalies and, 65
Cardiac outflow tract malformation, 59–102
  cardiac abnormalities, 64
  clinical perspectives, 60
  descriptive analyses, 68–69
    birthweight, gestational age, 72
    diagnosis, course, 69, 72
    gender, race, twinning, 72
    prevalence by time, season, area of residence, 68–69
    sociodemographic characteristics, 72
  developmental perspectives, 60–62
  diagnosis-specific analyses, 94–96
  environmental factors, 101–102
  epidemiologic perspectives, 62
  family history, 99
  genetic risk factors, 96–100
  heritable blood disorders, 99
  maternal diabetes, 101
  noncardiac anomalies, 64–68
  outflow tract anomaly, 95, 97, 98
  overview, 96
  potential risk factors, 79–96
    familial cardiac, noncardiac anomalies, 79
    genetic factors, environmental factors, 79
    lifestyle exposures, 88

    major groups of transposition, normal great artery outflow tract defects, 90–94
    maternal home, occupational exposures, 88–90
    maternal illnesses, 84
    maternal medications, 84
    maternal reproductive history, 79–84
    multivariate analysis, 88–90
    outflow tract anomalies, 91
    paternal medical exposures, 88
    univariate analysis, 79
  proband, 99
  study population, 63–68
Cardiology, cardiac morphogenesis, 386–387
Cardiomyopathy, 313–324
  atrioventricular septal defects, 115
  descriptive analyses, 316–319
    birthweight, gestational age, 317
    diagnosis, course, 316
    gender, race, twinning, 317
    prevalence by time, season, area of residence, 316
    sociodemographic characteristics, 317–319
  potential risk factors, 320–323
    familial cardiac, noncardiac anomalies, 320
    genetic, environmental factors, 320–323, 322
    multivariate analysis, 321–323
    univariate analysis, 320–321
  prevalence of, 341, 342
  study population, 313–316
    noncardiac anomalies, 314–316
Cardiovascular malformations, categorization, for risk factor analysis, 13–28
Carnitine deficiency, 397
  atrial septal defect, 270
Carpentry, as risk factor, 416
Case-control design, statistical methods, 29–37
Cataract, 428
  aortic stenosis, 203
  membranous ventricular septal defect, 139

Cataract (*continued*)
  pulmonic valve stenosis, 235
Categorization, cardiovascular malformations, risk factor analysis, 13–28, 26–27
  associated noncardiac anomalies, 27–28
  hierarchical order, 26
  mechanistic grouping, 26
  "pure" cardiac diagnoses, 26
Centrell syndrome, 408
Charcot-Marie syndrome, 422
  atrioventricular septal defects, 115
CHARGE association, 397
Chest x-rays, as risk factor, 416
Choanal atresia, 408
  atrial septal defect, 270
  outflow tract anomaly, 67
Classifications, cardiovascular malformations, previous, 24–26
Cleft lip
  laterality, looping defects, 51
  membranous ventricular septal defect, 139
  outflow tract anomaly, 67
  palate, 422
    patent arterial duct, 289
Cleft palate, 408, 425
  anal atresia, membranous ventricular septal defect, 130
  aortic stenosis, 203
  bicuspid aortic value, 213
  chalasia, membranous ventricular septal defect, 130
  membranous ventricular septal defect, 131
"Cleft sternum-hemangioma" association, 397
Clomiphene citrate
  coarctation of aorta, 200–201
  left-sided obstructive defects, 224, 225
  outflow tract anomaly, 97
Coarctation of aorta, 19, 427
  prevalence of, 341, 342
Cocaine use
  atrial septal defect, 279
  atrioventricular septal defects, 117
  laterality, looping defects, 52

membranous ventricular septal defect, 145
muscular ventricular septal defect, 161
Community concern, heart disease and, 4
Congenital deafness
  atrioventricular septal defects, 115
  laterality, looping defects, 51
Congenital heart disease, infant with, 337–357
  malformation triad, 344–345, 354–357
    cardiovascular malformations, 344–345
    noncardiac anomalies, 345–350
      chromosomal syndromes, 348
      small size at birth, 351–354
  prevalence, 339–344
Cooking with kerosene, coal, or wood, as risk factor, 416
Cornelia de Lange syndrome, atrioventricular septal defects, 108
Coronary artery anomaly, membranous ventricular septal defect, 138
Corticosteroids
  atrial septal defect, 279
  risk factors, 414
  total anomalous pulmonary venous return, 310
Costello syndrome, 397
Craniosynostosis, 408
  atrial septal defect, 270
  outflow tract anomaly, 66
Cystic fibrosis, 423
Cystic lung, atrial septal defect, 270
Cytomegaly, 397
Cytomegalovirus, atrial septal defect, 270

D
Dandy-Walker syndrome, atrioventricular septal defects, 115
De Lange syndrome, membranous ventricular septal defect, 130
Deafness, 422, 427
Defect of membranous ventricular septum, 19

Defects of laterality, looping, 41–57,
    45, 54
  descriptive analyses, 46–49
    birthweight, gestational age, 48
    diagnosis, course, 46
    gender, race, twinning, 46–48
    prevalence by time, season, area
        of residence, 46
    sociodemographic characteris-
        tics, 48–49
  environmental factors, 56
  genetic heterogeneity, 54–55
  maternal diabetes, 55–56
  multivariate analyses, 53
  potential risk factors, 50–53
    familial cardiac, noncardiac
        anomalies, 50
    genetic factors, environmental
        factors, 50–53
    univariates analyses, 50–53
  prevalence, 54
  socioeconomic status, 56
  study population, 42–45
    cardiac abnormalities, 42–44
    noncardiac anomalies, 44–45
Deficiency thyroxin binding globulin,
    203
Degreasing solvents
  bicuspid aortic valve, 218
  coarctation of aorta, 198
  hypoplastic left heart syndrome,
    186
  left-sided obstructive defects, 173
  pulmonic valve stenosis, 246
Dental x-rays, as risk factor, 416
Diabetes
  aortic stenosis, 209
  atrial septal defect, 279
  atrioventricular septal defects, 117
  bicuspid aortic valve, 218
  cardiomyopathy, 322
  coarctation of aorta, 198
  Ebstein's anomaly, 331
  hypoplastic left heart syndrome,
    186
  laterality, looping defects, 52
  left-sided obstructive defects, 173,
    224
    isolated/simplex subset, 174–176

  multivariate analysis, 174
  membranous ventricular septal de-
    fect, 145
  muscular ventricular septal defect,
    161
  outflow tract anomaly, 91, 92, 93, 95,
    97
  pulmonary valve atresia with intact
    ventricular septum, 256
  total anomalous pulmonary venous
    return, 310
  tricuspid atresia, 264
Diaphragmatic hernia, 408
  bicuspid aortic valve, 213
  membranous ventricular septal de-
    fect, 131
  outflow tract anomaly, 67
Diazepam, cardiac malformation, 367
DiGeorge syndrome, 398, 409, 410
  membranous ventricular septal de-
    fect, 130
  outflow tract anomalies and, 65
Dilated kidney, 426
Diuretics
  atrioventricular septal defects, 117
  risk factors, 414
Down syndrome, 422
  atrioventricular septal defects, 108
  bicuspid aortic valve, 213
  coarctation of aorta, 192
  hypoplastic left heart syndrome,
    180
  membranous ventricular septal de-
    fect, 130, 139
  prevalence by, 342
  prevalence of, 341
  pulmonic valve stenosis, 235
Drug manufacturing, as risk factor,
    417, 418
Dry cleaning solvents, as risk factor,
    416
Duodenal atresia, 408
  outflow tract anomaly, 66
Duplicate collecting system
  atrioventricular septal defects, 108
  membranous ventricular septal de-
    fect, 131
Duplicated right kidney, atrioventricu-
    lar septal defects, 115

Dysmorphology, genetics, 387–391
Dysplastic kidney
    atrioventricular septal defects, 115
    patent arterial duct, 289
Dysplastic right kidney, atrioventricu-
        lar septal defects, 115

E
Ebstein's malformation of tricuspid
        valve, 325–334
    descriptive analyses, 326–329
        birthweight, gestational age, 328
        diagnosis, course, 326
        gender, race, twinning, 326–328
        prevalence by time, season, area
            of residence, 326
        sociodemographic characteris-
            tics, 328
    potential risk factors, 329–332
        familial cardiac, noncardiac
            anomalies, 329–332, 331
            univariate, multivariate analy-
                sis, 329–332
    study population, 325–326
        cardiac abnormalities, 325
        noncardiac anomalies, 325
Ectopic kidney, membranous ventricu-
        lar septal defect, 131
Ehlers-Creveld syndrome, 398
Ehlers-Danlos syndrome
    membranous ventricular septal de-
        fect, 139
    pulmonic valve stenosis, 235
Electric heating, as risk factor, 416
Electric stove, as risk factor, 416
Ellis-Van Creveld syndrome, atrioven-
        tricular septal defects, 108
Encephalocele, muscular ventricular
        septal defect, 152
Epidemiologic case-control design,
        statistical methods, 29–37
    genetic component, evaluation o,
        36–37
    length of follow-up, 31
    quality control, data collection,
        33–34
    sampling of mild defects, for inter-
        view, 31
    statistical analysis, 35–36

    variables analyzed, 35
Epilepsy
    coarctation of aorta, 198
    left-sided obstructive defects, 173,
        175, 224, 225
    pulmonary valve atresia with intact
        ventricular septum, 256
Erythrocytosis, 423
Esophageal stenosis, laterality,
        looping defects, 45
Expectorants, risk factors, 414
Extreme cold, as risk factor, 417, 418
Extreme heat, as risk factor, 417, 418
Eye defects, hypoplastic left heart
        syndrome, 180

F
Facial dysmorphisms, membranous
        ventricular septal defect,
        131
Fancomi syndrome, membranous ven-
        tricular septal defect, 130
Fanconi anemia, 398
Fertility medications, outflow tract
        anomaly, 93, 98
Fetal alcohol syndrome, 398
    atrial septal defect, 270
    muscular ventricular septal defect,
        152
Fever, pulmonic valve stenosis, 246
Fireplace heating, atrioventricular
        septal defects, 117
Frequent use of fireplace for heat, as
        risk factor, 416
Friedreich ataxia, laterality, looping
        defects, 51
Frontometaphyseal dysplasia, 398

G
Gas heating, as risk factor, 416
Gas stove, as risk factor, 416
Gastrointestinal medications
    Ebstein's anomaly, 331
    muscular ventricular septal defect,
        161
    risk factors, 414
Gastroschisis
    cystic thymus, atrial septal defect,
        270

muscular ventricular septal defect, 152

General anesthesia
  hypoplastic left heart syndrome, 186
  outflow tract anomaly, 91, 93, 95, 97
  as risk factor, 416
Gestational diabetes, left-sided obstructive defects, 175
Glaucoma, pulmonic valve stenosis, 235
Goldenhar syndrome, 398, 408, 410
  atrial septal defect, 270
  atrioventricular septal defects, 108
  coarctation of aorta, 192
  membranous ventricular septal defect, 130
  muscular ventricular septal defect, 152
  outflow tract anomaly, 66
  patent arterial duct, 289
Goltz syndrome, 422
Growth hormone deficiency, muscular ventricular septal defect, 152
Guaifenesin, coarctation of aorta, 198

H
Hair dyes
  cardiac malformation, 367
  membranous ventricular septal defect, 145
  muscular ventricular septal defect, 161
  outflow tract anomaly, 91, 93, 95, 97
  pulmonic valve stenosis, 246
  as risk factor, 417, 418
Hemivertebrae, 409, 410
  atrial septal defect, 270
  atrioventricular septal defects, 115
  membranous ventricular septal defect, 131
Hemophilia, atrioventricular septal defects, 115
Hermivertebrae, 408
Holt-Oram syndrome, 398, 409
  atrial septal defect, 270
  atrioventricular septal defects, 108, 115
  membranous ventricular septal defect, 130

outflow tract anomalies and, 65
Holzgreve syndrome, 399, 410
  hypoplastic left heart syndrome, 180
  outflow tract anomalies and, 65
HOMAGE syndrome, 399
Horseshoe kidney, 426
House painting, indoors or outdoors, as risk factor, 416
Hydantoin embryopathy, 399
  atrioventricular septal defects, 108
Hydrocephaly, 408, 425
  atrial septal defect, 270
  laterality, looping defects, 51
  membranous ventricular septal defect, 139
  muscular ventricular septal defect, 152
  outflow tract anomaly, 66
  pulmonic valve stenosis, 235
Hydrocephaly-VACTERL syndrome, 399
Hydronephrosis, 426
  atrial septal defect, 270
  atrioventricular septal defects, 115
  bicuspid aortic valve, 213
  cleft palate, 409
  laterality, looping defects, 45
  membranous ventricular septal defect, 131
  muscular ventricular septal defect, 152
Hydroureter, 409
Hypertrophic clitoris, laterality, looping defects, 51
Hypoplastic cerebellar vermis, laterality, looping defects, 45
Hypoplastic left heart syndrome, 19
  prevalence of, 341, 342
Hypospadias, 408, 409, 426, 427
  aortic stenosis, 203
  atrioventricular septal defects, 108, 115
  membranous ventricular septal defect, 131
  outflow tract anomaly, 67
  patent arterial duct, 289
Hypothyroidism, 424
  membranous ventricular septal defect, 131

I
Ibuprofen
  atrioventricular septal defects, 117
  bicuspid aortic valve, 218
  cardiac malformation, 367
  left-sided obstructive defects, 224
  membranous ventricular septal de-
    fect, 145
  outflow tract anomaly, 91, 92, 95, 97
Ichthyosis, atrioventricular septal de-
    fects, 115
Imperforate anus, 427
  atrial septal defect, 270
Influenza
  aortic stenosis, 209
  atrial septal defect, 279
  atrioventricular septal defects, 117
  cardiomyopathy, 322
  coarctation of aorta, 198
  Ebstein's anomaly, 331
  hypoplastic left heart syndrome,
    186
  ibuprofen, outflow tract anomaly,
    91, 92, 95, 97
  laterality, looping defects, 52
  left-sided obstructive defects, 173
  muscular ventricular septal defect,
    161
  outflow tract anomaly, 91, 92, 95, 97
  pulmonary valve atresia with intact
    ventricular septum, 256
  total anomalous pulmonary venous
    return, 310
  tricuspid atresia, 264
Insulin, risk factors, 414
Intestinal atresia, 427
  laterality, looping defects, 45
  membranous ventricular septal de-
    fect, 131
Ionizing radiation
  aortic stenosis, 209
  atrial septal defect, 279
  atrioventricular septal defects, 117
  bicuspid aortic valve, 218
  cardiomyopathy, 322
  coarctation of aorta, 198
  Ebstein's anomaly, 331
  hypoplastic left heart syndrome,
    186

  laterality, looping defects, 52
  left-sided obstructive defects, 173
  muscular ventricular septal defect,
    161
  outflow tract anomaly, 91, 92, 93, 95,
    97, 98
  pulmonary valve atresia with intact
    ventricular septum, 256
  as risk factor, 417, 418
  total anomalous pulmonary venous
    return, 310
  tricuspid atresia, 264
Iris defect, 409
  outflow tract anomaly, 66
Ivemark syndrome, 399
  atrioventricular septal defects, 108

J
Jaw deformity, 424
Jejunal atresia, muscular ventricular
    septal defect, 152
Jewelry making, as risk factor, 417, 418

K
Kartagener syndrome, 399
  laterality, looping defects, 51
Keith, John D., contributions of, 9
Kerosene, coal, or wood heater, as
    risk factor, 416
Kidney agenesis, 408, 409
  outflow tract anomaly, 67
Kidney disease, laterality, looping de-
    fects, 51
Klippel-Feil syndrome, 399, 408, 409
  outflow tract anomaly, 66

L
Laboratory chemicals, as risk factor,
    417, 418
Laboratory viruses, as risk factor, 417,
    418
Larsen syndrome, 399
Laurence-Moon-Bardet-Biedl syn-
    drome, 399
Lazjuk syndrome, 400
  hypoplastic left heart syndrome,
    180
Left-sided lesions, 176, 192, 198, 213
  aortic stenosis, 202–211, 203, 209

birthweight, gestational age, 206
descriptive analyses, 204–207
diagnosis, course, 204
familial cardiac, noncardiac
    anomalies, 208
gender, race, twinning, 204–206
genetic, environmental factors,
    208–210
noncardiac anomalies, 203
potential risk factors, 208–210
prevalence by time, season, area
    of residence, 204
sociodemographic characteris-
    tics, 206–207
study population, 203–204
univariate analysis, 208–210
bicuspid aortic valve, 212–219, 218
    birthweight, gestational age, 214
    cardiac abnormalities, 212
    descriptive analyses, 212–215
    diagnosis, course, 213
    familial cardiac, noncardiac
        anomalies, 216–217
    gender, race, twinning, 214
    genetic, environmental factors,
        217–219
    noncardiac anomalies, 212
    potential risk factors, 216–219
    prevalence by time, season, area
        of residence, 212–213
    sociodemographic characteris-
        tics, 214
    study population, 212
    univariate analysis, 217–218
coarctation of aorta, 190–201,
    200–201
    birthweight, gestational age,
        195
    cardiac abnormalities, 192
    descriptive analysis, 193–196
    diagnosis, course, 193
    familial cardiac, noncardiac
        anomalies, 196–197
    gender, race, twinning, 193–195
    genetic, environmental factors,
        197–200
    multivariate analysis, 199–200
    noncardiac anomalies, 192
    potential risk factors, 196–200

prevalence by time, season, area
    of residence, 193
sociodemographic characteris-
    tics, 195
study population, 191–192
univariate analysis, 197–199
hypoplastic left heart syndrome,
    178–189, 180, 186
    birthweight, gestational age, 183
    cardiac abnormalities, 179
    descriptive analyses, 180–183
    diagnosis, course, 181
    environmental risk factors, 188
    familial cardiac, noncardiac
        anomalies, 183–185
    family history, 188
    gender, race, twinning, 181
    genetic, environmental factors,
        185–187
    maternal illnesses, 188
    multivariate analysis, 187
    noncardiac anomalies, 180
    potential risk factors, 183–187
    prevalence by time, season, area
        of residence, 180–181
    sociodemographic factors, 183
    study population, 179–180
    univariate analysis, 185
    variability of anatomic features,
        189
multiple anomalies/multiplex sub-
    set, 176
Left-sided obstructive lesions,
    165–225, 166–176, 173
    analysis plan, 167
    descriptive analyses, 220–223
    diagnostic considerations, 166–167
    diagnostic definition, 168
    genetic relationships among left-
        sided heart defects, 168–171
    infant characteristics, 171, 220
    potential risk factors, 172–176,
        224–225
        evaluation of effect of doubly as-
            certained families, 224–225
    race, gender, 220–223
    risk factor analysis of total group,
        168–172
    univariate analysis, 172–174

Limb-girdle muscular dystrophy, 422
Limb reduction defect
    laterality, looping defects, 51
    membranous ventricular septal de-
        fect, 131
Local contraceptives, risk factors, 414
Lung cysts, atrioventricular septal de-
        fects, 115

M
Macrodantin
    coarctation of aorta, 198, 200–201
    left-sided obstructive defects, 173,
        224, 225
Malformed heart, risk factor analysis
        and, 15–23
Malignant tumor on hip at birth, 428
Malrotation, outflow tract anomaly, 66
Marden-Walker syndrome, 400
    atrial septal defect, 270
Marfan syndrome, 422
Marijuana use
    outflow tract anomaly, 91, 92, 95, 97,
        98
    tricuspid atresia, 264
Meckel syndrome, 400, 409, 410
    membranous ventricular septal de-
        fect, 130
    outflow tract anomaly, 66
Mercury, as risk factor, 416, 418
Metronidazole
    cardiac malformation, 367
    membranous ventricular septal de-
        fect, 145
    outflow tract anomaly, 91, 93, 98
Microcephaly, 408, 410
    hypoplastic left heart syndrome, 180
    laterality, looping defects, 45
    membranous ventricular septal de-
        fect, 139
    outflow tract anomaly, 66
    pulmonic valve stenosis, 235
Microwave oven, as risk factor, 416
Milroy's lymphedema, 422
Miscarriages, outflow tract anomaly,
        93
Missing fingers, toes, 427
Multidisciplinary partnerships, 393
Muscular, prevalence of, 341

Mutations, microdeletions, 388–391

N
Nadas, Alexander S., contributions of,
        10–11
Narcotics, risk factors, 414
Neurofibromatosis, 424
Non-barbiturate tranquilizers, risk fac-
        tors, 414
Noonan syndrome, 400
    atrial septal defect, 270
    membranous ventricular septal de-
        fect, 130
    outflow tract anomalies and, 65
    pulmonic valve stenosis, 235
Normal heart, risk factor analysis and,
        14–15, 16, 17, 18

O
Oil heater, as risk factor, 416
Oligomeganephronia, laterality, loop-
        ing defects, 51
Omphalocele, 408, 428
    atrial septal defect, 270
    laterality, looping defects, 45
    membranous ventricular septal de-
        fect, 131
    outflow tract anomaly, 66
    pulmonic valve stenosis, 235
Opitz-Frias syndrome, 400
    pulmonic valve stenosis, 235
Oral contraceptives, risk factors, 414
Oro-facio digital syndrome, type 1, 400

P
Painting
    atrioventricular septal defects, 117
    coarctation of aorta, 198
    outflow tract anomaly, 95, 97, 98
    pulmonic valve stenosis, 246
    total anomalous pulmonary venous
        return, 310
Pancytopenia, bicuspid aortic valve,
        213
Parasympatholytics, risk factors, 414
Patent arterial duct, 285–299
    descriptive analyses, 288–292
        birthweight, gestational age, 290
        diagnosis, course, 288–290

gender, race, twinning, 290
    prevalence by time, season, area
        of residence, 288
    membranous ventricular septal de-
        fect, 138
    potential risk factors, 293–297
        familial cardiac, noncardiac
            anomalies, 293
        sociodemographic characteris-
            tics, 293
            genetic, environmental factors,
                294–296
                multivariate analysis, 296
                univariate analysis, 294–295
                maternal exposures during sec-
                    ond, third trimesters,
                    296–297
                potential risk factors, 297
                prevalence, relative frequency,
                    297
    prevalence of, 341
    study population, 287–288
        noncardiac anomalies, 287–288
Pectus excavatum, 427
    atrial septal defect, 270
    membranous ventricular septal de-
        fect, 131
    patent arterial duct, 289
Pena-Shokeir syndrome, 400
Pesticides
    as risk factor, 417, 418
    total anomalous pulmonary venous
        return, 310
Peters-plus syndrome, 401
    coarctation of aorta, 192
Phenothiazines
    risk factors, 414
    total anomalous pulmonary venous
        return, 310
Pilonidal cyst, atrioventricular septal
    defects, 115
Pituitary adenoma, 428
Plastics manufacturing, as risk factor,
    417, 418
Pol, syndrome, 422, 428
Polycystic kidney, 408, 423
    atrioventricular septal defects, 108
    hypospadias, membranous ventric-
        ular septal defect, 130

membranous ventricular septal de-
    fect, 131
outflow tract anomaly, 67
Polydactyly, 424
    atrioventricular septal defects, 115
    laterality, looping defects, 45
    membranous ventricular septal de-
        fect, 131, 139
    muscular ventricular septal defect,
        152
    outflow tract anomaly, 67
    pulmonic valve stenosis, 235
Pompe syndrome, 401
    membranous ventricular septal de-
        fect, 130, 139
Potter's syndrome, 427
Prader-Willi syndrome, 401
    atrial septal defect, 270
    muscular ventricular septal defect,
        152
Progesterone, outflow tract anomaly,
    91, 92, 95, 97
Progesterones
    outflow tract anomaly, 98
    risk factors, 414
Prune belly complex, 401
Prune belly syndrome, pulmonic valve
    stenosis, 235
Pseudotrisomy 13, 401
Pseudoxanthoma, membranous ven-
    tricular septal defect, 139
Pulmonary atresia, prevalence of, 341,
    342
Pulmonary valve stenosis, 22
    prevalence of, 341, 342
Pyloric stenosis, 408, 427
    aortic stenosis, 203
    atrial septal defect, 270
    laterality, looping defects, 51
    membranous ventricular septal de-
        fect, 131, 139
    muscular ventricular septal defect,
        152
    outflow tract anomaly, 66
    patent arterial duct, 289

R
Renal agenesis, membranous ventricu-
    lar septal defect, 139

Renal-hepatic-pancreatic dysplasia
 syndrome, 401
 hypoplastic left heart syndrome,
  180
Rhesus incompatibility, laterality,
 looping defects, 51
Rheumatic heart, 423
Right-sided obstructive defects,
 227–265, 264
 overview, 228–229
 prevalence of, 341
 pulmonary valve atresia, with intact
  ventricular septum, 249–257
  birthweight, gestational age, 252
  descriptive analyses, 250–254
  diagnosis, course, 251
  gender, race, twinning, 251–252
  potential risk factors, 254–257,
   256
   familial cardiac, noncardiac
    anomalies, 254
   genetic, environmental factors,
    254–257
  prevalence by time, season, area
   of residence, 250–251
  sociodemographic characteristics, 252
  study population, 249–250
   cardiac abnormalities, 249–250
   noncardiac anomalies, 250
 pulmonary valve stenosis, 230–248,
  231
  categorization of severity, 233
  descriptive analyses, 235–242
   birthweight, gestational age,
    240
   diagnosis, course, 236–237
   gender, race, twinning, 237–240
   prevalence by time, season,
    area of residence, 235–236
   sociodemographic characteris-
    tics, 240–242
  potential risk factors, 242–245
   familial cardiac, noncardiac
    anomalies, 242–244
   genetic, environmental factors,
    244–245, 246
    multivariate analysis, 245
    univariate analysis, 244–245

  recognition: changes by time,
   232–233
  study population, 233–235, 235
   cardiac abnormalities, 233–234
   noncardiac anomalies, 234
 tricuspid atresia with normally related great arteries, 258–265
  descriptive analyses, 259–262
   birthweight, gestational age,
    260
   diagnosis, course, 260
   gender, race, twinning, 260
   prevalence by time, season,
    area of residence, 259–260
   sociodemographic characteris-
    tics, 260–262
  potential risk factors, 262–265
   familial cardiac, noncardiac
    anomalies, 262
   genetic, environmental factors,
    262–265
  study population, 259
   cardiac abnormalities, 259
   noncardiac anomalies, 259
Risk factor analysis, 359–382
 cardiac, noncardiac abnormalities,
  familial aggregation of,
  364–366
 cardiac malformation, 367
  categorization of, 13–28
 diabetes, maternal, 368–370
 environmental risk factors, 367–382
 family history, cardiovascular malformations, 361–367
 fever, maternal, 370–371
 genetic heterogeneity, genetic liability, 366–367
 influenza, maternal, 370–371
 lead, 379
 lifestyle factors, 375–377
  alcohol, 376
  caffeine, 376–377
  cigarette smoking, 376
  hair dyes, 377
  recreational drugs, 376
 occupational exposures, 377–379
 paternal factors, 379–380
 pesticides, 379
 pharmaceuticals, 372–375

antitussives, 372
benzodiazepines, 374
clomiphene, 372–374
corticosteroids, 374
diazepam, 374
ibuprofen, 374
metronidazole, 374–375
salicylates, 375
reproductive history, maternal,
371–372
solvents, 377–379
Ritscher-Schinzel syndrome, 401
Roberts syndrome, 402, 409
coarctation of aorta, 192
outflow tract anomalies and, 65
Rowe, Richard D., contributions of, 10
Rubella syndrome, 402
atrial septal defect, 270

S
Scheueman syndrome, 426
Schinzel-Giedion syndrome, 402, 409
atrioventricular septal defects, 108
outflow tract anomalies and, 65
Sequestration of lung, patent arterial
duct, 289
Sexual ambiguity
membranous ventricular septal defect, 131
muscular ventricular septal defect,
152
Sickle cell disease, 422
atrioventricular septal defects, 115
laterality, looping defects, 51
membranous ventricular septal defect, 139
Skeletal x-rays, as risk factor, 416
Smith-Lemli-Opitz syndrome, 402, 409,
410
atrioventricular septal defects, 108
hypoplastic left heart syndrome,
180
outflow tract anomalies and, 65
Smoking
aortic stenosis, 209
atrial septal defect, 279
atrioventricular septal defects, 117
bicuspid aortic valve, 218
cardiomyopathy, 322

coarctation of aorta, 198
Ebstein's anomaly, 331
hypoplastic left heart syndrome, 186
laterality, looping defects, 52
left-sided obstructive defects, 173
muscular ventricular septal defect,
161
pulmonary valve atresia with intact
ventricular septum, 256
total anomalous pulmonary venous
return, 310
tricuspid atresia, 264
Soldering with lead
pulmonary valve atresia with intact
ventricular septum, 256
as risk factor, 416, 418
Solvents
cardiac malformation, 367
degreasing guns, motors, as risk factors, 416
Ebstein's anomaly, 331
outflow tract anomaly, 91, 92, 95, 97,
98
Spina bifida
atrial septal defect, 270
diaphragmatic hernia, atrial septal
defect, 270
membranous ventricular septal
defect, 131, 139
Spondylocostal dysplasia, 402
patent arterial duct, 289
Spondylolisthesis, 427
Stained glass crafts, as risk factor, 417
Stripping or sanding old paint, as risk
factor, 416
Sulfonamides, risk factors, 414
Sympathomimetics
coarctation of aorta, 198, 200–201
left-sided obstructive defects, 173,
175, 224, 225
risk factors, 414

T
Taussig, Helen Brooke, contributions
of, 8–9
Tetralogy of Fallot, 20
laterality, looping defects, 51
membranous ventricular septal
defect, 138

Tetralogy of Fallot (*continued*)
  prevalence by, 342
Textile dyes, as risk factor, 417, 418
Thalassemia, 423
  atrioventricular septal defects, 115
  membranous ventricular septal defect, 139
  muscular ventricular septal defect, 152
Thomas syndrome, 402
Thrombocytopenia, 408, 423
  membranous ventricular septal defect, 130, 139
  outflow tract anomaly, 67
Thyroglossal cyst, membranous ventricular septal defect, 139
Thyroid hemangioma, 428
Thyroid hormone
  pulmonic valve stenosis, 246
  risk factors, 414
Total anomalous pulmonary venous return, 303–312
  descriptive analyses, 306–308
    birthweight, gestational age, 307
    diagnosis, course, 306
    gender, race, twinning, 307
    prevalence by time, season, area of residence, 306
  potential risk factors, 309–312
    familial cardiac, noncardiac anomalies, 309
    genetic, environmental factors, 309–312
  study population, 303–306
    cardiac abnormalities, 304
    noncardiac anomalies, 304–306
Townes-Brocks syndrome, 402
  membranous ventricular septal defect, 130
  outflow tract anomalies and
Tracheoesophageal fistula
  atrial septal defect, 270
  atrioventricular septal defects, 108
  membranous ventricular septal defect, 130, 131
  muscular ventricular septal defect, 152
Tranquilizers, aortic stenosis, 209
Transposition of great arteries, 20

Treacher-Collins syndrome, laterality, looping defects, 51
Tricuspid valve atresia, 21
  prevalence of, 341, 342
Tuberous sclerosis, 408
  outflow tract anomaly, 67
Turner's syndrome, 422
  bicuspid aortic valve, 213
  coarctation of aorta, 192

U
Urethral stenosis, hypoplastic left heart syndrome, 180
Urinary tract infection
  atrial septal defect, 279
  laterality, looping defects, 52

V
Van de Woude syndrome, 403
  atrioventricular septal defects, 108
Varnishing
  atrioventricular septal defects, 117
  coarctation of aorta, 198
  Ebstein's anomaly, 331
  as risk factor, 416
Ventricular septal defect, 23, 123–163
  atrioventricular septal defects, 115
  laterality, looping defects, 51
  membranous type, 128–148, 145
    descriptive analyses, 132–136
      birthweight, gestational age, 135
      diagnosis, course, 132–133
      gender, race, twinning, 133–134
      prevalence by time, season, area of residence, 132
      similarity of small, moderate/large defects, 135
      sociodemographic characteristics, 136
    multivariate analysis, 140–145
    potential risk factors, 137–145
      familial cardiac, noncardiac anomalies, 137
      genetic, environmental factors, 137–145
      membranous, 138
      univariate analysis, 137–140
    study population, 128–131, 130

cardiac abnormalities, 128
noncardiac anomalies, 128–131
membranous ventricular septal defect, 138
muscular type, 149–163
changes in prevalence, risk factors, 162–163
descriptive analysis, 151–157, 152
birthweight, gestational age, 155
diagnosis, course, 154–155
gender, race, twinning, 155
prevalence by time, season, area of residence, 151–153
sociodemographic characteristics, 157
epidemiologic studies, 163
possible genetic risk factors, 162
potential risk factors, 158–159
familial cardiac, noncardiac anomalies, 158
genetic, environmental factors, 158–159
multivariate analysis, 159
univariate analysis, 159

study population, 149–151
associated cardiac abnormalities, 150–151
cardiac abnormalities, 149–150
noncardiac anomalies, 151
overview, 124–127, 147–148
prevalence of, 341, 342
Vertebral defects, outflow tract anomaly, 67
Vitiligo, 428
Von Willebr, syndrome, membranous ventricular septal defect, 139

W
Waardenburg syndrome, 403
Welding
atrioventricular septal defects, 117
cardiomyopathy, 322
as risk factor, 416, 418
Williams syndrome, 403
membranous ventricular septal defect, 130

X
Xanthine, risk factors, 414